UV and X-ray spectroscopy of astrophysical and laboratory plasmas draws interest from many disciplines. Contributions from international specialists are collected together in this book from a timely recent conference. In astrophysics, the Hubble Space Telescope, Astro 1 and ROSAT observatories are now providing UV and X-ray spectra and images of cosmic sources in unprecedented detail, while the Yohkoh mission recently collected superb data on the solar corona. In the laboratory, the development of ion-trap facilities and novel laser experiments are providing vital new data on high temperature plasmas. Recent innovations in the technology of spectroscopic instrumentation are discussed.

These papers constitute an excellent up-to-date review of developments in short-wavelength spectroscopy and offer a solid introduction to its theoretical and experimental foundations.

UV and X-Ray Spectroscopy of Astrophysical and Laboratory Plasmas

UV and X-Ray Spectroscopy of Astrophysical and Laboratory Plasmas

Proceedings from the
Tenth International Colloquium
held at
Berkeley, California
3–5 February 1992

Edited by

Eric H. Silver
Lawrence Livermore National Laboratory

and

Steven M. Kahn
University of California at Berkeley

CAMBRIDGE
UNIVERSITY PRESS

PUBLISHED BY THE PRESS SYNDICATE OF THE UNIVERSITY OF CAMBRIDGE
The Pitt Building, Trumpington Street, Cambridge, United Kingdom

CAMBRIDGE UNIVERSITY PRESS
The Edinburgh Building, Cambridge CB2 2RU, UK
40 West 20th Street, New York NY 10011–4211, USA
477 Williamstown Road, Port Melbourne, VIC 3207, Australia
Ruiz de Alarcón 13, 28014 Madrid, Spain
Dock House, The Waterfront, Cape Town 8001, South Africa

http://www.cambridge.org

First published 1993
First paperback edition 2003

A catalogue record for this book is available from the British Library

ISBN 0 521 43470 X hardback
ISBN 0 521 54816 0 paperback

CONTENTS

Contents

Contents

Tenth International Colloquium
UV and X-Ray Spectroscopy of Astrophysical and Laboratory Plasmas

Science Organizing Committee

S. Kahn, Chairman
F. Bely-Dubau
M. Bitter
C. Canizares
J. Culhane
G. Doschek
A. Dupree
A. Gabriel
W. Goldstein
M. Huber
M. Key
K. Koshelev
R. Mewe
H. Moos
D. Sampson
J. Schmitt
H. Schnopper
J. Schwob
P. Smith
T. Watanabe

Local Organizing Committee

E. Silver, Chairman
S. Allen
R. Falcone
S. Labov
J. Lemen
R. Malina
D. Matthews
K. Strong
J. Timothy
J. Underwood
A. Walker

List of Participants

L. Acton — Lockheed Palo Alto Research Laboratory, Palo Alto, California, USA
D. Alexander — University of Glasgow, Glasgow, Scotland
B. Aschenbach — Max Planck Institute, Garching, Germany
G. Athay — High Altitude Observatory, Boulder, Colorado, USA
G. Beadie — Brown University, Providence, Rhode Island, USA
P. Beiersdorfer — Lawrence Livermore National Laboratory, Livermore, California, USA
I. Beigman — P.N. Lebedev Physical Institute, Leninsky, Moscow, Russia
R. Benjamin — The University of Texas, Austin, Texas, USA
R. Bentley — Mullard Space Science Laboratory, Dorking, Surrey, United Kingdom
P. Bergamini — Stanford University, Stanford, California, USA
T. Berger — Stanford University, Stanford, California, USA
J. Bixler — Lawrence Livermore National Laboratory, Livermore, California, USA
W. Blair — Johns Hopkins University, Baltimore, Maryland, USA
S. Bliman — Universite Paris, Orsay, France
S. Bobashev — A.F. Ioffe Physiotechnical Institute, St. Petersburg, Russia
S. Bowyer — University of California, Berkeley, California, USA
L. Brewer — Lawrence Livermore National Laboratory, Livermore, California, USA
A. Brinkman — SRON-Utrecht, Utrecht, Netherlands
A. Brown — University of Colorado, Boulder, Colorado, USA
W. Brown — Lockheed Research Laboratory, Palo Alto, California, USA
R. Bruch — University of Nevada, Reno, Nevada, USA
C. Canizares — M.I.T., Cambridge, Massachusetts, USA
W. Cash — University of Colorado, Boulder, Colorado, USA
J. Castor — Lawrence Livermore National Laboratory, Livermore, California, USA
E. Chandler — Lawrence Livermore National Laboratory, Livermore, California, USA
M. Chen — Lawrence Livermore National Laboratory, Livermore, California, USA
K. Cheng — Lawrence Livermore National Laboratory, Livermore, California, USA
C. Cheng — U.S. Naval Research Laboratory, Washington, DC, USA
I. Coffey — JET Joint Undertaking, Abingdon, Oxon, England
M. Cornille — Observatoire de Paris, France
D. Cotton — University of California, Berkeley, California, USA
J. Culhane — Mullard Space Science Laboratory, Surrey, United Kingdom
C. Cunningham — Lawrence Livermore National Laboratory, Livermore, California, USA
S. Dalhed — Lawrence Livermore National Laboratory, Livermore, California, USA
J. Davila — NASA/Goddard Space Flight Center, Greenbelt, Maryland, USA
S. Deustua — Lawrence Livermore National Laboratory, Livermore, California, USA
D. Dewitt — Lawrence Livermore National Laboratory, Livermore, California, USA
D. Dietrich — Lawrence Livermore National Laboratory, Livermore, California, USA
J. Dubau — Observatoire de Paris, France
F. Bely-Dubau — Observatoire de la Cote d'Azur, France
J. Edelstein — University of California, Berkeley, California, USA
J. Edwards — Rutherford Appleton Laboratory, Chilton, United Kingdom
A. Faenov — Multicharged Ions Spectra Data Centerr, Moscow, Russia
P. Faucher — Observatoire de la Cote d'Azur, France
B. Feinberg — Lawrence Berkeley Laboratory, Berkeley, California, USA
U. Feldman — Naval Research Laboratory, Washington, DC, USA
D. Ferguson — W.J. Schafer Associates, Livermore, California, USA
M. Finkenthal — Hebrew University, Jerusalem, Israel
K. Flanagan — Harvard University, Cambridge, Massachusetts, USA
J. Fleischman — Columbia University, New York, New York, USA
M. Foord — Weizmann Institute, Rehovot, Israel
B. Monsignori-Fossi — Arcetri Astrophysical Observatory, Firenze, Italy
V. Foster — Queen's University, Belfast, Northern Ireland
D. Friart — Centre D'Etudes, France
P. Friedman — Columbia University, New York, New York, USA

D. Gallagher	University of Colorado, Boulder, Colorado, USA
J. Gauthier	Ecole Polytechnique, France
V. Gavrilov	Kurchatov Atomic Energy Institute, Russia
M. Giampapa	National Optical Astronomy Observatories, Tucson, Arizona, USA
W. Gladstone	University of California, Berkeley, California, USA
W. Goldstein	Lawrence Livermore National Laboratory, Livermore, California, USA
H. Gould	Lawrence Berkeley Laboratory, Berkeley, California, USA
B. Haisch	Lockheed Palo Alto Research Laboratory, Palo Alto, California, USA
W. Hallett	Oxford University, Oxford, United Kingdom
B. Hammel	Lawrence Livermore National Laboratory, Livermore, California, USA
L. Harra	Queen's University, Belfast, Ireland
S. Hawley	Lawrence Livermore National Laboratory, Livermore, California, USA
J. Henderson	Lawrence Livermore National Laboratory, Livermore, California, USA
M. Huber	ESA/ESTEC, Netherlands
C. Iglesias	Lawrence Livermore National Laboratory, Livermore, California, USA
S. Jaquemot	CEA Limeil-Valenton, Villeneuve St. Georges, France
E. Jenkins	Princeton University, Princeton, New Jersey, USA
T. Jernigan	NASA, Houston, Texas, USA
Y. Jung	NASA/Marshall Space Flight Center, Huntsville, Alabama, USA
J. Kaastra	SRON, Leiden, Netherlands
S. Kahn	University of California, Berkeley, California, USA
T. Kallman	NASA/Goddard Space Flight Center, Greenbelt, Maryland, USA
E. Rachlew-Kallne	Royal Institute of Technology, Stockholm, Sweden
Y. Karzhavin	University of Texas, Austin, Texas, USA
C. Keane	Lawrence Livermore National Laboratory, Livermore, California, USA
F. Keenan	Queen's University, Belfast, Northern Ireland
R. Kimble	NASA/Goddard Space Flight Center, Greenbelt, Maryland, USA
H. Kirby	Stanford University, Stanford, California USA
M. Klapisch	Naval Research Laboratoy, Washington, DC, USA
E. Knystautas	Universite de Laval, Quebec, Canada
K. Koshelev	Institute for Spectroscopy
Y. Ko	University of Maryland, Greenblet, Maryland, USA
J. Kohl	Harvard-Smithsonian Center for Astrophysics, Cambridge, Massachusetts, USA
M. Koike	Lawrence Berkeley Laboratory, Berkeley, California, USA
R. Kreplin	Naval Research Laboratory, Washington, DC, USA
S. Labov	Lawrence Livermore National Laboratory, Livermore, California, USA
M. Laming	Naval Research Laboratory, Washington, DC, USA
M. Landini	Florence University, Firenze, Italy
J. Lang	Rutherford Appleton Laboratory, Chilton, Didcot, United Kingdom
J. Larsen	Cascade Applied Sciences, Inc., Boulder, Colorado, USA
D. Leahy	University of Calgary, Alberta, Canada
D. Lechrone	NASA/Goddard Space Flight Center, Greenbelt, Maryland, USA
Y. Lee	Lawrence Livermore National Laboratory, Livermore, California, USA
R. Lee	Lawrence Livermore National Laboratory, Livermore, California, USA
M. LeGros	Lawrence Livermore National Laboratory, Livermore, California, USA
J. Lemen	Lockheed Palo Alto Research Laboratory, Palo Alto, California, USA
D. Liedahl	Lawrence Livermore National Laboratory, Livermore, California, USA
E. Liston	Belvedere, California
R. London	Lawrence Livermore National Laboratory, Livermore, California, USA
R. Malina	University of California, Berkeley, California, USA
G. Manzo	IFCAI del CNR, Palermo, Italy
T. Markert	M.I.T., Cambridge, Massachusetts, USA
R. Marrs	Lawrence Livermore National Laboratory, Livermore, California, USA
H. Mason	Cambridge University, Cambridge, United Kingdom
C. Mauche	Lawrence Livermore National Laboratory, Livermore, California, USA
J. McDonald	Lawrence Livermore National Laboratory, Livermore, California, USA
E. McGuire	Sandia National Laboratory, Albuquerque, New Mexico, USA
C. Mears	Lawrence Livermore National Laboratory, Livermore, California, USA

R. Mewe	SRON, Utrecht, Netherlands
J. Molitoris	Lawrence Livermore National Laboratory, Livermore, California, USA
W. Moos	Johns Hopkins University, Baltimore, Maryland, USA
G. Morris	Lawrence Livermore National Laboratory, Livermore, California, USA
H. Moseley	Goddard Space Flight Center, Greenbelt, Maryland, USA
T. Namioka	NASA/Goddard Space Flight Center, Greenbelt, Maryland, USA
J. Nash	Lawrence Livermore National Laboratory, Livermore, California, USA
J. Neff	Penn State University, University Park, Pennsylvania, USA
W. Neupert	NASA/Goddard Space Flight Center, Greenbelt, Maryland, USA
J. Nilsen	Lawrence Livermore National Laboratory, Livermore, California, USA
P. Nicolosi	University of Padova, Padova, Italy
E. Oks	Auburn University, Auburn, Alabama, USA
A. Osterheld	Lawrence Livermore National Laboratory, Livermore, California, USA
F. Paerels	University of California, Berkeley, California, USA
A. Panin	Bringham Young University, Provo, Utah, USA
W. Parkinson	Harvard Smithsonian Center for Astrophysics, Cambridge, Massachusetts, USA
N. Peacock	Culham Laboratory, Oxon, England
G. Peres	Observatorio Astronomico, Catania, Itlay
T. Perry	Lawrence Livermore National Laboratory, Livermore, California, USA
M. Pertsova	Stanford University, Stanford, Califonia, USA
R. Petre	NASA/Goddard Space Flight Center, Greenblet, Maryland, USA
T. Pfafman	University of California, Berkeley, California, USA
K. Phillips	Rutherford Appleton Laboratory, Didcot, United Kingdom
L. Piro	Istituto di Astrofisica Spaziale, Frascati, Italy
A. Ramsey	Princeton Plasama Physics Laboratory, Princeton, New Jersey, USA
A. Rasmussen	Columbia University, New York, New York, USA
J. Raymond	Harvard-Smithsonian Center for Astrophysics, Cambridge, Massachusetts, USA
K. Reed	Lawrence Livermore National Laboratory, Livermore, California, USA
F. Rogers	Lawrence Livermore National Laboratory, Livermore, California, USA
S. Rose	Rutherford Appleton Laboratory, Didcot, Oxon, United Kingdom
M. Rosen	Lawrence Livermore National Laboratory, Livermore, California, USA
J. Ryan	University of New Hampshire, Durham, New Hampshire, USA
J. Saba	NASA/Goddard Space Flight Center, Greenbelt, USA
U. Safronova	Russian Academy of Science, Russia
D. Sampson	Pennsylvania State University, Univeristy Park, Pennsylvania, USA
S. Sarlin	University of Colorado, Boulder, Colorado, USA
D. Schiminovich	Columbia University, New York, New York, USA
D. Schneider	Lawrence Livermore National Laboratory, Livermore, California, USA
H. Schnopper	Danish Space Research Institute, Denmark
U. Schuhle	Max-Planck-Institute, Katlenburg-Lindau, FRG
R. Scwartz	NASA/Goddard Space Flight Center, Greenblet, Maryland, USA
J. Schwob	The Hebrew University, Jerusalem, Israel
J. Scofield	Lawrence Livermore National Laboratory, Livermore, California, USA
H. Scott	Lawrence Livermore National Laboratory, Livermore, California, USA
J. Seely	Naval Research Laboratory, Washington, DC, USA
E. Silver	Lawrence Livermore National Laboratory, Livermore, California, USA
J. Silver	Oxford University, Oxford, England
D. Slater	Stanford University, Stanford, California, USA
O. Strand	Lawrence Livermore National Laboratory, Livermore, California, USA
J. Swenson	Lawrence Livermore National Laboratory, Livermore, California, USA
R. Thomas	NASA/Goddard Space Flight Center, Greenblet, Maryland, USA
T. Thomson	Lawrence Livermore National Laboratory, Livermore, California, USA
G. Timothy	Stanford University, Stanford, California, USA
A. Tlamicha	Czechoslovak Academy of Sciences, Ondrejov, Czechoslovakia
J. Underwood	Lawrence Berkeley Laboratory, Berkeley, California, USA
A. Van Teeseling	Sterrekundig Instituut, Utrecht, Netherlands
P. Viedler	University of California, Berkeley, California, USA
P. Vitello	Lawrence Livermore National Laboratory, Livermore, California, USA

K. Waljeski	NCR/ Naval Research Laboratory, Washington, DC, USA
A. Walker	Stanford University, Stanford, California, USA
F. Walter	State University of New York, Stony Brook, New York, USA
A. Wan	Lawrence Livermore National Laboratory Livermore, Calfiornia, USA
Z. Wang	University of Nevada, Reno, Nevada, USA
Z. Wang	Nanjing University, Najing, Republic of China
B. Wargelin	University of California, Berkeley, California, USA
T. Watanabe	National Astronomical Observatory, Mitaka, Japan
K. Widing	U.S. Naval Research Lab, Washington, DC, USA
P. Winkler	University of Nevada, Reno, Nevada, USA
K. Wong	Lawrence Livermore National Laboratory, Livermore, California, USA
J. Wyart	Centre Universitaire, France
Y. Yan	University of Nevada, Reno, Nevada, USA
P. Young	Lawrence Livermore National Laboratory, Livermore, California, USA
D. Zarro	NASA/Goddard Space Flight Center, Greenblet, Maryland, USA
H. Zhan	University of Nevada, Reno, Nevada, USA
Y. Zhang	University of Nevada, Reno, Nevada, USA
K. Ziock	Lawrence Livermore National Laboratory, Livermore, California, USA
A. Zigler	Hebrew University of Jerusalem, Jerusalem, Israel
A. Zwicker	Johns Hopkins University, Baltimore, Maryland, USA

PREFACE

The papers published with these Proceedings were presented in Berkeley at the tenth in a series of regular colloquia devoted to the interdisciplinary topic of ultraviolet and X-ray spectroscopy of laboratory and astrophysical plasmas. These conferences, which have been held roughly once every three years at various locations around the world, have provided a unique forum for astrophysicists, solar physicists, laboratory plasma experimentalists, and theoretical atomic and molecular physicists to collectively explore fundamental issues in short wavelength spectroscopy that are common to these diverse disciplines. Like its predecessors, the Berkeley conference was very well-attended; we had over 200 participants, far in excess of our initial expectations.

The colloquium came at an especially active time for this field. In particular, several prominent ultraviolet and X-ray satellite experiments had been launched within the year just prior to the meeting that collectively provided a wealth of new ultraviolet and X-ray spectroscopic data on cosmic sources, covering virtually all classes of astronomical systems. Various speakers presented some of the first results from the high resolution spectrograph on the *Hubble Space Telescope*, the high sensitivity far ultraviolet and X-ray spectrometers of the *ASTRO 1 Observatory*, the imaging X-ray spectrometer on the *ROSAT Observatory*, and the high resolution solar X-ray spectrometer on *Yohkoh*. We also heard of substantial progress in laboratory plasma research. The development of ion trap devices had brought about a revolution in laboratory investigations of atomic processes in highly charged atoms. X-ray laser experiments had not only yielded considerable insight into electron ion interactions in hot dense plasmas, but also demonstrated the tremendous versatility of laser plasmas as laboratory X-ray sources. Such measurements also motivated and led to refinements in the development of large-scale atomic and molecular codes. On the instrumental side, the design and development of the next series of very powerful short wavelength observatories had generated a large number of technological innovations in both dispersive and nondispersive spectroscopic instrumentation.

In addition to its scientific success, the conference proved to be a very pleasant affair. We were blessed with exceptional weather (even for Berkeley!), and the two major social events, the reception at the Santa Fe Bar and Grill and the banquet at the Exploratorium in San Francisco, were well-attended and very lively. In addition to the other members of the organizing committees, we are especially indebted to Marjorie Randell-Silver, Beth Saucier, Gloria Staude, Jan Wallace, Kerry O'Connor, Caryl Esteves, Susan Green, and Robin Weissberger for extensive help with the arrangements and for smooth operation of the meeting.

Finally, we would like to thank the National Aeronautics and Space Administration, the California Space Institute, the Lawrence Livermore National Laboratory, the University of California at Berkeley, the Lawrence Berkeley Laboratory, Stanford University, and the Lockheed Corporation for their financial support of the meeting. These funds were especially useful in enabling us to provide assistance in meeting travel expenses for selected attendees, particularly students and those from less well-endowed institutions. A widespread geographic distribution among conference participants was one of the goals of the organizing committees, and we believed that it was at least partially responsible for the intellectual vitality of the colloquium.

Steven M. Kahn
Eric H. Silver
Berkeley, California
July, 1992

Atomic Processes and Overview

UV and X–ray Spectroscopy in Astrophysics

Christopher F. McKee

Departments of Physics and Astronomy, University of California, Berkeley

ABSTRACT

Spectroscopy is an extremely powerful tool for measuring the physical conditions in astronomical plasmas, including the density, temperature, ionization, radiation intensity, and magnetic field. Application of spectroscopic analysis is more straightforward for diffuse plasmas, such as those in stellar coronae and the interstellar medium, since the effects of radiative transfer are less important there; this brief review focuses on such plasmas. The composition of stars and the interstellar medium is inferred from spectroscopy. Line profiles reveal the dynamics of astrophysical plasmas, and spectra calculated from global dynamical models can be compared with observation to obtain a more complete understanding of the dynamics. Both spectropolarimetry and the technique of "reverberation mapping" permit one to deduce the structure of astronomical sources on scales too small to be resolved directly. However, the potential of astrophysical spectroscopy can be realized only if accurate atomic data are available; dielectronic recombination is cited as an example.

1 THE POWER OF ASTROPHYSICAL SPECTROSCOPY

Astronomy is an unusual science in that it is based on observation instead of experiment. The astronomer must rely on analyzing the radiation that happens to reach the Earth from whatever source is being studied. There is a good deal of information in the distribution of this radiation on the sky and, for variable sources, its time evolution. However, the information content is multiplied manyfold when the radiation is broken down into different frequencies, its spectrum.

Here I shall present a brief and idiosyncratic look at some of the properties of astrophysical plasmas that can be inferred from spectroscopy, particularly spectroscopy of sufficiently high resolution that individual lines can be observed. A paper this brief is not a review and cannot be complete. In keeping with the topic of this conference, I shall focus on UV and X-ray spectroscopy, but I shall comment on techniques from other parts of the electromagnetic spectrum where appropriate. Furthermore, I shall concentrate on the spectroscopy of diffuse gases, such as stellar coronae and the interstellar medium (ISM), since then deductions from spectroscopic observations are less dependent on the effects of radiative transfer. First I discuss how spectroscopy can be used to infer the physical conditions in an astrophysical plasma, such as the density, temperature, and magnetic field. In some cases more subtle measurements are possible, so that, for example, the velocity distribution function of the emitting plasma is accessible to observation. The composition of astrophysical plasmas is determined by spectroscopic observations. Astrophysical plasmas are often violent, with flares, winds, and shock waves generating high velocities and temperatures, which are revealed by spectroscopy. Finally, it is possible to use spectroscopy to study the geometry of sources on very small scales by using spectropolarimetry or by using the technique of "reverberation mapping".

1.1 Physical Conditions

1.1.1 Density

The density of a plasma can be inferred from the intensity ratio of two lines of the same ion, at least one of which is not a permitted transition. For example, the relative populations of the $2\ ^3S$ and $2\ ^3P$ states in He–like ions are sensitive to the density, making the ratio of the forbidden and intercombination lines resulting from the decay of these states to the ground state a density diagnostic useful for stellar coronae (Gabriel and Jordan 1969). The excitation of Fe L–shell lines in X–ray photoionized nebulae has been studied by Liedahl $et\ al.$ (1992a), and they have shown that the collisional excitation of metastable levels in an Fe ion of charge $Z + 1$ affects the recombination spectrum of the Fe ion Z in a manner that permits the estimation of densities in the range $10^{12} - 10^{15}$ cm^{-3}. The density can also be estimated from absorption line observations, if one of the lines arises from an excited state.

If both emission and absorption line observations are available, it is even possible to estimate the density from observation of a single line: The intensity of the emission line can be written as $I_l = n_e N(X) C(T) h\nu/4\pi$, where $N(X)$ is the column density of the lower state of the emission line and $C(T)$ is the collisional excitation rate coefficient. Provided $C(T)$ is a weak function of temperature, as it generally is for photon energies $h\nu \lesssim kT$, one can infer the electron density n_e from observations of the emission line intensity $I_l \propto n_e N(X)$ and the absorption column density $N(X)$. Martin and Bowyer (1990) have applied this technique to estimate the density of the C IV emission region in the Galactic halo, finding $n_e \sim 0.01$ cm^{-3}.

1.1.2 Temperature

The temperature can be inferred from intensity ratios of permitted lines of the same ion. For example, transitions from the $2s2p^2$ excited states to the $2s^2 2p$ ground state in the B isoelectronic sequence provide a useful temperature diagnostic for the solar chromosphere (Zirin 1966). The dielectronic satellites of H–like and He–like ions are sensitive to the temperature (Gabriel 1972). Liedahl $et\ al.$ (1992b) have shown that the widths of the recombination continua in X–ray photoionized nebulae are relatively narrow and provide a good temperature diagnostic. Observations of line profiles in either emission or absorption give an upper limit on the temperature of the plasma, since bulk flows can also contribute to the line width.

1.1.3 Pressure

If both the density and temperature of the plasma can be determined, then the pressure is known. The pressure can thus be determined by using pairs of line ratios that depend on density and temperature, as described above. A good example of the determination of the pressure of an astronomical plasma is provided by the study of the resonant absorption lines from the excited fine structure states of neutral carbon in the cold phase of the ISM (Jenkins and Shaya 1979). Observation of absorption in the C I multiplets λ 1261, λ 1280, and λ 1329 showed that most of the cold gas in the ISM is at a pressure $P/k \lesssim 10^4$ K cm^{-3}, but that some is at pressures exceeding 10^5 K cm^{-3}.

1.1.4 Ionization

Comparison of the intensities of lines from different stages of ionization allows one to infer the degree of ionization in the plasma; when combined with an estimate of the temperature, one can infer whether the ionization is collisional [$kT \sim (0.1 - 0.3)E_i$, where E_i is the ionization potential of the dominant ion] or due to photoionization ($kT \ll E_i$). Collisionally ionized plasmas in stellar flares or in supernova remnants (SNR) are often not in ionization equilibrium. Deviations from ionization equilibrium in such plasmas can be measured from observations of the ratio of the forbidden [$1s^2 - 1s2s(^3S_1)$] to resonance [$1s^2 - 1s2p(^1P_1)$] lines in He–like ions (Vedder et al. 1986): the recombination contribution to the lines favors the forbidden line more than does the collisional excitation, so the forbidden/resonance ratio is reduced for ionizing plasmas. Vedder et al. used this diagnostic to demonstrate that the X–ray emitting plasma in the Cygnus Loop is not in ionization equilibrium.

The level of ionization can also be inferred from absorption observations. For example, Green et al. (1990) have used the first observation of the absorption edge of neutral helium in the ISM to infer the He I/ H I ratio in the local ISM from the EUV spectrum of the white dwarf G191-B2B; they found that this ratio is about 0.06 to within a factor of 1.5.

1.1.5 Local intensity of radiation

Atomic levels can be excited by radiation as well as by collisions, and this can often be used to infer the intensity of the local radiation field. Perhaps the most famous example of this procedure is the measurement of the temperature of the cosmic microwave background from absorption line observations of the ratio of the $R(0)$ and

$R(1)$ lines of the CN molecule at $\lambda 3874$ Å (e.g., Field and Hitchcock 1966). The excitation of the lower state of the $R(1)$ line is due almost entirely to the absorption of radiation by the $R(0)$ ground state at a wavelength $\lambda = 2.64$ mm. Meyer and Jura (1985) inferred a radiation temperature of 2.70 ± 0.04 K at this wavelength, in quite good agreement with the subsequent measurement by the Cosmic Background Explorer, 2.735 ± 0.06 K (Mather *et al.* 1990). The Bowen fluorescence mechanism provides another example: He II Lα at λ 304 excites a characteristic set of lines in O III in the near UV, so that measurement of the O III line intensities can be used to determine the intensity of He Lα. Schachter *et al.* (1990) have used observations of the Bowen lines in a sample of Seyfert galaxies to infer the properties of the emitting regions in these objects.

The level of ionization in photoionized plasmas depends on the ionization parameter $\Gamma = n_\gamma / n$ or, equivalently, $\Xi = u_\gamma / P$, where n_γ and u_γ are the number and energy density of ionizing radiation, respectively (e.g., Krolik *et al.* 1981). Observation of the level of ionization together with a determination of the density then allow one to infer the intensity of the ionizing radiation and thus the size of the source, a procedure commonly used to estimate the size of the broad emission line regions of active galactic nuclei. An ingenious variant of this approach has been developed by Bajtlik *et al.* (1988) to infer the intensity of the intergalactic ionizing radiation field at high redshift: They noted that the number of Lα absorption systems seen in quasar spectra was smaller at redshifts near that of the background quasar (the "proximity effect"). By attributing this decrease to the rise in the level of ionization as one approached the quasar, they were able to infer the distance out to which the quasar radiation dominated that of the background and thus the intensity of the background radiation itself. With this technique, it is possible in principle to measure the evolution of the intensity of the ionizing background with cosmic time.

1.1.6 Velocity distribution in shock fronts

Shocks generate non–Maxwellian velocity distributions, and for shocks in partially ionized gases this distribution is accessible to spectroscopic observation (Chevalier and Raymond 1978). The ionized component of the plasma is promptly heated and accelerated by collisionless processes (e.g., McKee and Draine 1991). The neutral component of the plasma can interact with the hot plasma only through collisions. Some of these collisions will excite the atoms, producing emission lines with the narrow profile characteristic of the ambient plasma; some of the collisions will ionize the atoms; and some will result in charge exchange, producing fast neutral atoms that emit lines with the profile appropriate to the shocked plasma. In a gas of cosmic abundances, the line emission from such shocks is predominantly in the Balmer lines of hydrogen, and as a result these shocks are often referred to as Balmer–dominated shocks.

Observation of the narrow and broad components of the emission lines from

Balmer–dominated shocks provides a number of important diagnostics of the shock (Raymond 1991): First, the profile of the broad lines gives a direct measure of the proton temperature. Next, the ratio of the narrow to the broad lines depends on the electron/ion temperature ratio behind the collisionless shock, which is not known theoretically. Some of the shock energy goes into accelerating cosmic rays, and in principle this fraction can be determined from observations of face–on shocks by comparing the velocity shift of the shocked gas (which is 3/4 the shock velocity) with the observed ion temperature. If the fraction of the shock energy going into the protons can be determined, then the shock velocity can be inferred from the temperature; observation of the proper motion of edge–on shocks can then be used to measure the distance to the shock. Measurement of these effects in the spectrum of helium would be extremely challenging, but useful.

1.1.7 Magnetic fields

Magnetic fields lead to a Zeeman splitting of $h\Delta\nu = 0.058(B/10^7 \text{ G})$ eV for an unpaired electron, where the normalization has been chosen to represent the typical fields found in magnetic white dwarfs (Angel 1977). These fields are generally measured in the optical region of the spectrum, but are accessible to UV spectroscopy as well. Much larger fields are found in neutron stars, and these can be measured only in the X–ray region of the spectrum since the cyclotron absorption resonance occurs at $11.6(B/10^{12} \text{ G})$ keV. For example, two gamma ray bursts have been observed with features at about 20 and 40 keV, and these observations have been interpreted as the first and second harmonics of the cyclotron resonance in a field of 1.7×10^{12} G (Murakami et al. 1988). These observations are consistent with the conventional view that gamma ray bursts arise from old neutron stars.

1.2 Abundances

Our knowledge of the composition of stars and the interstellar medium comes entirely from spectroscopy. In some cases, the composition can be measured by summing up the abundances in each ionization state, but this becomes more complicated if some of the atoms are in molecules or in grains. UV absorption line studies of the ISM provide the main source of information on the abundances in the ISM (Jenkins 1987). The difference between the observed abundances and the solar values is attributed to depletion onto grains. With this interpretation, there is no evidence for significant variations in abundances in the solar neighborhood. A direct confirmation of the result that interstellar abundances are in fact close to solar can be made by observing the K–shell absorption of an element, since the absorption cross section is almost unaffected by whether the atom is in a grain. *Einstein* observations of the Crab Nebula with the Focal Plane Crystal Spectrometer show that the oxygen abundance in the intervening ISM is indeed consistent with the solar value.

with the measurement giving an oxygen abundance 1.1 ± 0.3 times the solar value (Schattenburg and Canizares 1986).

Another technique for inferring the abundances in astrophysical plasmas is by direct comparison of the spectrum with a similar solar spectrum, thereby circumventing uncertainties in the atomic data. Applying this technique to the SNR Puppis A, Canizares and Winkler (1981) find that the strengths of the H and He–like lines of oxygen and neon compared to the expected lines of Fe XVII indicate that oxygen and neon are enhanced relative to iron by a factor $\sim 3 - 5$. They argue that this is not due to depletion onto grains, both because silicon does not appear to be depleted and because the high temperatures in the remnant are expected to destroy the grains; on the other hand, the results are consistent with the injection of about 3 $M_\odot$ of oxygen and neon by a supernova in a star of mass $\gtrsim 25 M_\odot$.

1.3 Dynamics

Line profiles are a direct manifestation of the dynamics of the emitting gas. P Cygni profiles in the spectra of early type stars indicate substantial mass loss in winds with velocities of order 3000 km s^{-1}. Such winds have a dramatic effect on the evolution of the star and on the ambient interstellar medium; in addition, they substantially complicate the interpretation of the underlying stellar spectrum (Kudritzki and Hummer 1990). Far more dramatic outflows are evident in the spectra of some quasars, the broad absorption line quasars, where line widths well in excess of 10^4 km s^{-1} are commonly observed (e.g., Turnshek *et al.* 1988). In many cases, the equivalent width of the emission is much less than that of the absorption, in contrast to the stellar case; this indicates that the high velocity outflow covers only part of the sky as seen from the quasar, and thus that a substantial fraction of quasars must have such outflows.

For the flows just described, the width of the line accurately portrays the velocity of the gas. This need not be the case, however, if the opacity in the line is large enough so that the line width is due to the damping wings. An interesting example of this is provided by the acceleration of Lα photons by shocks (Neufeld and McKee 1988). Shocks in atomic gas emit copious amounts of Lα radiation, which is effectively trapped by the large neutral hydrogen opacity on either side of the shock. Each time a Lα photon crosses the shock, it is blue–shifted by a process directly analogous to first order Fermi acceleration of cosmic rays in shocks. The Lα line profile of the radio galaxy 3C326.1 has a strong blue wing which Neufeld and McKee interpreted as being due to such acceleration. The peak of the profile of the accelerated photons occurs at a velocity of $270(N_{20}v_{s7})^{1/3}$ km s^{-1}, where N_{20} is the H I column density in units of 10^{20} cm^{-2} and v_{s7} is the shock velocity in units of 10^7 cm s^{-1}; widths substantially in excess of the shock velocity are possible. Low dust to gas ratios are required in order to ensure that the Lα photons survive.

It is also possible to use spectroscopy to study the dynamics of a system by developing a dynamical model of the system, calculating the resulting spectrum, and comparing with the observed spectrum. A good example of this procedure is provided by the modeling of the X-ray spectrum of Tycho's supernova remnant by Hamilton *et al.* (1986). They used the observed age, angular diameter, and X-ray flux as inputs to their model, and assumed that the supernova ejected a mass of 1.4 $M_\odot$. They used an analytic representation for the dynamics of the expanding SNR, including the blast wave in the ISM and a reverse shock expanding back into the ejecta of the supernova. By adjusting the composition of the ejecta, they were able to get good agreement with the *Einstein* spectra, and they inferred the distance to the remnant, the energy of the original explosion, the density of the ambient medium, and a number of parameters describing the SNR. This approach will become increasingly powerful as spectra of higher resolution become available.

1.4 Mapping the Source Geometry on Small Scales

1.4.1 Spectropolarimetry of NGC 1068

One of the remarkable applications of spectroscopy is in the measurement of the geometrical properties of the source on scales far too small to be spatially resolved. Some information on the structure of the source can be gleaned from measuring the polarization of the spectrum. A good example of the effectiveness of this procedure is the discovery that NGC 1068, a classical Seyfert 2 galaxy, is in fact a Seyfert 1 with an obscured broad emission line region (Antonucci and Miller 1985). Seyfert galaxies are characterized by bright nuclei with strong emission lines; those with line widths well in excess of 10^3 km s^{-1} are termed Seyfert 1, and those with narrower lines are termed Seyfert 2. The spectrum of NGC 1068 has the intense, narrow lines of a Seyfert 2 galaxy, but Antonucci and Miller showed that the *polarized* spectrum has the broad lines of a Seyfert 1. They interpreted this as indicating that the the broad emission line region is obscured by dust; the polarized flux was attributed to electron scattering by a plasma that has a clear line of sight both to us and to the central source, and that is confined to a region too small to be resolved optically.

A means of confirming this picture was proposed by Krolik and Kallman (1987): the same gas that we observe by its scattering in the optical region of the spectrum should also convert continuum X-rays to Fe Kα lines with a relatively large equivalent width. This prediction was soon confirmed by the *Ginga* satellite (Koyama *et al.* 1989). The efficiency of this process is enhanced by about a factor 2 due to resonant scattering of the continuum by the large number of Kα lines in the various stages of ionization of Fe (Band *et al.* 1990). In addition, Band *et al.* pointed out that strong Fe L-shell emission should be present. Such emission has recently been observed with BBXRT (Marshall *et al.* 1992); the emission is centered at 870 eV, with a width of 120 eV and an equivalent width of 340 eV. The quantitative interpretation of the X-ray spectrum of NGC 1068 should shed new light on the nature of Seyfert galaxies.

1.4.2 Reverberation mapping

A detailed picture of both the geometrical structure and the dynamics of a relatively compact, photoionized source can be obtained through the technique of reverberation mapping (Blandford and McKee 1982). Consider the broad emission line region of an active galactic nucleus (AGN) or the accretion disk of an X–ray binary, for example. Now imagine that the central source produces a brief flash of ionizing radiation, $L_c(t) = \delta(t)$. When this burst of radiation reaches an element of gas in the system, it will cause a brief change in the intensity of the emission line spectrum of the gas. The observer will see this emission shifted in both frequency and in time, with the frequency shift corresponding to the radial velocity of the emitting gas, v, and the time delay τ corresponding to the additional light travel time associated with the longer pathlength. Each emission line in a given source is thus characterized by a *transfer function* $\Psi(v, \tau)$ which gives the response of the line to a flash of ionizing radiation as a function of the delay time and of the frequency (measured in velocity units). The transfer function is the projection of the six dimensional phase space of the source (emissivity as a function of position and velocity) onto two dimensions: the spatial information is integrated over a paraboloid corresponding to the surface of constant delay time τ, and the velocity information is reduced to the radial velocity v. As such, it is impossible to completely reconstruct the source from the transfer function, but, particularly if the source has an underlying symmetry (spherical or axial, for example) it is possible to learn a great deal. Graphic depictions of the transfer function for different models have been presented by Welsh and Horne (1991).

Real sources generally do not oblige us with flashes of ionizing continuum radiation, and the line luminosity observed at a time t is an integral over the past history of the continuum emission of the source:

$$L_l(v,t) = \int_0^\infty \Psi(v,\tau) L_c(t-\tau) d\tau. \tag{1}$$

With an extensive set of observations of the continuum and line fluxes, this equation can be formally inverted using Fourier transforms to find the transfer function $\Psi(v,\tau)$ (Blandford and McKee 1982). A more robust method of evaluating the transfer function is to use the maximum entropy method, and this has been done by Krolik *et al.* (1991) using the extensive IUE data on the NGC 5548 (Clavel *et al.* 1991). The accuracy of the data gathered in this extensive campaign was in fact only adequate to determine the velocity–integrated transfer function, $\Psi(\tau) = \int \Psi(v,\tau) dv$. The high–ionization lines show a peak near zero time delay, and fall off to a lower value in 15–20 days. The low ionization line C III] $\lambda 1909$ shows a quite distinct behavior, with a peak in the transfer function at a lag of 25 days and a gradual roll off thereafter. Eventually, it should be possible to obtain the full transfer function $\Psi(v,\tau)$, and in that case Blandford and McKee have shown how the transfer function can be formally inverted to determine the emissivity and the moments of the velocity distribution of the emitting gas.

The technique of reverberation mapping should be particularly effective for X–ray sources, since the variability occurs on timescales of seconds to minutes for compact galactic X–ray sources and hours for AGN. For an accretion disk, it is possible to infer both the mass of the central object and the inclination of the disk (Blandford and McKee 1982; Stella 1990). AGN produce a greater observed flux of X–rays in a Schwarzschild crossing time than do galactic X–ray sources, and so they should be particularly amenable to this treatment (Stella 1990).

2 ATOMIC DATA

As the foregoing discussion has demonstrated, astrophysical spectroscopy is an extremely powerful tool for elucidating the physical processes occurring in astronomical sources of radiation as well as their geometrical and dynamical structure. However, this potential can be realized only if the atomic and molecular data used to interpret the observations are of sufficient quality. Here I shall illustrate this with a single example, that of dielectronic recombination rates.

The realization that dielectronic recombination is the dominant recombination process for incompletely ionized plasmas over a wide range of temperature was a major advance in our understanding of astrophysical plasmas (Burgess 1964). In this process, an incident electron is captured by an ion, forming a doubly excited state; usually, the ion autoionizes without recombining, but sometimes it radiates into a bound state. Since the original work of Burgess, a good deal of effort has gone into calculating more accurate dielectronic rates. In particular, Jacobs et al. (1980 and references therein) pointed out that it was essential to include autoionization to excited states of the recombining ion, a change which substantially reduced the dielectronic rate for a number of ions. Some of the transitions included by Jacobs et al. are energetically inaccessible, however (e.g., Smith et al. 1985). Using a quantum defect method similar to that of Jacobs et al., Romanik (1988, 1992) has calculated dielectronic rates of astrophysical interest by considering only energetically allowed transitions and by including captures to forbidden doubly excited states. Rates for the He, Li, Be, and Ne isoelectronic sequences were presented in the first paper; the rates for the remaining sequences between H and Ca are presented in the second. He has obtained good agreement with recent ab initio distorted wave calculations (e.g., Roszman 1987a,b). Fig. 1a shows a comparison between Romanik's (1992) results and Roszman's results for Fe XVIII; the discrepancy between the two may be due to the fact that Roszman used configuration–averaged energies in his calculation. In general, Romanik's results are intermediate in value between those of Jacobs et al. and those obtained with the original Burgess formula.

The values of the dielectronic recombination rates can have a substantial effect on the ionization equilibria of collisionally ionized plasmas. Romanik's (1992) results are compared with those of Shull and van Steenberg (1982a,b), who used the rates of Jacobs et al., in Fig. 1b. The temperature at which a given ion reaches its peak

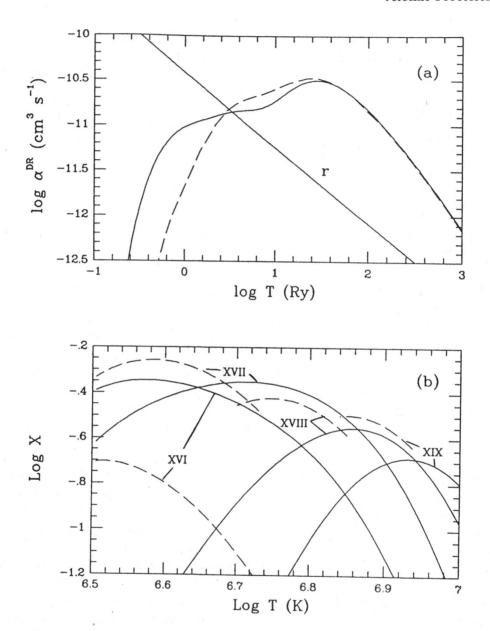

Figure 1. a) Dielectronic recombination rate coefficient for Fe XVIII. *Solid line* - results of Romanik (1992), *dashed line* - results of Roszman (1987b). Radiative recombination shown for comparison (r).

b) Ionization equilibrium calculation for Fe. *Dashed lines* - atomic data of Shull and vanSteenburg (1982a), *solid lines* - dielectronic data of Romanik (1990, 1992) substituted.

abundance is shifted upward by an amount up to 0.13 dex due to the improvements in the dielectronic rates; the value of the peak abundance can change by a larger factor, up to 0.35 dex.

This is but one example of the need for more accurate atomic data for the interpretation of astronomical spectra. The recent report of the Astronomy and Astrophysics Survey Committee (1991) highlighted this need, pointing out that the current level of support is inadequate and urging the federal funding agencies—NSF, NASA, and DOE—to support the laboratory and theoretical work needed for the interpretation of the wealth of data anticipated during the coming decade.

Acknowledgments. I wish to thank Carl Romanik for helpful remarks and for providing results in advance of publication; I also thank Richard Klein for comments that improved the presentation of the paper. My research is supported in part by NSF grant AST89–18573.

REFERENCES

Angel, R. 1977, *Ap. J.*, **216**, 1.

Antonucci, R. R. J., and Miller, J. S. 1985, *Ap. J.*, **297**, 621.

Bajtlik, S., Duncan R. C., and Ostriker, J. P. 1988, *Ap. J.*, **327**, 570.

Band, D. L., Klein, R. I., Castor, J. I., and Nash, J. K. 1990, *Ap. J.*, **362**, 90.

Blandford, R. D., and McKee, C. F. 1982, *Ap. J.*, **255**, 419.

Burgess, A. 1964, *Ap. J.*, **139**, 776.

Canizares, C. R., and Winkler, P. F. 1981, *Ap. J. (Letters)*, **246**, L33.

Chevalier, R. A., and Raymond, J. C. 1978, *Ap. J.*, **225**, L27.

Clavel, J. *et al.* 1991, *Ap. J.*, **366**, 64.

Field, G. B., and Hitchcock, J. L. 1966, *Phys. Rev. Letts.*, **16**, 817.

Gabriel, A. H. 1972, *M. N. R. A. S.*, **160**, 99.

Gabriel, A. H., and Jordan, C. 1969, *M. N. R. A. S.*, **145**, 241.

Green, J., Jelinsky, P., and Bowyer, S. 1990, *Ap. J.*, **359**, 499.

Hamilton, A. J. S., Sarazin, C. L., and Szymkowiak, A. E. 1986, *Ap. J.*, **300**, 713.

Jacobs, V. L., *et al.* 1980, *Ap. J.*, **239**, 1119.

Jenkins, E.B. 1987, in *Interstellar Processes*, ed. D. J. Hollenbach and H. A. Thronson (Dordrecht: Reidel), p. 533.

Jenkins, E. B., and Shaya, E. J. 1979, *Ap. J.*, **231**, 55.

Koyama, K., *et al.* 1989, *Pub. Astr. Soc. Japan*, **41**, 731.

Krolik, J. H. *et al.* 1991, *Ap. J.*, **371**, 541.

Krolik, J. H., and Kallman, T. R. 1987, *Ap. J. (Letters)*, **320**, L5.

Krolik, J. H., McKee, C. F., and Tarter, C. B. 1981, *Ap. J.*, **249**, 422.

Kudritzki, R. P., and Hummer, D. G. 1990, *Ann. Rev. Astr. Ap.*, **28**, 303.

Liedahl, D. A., Kahn, S. M., Osterheld, A. L., and Goldstein, W. H. 1992a, *Ap. J.*, submitted.

Liedahl, D. A., Kahn, S. M., Osterheld, A. L., and Goldstein, W. H. 1992b, in preparation.

Marshall, F. E., *et al.* 1992, in *Testing the AGN Paradigm*, ed. S. S. Holt (in press).

Martin, C., and Bowyer, S. 1990, *Ap. J.*, **350**, 242.

Mather, J. C., *et al.* 1990, *Ap. J. (Letters)*, **354**, L37.

McKee, C. F., and Draine, B. T. 1991, *Science*, **252**, 397.

Meyer, D. M., and Jura, M. 1985, *Ap. J.*, **297**, 119.

Murakami, T., *et al.* 1988, *Nature*, **335**, 234.

National Research Council, *The Decade of Discovery in Astronomy and Astrophysics* (Washington DC: National Academy Press) p. 68 ff.

Neufeld, D. A., and McKee, C. F. 1988, *Ap. J. (Letters)*, **331**, L87.

Raymond, J. C. 1991, *Pub. Astr. Soc. Pac.*, **103**, 781.

Romanik, C. J. 1988, *Ap. J.*, **330**, 1022.

Romanik, C. J. 1992, *Ap. J.*, submitted.

Roszman, L. J. 1987a, *Phys. Rev. A*, **35**, 2138.

Roszman, L. J. 1987b, *Phys. Rev. A*, **35**, 3368.

Schachter, J., Filippenko, A. V., and Kahn, S. M. 1990, *Ap. J.*, **362**, 74.

Schattenburg, M. L., and Canizares, C. R. 1986, *Ap. J.*, **301**, 759.

Shull, J. M., and van Steenberg, M. 1982a, *Ap. J. Suppl.*, **48**, 95.

Shull, J. M., and van Steenberg, M. 1982b, *Ap. J. Suppl.*, **49**, 351.

Smith, M. W., Raymond, J. C., Mann, J. B., and Cowan, R. D. 1985, *Ap. J.*, **298**, 898.

Stella, L. 1990, *Nature*, **344**, 747.

Turnshek, D. A., Foltz, C. B., Grillmair, C. J., and Weymann, R. J. 1988, *Ap. J.*, **325**, 651.

Vedder, P. W., Canizares, C. R., Markert, T. H., and Pradhan, A. K. 1986, *Ap. J.*, **307**, 269.

Welsh, W. F., and Horne, K. 1991, *Ap. J.*, **379**, 586.

Zirin, H. 1966, *The Solar Atmosphere* (Waltham: Blaisdell), p. 248.

X-ray Spectroscopy of Laboratory Plasmas

William H. Goldstein

L-Division, Lawrence Livermore National Laboratory

ABSTRACT

A biased overview is presented of x-ray spectroscopy of laboratory plasmas through a discussion of four lines of experimental investigation that are extremely active. These topics include laser-produced plasmas, specifically spectroscopic modeling and diagnostics in the late-time recombining phase; low density plasma sources, including EBITs and Tokamaks, for the study of isolated-ion atomic processes, particularly indirect excitation mechanisms; absorption spectroscopy of x-ray heated plasmas, with application to the measurement of heavy element opacities; and, finally, ultrashort pulse laser-produced plasmas. The discussion of each plasma source and application emphasizes more-or-less the trend towards better control and/or independent measurement of conditions in the plasma, and the concomitant reduction in ambiguities of analysis and heightened leverage of spectroscopic investigation. The choice of topics also reflects recent trends towards investigating systems of increasing complexity, including ionization transients and photoionized plasmas, near-solid and solid density plasmas, and the plasma spectroscopy of complex, heavy elements.

1 INTRODUCTION

Through the spectroscopic study of plasma created in the laboratory, we have the opportunity to access in a controlled and reproducible environment the atomic structure and dynamics of this ubiquitous state of matter. Spectroscopy provides a direct probe of macroscopic conditions in the plasma, including temperature, density, and charge-state distribution. The connection is made through the dependence of the emitted (or absorbed) spectrum on the average populations of the various atomic states accessible to an ion in the plasma, which in turn depend on the plasma conditions through electron, photon, and ion collision rates. Generally, though, the interpretation of spectra from laboratory sources is complicated by non-spectroscopic uncertainties arising from plasma inhomogeneities and gradients, line-of-sight integrations, and transport effects. Among the most important goals of the experimental study of plasma spectra is the ability to make precision measurements in well-characterized – preferably customized – homogenous plasma environments. In the last few years, new plasma sources and instrumental techniques, as well as increased computational sophistication have developed to realize this ideal over a large range of plasma densities.

In this overview, I will discuss four divers lines of spectroscopic investigation that are extremely active. The topics include laser-produced plasmas, specifically spectroscopic modeling and diagnostics in the late-time recombining phase; low density plasma sources, for the study of isolated-ion atomic processes, particularly indirect excitation; absorption spectroscopy of x-ray heated plasmas, with application to the measurement of heavy element opacities; and, finally, ultrashort pulse laser-produced plasmas. The discussion of each plasma source and application emphasizes more-or-less the trend towards better control and/or independent measurement of conditions in the plasma, and the concomitant reduction in ambiguities of analysis and heightened leverage of spectroscopic investigation. The choice of topics also reflects recent trends towards investigating systems of increasing complexity, including ionization transients and photoionized plasmas, near-solid and solid density plasmas, and the plasma spectroscopy of complex, heavy elements.

2 K-SHELL SPECTRA OF RECOMBINING LASER-PRODUCED PLASMAS

In highly transient, high density laser-produced plasmas, formed by the ablation of hot material from the surface of a solid irradiated by a high-power laser, the experimental problem has been to resolve the gradients and transients produced by the laser heating of the plasma on the one hand, and rapid expansion and cooling on the other. These plasmas are formed in the laser-driven implosion of ICF targets; they provide the amplification media for x-ray lasers, as well as useful sources of high-intensity, incoherent x-rays. The need for both time- and space-resolution has been recently met by the perfection of framing x-ray crystal spectrographs that are capable of taking 200 picosecond "snapshots" of the spatially imaged plasma plume. In experiments using framing cameras, the recombining phase of a laser-produced plasma has been resolved and its temperature and density successfully measured spectroscopically.

2.1 Some History

The first time-resolved line spectra of a laser-produced plasma were studied very soon after the development of high-sensitivity, picosecond resolution x-ray streak cameras. In 1980, Key *et al.* measured heliumlike aluminum resonance and intercombination lines with the goal of determining the electron density. The spectra were time-resolved but spatially integrated. The time-resolution allowed them to observe, for the first time, a late-time, "supercooled" recombination phase, characterized by a departure of the density-sensitive resonance-to-intercombination line strength ratio from that expected assuming self-similar planar expansion.

In 1979, Boiko et al., had inferred similar conclusions based on time-integrated but spatially resolved line strength ratio measurements far from the target surface. Boiko continued this line of research during the first half of the 1980's (Boiko et al. 1983, 1984), though without time-resolution, and with uncontrolled radial gradients in his plasmas. Boiko developed a quasisteady-state model (Boiko et al. 1984) to analyze his spectra that did not assume collisional-radiative equilibrium, but, owing to a dearth of collisional-radiative data, was limited to treating coronal conditions.

Nakano and Kuroda (1983) sharply criticized the previous work, claiming that a fully time-dependent spectroscopic model was required to correctly analyze laser-produced plasma spectra. If they were correct, the utility of spectroscopic diagnostics would be seriously compromised in this application, since a time-dependent model would require temperature and density histories as input, rather than provide these parameters as results of the analysis.

In 1983, Burkhalter et al. introduced a new experimental technique, microdot spectroscopy, designed to minimize spatial gradients, two-dimensional effects, and optical depth uncertainties from the analysis of laser-produced plasma spectra. As shown in Fig. 1, the method leads to a small, confined, one-dimensional blow-off plasma of the dot material. The plasma is spatially imaged along its plume to spectroscopically resolve small isolated volumes of plasma. In long pulse (3-4 ns), time-integrated experiments, they found good agreement between temperatures and densities determined by hydrodynamic simulation, and by collisional-radiative equilibrium line ratios.

The introduction of microdot spectroscopy was a major advance in

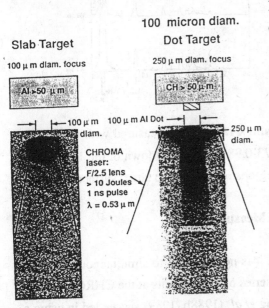

Fig. 1—X-ray pinhole pictures comparing the blowoff plasma of an aluminum slab (left) and microdot (right). From Young et al. (1988a).

controlling the conditions that are spectroscopically observed in laser-produced plasmas, but there remained questions in the original work owing to the absence of time-resolution. In particular, measurements that were integrated over the duration of the x-ray emission were compared with snapshots of plasma conditions at a particular time provided by the simulation. Subsequently, Kauffman, Lee and Estabrook (1987) obtained time-resolved spectra from sulfur microdots with a streak camera, but without imaging along the plasma plume. Their results echoed the time-resolved experiment of Key *et al.* (1980) in detecting significant disagreements between late-time emission measurements and simulations.

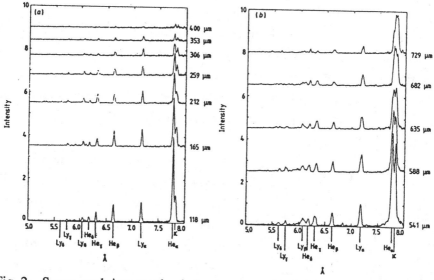

Fig. 2—Space- and time-resolved aluminum x-ray spectra measured with the FCXS. The exposures are (a) 0.23-0.48 ns and (b) 0.73-0.98 ns after shutdown of the 1.0 ns laser heating pulse. The spectra represent plasma regions relative to the target surface.

2.2 Space-, Time- and Spectrally-Resolved Measurements

Of course, with film and a streak camera it is not possible to simultaneously obtain spectral, spatial and time resolution. But in a series of experiments at the CHROMA laser at KMS Fusion, Inc., in 1987 and 1988, Young *et al.* (1988b, 1989) succeeded in using a Framing Crystal X-ray Spectrometer (FCXS) to take 200 ps long exposures of a spatially imaged aluminum microdot plasma plume. As shown in Fig. 2, this data revealed striking spatial and temporal variations in the emission late in time. Two general features are especially noteworthy. First, the comparable and, and, in fact, increasing, strength in Fig.

2(b) of the spin-forbidden $1s^2$-$1s2p(^3P_1)$ intercombination (IC) line relative to the resonant $1s^2$-$1s2p(^1P_1)$ (He_α) line of heliumlike aluminum. The ratio of these lines is strongly dependent on density, with a prominent forbidden line implying lower density. If these lines are formed by collisional excitation, the low density limit of the IC/He_α ratio is around .7. But if they are formed primarily by recombination (and cascades), the ratio exceeds 1.0 for densities below 5×10^{19} cm^{-3}. Thus, the data clearly indicates that the plasma is recombining late in time. The second highlight in the heliumlike spectrum is the inverted order of intensities in the He_β, He_γ and H_δ lines observed at distances of greater than about 600 μm from the target surface. This behavior indicates an excess of population in higher Rydberg levels and population inversions between metastable levels in the n=4 and n=5 manifolds and the resonantly decaying $1s3p(^1P_1)$. These inversions are also characteristic of a recombining plasma.

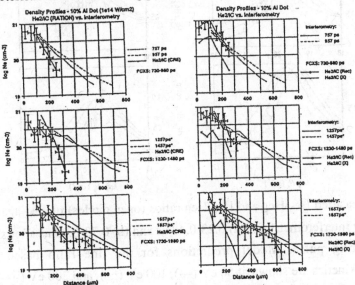

Fig. 3—Interferometrically measured densities compared with spectroscopic densities from the IC/He_α ratio in several models.

In the same experiment, model-independent measurements of electron density and temperature were obtained from holographic interferometry (Busch *et al.* 1985) and free-bound continuum emission (Young *et al.* 1988b), respectively. With independently measured temperatures and densities, it was finally possible to compare various spectroscopic models as diagnostics for these parameters, without the complications of spatial and temporal gradients (Young *et al.* 1989). In Fig. 3 interferometrically measured densities are compared with the spectroscopic densities from the IC/He_α ratio in several models. Both a collisional radiative equilibrium (CRE) model and a model assuming line formation by collisional excitation alone substantially underestimate the density. On the other hand, a quasisteady state model assuming line formation by recombination shows excellent agreement with the interferometry. This model also explained the He_γ/He_β and He_δ/He_β ratios, as shown in Fig. 4. The sub-100 eV temperature indicated by this analysis was in agreement with that obtained from the slope of free-bound continuum emission. Thus, given suitably resolved data and the right model, spectrsoscopic diagnostics can be accurate and useful, even in the highly transient environment of a laser-produced plasma.

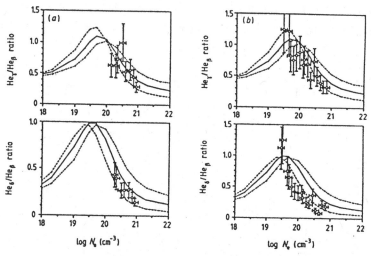

Fig. 4—Measured He_γ/He_β and He_δ/He_β ratios (open circles) plotted against density as measured by interferometry for (a) 0.23-0.48 ns and (b) 0.73-0.98 ns after shutdown of the laser heating pulse. Model predictions for the line ratio based on purely recombination kinetics are shown at 50 eV (---), 100 eV (—), and 200 eV (·····).

3 NEONLIKE KINETICS IN AN EBIT

At low densities, "customizing" the plasma has meant eliminating the transport issues and line-of-sight integrations that compromise the utility of tokamak spectra, and, in effect, eliminating the plasma itself as the perturbing environment, replacing it by a narrow, well-defined, electron beam. In an Electron Beam Ion Trap (EBIT), this beam serves not only to strip arbitrary elements to any charge state, and to collisionally populate excited states with high specificity, but also confines the ions over long time periods for precision study of their spectra (Levine *et al.* 1988; Marrs *et al.* 1988). The

EBIT provides a unique capability for measuring atomic structure and cross sections in the isolated-ion limit, and for efficiently expanding the experimental data base of atomic energy levels, oscillator strengths and transition cross sections. Among recent EBIT applications have been precise wavelength determinations for neonlike radiative transitions in elements as heavy as thorium (Beiersdorfer 1991a); measurements of electron collisional transition rates (Marrs *et al.* 1988), including resonance contributions (Beiersdorfer *et al.* 1990); dielectronic recombination rates in heliumlike (Knapp *et al.* 1989) and neonlike ions (Schneider *et al.* 1992); and the observation of highly forbidden magnetic octupole decays in nickellike uranium (Beiersdorfer *et al* 1991b). The precision and flexibility of the EBIT have been instrumental in evaluating the feasibility of a variety of proposed resonantly photopumped x-ray laser schemes (Beiersdorfer *et al* 1991c).

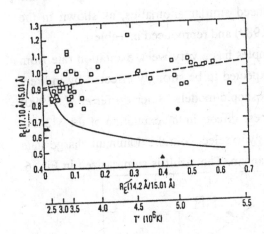

Fig. 5—Energy flux ratio of the 17.10 and 15.01 Å neonlike iron lines plotted as a function of the temperature sensitive ratio of fluorinelike and neonlike resonance lines. The solid line represents theoretical calculations by Smith *et al.* (1984); while measurements are shown as open squares.

3.1 Anomalies in Neonlike Line Ratios

In a recent experiment, EBIT spectroscopy has helped clarify the importance of various indirect line formation processes involved in an important temperature and density dependent line intensity in neonlike ions. These indirect processes – resonance excitation, ionization and recombination to excited states, and cascades – are difficult andpainful to include accurately in spectroscopic models, but their absence can severely compromise the diagnostic utility of spectral lines.

Neonlike x-ray spectra have been intensely studied for both astrophysical and laboratory applications. In the solar corona, the ratio of the FeXVII magnetic quadrupole (M2) line at 17.01 Å, $2p^6$-$2p^53s(^3P_2)$, and resonance line at 15.01 Å,

TABLE 1

PREDICTED AND EXPERIMENTAL INTENSITY RATIOS OF LINES M2 AND 3G

Element	M2/3G		Te
	Experiment	Theory	(keV)
Ge	0.83	0.45	1.5
Se	0.80	0.41	2.0
Ag	0.74	0.38	3.0

$2p^6-2p^53d(^1P_1)$, is predicted to be highly temperature sensitive. A comparison between data from a range of solar regions and model predictions was carried out by Rugge and McKenzie (1985), with the disappointing results shown in Fig. 5.

Tokamak measurements involving the same M2 line in germanium, selenium and silver, and a $2p^6-2p^53s$ line (3G) have evinced similar anomalies, as shown in the comparisons collected by Beiersdorfer *et al.* (1989) and reproduced in Table 1.

The M2 transition common to these examples has a very week excitation rate from the neonlike ground state, and is therefore expected to be strongly effected by indirect processes not typically included in the spectroscopic models. Beiersdorfer *et al.* (1990) used an EBIT to systematically demonstrate these effects in the excitation of the M2 line in barium, by adjusting the electron beam energy to select both the dominant charge state in the trap, and the dominant excitation mechanism. The result is summarized in Fig. 6.

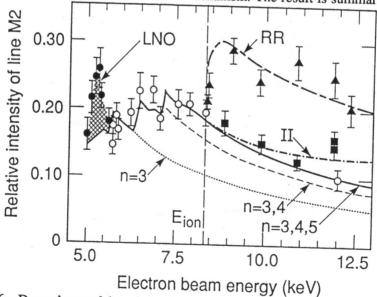

Fig. 6—Dependence of the relative intensity of the M2 line on EBIT beam energy.

Below the ionization limit, the analysis shows that the M2 line is predominantly formed through resonant excitation below about 6 keV (solid circles), and by cascades from the n=3,4 and 5 manifolds above this energy (open circles). The importance of cascades is demonstrated by the comparison of the data to spectroscopic models that included the

n=3 neonlike levels (dotted line), the n=3 and n=4 levels (dashed line) and the n=3,4 and 5 levels (solid line), with agreement obtaining over the full range of energies only for the latter. Above the ionization limit, it was possible to study line

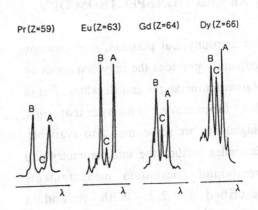

Fig. 7—4s-4p$_{3/2}$ spectra of rare earths in the TEXT tokamak. A: copperlike; B: zinklike; C: galliumlike.

formation by the additional processes of recombination of fluorinelike ions (solid triangles) and innershell ionization of sodiumlike ions (solid squares).

3.3 Tokamak Spectroscopy and Indirect Processes

Tokamak spectroscopy has recently elucidated the role of another indirect process, excitation-autoionization, and its dramatic effect on ionization balance (Mandlebaum *et al* 1990). As shown in Fig. 7, when rare earth atoms were introduced into the steady-state phase of the standard discharge in the TEXT tokamak, the relative intensities of 4s-4p lines in copperlike, zinklike and galliumlike ions exhibited a strong systematic variation with Z. The apparent explanation for this trend is the progressive loss of the galliumlike indirect ionization channel

$$3d^{10}4s^24p + e^- \rightarrow 3d^94s^24p4l + e^- \rightarrow 3d^{10}4s4p, \ 3d^{10}4s^2 + 2e^-.$$

The first step in this process is impact excitation. The second is autoionization, and proceeds only if the intermediate $3d^94s^24p4l$ level is above the ionization limit. As shown in Fig. 8, the availability of this channel, and thus the ionization balance, is strongly dependent on Z.

4 ABSORPTION SPECTROSCOPY

For astrophysical plasmas, spectroscopy frequently provides the only constraint on plasma temperatures and densities. But in the laboratory, non-spectroscopic diagnostics are sometimes also available. Examples include the interferometry and free-bound continuum measurements described in 2.2. With redundant measurements of conditions, it becomes possible to use plasma experiments to test the atomic models that, for example, enter radiation flow simulations. Recent experiments have succeeded in measuring the absorption spectrum of heavy elements under known plasma conditions as a test of the predictive capability of LTE models for the opacity of high density plasma.

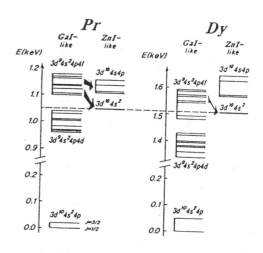

Fig. 8—Energy diagram for configurations relevant to excitation autoionization in galliumlike ions. Lines within configurations represent jj coupling relativistic subconfigurations. Arrows indicate autoionization channels

These experiments used the technique of absorption, or backlighter, spectroscopy, a method that has also been recently used as a direct probe of ionization balance (Abdallah and Clark 1990), to measure Stark broadening of high-n lines (Springer $et\ al.$ 1991), and as a diagnostic for growth rates of hydrodynamic instabilities (see, for example, Goldsack $et\ al.$ 1991; Molitoris $et\ al.$ 1991). To measure the frequency dependent opacity of a sample, it is necessary to record simultaneously the unattenuated backlighter spectrum, the spectrum transmitted through the sample, and the self-emission of the sample, which must be subtracted from the signal. This is accomplished through a technique called point projection spectroscopy, described originally by Lewis and McGlinchey (1985) and first applied to opacity measurements by Davidson $et\ al.$ (1988). The point-projection method is shown schematically in Fig. 9.

In opacity experiments, the necessity of dealing with a homogenous, gradient-free volume of well-characterized plasma is exacerbated by the additional requirement that it be large enough to produce measurable absorption of the backlighter, and be in LTE. Therefore, the plasma is created by bathing the sample with x-rays, rather than ablating it with a high-power laser. The x-rays heat the target more uniformly, at a higher density and to lower temperature. The result is an isothermal plasma in LTE that expands uniformly and, if it has the right aspect ratio, essentially in one-dimension.

Since the ionization balance, and thus the opacity, of LTE plasma is governed by Saha-Boltzmann statistics, models are highly sensitive to temperature and density, and it is essential to independently measure them to rather high precision. This step has only recently been achieved in measurements of niobium opacity at NOVA by Perry *et al.* (1991) and Springer *et al.* (1991). In these experiments, schematically shown in Fig. 10, the density was determined by spectroscopically radiographing the expanding sample side-on to measure its physical extent. The temperature was obtained by adding aluminum to the niobium target, and using ratios of absorption lines from the lithiumlike through carbonlike charge-states as a thermometer. This procedure was not entirely satisfactory, since it actually relies on a model for the aluminum opacity. But this model is expected to be considerably more accurate than that for the much more complex niobium component. Using these diagnostics, the density was determined to within 20%, while the temperature was narrowed down to 47±3 eV. Comparisons between the measured niobium transmission and predictions using the super transition array (STA) model of Bar-Shalom *et al.* (1989, 1990, 1991) are shown in Figs. 11 and 12. It should be noted that under these plasma conditions, the niobium is stripped only into its N-shell. Thus the features observed are not generally individual spectral lines, but quasi-continua

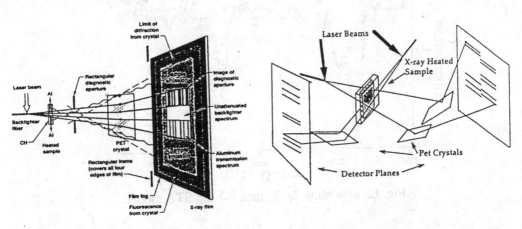

Fig. 9—The point-projection method for simultaneous measurement of unattenuated backlighter and transmitted spectrum.

Fig. 10—Experimental setup showing side-on radiography, and crystals for recording both niobium and aluminum transmission spectra.

built up out of a tremendous number of unresolved transition arrays or UTAs (Bauche *et al* 1988). The agreement between data and prediction for the structure of these quasi-continua, shown most dramatically in Fig. 12, is a spectacular validation of the opacity model

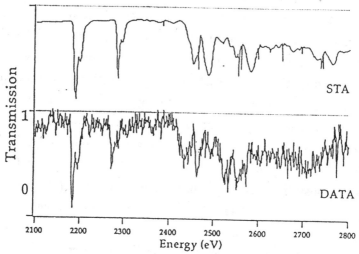

Fig. 11—Niobium transmission data and STA model.

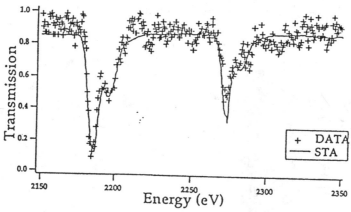

Fig. 12—Niobium 2p-3d data (+) and STA model.

5 ULTRASHORT PULSE LASER-PRODUCED PLASMA SPECTROSCOPY

Ultrashort pulse (USP) laser technology, providing subpicosecond pulse lengths and power densities of up to 10^{19} W/cm^2 from table-top systems, has significantly expanded the field of laser-produced plasma research. Applications presently under investigation include x-ray lasers (Burnett and Corkum 1989), ultrashort pulse x-ray flashlamps (Murnane *et al.* 1989), and the study of the laser-matter interaction at high density and temperature.

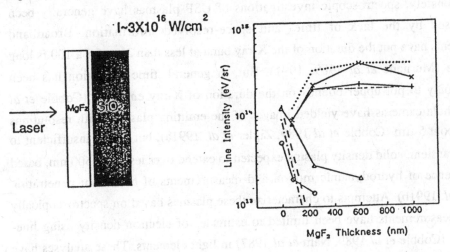

Fig. 13—Line intensities of silicon and magnesium $L\alpha$ (o,Δ) and $He\beta$ (+,$\times$) as a function of MgF_2 thickness on SiO_2 substrates.

With these systems it has become feasible to study high temperature plasma without the complication of hydrodynamic motion. The laser energy is absorbed in, and heats, a skin depth of solid material, and the resulting plasma remains at solid density for the duration of the interaction. At the end of the laser pulse, owing to the very high density, the plasma recombines rapidly through enhanced three-body processes, while heat is quickly conducted into the cool solid. The result is a very small volume of hot, short-lived, solid density plasma that cools before it can disassemble into an inhomogenous plasma plume, replete with gradients and hydrodynamic effects. The opportunities presented by such sources for studying spectroscopy in a well-defined, solid density plasma environment, including line shapes, plasma polarization shifts, continuum lowering, radiation transport and solid-state effects, are matched only by the uncertainties of how to experimentally resolve in time a plasma that lasts less than half a picosecond, and how, theoretically to describe energy absorption in a solid at irradiation intensities exceeding 10^{17} Watts/cm^2. The latter question raises the problem of treating atomic processes in strong fields. For example, at 10^{17} Watts/cm^2 the electric field strength exceeds 10^9 Volts/cm, roughly the field that binds the electron in a hydrogen atom. Under such extreme conditions, the huge ponderomotive forces and relativistic electron motion can give rise to multielectron ionization, while the intense laser field can lead to multiphoton ionization. Unfortunately, the influence of a plasma environment on these many-body phenomena has yet to be determined.

Unfortunately, spectroscopic investigations of USP plasmas have generally been compromised by the lack of time- and space-resolved information. Broadband measurements have put the duration of the X-ray burst at less than 2 ps, for a 100 fs long laser pulse (Murnane *et al* 1989, 1991). But, in general, time-resolution has been sufficient only to put upper bounds on the duration of X-ray emission (Cobble *et al* 1989). Pinhole cameras have yielded images of the emitting plasmas with resolutions down to about 5 μm (Cobble *et al* 1989, Zigler *et al.* 1991a), but this is insufficient to resolve a transient, solid density plasma expected to extend over less than 500 nm, based on the absence of hydrodynamic motion and measurements of heat front penetration (Zigler *et al.* 1991b). Attempts to characterize these plasmas based on spectroscopically resolved measurements have been limited to estimates of electron density using line-broadening (Cobble *et al* 1989, Nam *et al* 1987) in light elements. These analyses have encountered ambiguities owing to the long time-scales involved in the atomic kinetics, and have involved short-pulse energy deposition in plasmas preformed by an ASE prepulse and consequently emitting at much lower than solid density(Cobble *et al* 1989; Nam *et al* 1987, 1990).

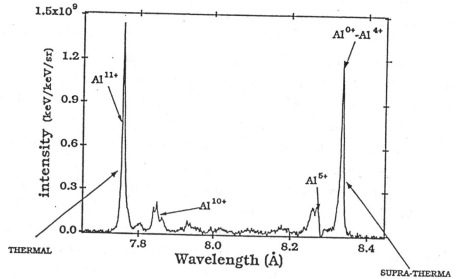

Fig. 14—Aluminum 1s-2p spectrum from an USP laser-produced plasma showing thermal emission from the heliumlike and lithiumlike charge states, as well as fluorescent lines from low ionization stages.

However, spectroscopic experiments have been successful in measuring the energy penetration depth in solid targets of the heat front generated by an USP laser . By varying the thickness of a MgF_2 layer on a SiO_2 substrate and measuring the $L\alpha$ and $He\beta$ yield of the magnesium and silicon, Zigler et al. (1991b) found that the target was heated to a depth of approximately 1000 Å. These results are summarized in Fig. 13.

Bennattar et al (1992) have found spectroscopic evidence for the simultaneous existence of a thermal plasma at the surface of an USP heated sample, and a deeper, cool region emitting fluorescent x-rays from impact ionization by suprathermal electrons. A typical 1s-2p aluminum spectrum is reproduced in Fig. 14, showing thermal emission from the heliumlike and lithiumlike charge states, as well as fluorescent lines from low ionization stages. From the latter, it was possible to determine the temperature and flux of the suprathermal electrons.

These two early examples point to the potential of spectroscopic measurements for helping to analyze and understand the new plasma regimes accessed in USP laser-produced plasmas.

6 REFERENCES

Abdallah Jr., J. and Clark, R. E. H. 1991 J. Appl. Phys. **69** 23.

Bar-Shalom, A., Oreg, J., Goldstein, W. H., Shvarts, D., and Zigler, A. 1989 Phys. Rev. A **40** 3183.

Bar-Shalom, A., Oreg, J., and Goldstein, W. H. 1991, in *Radiative Properties of Hot Dense Matter*, eds. W. Goldstein, C. Hooper, J. Gauthier, J. Seely and R. Lee (Singapore: World Scientific), p. 163.

Bar-Shalom, A., Oreg, J., and Goldstein, W. H. 1992, in *Atomic Processes in Plasmas*, ed. E. Marmar (New York: AIP), in press.

Bauche, J., Bauche-Arnoult, C., and Klapisch, M 1988 Adv. in At. Mol. Phys. **23** 131.

Beiersdorfer, P., Von Goeler, S., Bitter, M., and Hill, K. W. 1989 Nucl. Instrum. and Meth. **B43** 347.

Beiersdorfer, P., et al. 1990 Phys. Rev. Lett. **65** 1995.

Beiersdorfer, P. 1991a Nucl. Instrum. and Meth. **B56/57** 1144.

Beiersdorfer, P., Osterheld, A. O., Scofield, J., Wargelin, B., and Marrs, R. E. 1991b Phys. Rev. Lett. **67** 2272.

Beiersdorfer, P., et al. 1991c preprint; Phys. Rev. A, in press.

Bennattar, R,. et al. 1992 Optics Comm. 88 **376**.

Boiko, V. A., *et al.* 1979 J. Phys. B **12** 213.

Boiko, V. A., *et al.* 1983 J. Phys. B **16** L77.

Boiko, V. A., Skobelev, I. Yu., and Faenov, A. Ya. 1984 Sov. J. Plasma Phys. **10** 82.

Burkhalter, P. G., *et al.* 1983 Phys Fluids **26** 3650.

Burnett, N. H. and Corkum, P. B. 1989 J. Opt. Soc. Am. B **6** 1195.

Busch, Gar. E., Shepard, C. L., Siebert, L. D., and Tarvin, J. A. 1985 Rev. Sci. Instrum. **56** 879.

Cobble, J. A., *et al.* 1989 Phys. Rev. A **39** 454.

Davidson, S. J., Foster, J. M., Smith, C. C., Warburton, K. A., and Rose, S. J. 1988 Appl. Phys. Lett. **52** 847.

Goldsack, T. J., *et al.* 1991, in the Proceedings of the 3rd International Workshop on The Physics of Compressible Turbulent Mixing, Royaumont, France, in press.

Kauffman, R. L., Lee, R. W., and Estabrook, K. 1987 Phys. Rev. A **35** 4286.

Key, M. H., *et al.* 1980 Phys. Rev. Lett. **44** 1669.

Knapp, D. A., *et al.* 1989 Phys. Rev. Lett. **62** 2104.

Levine, M. A., Marrs, R. E., Henderson, J. R., Knapp, D. A., and Schneider, M. B. 1988 Phys. Scripta **T22** 157.

Lewis, C. L. S. and McGlinchey, J. 1985 Opt. Commun. **53** 179.

Mandlebaum, P., *et al.* 1990 Phys. Rev. A **42** 4412.

Marrs, R. E., Levine, Knapp, D. A., M. A., Henderson 1988 Phys. Rev. Lett. **60** 1715.

Molitoris, J. D., *et al.* 1991, in the Proceedings of the 3rd International Workshop on The Physics of Compressible Turbulent Mixing, Royaumont, France, in press.

Murnane, M. M., Kapteyn, H. C., and Falcone, R. W. 1989 Phys. Rev. Lett. **62** 155.

Murnane, M. M., Kapteyn, H. C., Rosen, M. D., and Falcone, R. W. 1991 Science **251** 531.

Nam, C. H., *et al.* 1987 Phys. Rev. Lett. **59** 2427.

Nam, C. H., Tighe, W., Valco, E., and Suckewer, S. 1990 Appl. Phys. B **50** 275.

Noboru, N. and Kuroda, H 1983 Phys. Rev. A **29** 3447.

Perry, T. S., *et al.* 1991 Phys. Rev. Lett. **67** 3784.

Rugge, H. R. and McKenzie, D. L. 1985 Ap. J. **297** 338.

Schneider, M. B., *et al.* 1992 Phys. Rev. A **45** R1291.

Smith, B. W., Raymond, J. C., Mann, J. B., and Cowan, R. D. 1984 Ap. J. **298** 898.

Springer, P. T. *et al.* 1991, in *Radiative Properties of Hot Dense Matter*, eds. W. Goldstein, C. Hooper, J. Gauthier, J. Seely and R. Lee (Singapore: World Scientific), p. 42.

Springer, P. T., *et al.* 1992, in *Atomic Processes in Plasmas*, ed. E. Marmar (New York: AIP), in press.

Young, B. K. F., Stewart, R. E., Charatis, G., and Busch, Gar. E. 1988a, in *X Rays from Laser Plasmas*, ed. M. Richardson, Proc. SPIE 831, p. 18.

Young, B. K. F., Stewart, R. E., Cerjan, C. J., Charatis, G., and Busch, Gar. E. 1988b Phys. Rev. Lett. **61** 2851.

Young, B. K. F., *et al.* 1989 J. Phys. B **22** L533.

Zigler, A., *et al.* 1991a, Appl. Phys. Lett. **59**, 777.

Zigler, A., *et al.* 1991b, Appl. Phys. Lett. **59**, 534.

METHODS OF CALCULATING ELECTRON-ATOM CROSS-SECTIONS FOR KINETICS APPLICATIONS

I. L. Beigman

P. N. Lebedev Physical Institute, Leninsky pr., 53, Moscow

1 INTRODUCTION

There are various problems associated with the modern theory of electron–atom collisions. For example, there is the problem of calculating selected cross-sections with enough accuracy to check the theory. Another problem deals with the calculation of line intensities observed in plasmas. There are two parts to the line calculations:

1. atomic constants and rate coefficients;
2. the composition and solution of the kinetic equations.

Analysis shows that, in addition to the rate coefficients directly connected with a specific transition, there are many other rate coefficients and atomic constants that influence the selected line intensity (see Figure 1). The methods for calculating this rather large array of cross-sections is the subject of this paper. The methods considered here:

1. are relatively simple to implement, especially with limited computer resources:
2. are accurate to about 30–50%;
3. make it possible for the results to be conveniently represented.

These requirements are necessary primarily because the number of rate coefficients is large. Furthermore, the usual representation, a tabulation of numerical results for cross-sections or rate coefficients, is not convenient for the single solution to the kinetic problem. We need to know the dependence on different parameters, the possible limitation of the number of levels (or ions), and the influence of the errors of atomic constants on the line intensity of interest. An analytical representation (may be an

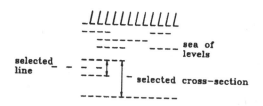

Fig. 1 – Atomic diagram.

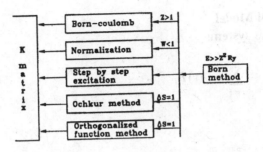

Fig. 2 – The methods based on the perturbation theory.

approximation formula), therefore, is preferable. Below we consider this question in more detail.

There are a number of methods that satisfy the requirements above: classical, semiclassical, and methods based on the perturbation theory of the electron–atom interaction, called the generalized Born approach. The Born method is valid at large energies, but low energies are of primary importance for our applications. The generalized Born method is an extrapolation from large energy to small energies taking into account various physical effects (see Figure 2). All these effects may be described by the K-matrix method.

2 K-MATRIX METHOD (LOW EXCITED LEVELS)

2.1 Background

The K-matrix method was introduced in the theory of electron–atom collisions by Seaton many years ago (Seaton, 1961). He defined the K-matrix by the relation:

$$S = (I - iK)/(I + iK); \quad T = K/(I + iK)$$

where S is the usual scattering matrix, T is the matrix of transition, and the K-matrix is calculated with the help of perturbation theory. In this case, matrix elements of K are proportional to matrix elements of the interaction potential. The K-matrix is, therefore, hermitian and the S-matrix is automatically unitary.

To illustrate the features of this old method, we consider the system of the close-coupling equations with well potentials, which may be solved analytically. Comparisons of the K-matrix method with the analytical solution are presented. Next, this method is applied to the process of ionization. Here we set the diagonal elements of interaction to zero, which is an effective way to compensate for the polarization interaction of the atom with an external electron.

2.2 Potential Well Model

Let us consider the system:

$$\frac{d^2}{dr^2}F_i + k_i^2 F_i = \sum_{i \neq i'} h_{ii'}(r) F_{i'}$$

where

$$h_{ii'}(r) = \begin{cases} h_{ii'} \text{ if } r \leq a \\ 0, \quad \text{if } r > a \end{cases}$$

with boundary conditions:

$$F_{ii'} \xrightarrow[r \to \infty]{} \begin{cases} (k_i)^{-1/2}[\delta_{ii'} \sin(k_i r) + T_{ii'} \exp(ik_i r, \ k_i > 0 \\ 0 \qquad\qquad\qquad\qquad\qquad , \ k_i > 0 \end{cases}$$

First, we consider a two-level system. It is clear that at $h \to 0$ perturbation theory is valid, and the K-matrix method must coincide with the exact result. In the opposite case, $h \gg k^2$, the close-coupling system gives:

$$T \to \frac{-k[\sin(h^{1/2}a) - \cos(h^{1/2}a)]}{h^{1/2}[\cos(h^{1/2}a) - (2ik/h)\sin(h^{1/2}a)] + \cos(h^{1/2}a)} \sim h^{-1/2}$$

the K - matrix :

$$T = -\frac{4k^2}{h[\sin(2ka) - 2ka)} \sim \frac{1}{h}$$

Thus, in a strong interaction region, the K-matrix behaves opposite to the usual perturbation theory decrease of the transition amplitude.

Next we consider a three-level system (see Figure 3). There is a nontrivial moment at small h. If $h_{13} = 0$, $T_{13} \sim h_{12} h_{13}/k^4$ for both cases of the K-matrix and the exact solution,

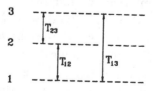

Fig. 3 – Three-level diagram.

this means that the K-matrix method qualitatively takes into account the step-by-step transition in electron–atom collisions.

The case of large h is shown in the table:

$$(h_{23} \gg |h_{12}| + |h_{13}| + \Sigma k_i^2)$$

	T_{12}	T_{13}	T_{23}
Exact solution	$\sim h_{13}/h_{23}$	$\sim h_{13}/h_{23}$	$\sim (h_{23})^{-1/2}$
K-matrix	$\sim h_{13}/h_{23}$	$\sim h_{12}/h_{23}$	$\sim (h_{23})^{-1}$

Again we see the qualitative agreement. It should be noted that the strong interaction between levels 2 and 3 decreases the amplitude of transitions from 1 to 2 and 3. We call it the supressing effect.

2.3 Some Examples

Figure 4 shows the cross-section of the transition 3s–3p in sodium (Na). We see the satisfactory agreement with experimental data and close-coupling method. Here the main effect is normalization.

Figure 5 shows the cross-section of the transition 4s–3d in potassium (K). We see that the step-by-step effect, scheme '4s–4p–3d', gives the cross-section above the Born one. The included levels 5p, 4d, 4f decrease the result (suppressing effect), and the cross-

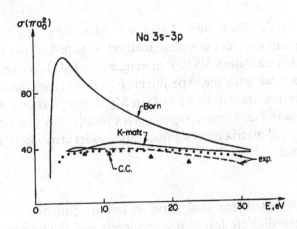

Fig. 4 – The cross-section of the transition 3s–3p in Na. Exp: ••• Enemark, E. A. and Gallagher, A. C. 1972, *Phys. Rev. A* **6**, 192; – – – Phelps J. O. and Lin, C. C. 1981, *Phys. Rev. A* **24**, 1299.

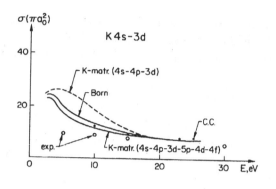

Fig. 5 – The cross-section of the transition 4s–3d in K. Exp: •••. Close coupling: ooo Phelps, J. O., Solomon, J. E., Korff, D. L. and Lin, C. C., (15 levels) 1979, *Phys. Rev. A* **20**, 1418.

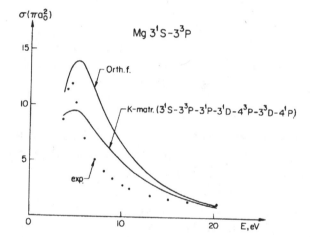

Fig. 6 – The cross-section of the transition 3¹S³–3³P in Mg. Exp: ••• Aleksachin, I. S., Zapesohnnyi, I. P., Garga, I. I. and Starodub, V. P. 1973, *Opt. & Spectr.* **34**, 1054.

section is less than the Born one. Again we see satisfactory agreement with the experimental data and the close-coupling method. Figure 6 shows the cross-section of the intercombination transition 3¹S–3³P in magnesium (Mg). We see that the *K*-matrix improved the agreement with the experiment. Figure 7 shows the cross-section of the other intercombination transition 3¹S–3³D in Mg. Figure 8 shows the cross-section of the transition 3s–4s in Na. Close-coupling gives results that practically coincide with the Born results. The *K*-matrix gives significantly smaller cross-sections which are closer to experimental data.

2.4 Ionization

The generalized *K*-matrix for ionization is infinite-dimensional and involves the interactions between discrete levels, discrete levels and the continuum, and between continuum levels. We consider a simple model that does not include the diagonal elements nor the elements of continuum–continuum interactions:

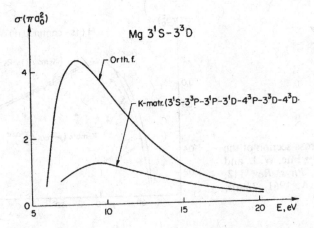

Fig. 7 – The cross-section of the transition 3^1S-3^3D in Mg.

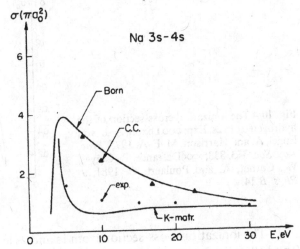

Fig. 8 – The cross-section of the transition 3s–4s in Na. Exp: ••• Phelps, J. O. and Lin, C. C. 1981, *Phys. Rev. A* **24**, 1299. Close coupling (3s–3p–4s–3d–4p): ▲▲▲ Korff, D. L., Chung, S and Lin, C. C. 1973, *Phys. Rev. A* **7**, 545.

$$K = \begin{vmatrix} H_{ab} & Y_{a\beta} \\ Y_{ab} & 0 \end{vmatrix}$$

The Latin indices indices denote the discrete spectra and Greek ones denote the continuum. $Y_{a\beta}$ is the matrix element of the rectangular matrix describing the connection between discrete levels and states of the continuum. For matrix element $T_{a\alpha}$ the transition from level a to continuum level α the expression

$$T = \tfrac{1}{2}i(S - I) = (I + iH + P)^{-1} \cdot Y,$$

where $P_{ab} = Y_{a\alpha} Y_{\alpha b}$, may be obtained.

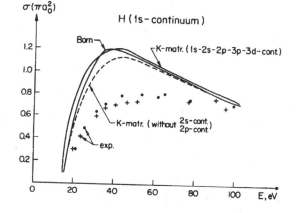

Fig. 9 – The ionization cross-section of the hydrogen from 1s. Exp: ••• Fite, W. L. and Brackmann, R. T. 1958, *Phys. Rev.* **112**, 1141; +++ Boksenberg, A. 1961, *Thesis*, Univ. of London.

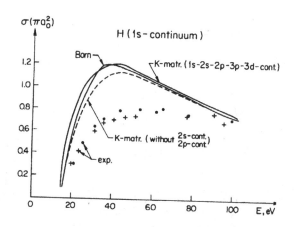

Fig. 10 – The ionization cross-section of the hydrogen from 2s. Exp: ooo Dixon, A. J., von Engel, A. and Harrison, M. F. A. 1975, *Proc. Roy. Soc.* **343**, 333; ×××Defrance, P., Clayes, W., Cornet, A. and Poulaert, G. 1981, *J. Phys. B* **14**.

The ionization cross-section from 1s and 2s levels of hydrogen is shown in Figures 9 and 10. It can be seen that the *K*-matrix cross-section is slightly closer to experimental findings, but, on the whole, the difference between experiment and calculation is significant. It is possible that the continuum–continuum interactions, which are neglected, play a principal role.

3 ANALYTICAL APPROACH (HIGHLY EXCITED LEVELS)

Here we consider the cross-sections averaged over all angular quantum numbers n–n'.

3.1 Dipole and Impulse Approximation

The simplest approach suggested by Seaton thirty years ago (Seaton, 1962) is the dipole approximation. In this approximation the interaction potential is

$$U = e^2 \frac{R\langle n' | \mathbf{r} | n \rangle}{R^3},$$

where R is the distance between electron and atom and the matrix element $\langle n' | \mathbf{r} | n \rangle$ defines the strength of the interaction. The cross-section is proportional to

$$\sigma \sim \pi a_0^2 \frac{n^4}{\Delta n^4} \frac{\ln E}{E}; \quad \Delta n = |n' - n|.$$

The specific dependence on quantum numbers is given by the matrix element and the dependence on energy is given by the shape of the potential. The dipole interaction dominates at long distances and corresponds to slow changes of orbital parameters. The alternative impulse approach has been known for many years. It removes the interactions of the colliding electrons with atomic core and is characterized by sharp changes in the orbital parameters. The cross-section of energy transfer is given by Stabler's formula (Stabler, 1964):

$$\frac{d\sigma}{d\Delta\varepsilon} = (\pi E_{min}^{0.5}) \, e^4 (1/\Delta\varepsilon^2 + 4E_{min}/3\Delta\varepsilon^3)/(E \, E_n^{0.5}),$$

where $\Delta\varepsilon$ is the transfer energy, E is the energy of external particle and E_{min} is the smallest of the four initial and final electron energies. The corresponding cross-section is proportional to

$$\sigma \sim \pi a_0^2 \frac{n^4}{\Delta n^3} \frac{1}{E}.$$

3.2 Born Approximation

The sequential perturbation theory on electron–atom interactions gives (Bergman, Urnov and Shevelko, 1970):

$$\sigma_{n-n'} = (8\pi/k^2)(1/n^2) \int_{k-k'}^{k+k'} f(q) \, q \, dq$$

$$f(q) = \sum_{lm} |\langle n \, l \, m | e^{iqr} | n' \, l' \, m' \rangle|^2.$$

The generalized oscillator strength, $f(q)$, may be expressed through hypergeometrical functions. It is a very cumbersome expression.

$$f(q) = \frac{1}{q} \int_0^q A(q')\, dq',$$

$$A(q) = (nn')^{-2} \operatorname{Re}\{I'(-n+1, -n'+1)\, I'(-n, -n') - I'(-n+1, n')I'(-n, -n'+1)$$

$$-\frac{1}{6}\frac{d^2}{dq^2}[I(-n+1, -n'+1)\, I(-n, -n') - I(-n+1, -n')\, I(-n, -n'+1)]\}$$

$$I(w, w') = \frac{\lambda^{w+w'-1}}{(\lambda - p)^w (\lambda - p')^{w'}}\, F\left(w, w', 1, \frac{pp}{(\lambda - p)\,(\lambda - p')}\right),$$

where $p = 1/n$, $p' = 1/n$, $\lambda = (p+p'+iq)/2$ and $F(a, b, 1, x)$ is a hypergeometric function. The symbol $\Gamma(w, w')$ stands for dI/dq, and I is the complex conjugate of I. In the asymptotic case $n \gg 1$, this formula becomes simpler and may be expressed through Bessel functions. At first, we consider the limiting case $n \gg dn \gg 1$. In this case the Born cross-section has the form:

$$\sigma_{n-n'} = \frac{\pi a_0^2}{Z^4}\, \frac{8}{n^2}\, \frac{Z^2 Ry}{E}\left[\frac{(\varepsilon \varepsilon')^{3/2}}{(\Delta \varepsilon)^4}\, \ln(E/E_n) + \frac{(\varepsilon')^{3/2}}{(\Delta \varepsilon)^2}\left(\frac{4}{3}\frac{1}{\Delta \varepsilon} + \frac{1}{\varepsilon}\right)\right].$$

We see that the first term corresponds to the dipole approximation and the second term corresponds to the impulse approximation. Thus the sequential perturbation theory consists of both alternative approaches. Now we show the general Born formula for cross-section:

$$\sigma_{n-n'} = \frac{\pi a_0^2}{Z^4}\, \frac{8}{n^2}\, \frac{Z^2 Ry}{E}$$

$$\times \left[\left(1 - \frac{0.25}{\Delta n}\right)\frac{(\varepsilon \varepsilon')^{3/2}}{(\Delta \varepsilon)^4}\, \ln(E/E_n) + \left(1 - \frac{0.6}{\Delta n}\right)\frac{E/E_n}{(1+E/E_n)}\frac{(\varepsilon')^{3/2}}{(\Delta \varepsilon)^2}\left(\frac{4}{3}\frac{1}{\Delta \varepsilon} + \frac{1}{\varepsilon}\right)\right]$$

3.3 Semiclassical Approach

It is known that the Born approximation as a rule is not valid for energies about 1/5–1/30 of ionization potential. This is the important region for applications. In the early 1970s, three independent approaches to this problem were suggested: 'The correspondence principle' by Percival and Richards (1970); 'The model of equidistant levels' by Presnyakov and Urnov (1970); and 'Semiclassical description of the Rydberg electron' by Beigman, Vainshtein and Sobelman (1969).

Richards showed that all three approaches are equivalent (Richards, 1972). There is not enough space to discuss this question in detail. Briefly though, our approaches describe the highly excited electron with the help of the classical function of action. The methods of classical mechanics, in particular, and perturbation theory in the variable of action have been used to find the additions to action induced by collisions with charged particles. The final results may be presented in the form:

$$\sigma^Q = \sigma^B\, F(E, Z\Delta n); \quad F(E, Z\Delta n) = \frac{\ln\{1 + (vn/Z\Delta n)/[1.25/vZ\Delta n]^{3/2}}{\ln[1 + (vn/Z\Delta n)]}.$$

3.4 Intercombination Transitions

Now, we consider intercombination transitions averaged over angular quantum numbers. Relative concentrations of highly excited atoms with different spins in a plasma are determined by such transitions, and therefore cross-sections of the pertinent transitions are significant for some kinetics problems. First we discuss a classical approach. From the classical point of view, the intercombination transition consists of the capture of an external electron and the ejection of the atomic one (see Figure 11). The use of the classical impulse approximation is justified owing to the large energy transfer in this process:

$$\Delta E = k^2 + 1/(n')^2$$

In accordance with the classical cross-section of energy transfer mentioned above, we have:

$$\sigma_{n-n'} = (8\pi/k^2)(E_{min}/E_n)^{0.5}\frac{1}{n'^3}\{1/k^2 + 1/n'^2) + 4E_{min}/[3(k^2 + 1/n'^2)^3]\}.$$

Another approach is based on the Ochkur approximation. States with large orbital quantum numbers and small quantum defects give the main contribution to the cross-section, and therefore we may use the hydrogen wavefunctions for matrix elements:

$$\sigma_{n-n'} = (8\pi/k^2 Z^2)\, F/(k^2 + 1/n^2)^2 (1/n^2)\int_{k-k'}^{k+k'} f(q)q\, dq, \quad F = (2S+1)/2(2S_p + 1),$$

where S, S_p are spins of the atom after collision and the atomic core, respectively.

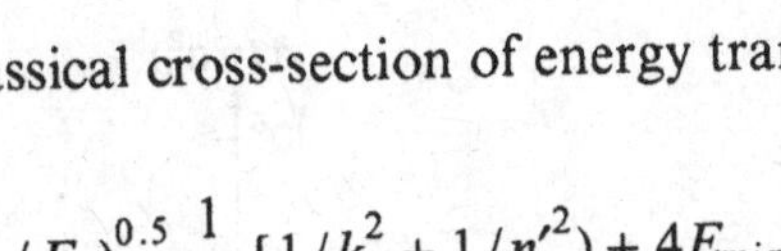

Fig. 11. The exchange diagram

There are some differences between methods of calculating $f(q)$ for the usually allowed transitions discussed above and the intercombination ones. In the first case, the region of momentum transfer, $q \sim 1/n^2$, gives the main contribution. In the intercombination case, the range of comparatively large $q \sim 1/n$ dominates. Since there are no good asymptotical expansions of hypergeometrical functions for such arguments, this case is more difficult. Figures 12 and 13 show the comparison between classical and Ochkur approximations for some helium transitions.

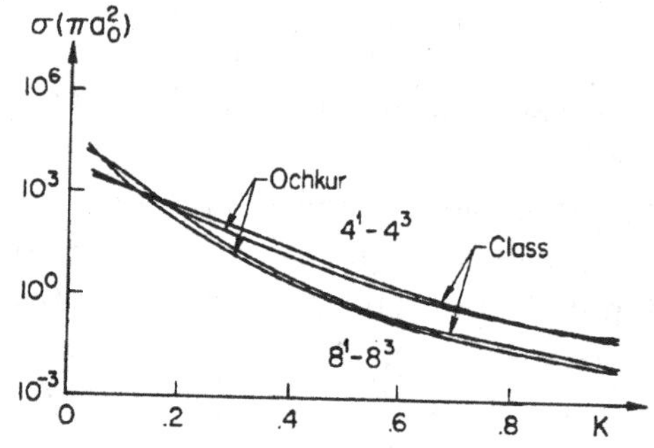

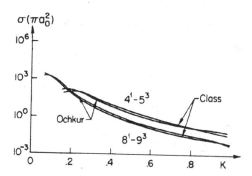

Fig. 12 – The cross-sections of the intercombination transitions averaged over orbital quantum numbers in He ($\Delta n = 0$).

Fig. 13 – The cross-sections of the intercombination transitions averaged over orbital quantum numbers in He ($\Delta n = 1$).

4 CONCLUSIONS

1. The K-matrix method is a convenient and reasonable method for computer calculations.

2. There is an analytical description for the cross-sections of the transitions between highly excited states induced by electron impact.

3. The direct experimental data about cross-sections between highly excited levels with $\Delta n \geq 1$ are absent.

5 REFERENCES

Seaton, M. J. 1961, *Proc. Phys. Soc.* **77**, 184.

Vainshtein, L. A. and Presnyakov, L. P. 1968, *ZhETF* **55**, 137 (Engl. transl.: 1969 *JETP* **28**, 156).

Ochkur, V. I., *ZhETF* **45**, 735 (Engl. transl.: 1964 *JETP* **18**, 503).

Beigman, I. L. and Vainshtein, L. A. 1967, *ZhETF* **52**, 185 (Egnl. transl.: 1967, *JETP* **25**, 119.

Beigman, I. L., Gaisinsky, I. M. and Rumyantzev, N. M. 1988, *Council of Spectoscopy*, 28.

Seaton, M. J. 1962, *Proc. Phys. Soc.* **79**, 1105.

Stabler, R. C. 1964, *Phys. Rev. A* **133**, 1268.

Beigman, I. L., Urnov, A. M. and Shevelko, V. P. 1970, *ZhETF* **58**, 1825 (Engl. transl.: *JETP* **31**, 918).

Percival, I. C and Richards, D. 1970, *J. Phys. B* **3**, 1035.

Presnyakov, L. P. and Urnov, A. M. 1970, *J. Phys. B* **3**, 1267.

Beigman, I. L., Vainshtein, L. A. and Sobel'man, I. I. 1969, *ZhETF* **57**, 1703 (Engl. transl.: *JETP* **30**, 920).

Richards, D. 1972, *J. Phys. B* **5**, L53.

Line Ratio Diagnostics for Astrophysical Plasmas

F. P. Keenan

Department of Pure and Applied Physics, The Queen's University of Belfast, Belfast BT7 1NN, Northern Ireland

ABSTRACT

In this paper the conditions under which emission (or absorption) line intensity ratios are sensitive to variations in the physical conditions of a plasma, such as electron temperature and density, are discussed. More importantly, a bibliography is provided of the most reliable line ratio diagnostic calculations currently available for application to the IR, optical, UV and X-ray spectra of a wide range of astrophysical sources. These include the solar chromosphere, transition region and corona, late-type stellar atmospheres, gaseous nebulae (such as H II regions and planetary nebulae), active galactic nuclei, supernova remnants and the interstellar medium.

1 INTRODUCTION

Emission lines are frequently observed in the infrared, optical, ultraviolet and X-ray spectra of astrophysical plasmas, including the solar transition region/corona [21,149,156,167], early and late-type stellar atmospheres [54,159], planetary nebulae [117], H II regions [43], symbiotic stars [158], active galactic nuclei [146] and solar system objects, such as the Io Torus [136]. These transitions may be used to derive information on the plasma parameters, such as electron temperature (T_e) and density (N_e), through line intensity ratios (which are usually called diagnostic line ratios). In Section 2 we discuss under what conditions line ratios may be sensitive to variations in the plasma parameters, while in Section 3 we give a bibliography of the most reliable diagnostic calculations currently available for application to the spectra of astrophysical plasmas.

2 THEORY

Consider a set of n levels for a given ion in a plasma where the principle population and de-population mechanisms are collisions with electrons and spontaneous radiative de-excitation. The change in population dN_i/dt of a level i is then given by

$$\frac{dN_i}{dt} = N_e \sum_{j=1}^{n} N_j C_{ji} - N_e N_i \sum_{j=1}^{n} C_{ij} + \sum_{j=i}^{n} N_j A_{ji} - N_i \sum_{j=1}^{i} A_{ij} \qquad (1)$$

where the first and second terms are the collisional rates in and out of level i, respectively, the third and fourth terms are the radiative rates in and out of the level, C_{ij} is the electron collisional rate from level $i \rightarrow j$ and unit N_e, and A_{ij} is the spontaneous radiative de-excitation rate from $i \rightarrow j$. For a stationary plasma, $dN_i/dt = 0$ and hence

$$N_i = \frac{N_e \sum_{j=1}^{n} N_j C_{ji} + \sum_{j=i}^{n} N_j A_{ji}}{N_e \sum_{j=1}^{n} C_{ij} + \sum_{j=1}^{i} A_{ij}} \qquad (2)$$

where $i = 1, \ldots, n$. The level populations are related to the total volume density of the ionization stage N_{ion} by

$$N_{ion} = \sum_{i=1}^{n} N_i \qquad (3)$$

2.1 Low values of N_e

If level i has an allowed transition to the ground state (i.e. A_{i1} is large), then the first term in the denominator of (2) ($N_e \sum_{j=1}^{n} C_{ij}$) is negligible. Also for low N_e the level populations of the excited levels will be very small compared with the ground state, and hence the second term in the numerator ($\sum_{j=i}^{n} N_j A_{ji}$) becomes negligible. Hence we get the coronal approximation [37]

$$N_i = \frac{N_e N_1 C_{1i}}{A_{i1}} \qquad (4)$$

where $1 =$ ground state. The line intensity is therefore

$$I_{i1} = E_i N_i A_{i1} = E_i N_e N_1 C_{1i} \qquad (5)$$

where E_i is the energy of level i relative to the ground state, and is directly proportional to the collisional excitation rate, but is independent of the A-value.

2.2 High values of N_e

The radiative terms in (2) become negligible and

$$N_i = \frac{\sum_{j=1}^{n} N_j C_{ji}}{\sum_{j=1}^{n} C_{ij}} \qquad (6)$$

The relation between inverse collisional rates then gives the thermodynamic equilibrium population distribution

$$\frac{N_j}{N_i} = \frac{g_j}{g_i} \, exp(-E_{ji}/kT_e) \qquad (7)$$

where E_{ji} is the energy difference of the levels and g is the level degeneracy. The line intensity is therefore

$$I_{i1} = E_i N_i A_{i1} = E_i N_1 \frac{g_i}{g_1} A_{i1} \ exp(-E_i/kT_e) \tag{8}$$

Hence the line intensity is directly proportional to the A-value, and is independent of the collision rate.

2.3 T_e-diagnostics

Consider two levels i and j for which the principal rates are spontaneous radiative de-excitation and electron impact excitation from the ground state (i.e. we have no metastable levels). Then the coronal approximation gives for the emission line ratio R

$$R = \frac{I_{j1}}{I_{i1}} = \frac{E_j}{E_i} \frac{C_{1j}}{C_{1i}} \tag{9}$$

We may write C_{1j} as

$$C_{1j} = \frac{8.63 \times 10^{-6}}{g_1 \sqrt{T_e}} \Upsilon_{1j} \ exp(-E_j/kT_e) \tag{10}$$

where Υ_{1j} is the effective collision strength, which is a slowly varying function of T_e. Hence we have

$$R = \frac{\Upsilon_{1j}}{\Upsilon_{1i}} \frac{E_j}{E_i} \ exp(-(E_j - E_i)/kT_e) \tag{11}$$

so that from the observed value of R we may derive T_e. However note that $(E_j - E_i)$ needs to be large for R to be sensitive to variations in T_e, so that the relevant emission lines are often well separated in wavelength.

2.4 N_e-diagnostics

In this instance we need two lines $1 - i$ and $1 - k$, where i has a small radiative decay rate (i.e. is a metastable level), and can be depopulated by electron collisions to another level m with collisional loss rate C_{im}. Hence the population of level i is given by

$$N_i(A_{i1} + N_e C_{im}) = N_e N_1 C_{1i} \tag{12}$$

as we can neglect the $N_m A_{mi}$ term as N_m will be small. The line intensity ratio is therefore given by

$$R = \frac{I_{k1}}{I_{i1}} = \frac{E_k}{E_i} \frac{N_e N_1 C_{1k}}{N_i A_{i1}}$$

$$= \frac{E_k}{E_i} \frac{N_e N_1 C_{1k}}{N_e N_1 C_{1i} A_{i1}} (A_{i1} + N_e C_{im})$$

$$= \frac{E_k}{E_i} \frac{C_{1k}}{C_{1i}} \left(1 + \frac{N_e C_{im}}{A_{i1}}\right) \tag{13}$$

If $N_e C_{im} \ll A_{i1}$ then R is independent of N_e (coronal approximation), but if $N_e C_{im} \gtrsim A_{i1}$ then R is sensitive to variations in N_e. The presence of the C_{1k}/C_{1i} term in (13) implies that R will also be T_e–sensitive, particularly when $(E_k - E_i)$ is large.

The magnitude of A_{i1} (i.e. the type of transition) dictates the density range over which R is N_e–sensitive. In general, the types of transitions may be divided into three groups, namely

- Intercombination transitions in low atomic number species (e.g. $2s2p\ ^3P_1 \rightarrow 2s^2$ ^1S in O V) or forbidden transitions in high atomic number species (e.g. $2s^2 2p^2$ $^3P_1 \rightarrow 2s^2 2p^2\ ^3P_0$ in Ca XV) have $A_{i1} \simeq 10^2 - 10^4\ s^{-1} \Rightarrow N_e$ range $\simeq 10^8 - 10^{13}\ cm^{-3}$ (applicable, for example, to the solar transition region/corona and late-type stellar atmospheres).

- Forbidden transitions in low atomic number species (e.g. $3s3p\ ^3P_2 \rightarrow 3s^2\ ^1S$ in Si III) have $A_{i1} \simeq 10^{-3} - 1\ s^{-1} \Rightarrow N_e$ range $\simeq 10^2 - 10^7\ cm^{-3}$ (applicable to gaseous nebulae such as H II regions and planetary nebulae).

- Forbidden transitions within the ground term of low atomic number species (e.g. $2s^2 2p\ ^2P_{3/2} \rightarrow 2s^2 2p\ ^2P_{1/2}$ in C II) have $A_{i1} \simeq 10^{-6} - 10^{-3}\ s^{-1} \Rightarrow N_e$ range $\leq 100\ cm^{-3}$ (applicable to the interstellar medium and supernova remnants).

3 BIBLIOGRAPHY OF DIAGNOSTIC CALCULATIONS

In Tables 1 – 13 we list the most reliable diagnostic calculations currently available for astrophysical plasmas. These cover the spectral range from the infrared (the longest wavelength considered is $\sim 200\ \mu m$ [153]) down to the X-ray at ~ 1 Å [15], and include not only emission line diagnostics, but also absorption diagnostics applicable to the interstellar medium, as illustrated by for example [1,51]. Excluded from the tables are diagnostics that involve recombination/satellite lines, which are discussed in the excellent reviews by Doschek [23] for solar observations and Aller [2] for gaseous nebulae spectra. Nor do we list diagnostics that involve lines from different species, which are rather limited in number and are summarized in references [25,88,91,93] and [38,153,161] for solar/stellar and gaseous nebulae plasmas, respectively.

Species are listed in the tables by isoelectronic sequence, apart from ions of Fe and Ni which are considered separately. In brackets after each reference is one or more of the letters S, G and I. These indicate the approximate density range over which the diagnostic calculations are applicable, with S referring to solar-type plasmas ($N_e \gtrsim 10^8\ cm^{-3}$), G to gaseous nebulae ($10^2 \lesssim N_e \lesssim 10^7\ cm^{-3}$) and I to the interstellar medium ($N_e \lesssim 10^2\ cm^{-3}$). Note that these divisions are *very* approximate, and that many of the diagnostics are applicable over a large density range, which is illustrated by the fact that many of the references are followed by more than one letter.

In many instances a species may have more than one reference listed for a particular plasma type. This indicates that the species has several different diagnostic line ratios, usually spanning different wavelength regions, which are not covered in a single paper. For example, there are O III diagnostics for gaseous nebulae involving infrared [69], optical [70] and ultraviolet lines [68], while similarly for solar-type plasmas there are O III UV [72], EUV [104] and XUV [9] diagnostics.

Finally, I would like to apologize in advance to anyone whose work has been omitted from the tables. For the future, I would be very grateful if researchers would inform me of any references that I have omitted, and also of any of their more recent diagnostic work that could be included in future versions of these tables.

Acknowledgements: I would like to thank all those who sent me reprints/preprints of their work for inclusion in this paper, including L. Aller, A. Bhatia, R. Clegg, G. Doschek, U. Feldman, C. Jordan, S. Kastner, P. McWhirter, M. Malkan, H. Mason, H. Nussbaumer, D. Osterbrock, S. Pottasch and R. Rubin.

REFERENCES

1. Allen, M. M., Snow, T. P., Jenkins, E. B. 1988. *Ap. J.* **355**, 130.
2. Aller, L. H. 1990. *Publ. Astr. Soc. Pacif.* **102**, 1097.
3. Bhatia, A. K., Fawcett, B. C., Lemen, J. R., Mason, H. E., Phillips, K. J. H. 1989. *M.N.R.A.S.* **240**, 421.
4. Bhatia, A. K., Feldman, U., Seely, J. F. 1989. *Atom. Data Nucl. Data Tables* **32**, 435.
5. Bhatia, A. K., Kastner, S. O. 1980. *Solar Phys.* **65**, 181.
6. Bhatia, A. K., Kastner, S. O. 1985. *Solar Phys.* **96**, 11.
7. Bhatia, A. K., Kastner, S. O. 1988. *Ap. J.* **332**, 1063.
8. Bhatia, A. K., Kastner, S. O. 1992. *Ap. J. Suppl.* (in press).
9. Bhatia, A. K., Kastner, S. O., Behring, W. E. 1982. *Ap. J.* **257**, 887.
10. Bhatia, A. K., Mason, H. E. 1980. *M.N.R.A.S.* **190**, 925.
11. Bhatia, A. K., Mason, H. E. 1980. *Astr. Ap.* **83**, 380.
12. Bhatia, A. K., Mason, H. E. 1981. *Astr. Ap.* **103**, 324.
13. Bhatia, A. K., Mason, H. E. 1983. *Astr. Ap. Suppl.* **52**, 115.
14. Bhatia, A. K., Mason, H. E. 1986. *Astr. Ap.* **155**, 417.
15. Bitter, M., *et al.* 1991. *Phys. Rev. A* **44**, 1796.
16. Bromage, B. J. I., Phillips, K. J. H., Keenan, F. P., McCann, S. M. 1989. *Solar Phys.* **124**, 289.
17. Brown, W. A., Bruner, M. E., Acton, L. W., Mason, H. E. 1986. *Ap. J.* **301**, 981.
18. Butler, K., Mendoza, C. 1984. *M.N.R.A.S.* **208**, 17P.
19. Cornille, M., Dubau, J., Loulerge, M., Mason, H. E. 1992. *Astr. Ap.* (in press).

20. Czyzak, S. J., Keyes, C. D., Aller, L. H. 1986. *Ap. J. Suppl.* **61**, 159.
21. Dere, K. P. 1978. *Ap. J.* **221**, 1062.
22. Dere, K. P., Mason, H. E., Widing, K. G., Bhatia, A. K. 1979. *Ap. J. Suppl.* **40**, 341.
23. Doschek, G. A. 1990. *Ap. J. Suppl.* **73**, 117.
24. Doschek, G. A., Feldman, U., Bhatia, A. K. 1991. *Phys. Rev. A* **43**, 2565.
25. Doschek, G. A., Feldman, U., Mason, H. E. 1979. *Astr. Ap.* **78**, 342.
26. Doyle, J. G., Dufton, P. L., Keenan, F. P., Kingston, A. E. 1983. *Solar Phys.* **89**, 243.
27. Doyle, J. G., Keenan, F. P., Harra, L. K., Aggarwal, K. M., Tayal, S. S. 1992. *Astr. Ap.* (in press).
28. Dufton, P. L., Hibbert. A., Keenan, F. P., Kingston, A. E., Doschek, G. A. 1986. *Ap. J.* **300**, 448.
29. Dufton, P. L., Hibbert. A., Kingston, A. E., Doschek, G. A. 1982. *Ap. J.* **257**, 338.
30. Dufton, P. L., Kingston, A. E. 1985. *Ap. J.* **289**, 844.
31. Dufton, P. L., Kingston, A. E., Doyle, J. G., Widing, K. G. 1983. *M.N.R.A.S.* **205**, 81.
32. Dufton, P. L., Kingston, A. E., Widing, K. G. 1990. *Ap. J.* **353**, 323.
33. Dufton, P. L., et al. 1991. *M.N.R.A.S.* **253**, 474.
34. Dwivedi, B. N., Gupta, A. K. 1991. *Solar Phys.* **135**, 415.
35. Dwivedi, B. N., Raju, P. K. 1980. *Solar Phys.* **68**, 111.
36. Eaton, J. A., Johnson, H. R. 1988. *Ap. J.* **325**, 355.
37. Elwert, G. 1952. *Z. Naturforsch* **7A**, 432.
38. Feibelman, W. A., Aller, L. H. 1987. *Ap. J.* **319**, 407.
39. Feldman, U. 1992. *Ap. J.* (in press).
40. Feldman, U., Doschek, G. A. 1977. *J. Opt. Soc. Am.* **67**, 726.
41. Feldman, U., Doschek, G. A., Widing, K. G. 1978. *Ap. J.* **219**, 304.
42. Feldman, U., Seely, J. F., Bhatia, A. K. 1985. *Atom. Data Nucl. Data Tables* **32**, 305.
43. Fich, M., Silkey, M. 1991. *Ap. J.* **366**, 107.
44. Gabriel, A. H., Bely-Dubau, F., Faucher, P., Acton, L. W. 1991. *Ap. J.* **378**, 438.
45. Garstang, R. H., Robb, W. D., Rountree, S. P. 1978. *Ap. J.* **222**, 384.
46. Hayes, M. A., Nussbaumer, H. 1983. *Astr. Ap.* **124**, 279.
47. Hayes, M. A., Nussbaumer, H. 1984. *Astr. Ap.* **134**, 193.
48. Hayes, M. A., Nussbaumer, H. 1984. *Astr. Ap.* **139**, 233.
49. Heroux, L., Cohen, M. 1971. *Phil. Trans. R. Soc. London A* **270**, 99.
50. Ho, Y. K., Henry, R. J. W. 1984. *Ap. J.* **284**, 435.
51. Howarth, I. D., Phillips, A. P. 1986. *M.N.R.A.S.* **222**, 809.
52. Hutcheon, R. J., McWhirter, R. W. P. 1973. *J. Phys. B* **6**, 2668.
53. Jordan, C. 1967. *Solar Phys.* **2**, 441.

54. Jordan, C. 1988. *J. Opt. Soc. Am. B* **5**, 2252.
55. Jordan, C., Veck, N. J. 1982. *Solar Phys.* **78**, 125.
56. Judge, P. G. 1986. *M.N.R.A.S.* **221**, 119.
57. Judge, P. G., Carpenter, K. G., Harper, G. M. 1991. *M.N.R.A.S.* **253**, 123.
58. Judge, P. G., Jordan, C. 1991. *Ap. J. Suppl.* **77**, 75.
59. Kastner, S. O., Bhatia, A. K. 1984. *J. Quant. Spectrosc. Radiat. Transfer* **32**, 249.
60. Kastner, S. O., Bhatia, A. K. 1984. *Ap. J.* **287**, 945.
61. Kastner, S. O., Bhatia, A. K. 1986. *Ap. J.* **309**, 883.
62. Kastner, S. O., Bhatia, A. K. 1989. *Ap. J. Suppl.* **71**, 665.
63. Keenan, F. P. 1988. *Solar Phys.* **116**, 279.
64. Keenan, F. P. 1989. *Ap. J.* **339**, 591.
65. Keenan, F. P. 1990. *Solar Phys.* **126**, 311.
66. Keenan, F. P. 1991. *Solar Phys.* **131**, 291.
67. Keenan, F. P. 1992. *Astr. Sp. Sci.* (in press).
68. Keenan, F. P., Aggarwal, K. M. 1987. *Ap. J.* **319**, 403.
69. Keenan, F. P., Aggarwal, K. M. 1988. *J. Astr. Ap.* **9**, 237.
70. Keenan, F. P., Aggarwal, K. M. 1989. *J. Astr. Ap.* **10**, 147.
71. Keenan, F. P., Aggarwal, K. M. 1989. *Solar Phys.* **122**, 1.
72. Keenan, F. P., Aggarwal, K. M. 1989. *Ap. J.* **344**, 522.
73. Keenan, F. P., Aggarwal, K. M. 1990. *Ap. J.* **350**, 262.
74. Keenan, F. P., Aggarwal, K. M., Berrington, K. A., Widing, K. G. 1988. *Ap. J.* **327**, 473.
75. Keenan, F. P., Berrington, K. A. 1985. *Solar Phys.* **99**, 25.
76. Keenan, F. P., Berrington, K. A. 1988. *Ap. J.* **333**, 806.
77. Keenan, F. P., Berrington, K. A., Burke, P. G., Kingston, A. E., Dufton, P. L. 1984. *M.N.R.A.S.* **207**, 459.
78. Keenan, F. P., Burke, P. G., Hibbert, A., Mohan, M., Reid, R. H. G. 1990. *Phys. Scripta* **41**, 257.
79. Keenan, F. P., Burke, V. M., Aggarwal, K. M. 1991. *Ap. J.* **371**, 636.
80. Keenan, F. P., Conlon, E. S., Harra, L. K., Aggarwal, K. M., Widing, K. G. 1992. *Ap. J.* (in press).
81. Keenan, F. P., Conlon, E. S., Harra, L. K., Burke, V. M., Widing, K. G. 1992. *Ap. J.* **385**, 381.
82. Keenan, F. P., Conlon, E. S., Harra, L. K., Widing, K. G. 1992. *Ap. J.* (in press).
83. Keenan, F. P., Cook, J. W., Dufton, P. L., Kingston, A. E. 1989. *Ap. J.* **340**, 1135.
84. Keenan, F. P., Cook, J. W., Dufton, P. L., Kingston, A. E. 1992. *Ap. J.* (in press).
85. Keenan, F. P., Doyle, J. G. 1988. *Solar Phys.* **115**, 229.
86. Keenan, F. P., Doyle, J. G. 1990. *Solar Phys.* **128**, 345.

87. Keenan, F. P., Doyle, J. G., Tayal, S. S., Henry, R. J. W. 1991. *Solar Phys.* **135**, 353.
88. Keenan, F. P., Dufton, P. L., Aggarwal, K. M., Kingston, A. E. 1988. *Ap. J.* **324**, 1068.
89. Keenan, F. P., Dufton, P. L., Boylan, M. B., Kingston, A. E., Widing, K. G. 1991. *Ap. J.* **373**, 695.
90. Keenan, F. P., Dufton, P. L., Kingston, A. E. 1986. *Astr. Ap.* **169**, 319.
91. Keenan, F. P., Dufton, P. L., Kingston, A. E. 1987. *M.N.R.A.S.* **225**, 859.
92. Keenan, F. P., Dufton, P. L., Kingston, A. E. 1989. *Solar Phys.* **123**, 33.
93. Keenan, F. P., Dufton, P. L., Kingston, A. E. 1990. *Ap. J.* **353**, 636.
94. Keenan, F. P., Feibelman, W. A., Berrington, K. A. 1992. *Ap. J.* (in press).
95. Keenan, F. P., Harra, L. K., Aggarwal, K. M., Feibelman, W. A. 1992. *Ap. J.* **385**, 375.
96. Keenan, F. P., Hibbert, A., Burke, P. G., Berrington, K. A. 1988. *Ap. J.* **332**, 539.
97. Keenan, F. P., Johnson, C. T., Kingston, A. E. 1988. *Astr. Ap.* **202**, 253.
98. Keenan, F. P., Johnson, C. T., Kingston, A. E., Dufton, P. L. 1985. *M.N.R.A.S.* **214**, 37P.
99. Keenan, F. P., Kingston, A. E., Aggarwal, K. M., Widing, K. G. 1986. *Solar Phys.* **103**, 225.
100. Keenan, F. P., Kingston, A. E., Dufton, P. L., Doyle, J. G., Widing, K. G. 1984. *Solar Phys.* **94**, 91.
101. Keenan, F. P., Kingston, A. E., McKenzie, D. L. 1985. *Ap. J.* **291**, 855.
102. Keenan, F. P., Lennon, D. J., Johnson, C. T., Kingston, A. E. 1986. *M.N.R.A.S.* **220**, 571.
103. Keenan, F. P., McCann, S. M. 1987. *Solar Phys.* **109**, 31.
104. Keenan, F. P., McCann, S. M., Aggarwal, K. M., Widing, K. G. 1989. *Solar Phys.* **122**, 7.
105. Keenan, F. P., McCann, S. M., Widing, K. G. 1988. *Solar Phys.* **117**, 69.
106. Keenan, F. P., McCann, S. M., Widing, K. G. 1990. *Ap. J.* **363**, 315.
107. Keenan, F. P., McKenzie, D. L., McCann, S. M., Kingston, A. E. 1987. *Ap. J.* **318**, 926.
108. Keenan, F. P., Norrington, P. H. 1991. *Ap. J.* **368**, 486.
109. Keenan, F. P., Phillips, K. J. H., Harra, L. K., Conlon, E. S., Kingston, A. E. 1992. *Ap. J.* (in press).
110. Keenan, F. P., Tayal, S. S., Henry, R. J. W. 1990. *Solar Phys.* **125**, 61.
111. Keenan, F. P., Tayal, S. S., Kingston, A. E. 1984. *Solar Phys.* **92**, 75.
112. Keenan, F. P., Tayal, S. S., Kingston, A. E. 1984. *Solar Phys.* **94**, 85.
113. Keenan, F. P., Widing, K. G., McCann, S. M. 1989. *Ap. J.* **338**, 563.
114. Keenan, F. P., *et al.* 1991. *Ap. J.* **379**, 406.
115. Keenan, F. P., *et al.* 1991. *Ap. J.* **382**, 349.
116. Keenan, F. P., *et al.* 1992. *Ap. J.* **384**, 385.

117. Keyes, C. D., Aller, L. H., Feibelman, W. A. 1990. *Publ. Astr. Soc. Pacif.* **102**, 59.
118. Kunc, J. A. 1988. *J. Appl. Phys.* **63**, 656.
119. Lang, J., Dufton, P. L., Kingston, A. E. 1986. *Solar Phys.* **105**, 313.
120. Lennon, D. J., Dufton, P. L., Hibbert, A., Kingston, A. E. 1985. *Ap. J.* **294**, 200.
121. Ljepojevic, N. N., Hutcheon, R. J., McWhirter, R. W. P. 1984. *J. Phys. B* **17**, 3057.
122. Ljepojevic, N. N., McWhirter, R. W. P., Volonté, S. 1985. *J. Phys. B* **18**, 3285.
123. Loulergue, M., Mason, H. E., Nussbaumer, H., Storey, P. J. 1985. *Astr. Ap.* **150**, 246.
124. Lynch, J. P., Kafatos, M. 1991. *Ap. J. Suppl.* **76**, 1169.
125. McCann, S. M., Keenan, F. P. 1987. *Solar Phys.* **112**, 83.
126. McCann, S. M., Keenan, F. P. 1988. *Ap. J.* **328**, 344.
127. McKenzie, D. L., *et al.* 1992. *Ap. J.* **385**, 378.
128. McWhirter, R. W. P., MacNeice, P. J. 1987. *Solar Phys.* **107**, 323.
129. Malkan, M. A. 1983. *Ap. J. Lett.* **264**, L1.
130. Mason, H. E. 1975. *M.N.R.A.S.* **171**, 119.
131. Mason, H. E., Bhatia, A. K. 1983. *Astr. Ap. Suppl.* **52**, 181.
132. Mason, H. E., Bhatia, A. K., Kastner, S. O., Neupert, W. M., Swartz, M. 1984. *Solar Phys.* **92**, 199.
133. Mason, H. E., Doschek, G. A., Feldman, U., Bhatia, A. K. 1979. *Astr. Ap.* **73**, 74.
134. Mason, H. E., Storey, P. J. 1980. *M.N.R.A.S.* **191**, 631.
135. Mendoza, C., Zeippen, C. J. 1982. *M.N.R.A.S.* **198**, 127.
136. Moos, H. W., *et al.* 1991. *Ap. J. Lett.* **382**, L105.
137. Nussbaumer, H. 1976. *Astr. Ap.* **48**, 93.
138. Nussbaumer, H. 1986. *Astr. Ap.* **155**, 205.
139. Nussbaumer, H., Schild, H. 1981. *Astr. Ap.* **99**, 177.
140. Nussbaumer, H., Storey, P. J. 1978. *Astr. Ap.* **70**, 37.
141. Nussbaumer, H., Storey, P. J. 1979. *Astr. Ap.* **71**, L5.
142. Nussbaumer, H., Storey, P. J. 1980. *Astr. Ap.* **89**, 308.
143. Nussbaumer, H., Storey, P. J. 1982. *Astr. Ap.* **110**, 295.
144. Nussbaumer, H., Storey, P. J. 1982. *Astr. Ap.* **115**, 205.
145. Nussbaumer, H., Storey, P. J. 1988. *Astr. Ap.* **193**, 327.
146. Osterbrock, D. E., Shaw, R. A, Veilleux, S. 1990. *Ap. J.* **352**, 561.
147. Osterbrock, D. E., Tran, H. D., Veilleux, S. 1992. *Ap. J.* (in press).
148. Phillips, K. J. H., Keenan, F. P., Harra, L. K., McCann, S. M. 1992. *Ap. J.* (in press).
149. Phillips, K. J. H., *et al.* 1982. *Ap. J.* **256**, 774.
150. Pradhan, A. K. 1982. *Ap. J.* **263**, 477.
151. Pradhan, A. K., Shull, J. M. 1981. *Ap. J.* **249**, 821.

152. Raymond, J. C., Smith, B. W. 1986. *Ap. J.* **306**, 762.
153. Rubin, R. H. 1989. *Ap. J. Suppl.* **69**, 897.
154. Rugge, H. R., McKenzie, D. L. 1985. *Ap. J.* **297**, 338.
155. Saha, H. P., Trefftz, E. 1982. *Astr. Ap.* **116**, 224.
156. Sandlin, G. D., Bartoe, J.-D. F., Tousey, R., Van Hoosier, M. E. 1986. *Ap. J. Suppl.* **61**, 801.
157. Saraph, H. E., Seaton, M. J. 1970. *M.N.R.A.S.* **148**, 367.
158. Schmid, H. M., Schild, H. 1990. *M.N.R.A.S.* **246**, 84.
159. Shore, S. N., Sanduleak, N. 1984. *Ap. J. Suppl.* **55**, 1.
160. Smeding, A. G., Pottasch, S. R. 1979. *Astr. Ap. Suppl.* **35**, 257.
161. Spinoglio, L., Malkan, M. A. 1992. *Ap. J.* (in press).
162. Stanghellini, L., Kaler, J. B. 1989. *Ap. J.* **343**, 811.
163. Tayal, S. S., Henry, R. J. W. 1988. *Ap. J.* **329**, 1023.
164. Tayal, S. S., Henry, R. J. W., Keenan, F. P., McCann, S. M., Widing, K. G. 1989. *Ap. J.* **343**, 1004.
165. Tayal, S. S., Henry, R. J. W., Keenan, F. P., McCann, S. M., Widing, K. G. 1991. *Ap. J.* **369**, 567.
166. Vernazza, J. E., Mason, H. E. 1978. *Ap. J.* **226**, 720.
167. Vernazza, J. E., Reeves, E. M. 1978. *Ap. J. Suppl.* **37**, 485.
168. Zeippen, C. J., Butler, K., Le Bourlot, J. 1987. *Astr. Ap.* **188**, 251.

Table 1. Diagnostics for H-like ions.

Species	Reference
C VI	121(S)
N VII	52(S)
O VIII	52(S)
Ne X	52(S)
Mg XII	128(S)
Si XIV	122(S)
S XVI	122(S)
Ar XVIII	121(S)
Ca XX	122(S)

Table 2. Diagnostics for He-like ions.

Species	Reference
C V	150(S)
N VI	151(S)
O VII	44(S,I), 101(S), 111(S)
Ne IX	107(S), 112(S)
Mg XI	16(S), 109(S)
Al XII	16(S), 103(S)
Si XIII	125(S)
S XV	126(S)
Ar XVII	148(S)
Ca XIX	55(S)

Table 3. Diagnostics for Li-like ions.

Species	Reference
C IV	81(S), 118(S)
N V	118(S)
O VI	118(S)
Ne VIII	118(S)
Mg X	49(S)

Table 4. Diagnostics for Be-like ions.

Species	Reference
C III	8(S), 75(S), 77(S,G), 94(G)
N IV	65(S), 77(S,G), 139(G)
O V	26(S), 115(S)
Ne VII	66(S), 106(S), 119(S)
Mg IX	82(S)
Si XI	100(S)
S XIII	105(S)
Ar XV	113(S)
Ca XVII	13(S), 31(S)

Table 5. Diagnostics for B-like ions.

Species	Reference
C II	47(S,G), 48(S,G), 102(I), 120(S,G)
N III	20(G), 59(S), 60(S,G), 141(S), 160(I), 166(S)
O IV	46(G), 59(S), 144(S,G), 166(S)
Ne VI	124(G), 166(S)
Mg VIII	35(S), 124(G)
Al IX	34(S)
Si X	35(S), 124(G), 155(S)
S XII	124(G), 166(S)
Ar XIV	22(S)
Ca XVI	22(S)

Table 6. Diagnostics for C-like ions.

Species	Reference
C I	53(S), 64(I)
N II	20(G), 153(G), 160(I)
O III	9(S), 62(S,G), 68(G), 69(G), 70(G), 72(S), 73(G), 104(S)
Ne V	71(S), 79(G), 80(S)
Mg VII	99(S)
Si IX	17(S), 99(S)
S XI	61(S)
Ar XIII	22(S)
Ca XV	14(S), 74(S), 130(S)

Table 7. Diagnostics for N-like ions.

Species	Reference
N I	124(S,G), 153(G)
O II	124(S,G), 129(G), 162(G)
Ne IV	7(S), 20(G)
Mg VI	10(S), 124(S,G)
Si VIII	10(S), 124(S,G)
S X	10(S), 124(S,G)
Ar XII	10(S)
Ca XIV	10(S)

Table 8. Diagnostics for O- and F-like ions.

Species	Reference
O I	56(S), 76(I), 124(S,G), 153(G)
Ne III	18(G), 124(S,G), 153(G)
Mg V	124(S,G)
Si VII	124(S,G)
S IX	124(S,G)
Ne II	124(S,G)
Mg IV	124(S,G)
Si VI	124(S,G)
S VIII	124(S,G)
Ar X	40(S)

Table 9. Diagnostics for Ne- and Na-like ions.

Species	Reference
Si V	6(S)
S VII	6(S)
Ar IX	6(S)
Ca XI	6(S)
Mg II	40(S)
Al III	90(S)
Si IV	85(S), 90(S)
S VI	63(S)
Ca X	40(S)

Table 10. Diagnostics for Mg-, Al- and Si-like ions.

Species	Reference
Al II	27(S), 95(G)
Si III	83(S), 92(S), 94(G), 138(S,G)
S V	28(S), 86(S)
Si II	30(S), 33(S), 57(S,G), 84(S), 98(I)
S IV	29(S)
S III	50(G), 124(S,G), 153(G)

Table 11. Diagnostics for P- and S-like ions.

Species	Reference
S II	20(G), 67(G), 153(G)
Cl III	20(G)
Ar IV	168(G)
K V	135(G), 157(G)
S I	56(S)
Ar III	97(G), 153(G)

Table 12. Diagnostics for ions of Fe.

Species	Reference
Fe II	36(S), 58(S), 96(G,I), 142(G,I), 145(G,I)
Fe III	116(S,G)
Fe VI	45(S,G), 140(G)
Fe VII	108(S,G)
Fe IX	39(S), 41(S)
Fe X	124(G), 137(S)
Fe XI	22(S), 124(G)
Fe XII	22(S), 87(S), 110(S), 124(G), 163(S), 164(S), 165(S)
Fe XIII	22(S), 124(G)
Fe XIV	17(S), 89(S), 114(S), 124(G)
Fe XV	32(S)
Fe XVI	19(S), 40(S)
Fe XVII	24(S), 152(S), 154(S)
Fe XVIII	40(S), 127(S)
Fe XIX	3(S), 123(S)
Fe XX	11(S), 131(S)
Fe XXI	133(S)
Fe XXII	134(S)
Fe XXIII	12(S)
Fe XXIV	132(S)
Fe XXV	150(S)
Fe XXVI	122(S)

Table 13. Diagnostics for ions of Ni.

Species	Reference
Ni II	143(G)
Ni III	147(G)
Ni XVII	5(S)
Ni XIX	4(S)
Ni XX	78(S)
Ni XXI	42(S)
Ni XXII	42(S)
Ni XXIII	42(S)
Ni XXIV	42(S)
Ni XXV	42(S)
Ni XXVII	15(S)
Ni XXVIII	52(S)

X-RAY SPECTROSCOPY WITH EBIT

P. Beiersdorfer[1], R. Cauble[1], S. Chantrenne[2], M. Chen[1], N. DelGrande[1],
D. Knapp[1], R. Marrs[1], A. Osterheld[1], K. Reed[1], M. Schneider[1],
J. Scofield[1], B. Wargelin[3], K. Wong[1], D. Vogel[1], R. Zasadzinski[1]

[1]Lawrence Livermore National Laboratory, University of California,
  Livermore, CA 94550
[2]Hewlett Packard Co., Silicon Process Laboratory; Palo Alto, CA 94303
[3]Space Sciences Laboratory, University of California, Berkeley, CA 94720

ABSTRACT

X-ray spectroscopy with the Livermore electron beam ion traps provides data on a wide range of atomic physics issues including ionization, recombination, and excitation cross sections, identification of forbidden transitions, and contributions from relativity and quantum electrodynamics to the transition energies. Here we briefly discuss the source characteristics and x-ray instrumentation, and report measurements of the excitation cross sections of the $K\alpha$ transitions in heliumlike Ti^{20+} as a function of beam energy. The measurements allow detailed comparisons with theoretical predictions of the direct electron-impact excitation cross sections, resonance-excitation contributions, and the electron temperature dependence of the ratio of triplet and singlet lines. The results demonstrate the importance of such measurements for increasing the reliability of x-ray diagnostics of laboratory and astrophysical plasmas.

1 INTRODUCTION

The purpose of x-ray spectroscopy is two-fold. First, x-ray spectroscopy is a tool to determine the basic properties of highly charged ions such as their structure and the cross sections for excitation, ionization, and recombination with electrons. Second, x-ray spectroscopy provides information about the medium embedding highly charged ions and thus is widely employed in the diagnosis of astrophysical, x-ray laser, and fusion plasmas. The two missions of x-ray spectroscopy are evidently interdependent, as a clear understanding of atomic properties is prerequisite for inferring plasma parameters from spectroscopic observations, while, conversely, our understanding of atomic processes has been greatly refined by the use of x-ray spectroscopy for diagnostic purposes.

The electron beam ion traps (EBIT) at Lawrence Livermore Laboratory were specifically designed for studying the properties of highly charged ions by way of x-ray spectroscopy. EBIT uses a mono-energetic electron beam to successively ionize and excite ions, presently up to U^{86+} (Marrs et al. 1988; Levine et al. 1989). The processes of ion generation and excitation are thus the inverse of the beam-foil method of accelerators.

Since the ions, trapped radially by the electric field of the electron beam and axially by the bias potential of two cylindrical electrodes, are virtually at rest, x-ray measurements are unaffected by Doppler shifts. In addition, ballistic injection of neutral nitrogen (Schneider *et al.* 1989) is used to continuously cool the ions and to maintain their temperature below a level at which Doppler broadening would affect the measurements.

X-ray spectroscopy on EBIT can thus fulfil its mission of determining the properties of highly charged ions unimpeded by difficulties encountered in other experimental arrangements. Moreover, since the energy of the electron beam can be precisely controlled, the excitation process of interest can be selected without interference from competing processes and the electron-ion interaction cross section can be determined for individual excitation processes.

A wide range of atomic physics issues have been investigated with x-ray spectroscopy on EBIT. These include the study of x-ray line excitation by direct electron collisions (Marrs *et al.* 1988) and by indirect processes, such as resonance excitation (Beiersdorfer *et al.* 1990c), the measurement of relativistic and quantum electrodynamical shifts in the structure of very highly charged ions such as heliumlike Ge^{30+} (MacLaren *et al.* 1992), neonlike Yb^{60+} (Beiersdorfer *et al.* 1990a), and sodiumlike Pt^{67+} (Cowan *et al.* 1991), the identification of highly forbidden transitions such as magnetic octupole decay in nickellike Th^{62+} and U^{64+} (Beiersdorfer *et al.* 1991b), and measurements of line overlap for resonant photo-pumping of x-ray lasing transitions (Beiersdorfer *et al.* 1992b). In addition, a variety of cross sections have been measured with x-ray spectroscopy on EBIT. These include those of dielectronic recombination in heliumlike Ni^{26+} (Knapp *et al.* 1989) and neonlike Au^{69+} (Schneider *et al.* 1992), of K-shell ionization of lithiumlike Cr^{21+} (Vogel *et al.* 1991), of L-shell ionization of lithiumlike Ba^{53+} (Wong *et al.* 1991), and of electron-impact excitation of the heliumlike transition metals Ti^{20+}, V^{21+}, Cr^{22+}, Mn^{23+}, and Fe^{24+} (Beiersdorfer *et al.* 1992a).

In the following we concentrate our discussion on measurements of the energy dependence of the electon-impact excitation cross sections of heliumlike Ti^{20+}. The x-ray spectra of heliumlike ions play an important role in the diagnostics of fusion and astrophysical plasmas, and detailed experimental investigations of the excitation processes are necessary for the proper interpretation of the x-ray spectra from these ions. This is emphasized by significant differences between our experimental results and calculations. Before detailing these measurements, however, we first present a brief overview of the x-ray instrumentation on EBIT and its operation.

2 X-RAY INSTRUMENTATION AND OPERATION

EBIT is designed as an x-ray source with direct, line-of-sight access to ions in the trap. Because of the 70-μm width of its electron beam EBIT represents a line source and thus is well suited for deployment of flat-crystal or curved-crystal spectrometers in the von Hámos geometry. The von Hámos geometry provides focusing of x rays in the non-dispersive direction. As a result, its efficiency surpasses that of a flat-crystal spectrometer

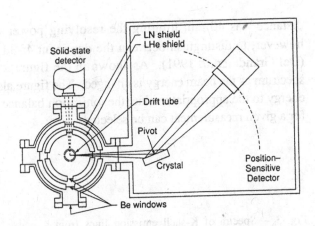

Fig. 1 -- Layout of EBIT and flat-crystal spectrometer. The electron beam direction is out of the page.

(Beiersdorfer *et al.* 1990b), and, by employing large-radius crystals, very high resolving powers can be attained. For example, using a Si(220) crystal with a 75-cm radius of curvature resulted in a resolving power $\lambda/\Delta\lambda = 4500$ (Beiersdorfer 1991). Flat-crystal spectrometers, by contrast, are easier to operate and to align to different Bragg angles. In Fig. 1 we show the layout of EBIT's flat-crystal spectrometer. The spectrometer operates *in vacuo* and uses a 1-D position-sensitive detector with Xe-CH4 detection gas. A 5-mil beryllium window is used to separate the spectrometer vacuum (10^{-5} torr) from EBIT's ($\leq 10^{-9}$ torr) and thus to prevent a degradation of the ionization balance resulting from an increased neutral background in the trap. In a recent modification the detector window and the EBIT beryllium windows were replaced by 4-μm polypropylene, allowing an extension of the useful range of the instrument to energies below 1000 eV.

Representative spectra of M-shell transitions in near nickel-like uranium ions recorded with a flat-crystal spectrometer are shown in Fig. 2. The EBIT-crystal-detector

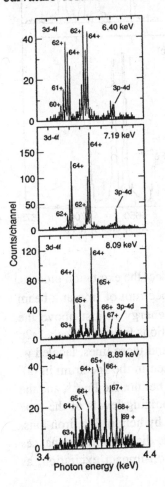

Fig. 2 -- Spectra of the 3d-4f x-ray transitions in ten near nickel-like charge states U^{60+} -- U^{69+}. The spectra are recorded with EBIT's flat-crystal spectrometer at different beam energies between 6.40 and 8.89 keV.

distance was 40 cm, limiting the resolving power to $\lambda/\Delta\lambda = 800$. This is sufficient, however, to distinguish between the dominant 4f-3d transitions in various ionic species (Del Grande *et al.* 1991). As shown in the figure, different ionic species dominate the spectrum as the beam energy is changed. The figure also illustrates that by setting the beam energy to the appropriate value, the ionization balance among the target ions most suitable for a given measurement can be selected.

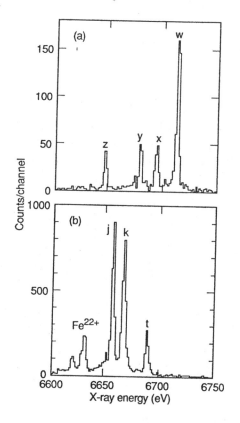

Fig. 3 -- Spectra of K-shell emission lines from heliumlike iron recorded at different electron beam energies: (a) 6.76 keV, (b) 4.68 keV. The first energy is 60 eV above the threshold for electron-impact excitation of the heliumlike K-shell levels. This energy region is free of resonances, and all lines seen are excited directly by electron collisions. The heliumlike transitions are labeled w, x, y, and z and represent the transitions from upper levels $1s2p\ {}^1P_1$, $1s2p\ {}^3P_2$, $1s2p\ {}^3P_1$, and $1s2s\ {}^3S_1$ to the ${}^1S_0$ ground level, respectively. At the latter energy, electrons dielectronically recombine with the heliumlike iron ions and excite the satellite transitions shown. j, k, and t denote the lithiumlike transitions $1s2p^2\ {}^2D_{5/2}$ - $1s^22s\ {}^2P_{3/2}$, $1s2p^2\ {}^2D_{3/2}$ - $1s^22s\ {}^2P_{1/2}$, and $1s2s2p\ {}^2P_{1/2}$ - $1s^22s\ {}^2S_{1/2}$, respectively.

By choosing the appropriate beam energy we can also select the excitation process. This is illustrated in Figs. 3(a) and (b), which show the Kα spectrum of iron at a beam energy of 6.76 keV and 4.68 keV, respectively. The first energy is just above the threshold for electron-impact excitation of the heliumlike transitions from the levels $1s2p$ ${}^1P_1$, $1s2p\ {}^3P_2$, $1s2p\ {}^3P_1$, and $1s2s\ {}^3S_1$ to the ${}^1S_0$ ground level, and these lines, labeled w, x, y, and z in the notation of Gabriel (1972), are prominently seen in the spectrum in (a). The second energy is amidst the KLL dielectronic resonances of heliumlike Fe^{24+}, and the lines seen are produced in the radiative stabilization of autoionizing levels of the type $1s2\ell2\ell'$ populated in the dielectronic capture of beam-electrons by helium-like iron ions. By alternating between the two beam energies the strength of the dielectronic resonances can then be measured relative to the cross section for electron-impact excitation, as discussed in detail by Beiersdorfer *et al.* (1992c).

EBIT's x-ray instrumentation also includes solid-state detectors. These provide a highly efficient means for x-ray observation due to their near 100% quantum efficiency, as well as a wide energy coverage. This allows the simultaneous recording of the x-ray emission due to electron excitation and due to non-resonant radiative electron capture. For instance, the spectrum of the x-ray emission from neonlike Au[69+] recorded with a Ge detector, shown in Fig. 4, covers the energy band from 2 to 28 keV and includes features from 3-4, 3-5, and 3-6 transitions and from 2-3, 2-4, and 2-5 transitions. Features due to non-resonant radiative electron capture into the n=3, 4, 5, and 6 shells are found at energies exceeding that of the electron beam. The cross section of the latter process, the inverse of photoionization, can be calculated with little uncertainty. Hence, normalization of cross sections to those of radiative recombination by comparing the relative x-ray intensities resulting from the respective processes has become a standard procedure in EBIT measurements.

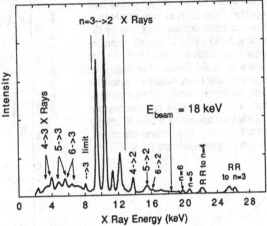

Fig. 4 -- X-ray emission from neonlike Au[69+] interacting with an 18-keV electron beam recorded with a Ge detector. The energy of the photon emitted in the radiative capture of a beam electron equals that of the captured electron plus the binding energy of the recombined ion. As a result, features due to radiative recombination (RR) are situated above the beam energy. Features due to electron-impact excitation are situated below the beam energy.

3 ELECTRON-IMPACT EXCITATION CROSS SECTIONS OF HELIUMLIKE Ti[20+]

Kα radiation from heliumlike ions, especially those of the transition metals, is observed in a wide variety of high-temperature plasmas. It is the primary ion temperature diagnostic for tokamaks (Bitter *et al.* 1982), and it is used to infer electron temperatures and bulk plasma motion of solar flares (Doschek *et al.* 1980; Seely, Feldman, and Doschek 1987). Evidently, a clear understanding of the excitation processes of heliumlike ions is paramount to their use in plasma diagnostics.

We have measured the excitation function of heliumlike Ti[20+] with EBIT by recording high-resolution crystal spectra and determining the individual intensities of lines w, x, y, and z as a function of electron beam energy. Adjusting for crystal efficiency and polarization effects and normalizing the data to the cross section for radiative electron capture, we infer the electron-impact excitation cross section for each heliumlike transition

(Chantrenne *et al.* 1992). The results are shown in Fig. 5. Here, the error bars represent statistical uncertainties; absolute uncertainties are ±18% for *w*, ±14% for *x*, and ±11% for *y* and *z*, as indicated by the error bars at 8200 eV.

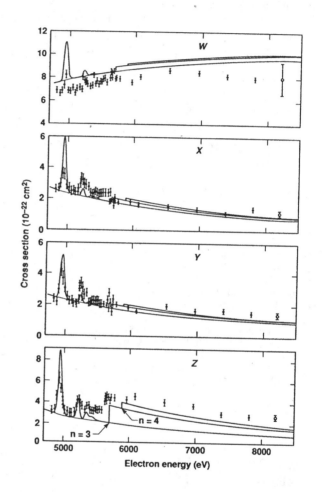

Fig. 5 -- Excitation cross sections of *w*, *x*, *y*, and *z* in heliumlike Ti^{20+} as a function of electron energy. Error bars represent statistical uncertainties. Absolute uncertainties are given by the error at 8200 eV. Calculated electron-impact excitation cross sections are shown as solid lines. The step-like increases are from radiative cascades from higher n shells. Enhancements from above-threshold resonances are modeled by assuming a 50-eV energy spread among the beam electrons. Note the suppressed zero for *w*.

The data are compared to results, shown as solid lines in Fig. 5, from a fully relativistic distorted wave code of Zhang, Sampson, and Clark (1990), which was developed specifically for highly charged ions. Radiative redistribution within the n=2, n=3, and n=4 complex is included. Redistribution has the most profound effect on *z*, which receives thirty percent of the $1s2p\ ^3P_2$ level population, all of the $1s2p\ ^3P_0$ level population of those titanium isotopes with zero nuclear spin, and much of the 1s3l and 1s4l level population. Enhancement of *z*'s excitation cross section due to radiative redistribution from the n=3 levels is clearly marked in the data by a step-like increase at 5580 eV, i.e., at the threshold for excitation of the n=3 shell. The step is also seen in the calculated cross sections. A smaller step occurs at the threshold for excitation of the n=4 shell. Similar steps in the data cannot be seen for the other lines. Indeed, radiative redistribution from the n=3 and n=4 levels is predicted to play a much smaller role for w, x, and y. Table 1 gives an overview of the predicted contributions to the effective electron-impact cross sections of w, x, y, and z from the n=3 and n=4 levels.

TABLE 1. Partial cross sections for feeding w, x, y, and z in heliumlike Ti^{20+} via radiative cascades from the n=3,4 shells. Values are given in units of 10^{-23} cm^2 and are expressed as a percentage of the total cross section for excitation of a given shell in the parentheses below. $K\beta_1$ and $K\gamma_1$ denote the transitions from levels 1s3p $^1P_1$ and 1s4p $^1P_1$ to ground, respectively.

beam energy	$K\beta_1$	x	y	z	w
		n = 3 excitation			
5700 eV	10.91 (29.2%)	2.63 (7.0%)	1.99 (5.3%)	16.32 (43.6%)	4.85 (13.0%)
7000 eV	13.11 (42.5%)	1.31 (4.3%)	1.28 (4.2%)	8.59 (27.9%)	4.45 (14.4%)
8500 eV	14.35 (53.0%)	0.79 (2.9%)	0.91 (3.4%)	5.16 (19.1%)	4.15 (15.3%)

beam energy	$K\gamma_1$	x	y	z	w
		n = 4 excitation			
5900 eV	3.58 (24.6%)	1.28 (8.8%)	0.89 (5.8%)	5.92 (40.6%)	1.27 (8.7%)
7000 eV	4.29 (35.8%)	0.79 (6.6%)	0.55 (4.6%)	3.74 (31.2%)	1.17 (9.8%)
8500 eV	4.80 (47.0%)	0.47 (4.6%)	0.36 (3.5%)	2.24 (21.97%)	1.10 (10.8%)

The data in Fig. 5 also include features from above-threshold resonances. The strongest are the KMM resonances, i.e., the process

$$1s^2 + e^- \rightarrow 1s3\ell3\ell' \rightarrow 1s2\ell + e^-, \tag{1}$$

situated at 4950 eV, followed by the KMN, KMO, and weaker resonances at higher beam energies. These are discussed in more detail below.

Ignoring the contribution from above-threshold resonances, the distorted-wave calculations agree well with our measurements of x, y, and z. In the case of w the measured cross sections are slightly smaller than calculated. (Note, though, the suppressed zero for w in Fig. 5.) Inclusion of radiative redistribution from n $\geq$ 3 levels in the theoretical cross sections only increases the discrepancy with the measurements. Slightly smaller cross sections were also measured for w at threshold in heliumlike V^{21+}, Cr^{22+}, Mn^{23+}, and Fe^{24+} (Beiersdorfer et al. 1992a). Whether this is indeed significant or the result of some, yet-to-be-identified systematic effect in our measurement is under investigation.

To compare the measured enhancements in the excitation cross sections from above-threshold resonances with theory, a least-squares fitting routine was used to fit the resonance peaks and to determine their area. The results, in units of cm^{2}eV, provide the strength of the resonances, as discussed by Beiersdorfer *et al.* (1990c). In Table 2 we list the experimental values for the KMM resonances and compare them to calculations performed using the procedures of Chen (1985). Rather poor agreement is found.

TABLE 2. Comparison of measured and calculated strengths of the KMM resonance contributions to the excitation of *w*, *x*, *y*, and *z* in heliumlike Ti^{20+}. Strengths are given in units of 10^{-20} cm^{2}eV.

	experiment	theory	ratio
w	0.6	1.5	2.4
x	0.8	1.7	2.1
y	1.5	1.5	1.0
z	2.2	3.2	1.5

Because the cross section of *w* increases over a wide range of electron energies, while that of the triplet lines decreases, the lines provide a means for determining the electron temperature of high-temperature plasmas. For this it is customary to define the temperature-sensitive line ratio $G = (x+y+z)/w$. Theoretical predictions for G over a wide range of electron temperatures have been reported for titanium by several authors (Bitter *et al.* 1985, Lee *et al.* 1985). Typically these calculations include the effects of radiative redistribution from levels $n \geq 3$ and above-threshold resonances.

To compare our data with the calculations for G we have calculated Maxwellian-averaged excitation rates. Our cross section data extend up to about 1.7 times threshold. This is sufficiently high to calculate excitation rates for electron temperatures as high as 1 keV. To calculate excitation rates for higher temperatures, we have extrapolated the data according to their predicted high energy behavior, i.e., a $1/E^2$ falloff for the triplet lines and a $\ln(E)/E$ falloff for the singlet line. For example, at an electron temperature of 1 keV the contributions from the extrapolated regions are 3%; at 2.5 keV they are 18% of the inferred excitation rate for line *z*.

Figure 6 compares the values of G inferred from our data with those reported by Lee *et al.* (1985) and by Bitter *et al.* (1985). The original calculations by Lee *et al.* did not include contributions from resonance excitation; however, for better comparison, we have augmented their values to approximate the inclusion of resonances. The resonance contribution is small and adds 11% to G at 500 eV and 10% at 1 keV. By contrast, radiative redistribution from levels $n \geq 3$ add 27% to G at 500 eV and 44% at 1 keV, as determined from our data. The calculations reported by Bitter *et al.* (1985) and Lee *et al.* (1985) differ most strongly in their predictions of the effect of radiative redistribution. Bitter *et al.* predict a more than six-fold increase in *z*'s intensity at 1.7 keV from radiative redistribution among the $n \geq 2$ levels; Lee *et al.* predict only a two-fold increase. As a consequence, the latter predictions for G are lower than those of Bitter *et al.* and are lower than those inferred from our data.

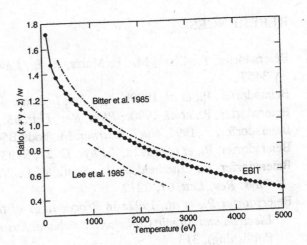

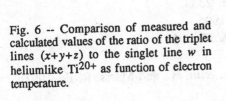

Fig. 6 -- Comparison of measured and calculated values of the ratio of the triplet lines $(x+y+z)$ to the singlet line w in heliumlike Ti^{20+} as function of electron temperature.

Differences in the theoretical predictions for G translate to differences in the temperature inferred from a measured value of G. For example, if a value G=0.80 were observed in a hypothetical measurement, a plasma temperature of 1.1 keV would be inferred from the predictions reported by Lee *et al.* (1985), while a temperature of 2.3 keV would result from the predictions of Bitter *et al.* (1985). From our data we infer a temperature of 2.1 keV.

4 SUMMARY

X-ray spectroscopy with EBIT has made possible the investigation of atomic processes in detail hitherto impossible in plasma sources. A wide variety of atomic processes can be studied that range from atomic structure determinations to excitation, ionization, and recombination cross sections. To illustrate the uses of EBIT x-ray spectroscopy we presented measurements of the above-threshold excitation of heliumlike Ti^{20+}. Because individual excitation processes are resolved from each other, the EBIT x-ray data can test theory in great detail. Good overall agreement, for example, was found with calculations of direct electron-impact excitation, while the measured values of above-threshold excitation resonances differ significantly from predictions. Such detailed comparisons with theory are essential for improving the utility of x-ray spectra in plasma diagnostics.

ACKNOWLEDGMENTS

The continued support of Dr. M. Eckart and Dr. A. Hazi is greatly appreciated. This work was performed by the Lawrence Livermore National Laboratory under the auspices of the Department of Energy under contract W-7405-ENG-48.

REFERENCES

Beiersdorfer, P., Chen, M. H., Marrs, R. E., Levine, M. A. 1990a, *Phys. Rev. A* **41**, 3453.

Beiersdorfer, P., *et al.* 1990b, *Rev. Sci. Instrum.* **61**, 2338.

Beiersdorfer, P., *et al.* 1990c, *Phys. Rev. Lett.* **65**, 1995.

Beiersdorfer, P. 1991, *Nucl. Instrum.* Methods B**56/57**, 1144.

Beiersdorfer, P., *et al.* 1991a, *Z Phys. D* **21**, S193.

Beiersdorfer, P., Osterheld, A. L., Scofield, J., Wargelin, B., and Marrs, R. E., 1991b, *Phys. Rev. Lett.* **67**, 2272.

Beiersdorfer P., *et al.*, 1992a in *Proceedings of the XVII Int. Conf. on the Physics of Electron and Atomic Collisions, Brisbane, Australia, July 10-16, 1991*, (London: IOP Publishing), 313.

Beiersdorfer, P., *et al.* 1992b, *Phys. Rev.* A, in press.

Beiersdorfer, P., *et al.* 1992c, *Phys. Rev.* A, submitted.

Bitter, M., *et al.* 1982 in *Temperature, its Measurement and Control in Science and Industry*, **Vol. 5**, edited by S. F. Schooley (New York: American Institute of Physics), 693.

Bitter, M., *et al.* 1985, *Phys. Rev. A* **32**, 3011.

Chantrenne, S., Beiersdorfer, P., Cauble, R., and Schneider, M. B. 1991, *Phys. Rev. Lett.*, submitted.

Chen, M. H. 1985, *Phys. Rev. A* **31**, 1449.

Cowan, T. E., *et al.* 1991, *Phys. Rev. Lett.* **66**, 1150.

DelGrande, N. K., *et al.* 1991, *Nucl. Instrum. Methods Phys. Res.* B **56/57**, 227.

Doschek, G. A., Feldman, U., Kreplin, R. W., and Cohen, L. 1980, *Astrophys. J.* **239**, 725.

Gabriel, A. H. 1972, *Mon. Not. R. Astron. Soc.* **160**, 99.

Knapp, D. A., *et al.* 1989, *Phys. Rev. Lett.* **62**, 2104.

Lee, P., Lieber, A. J., Chase, R. P., and Pradhan, A. K. 1985, *Phys. Rev. Lett.* **55**, 386.

Levine, M. A., *et al.* 1989, *Nucl. Instrum. Methods, Phys. Res. B* **43**, 431.

MacLaren, S., *et al.* 1990, *Phys. Rev. A.* **45**, 329.

Marrs, R. E., Levine, M. A., Knapp, D. A., and Henderson, J. R. 1988, *Phys. Rev. Lett.* **60**, 1715.

Schneider, M. B., *et al.* 1989 in *International Symposium on Electron Beam Ion Sources and their Applications - Upton, NY 1988*, AIP Conference Proceedings No. 188, edited by A. Herschcovitch (New York: American Institute of Physics), p. 158.

Schneider, M. B., *et al.* 1992, *Phys. Rev. A* **45**, R1291.

Seely, J. F., Feldman, U., and Doschek, G. A. 1987, *Astrophys. J.* **319**, 541.

Vogel, D. A., Beiersdorfer, P., Marrs, R., Wong, K., and Zasadzinski, R. 1991, *Z Phys. D* **21**, S193.

Wong, K., Beiersdorfer, P., Marrs, R., Vogel, D., and Levine, M. 1991, *Z. Phys. D* **21**, S197.

Zhang, H. L., Sampson, D. H., Clark, R. E. H. 1990, *Phys. Rev. A* **41**, 198.

X-Ray Transfer*

John I. Castor

Lawrence Livermore National Laboratory

ABSTRACT

The computational techniques for treating radiative transfer in general, and x-ray transfer in particular, are reviewed, with emphasis on the difficult problems associated with systems that are not in local thermodynamic equilibrium. Some special aspects of x-ray transfer are mentioned. The computer code ALTAIR, developed at LLNL to solve such problems, is described briefly, with an example of x-ray fluorescence in a Seyfert galaxy. Some of the prospects for experimental tests of x-ray radiative transfer theory are considered.

1 INTRODUCTION

Applications of x-ray transfer—situations in which x-radiation is emitted and absorbed or scattered by an object—range from astronomical systems like the sun and the compact companions of some evolved stars to plasmas that are created on the earth in various ways. The common feature is electron temperatures in the range of hundreds of eV—millions of kelvins. Radiative transfer becomes an issue whenever the column density ($\int \rho dz$) of the object is large enough for there to be appreciable optical depth. The objective of doing radiative transfer calculations is to enable the measured x-ray spectra to be used as diagnostics of the system, as in astrophysical and ICF plasmas, or as part of a design in which the x-ray output itself is useful, as in the soft x-ray laser. Since many or most systems that emit x-rays are not in local thermodynamic equilibrium (LTE), the difficulties of treating radiative transfer in non-LTE must be faced.

For the most part x-ray transfer can be treated using the same mathematical techniques as in the transfer of optical and UV light, for example. The concepts of optical depth, source function, isotropic scattering, *etc.*, can all be applied. The second part of the talk will review the standard radiative transfer computational methods developed for those other spectral regions.

Some problems arise only in the x-ray region, or are particularly troublesome

---

* This work was performed under the auspices of the U. S. Department of Energy by the Lawrence Livermore National Laboratory under Contract No. W-7405-ENG-48.

there. These include the tendency for the x-ray lines to be densely clustered in frequency (unresolved transition arrays), and for many of the lines to arise from inner-shell de-excitation, as may occur following inner-shell ionization or dielectronic recombination. The atomic physics needs careful consideration in these cases, but the difficulty for radiative transfer is in treating the large number of blended lines. The frequency redistribution in Compton scattering, a major effect at the high temperatures and high frequencies appropriate to x-ray systems, is another computational problem. Since some x-ray emitting objects are at solid density or higher, a variety of dense plasma effects enters, some of which can be treated only poorly at present.

2 COMMON RADIATIVE TRANSFER TECHNIQUES

2.1 LTE Case—Formal (Lambda) Solution

It is supposed that the atomic level populations are given—perhaps by the Boltzmann and Saha relations—and thereby the absorptivity k and emissivity j of x-rays, and the intensity of x-rays is to be found. The *source function* $S = j/k$ reduces to the Planck function B_ν. A simple integration of the equation of transfer is all that is required. (The notation of radiative transfer, and many of the solution techniques, are described by Mihalas 1978.)

The integral methods use an integral relation such as $J(\tau) = \int S(\tau')K(\tau' - \tau)d\tau'$ for the mean intensity J, which is done by a numerical quadrature. A review of this kind of method, in the case of spherical symmetry, is given by Avrett and Loeser (1984). Special quadrature schemes may be designed for standard kernel functions $K(\tau)$. The use of a one- or two-point quadrature scheme is the basis of Scharmer's (1981, 1984) approximate method.

If the source function S is taken through the integral sign, the integral formulation leads to a relation $J = (1 - p_{esc})S$ between J and the local value of S; this forms the basis of the escape probability method. Here p_{esc} is an integral of K, $\int_\tau^\infty K(\tau')d\tau'$ for semi-infinite slab geometry, and is easily approximated. A similar relation gives the total flux leaving the problem in terms of an integral of p_{esc} times S. By using a single representative point to describe the whole system, and an average of p_{esc}, the flux can be approximated using the information from the single representative point alone. A comprehensive review of the escape probability method is given by Rybicki (1984).

In Feautrier's (1964) method the transfer equation is manipulated into the form of a second-order differential equation at each $\mu = \cos\theta > 0$ for $u_\nu \equiv (I(\mu) + I(-\mu))/2$. (The angle θ is the inclination of a ray to the normal or radial direction.) Or, with the specification (iteratively) of the Eddington factor $\int d\mu\mu^2 I(\mu)/\int d\mu I(\mu)$, a single ODE for J_ν is obtained (Mihalas 1978, §6-3). Either form lends itself to a solution by a tridiagonal discretization of the ODE. The time compares very favorably with that for the integral method, as the tri-diagonal system is solved with a single forward+reverse sweep.

Alternatively, the 1st order transfer equation itself can be solved in a single sweep, but special care must be taken in the numerical differencing to retain 2nd-order accuracy and positivity. The discontinuous finite-element method is a good way to do this (Reed and Hill 1973; Castor, Dykema and Klein 1992). This formulation has advantages over Feautrier's in cases of non-isotropic opacity (*e.g.*, velocity fields) or angle-dependent scattering.

2.2   Non-LTE — Rate Equations — Line Scattering

The Boltzmann and Saha relations are not applicable in non-LTE, and the atomic populations must be found by solving the appropriate kinetic equations. From these, the absorptivity and emissivity needed for the x-ray transfer can be computed. The coupling between atomic populations and radiation field is strong (stiff), and these two elements must be solved for simultaneously by a process of iteration. The simplest method is to solve the two alternately (lambda iteration) until convergence is attained, but iteration counts are known to increase with the optical depth, so that for optical depths more than a few, lambda iteration is impractical. (Collision-dominated problems that are, in fact, close to LTE are an exception.) A variety of methods for solving non-LTE problems have been used.

The concept of line scattering applies to just those problems that are radiation dominated and for which lambda iteration converges very poorly. The typical case is a resonance line, where absorption of a photon in the line is followed with a very high probability ($1 - \epsilon$, where ϵ can be quite small, perhaps 10^{-4} or less) by the emission of another photon in the same line, although usually in a different direction and at a somewhat different frequency. This is the elementary scattering process, and if the medium has a sufficiently large optical depth the process recurs over and over until of order $1/\epsilon$ scatterings have occurred, and some other process intervenes. We say that the line radiation is trapped, and its intensity builds up to a value $O(1/\epsilon)$ times as strong as that due to the primary sources, if there were no scattering. Certain numerical errors made in computing the radiative transfer, even though they are as small as ϵ, can cause a large error to be made in the value of the trapped intensity, and thus some care is needed in the numerical technique.

2.3   ETLA Theory

The photoabsorption rate in a bound-bound transition is proportional to a profile-weighted average, $\bar{J}$, of the radiation intensity over angle and frequency: $\bar{J} = \int d\nu \phi_\nu J_\nu = \int d\nu \int_0^1 d\mu \phi_\nu u_\nu$. If the kinetic equations are solved for the populations N of the upper and lower levels u and ℓ of the transition, and then substituted into the expression for the line source function $S = (2h\nu^3/c^2)/(g_u N_\ell/g_\ell N_u - 1)$, S is found to depend on $\bar{J}$ linearly: $S = (1 - \epsilon)\bar{J} + \epsilon B^*$ (*cf.*, Mihalas 1978, §12-1; Vernazza, Avrett and Loeser 1981). The line scattering problem is to solve the transfer equation self-consistently with this relation for S, which depends on the intensity to be found. The ETLA method for multi-level non-LTE problems repeatedly solves all the line scattering problems, using in each case current estimates of ϵ and B^*, then builds new radiative rates from the $\bar{J}$ values obtained, after which the kinetic equations can be solved and improved values of ϵ and B^* obtained. It is found to be helpful to parameterize the radiative rates in terms of the net radiative bracket $z = 1 - \bar{J}/S$ (Avrett and Loeser 1987); this is analogous to the escape probability, but is iterated to self-consistency in the ETLA method.

2.4 Methods for ETLA Line Scattering

A single line scattering problem, that is, to solve the equation of transfer simultaneously with $S = (1 - \epsilon)\bar{J} + \epsilon B^*$, is a linear problem, and amenable to a variety of techniques.

In the integral equation method $\bar{J}$ can be replaced by the formal quadrature over optical depth of K times the unknown S. This leads to a full system of linear equations for S, with a dimension equal to the number of spatial mesh points. The costs of finding all the necessary K values and solving the full linear system are significant. Some care with the quadrature is needed to successfully treat the large-τ regime.

In Feautrier's (1964) original method for line scattering the tri-diagonal systems for u_ν at each frequency are taken together, with S replaced in terms of the $\bar{J}$'s, leading to a grand system of equations for all the u_ν's. These can be grouped together by spatial point, leading to a block tridiagonal structure. The solution is by a block version of the usual elimination method for tri-diagonal systems of equations. This method is intrinsically robust when $\tau \gg 1$, and no values of K are needed, but it can still be costly.

In Rybicki's (1971) variation of Feautrier's method the equations are grouped by frequency, leading to a block-diagonal structure bordered by horizontal and vertical stripes (from the $S = (1 - \epsilon)\bar{J} + \epsilon B^*$ relation, kept separate). The diagonal blocks are themselves tri-diagonal matrices. The intensities are eliminated using the inverses of the tri-diagonal blocks, leading to a full system for the S's, as in the integral method. The tri-diagonal inverses are less costly than the kernel quadrature calculations, however.

Lambda iteration converges slowly because of the poor coupling between neighbor zones at large optical depth. The ALI (for accelerated lambda iteration, also known as AOI for approximate operator iteration) methods apply a preconditioning step based on an approximate lambda operator. For preconditioning with a diagonal operator $1 - p_{esc}^z$, the $(n+1)$st iterate S_{n+1} is obtained from the nth iterate S_n and the formal solution $\bar{J}_n$ based on S_n by:

$$[1 - (1 - p_{esc}^z)(1 - \epsilon)]S_{n+1} = (1 - \epsilon)\left[\bar{J}_n - (1 - p_{esc}^z)S_n\right] + \epsilon B^* .$$

(See Werner and Husfeld 1985; Olson, Auer and Buchler 1986; Klein, *et al.* 1989; also the review by Rybicki 1991.) Here p_{esc}^z is approximately the escape probability from the local zone; its exact definition varies with the implementation. The iteration relation is very simple, and most of the cost is obtaining $\bar{J}_n$, as in lambda iteration; no large matrices need to be set up or solved. The diagonal operator $1 - p_{esc}^z$ may be replaced with operators coupling 3 or 5 spatial points, giving faster convergence at the expense of more computation. The convergence in all cases can be further improved by applying an acceleration technique such as Ng's method or Orthomin. These acceleration methods are reviewed by Auer (1991). The final result is convergence in 15–30 iterations, independent of optical thickness.

In the escape probability approximation (see Rybicki 1984), $\bar{J}$ is simply replaced by $(1 - p_{esc})S$ (possibly plus an external source term), so a single algebraic equation suffices to obtain $\bar{J}$. This result gives the correct average S and $\bar{J}$ taken over the whole of a homogeneous problem, but is substantially in error in the boundary layers. A so-called second-order form of the escape probability theory for homogeneous semi-infinite media leads to a first-order differential equation for $\bar{J}$. This is easily solved, and the result is nowhere in error by more than about 20%. The extension of the second-order escape probability theory to more complicated systems is an area of ongoing work.

2.5 Other Multi-level Methods

The ETLA method is a type of functional iteration, in which the large set of nonlinear equations for the intensities and level populations is solved by repeatedly applying some relatively simple operations. An alternative is the Newton-Raphson or linearization method of iteration in which a very large linear system is repeatedly solved for the corrections to the current estimates of the unknowns. Different ways of solving this system distinguish methods of this type.

The Mihalas and Auer approach (Mihalas 1978, §12-3) is to solve the system directly, by casting it in block tri-diagonal form, with all the unknowns at one space point grouped together. The tri-diagonal structure stems from the use of the Feautrier method for the radiative transfer. The size of the blocks makes this method impractical for problems with many lines and levels. Auer and Heasley (1976; see also Mihalas 1978, §12-3) used a variation in which the corrections to the intensities are expressed, using the perturbed transfer equation, in terms of the corrections to the level populations, and these, in turn, are eliminated, using the perturbed kinetic equations, to give a system of equations for the perturbations to the net radiative rates in the various transitions. The matrix is blocked with the different zones for one transition grouped together, as in Rybicki's method for single lines. The block sizes are still large: the number of spatial zones. The block SOR (successive over-relaxation, Ralston 1965) method is used to treat coupling between transitions.

When the radiative transfer is approximated using the local escape probability method for every line, the only remaining unknowns are the level populations, which obey a separate system of non-linear equations for each space point. The Newton-Raphson method may be used to solve these, although careful attention to limiting the corrections may be necessary—a general feature of linearization methods.

The extension of the ALI technique to multi-level problems is based on its analogy to the escape probability method: $\bar{J}$ is replaced by $\Lambda^* S +$ (inhomogeneous term), where Λ^* often has the form $1 - p_{esc}^z$ in terms of the escape probability from the local zone. The inhomogeneous term is adjusted to make the substitution consistent with the last formal solution of the transfer equation. With all the values of $\bar{J}$ replaced in this way, the kinetic equations are reduced to the same form as in the escape probability method, and, as in that case, can be solved with the Newton-Raphson method, zone-by-zone. This method has been reviewed recently by Dreizler and Werner (1991). It is not absolutely necessary to use the self-consistent p_{esc}^z defined by the unknown populations; lagged or relaxed values may also be used, obviating the need for the non-linear solution (Hummer and Rybicki 1991).

3 SPECIAL PROBLEMS FOR X-RAYS

The foregoing material applies equally for optical, UV and x-ray transfer. Some special considerations for x-ray transfer will be taken up next.

X-ray spectra are usually of highly ionized species, and the energy level splittings associated with electron correlations are small compared with the single-electron transition energies. This makes the lines bunch into unresolved transition arrays. A special case of this is the cluster of dielectronic satellite lines surrounding a resonance line, formed by the various spectator states that a resonantly-captured electron may occupy. A similar spectrum may be produced by inner-shell ionization as well. The difficulty this poses is that the various lines are not at all isolated from each other, a simplifying assumption that is often made for optical radiative transfer. The non-LTE algorithms must allow for many overlapping lines to successfully treat x-ray transfer. Dielectronic recombination can not be treated as a one-way process, either. The reverse rate, which depends on the radiative transfer, must be included in the branching calculation.

Since for x-ray-emitting plasmas neither kT nor $h\nu$ is completely negligible compared with mc^2, the Thomson scattering approximation of the Compton effect, which neglects the frequency shift in scattering, is not good enough. Modest thicknesses produce Compton y parameters ($y \equiv (4kT/mc^2)\max(\tau, \tau^2)$) large enough to appreciably distort the spectrum. Even one Compton scattering displaces and broadens a line by a large amount compared with its width. An accurate inclusion of Compton redistribution, which is difficult to calculate and depends sensitively on the angular distribution of the radiation, has not yet been achieved in radiative transfer codes except in those using the Monte-Carlo method. Previous work has concentrated on the Fokker-Planck approach, which is unsuitable for problems including lines. New developments address the case of accurate angle-averaged frequency redistribution, which permit line scattering and Compton scattering to be treated at the same time.

There are a few problems that do not affect optical and UV transfer very much, but which are severe in x-ray sources approaching solid density:

The Debye-Hückel theory of ionization potential lowering works well for the low-density gaseous materials that are the usual source for optical and UV spectra; the reduction of the ionization energy is by a small fraction of kT, and this converts to a yet-smaller fraction of the ionization potential itself. Only Rydberg levels much higher than those of spectroscopic interest are affected. But some of the x-ray sources are dense enough that the plasma coupling parameter $\Gamma = Ze^2/r_0 kT$ approaches unity, and the reduction in ionization potential is significant compared both with kT and with the IP itself. Only the valence levels remain bound to the ion, and the spectrum becomes very sparse. Yet there may be circumstances in which, besides these effects, the plasma still is radiation-dominated and the methods of non-LTE should be used. The result is that the energy-level data, the statistical weights, and the photoionization cross sections become dependent on plasma conditions, which invalidates most or all the solution methods that have so far been employed for non-LTE problems. This problem is only now being first addressed.

The non-LTE radiative transfer codes are required to repeatedly evaluate the line shape for each of the hundreds of lines being treated. The codes also require the cross section for an individual transition between specific atomic states in the atom. The theory of line shapes for dense plasmas provides the total emissivity as a function of frequency, and the association of parts of it with individual atomic states is inappropriate. Or, rather, the atomic states of the isolated ion do not form a good description of the radiative transitions in the dense plasma, owing to Stark mixing. A suitable theory of the line shape in a non-LTE dense plasma does not seem to exist. The best that may be hoped at present is to make an *ad hoc* division of the line shape among the isolated-ion transitions, and adopt these as the profile functions for the non-LTE calculation. This has not so far been done.

Radiative transfer theory is based on a rare-gas idealization of the dielectric function of the plasma. It is assumed that the index of refraction departs only slightly from unity, that the mean free path is large compared with the wavelength, and that the scatterers are many wavelengths apart and not ordered in space, *etc.* For x-ray transfer in solid-density materials some of these assumptions may break down, and the methods of light scattering in condensed matter must be used. It is not known how to incorporate non-LTE into that picture.

4 X-RAY CALCULATIONS WITH THE ALTAIR CODE

The computer code ALTAIR is a general purpose non-LTE radiative transfer and hydrodynamics tool that has been developed at Lawrence Livermore over the last several years (Klein, *et al.* 1989; Castor, *et al.* 1992). In its one-dimensional version it can solve the time-dependent kinetics for species with more than 1000 energy levels, while self-consistently calculating the radiative transfer in hundreds of spectral lines and photoionization continua, in addition to computing Lagrangian hydrodynamics and energy balance. The code is completely general in terms of the atomic data, since these are supplied in a separate data set that is prepared for each problem of interest. The methods used are the ETLA method for the multi-level problem, in which ALI is used to accelerate the discontinuous finite-element formal iteration for each separate line. Orthomin is used in conjunction with the ALI, and the overall ETLA iteration is accelerated using the Chebyshev method (Manteuffel 1977, 1978). The kinetic equations for the level populations are solved in an iterative fashion, using again the Chebyshev method. Details of these methods are given by Castor, *et al.* (1992).

ALTAIR's capability for computing x-ray transfer is illustrated by an example drawn from Castor, *et al.* (1992) of the K-shell iron fluorescence from a cloud located near a hypothetical active galaxy nucleus. (This calculation was originally part of a study of the Seyfert 2 galaxy NGC 1068 by Band, *et al.* 1990.) The cloud is at a density of $3 \times 10^4 \, cm^{-3}$ and a temperature of 10^5 kelvins. It is 10^{18} cm thick, and located about 10^{19} cm from the center of the galaxy. The atomic model for Fe consists of the Na-like to bare ions, and includes 546 energy levels, mostly computed with Grant's (1980) Dirac-Fock atomic structure code. The 4027 bound-bound radiative transitions are either from the Dirac-Fock code, or hydrogenic values are used. The electron collision data are mostly computed with the Coulomb-Born-Oppenheimer method of Golden, *et al.* (1981). A total of 1409 Auger decay rates following Chen (1985) were included, in addition to dielectronic recombination data from Romanik (1988, 1992).

The diffuse fluorescence spectrum from the non-illuminated side of the cloud is shown in Figure 1, for the region around the iron K lines. The extremely weak H-like lines are at the far right, and the strongest line in the center is the He-like intercombination line. The other lines represent the species of Li-like to C-like. It is significant that all the lines in the figure were given a full non-LTE line scattering radiative transfer treatment.

5 EXPERIMENTAL TESTS OF X-RAY TRANSFER

The primary problem of testing radiative transfer theory is finding a source with well-characterized structure and with physical properties that are known with sufficient accuracy to allow a significant test to be made. None of the possible environ-

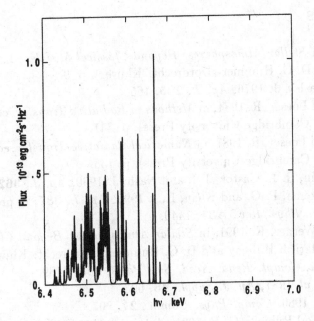

Fig. 1—ALTAIR calculation of the spectrum in the iron K-line region of the flux diffusely transmitted through a slab located near the nucleus of an active galaxy. The Compton-scattered background and the direct transmission are suppressed.

ments is perfect in this respect.

Astronomical x-ray sources span wide ranges in physical properties, which is useful for testing theories in many different regimes. The ample spatial scale means that fringing effects—walls, *etc.*—are not a problem. The drawbacks are (1) spatial structure that is often complex, and in some cases unknown, and (2) no direct measurements of physical conditions. High signal levels in some cases mean that very high resolution diagnostics may be used.

For laser-produced plasmas a number of excellent diagnostics exists to measure the physical conditions, and controlled fabrication allows simple geometries to be created, with well-known parameters. The severe size constraints limit the regimes that can be explored, and limit the density range to 2–3 orders of decompression from solid density. The short time scale and throughput constraints limit the attainable resolution. The moderately large density means that dense plasma effects can become an issue.

Plasma machines deal with plasmas that are geometrically much larger than the laser-produced plasmas, and are correspondingly easier to measure and characterize. The plasma instabilities make inhomogeneity a worse problem than with laser-produced plasmas. The density is limited to rather smaller values than in the former case. Large optical thicknesses may be difficult to obtain.

REFERENCES

Auer, L. 1991, in *Stellar Atmospheres: Beyond Classical Models*, ed. L. Crivellari, I. Hubeny and D. G. Hummer (Dordrecht: Kluwer), p. 9.

Auer, L. and Heasley, J. 1976, *Ap. J.*, **205**, 165.

Avrett, E. H. and Loeser, R. 1984, in *Methods in Radiative Transfer*, ed. W. Kalkofen (Cambridge: Cambridge University Press), p. 341.

Avrett, E. H. and Loeser, R. 1987, in *Numerical Radiative Transfer*, ed. W. Kalkofen (Cambridge: Cambridge University Press), p. 135.

Band, D. L., Klein, R. I., Castor, J. I. and Nash, J. 1990, *Ap. J.*, **362**, 90.

Castor, J. I., Dykema, P. G. and Klein, R. I. 1992, *Ap. J.*, **387**, in press.

Chen, M. H. 1985, *Phys. Rev.*, **A31**, 1449.

Dreizler, S. and Werner, K. 1991, in *Stellar Atmospheres: Beyond Classical Models*, ed. L. Crivellari, I. Hubeny and D. G. Hummer (Dordrecht: Kluwer), p. 155.

Feautrier, P. 1964, *Compt. Rend. Acad. Sci. Paris*, **258**, 3189.

Golden, L. B., *et al.* 1981, *Ap. J. Suppl.*, **45**, 603.

Grant, I. P. *et al.* 1980, *Comp. Phys. Comm.*, **21**, 207.

Hummer, D. G. and Rybicki, G. B. 1991, *Astr. Ap.*, **245**, 171.

Klein, R. I., Castor, J. I., Greenbaum, A., Taylor, D. and Dykema, P. G. 1989, *J. Quant. Spectrosc. Rad. Transf.*, **41**, 199.

Manteuffel, T. A. 1977, *Numer. Math.*, **28**, 307.

Manteuffel, T. A. 1978, *Numer. Math.*, **31**, 183.

Mihalas, D. 1978, *Stellar Atmospheres* (2nd ed., San Francisco: W. H. Freeman and Co.)

Olson, G. L., Auer, L. H. and Buchler, J.-R. 1986, *J. Quant. Spectrosc. Rad. Transf.*, **35**, 431.

Ralston, A. 1965, *A First Course in Numerical Analysis* (New York: McGraw-Hill), p. 438.

Reed, W. H. and Hill, T. R. 1973, Los Alamos National Laboratory report LA-UR-73-479.

Romanik, C. J. 1988, *Ap. J.*, **330**, 1022.

Romanik, C. J. 1992, *Ap. J.*, submitted.

Rybicki, G. B. 1971, *J. Quant. Spectrosc. Rad. Transf.*, **11**, 589.

Rybicki, G. B. 1984, in *Methods in Radiative Transfer*, ed. W. Kalkofen (Cambridge: Cambridge University Press), p. 21.

Scharmer, G. B. 1981, *Ap. J.*, **249**, 720.

Scharmer, G. B. 1984, in *Methods in Radiative Transfer*, ed. W. Kalkofen (Cambridge: Cambridge University Press), p. 173.

Vernazza, J. E., Avrett, E. H. and Loeser, R. 1981, *Ap. J. Suppl.*, **45**, 635.

Werner, K. and Husfeld, D. 1985, *Astr. Ap.*, **148**, 417.

Laboratory Plasma Experiments Using X-ray Heating

J. Edwards

Blackett Laboratory, Imperial College of Science Technology and Medicine, London SW7 2BZ, U.K.

ABSTRACT

Laboratory experiments which generate and investigate plasmas by irradiating targets with soft-X-rays produced in high-power laser laboratories are discussed. Particular emphasis is placed on research utilizing recent developments in experimental design which allow experiments to be performed in well defined and well controlled plasmas. These experiments are motivated by large research programmes such as inertial confinement fusion but also have some relevance to astrophysics.

1 INTRODUCTION

Laser-plasma (LP) research has experienced intense activity during the past twenty years. The motivation for much of this has been the prospect of achieving controlled fusion in the laboratory by using high-power lasers to drive the implosion of small, spherical targets containing nuclear fuel. There are currently two methods using high-power lasers which are considered promising candidates for achieving this goal. In the first, intense laser pulses are used to uniformly irradiate the spherical shell of the fuel pellet, directly driving the implosion. In the second, the laser light is first converted into intense soft-X-ray pulses which are then used to uniformly drive the target. Both systems are very complex and require a large amount of interacting physics to describe them. It is not surprising then that much of the considerable effort and resources devoted to LP research has been directed towards understanding both target behaviour (and performance) under these conditions and the underlying physical processes governing it.

The conditions generated in LPs are such that they are sources of copious XUV and soft-X-ray emission. This fact has been exploited extensively to diagnose the temperatures and densities of plasmas and their behaviour. X-ray spectroscopic diagnostic techniques used in LPs have been reviewed recently (Hauer, Delamater and Koenig 1991). However, the large temperature and density gradients generated in laser plasmas (Herbst et al. 1982), and the large spatial nonuniformities introduced by the initial laser-target interaction (Desselberger et al. 1992), make them unsuitable for measuring those plasma properties which are very sensitive to plasma conditions, or, in particular, those which are required for rigorous testing of theoretical calculations.

Recent experimental developments now present the possibility of making measurements of plasma properties in systems which have been demonstrated to be well defined (in terms of temperature and density), uniform, controllable and even reproducible. They are produced by heating thin foils with intense soft-X-ray pulses generated from separate laser-irradiated X-ray converter targets. This area of research is growing rapidly and many experiments using this technique have recently been performed. Much of the effort to date has been concentrated towards characterising the X-ray heating sources and plasmas they produce (Edwards et al. 1990; Back et al. 1990; Lee et al. 1992). Presently, useful measurements can be made for densities below about one tenth of solid and temperatures below approximately 50eV. Recent experiments include measurements of: LTE opacities of low and intermediate atomic number plasmas (Davidson et al. 1988; Perry et al. 1991; Foster et al. 1991); lineshapes of He-like lines in a magnesium plasma (Springer et al. 1990); selective broadband photopumping of excited states and/or the continuum in a partially ionized aluminium plasma (Smith et al. 1990); and preliminary measurements necessary for radiation-hydrodynamics studies (Back et al. 1990; Lee et al. 1992). In this paper we give a brief description of the basic experimental set-up used and of some of the experiments performed.

The direct relevance of these experiments to astrophysical plasmas is limited because in general astrophysical plasmas reside in very different temperature and density regimes. However, the type of problems which we wish to tackle experimentally in the laboratory (e.g. radiative-transfer, opacities) are often of a very similar nature to those which occur in astrophysics. Laboratory experiments which have relevance to astrophysics have been reviewed recently by Rose (Rose 1991).

2 BASIC EXPERIMENTAL SYSTEM FOR X-RAY HEATING

2.1   Basic Set-Up

The requirements of an experimental system for X-ray heating are as follows: that the sample be isolated both from the undesired effects of direct laser irradiation and from interaction with the X-ray converter plasma; that a large X-ray flux from the converter be available so that as large as possible a region of the (ρ, T) plane can be examined; and that the design of the sample be such that gradients in it are kept to a minimum. A schematic of how this can be achieved is shown in figure 1.

In brief, the laser light is incident on the X-ray converter and the soft-X-rays emitted to the rear of this foil are used to heat the sample. The converter is sufficiently thick that no laser light reaches the sample. Typically, X-ray conversion efficiencies up to 20% are found from the rear of the targets depending on the laser parameters and target thickness (Davidson 1988a; Edwards et al. 1990; Back et al. 1990; Bruneau et al. 1991; Lee et al. 1992). The converter and sample are sufficiently separated that they do not interact during the time of observation. The sample is tamped to reduce gradient formation within it to a minimum. Sample design has been investigated recently by Back (Back et al. 1990).

Finally, it is noted that significantly larger X-ray fluxes can be obtained in enclosed targets (hohlraums) which can be used to heat or drive samples (Sigel et al. 1990; Remmington et al. 1991).

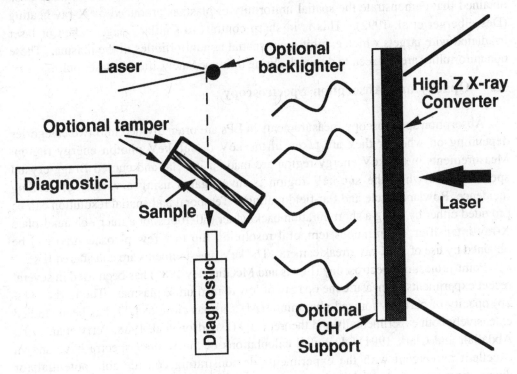

Figure 1. Schematic of set-up for experiments using X-ray heating.

3   EXPERIMENTS USING X-RAY HEATING

X-ray heating has, of course, been used extensively in government laboratories for fusion studies. Much of this work has been conducted in closed geometries and information pertaining to this work is not available.

3.1   Spatial Uniformity Measurements

Recently, several experiments have been performed using a novel monochromatic XUV imaging system (Desselberger et al. 1991; 1992). The method uses a multi-layered spherical mirror to image emission from a plasma. Images of emission from a separate backlighter in the target plane can also be obtained so that target plasma is seen as absorbing regions on the image. Images can be obtained at wavelengths down to about 30 angstroms depending on the optical properties of the multi-layers. It has been demonstrated that the spatial resolution of the instrument can be better than 1μm in the object plane.

Images of the self-emission and backlit images of X-ray heated wires have been obtained that demonstrate the spatial uniformity of plasmas produced by X-ray heating (Desselberger et al. 1992). This is in sharp contrast to similar images taken of laser irradiated wire targets which exhibit gross spatial nonuniformities in the plasma. These nonuniformities are not seen to diminish during or immediately after the laser pulse.

3.2   Opacity and Absorption Spectroscopy

Absorption spectroscopy measurements in LPs are often divided into two categories depending on whether they are taken in the keV or sub-keV photon energy region. Measurements in the keV energy region are made using flat and curved Bragg crystal spectrometers while the sub-keV region is investigated using a variety of grazing incidence, Rowland circle and flat-field gratings. Temporal and spatial resolution can be provided either by using a short duration backlighter (50-100ps) or a micro-channel-plate X-ray intensifier. Continuous temporal resolution (up to a few picoseconds) can be obtained by use of an X-ray streak camera. The keV measurements are considered first.

Point-projection spectroscopy (Lewis and McGlinchey 1985) has been used in several recent experiments to measure the opacity of low to medium Z plasmas. The K-shell (1s-2p) opacity of partially ionized aluminium (Al) plasmas close to LTE has been studied extensively both experimentally and theoretically (Davidson et al. 1988; Perry et al. 1991; Abdallah and Clark 1991). Detailed calculations of the K-shell spectra have shown excellent agreement with the experiments demonstrating considerable potential for diagnostic purposes. Indeed, Al has been doped into a tamped niobium target to act as a diagnostic of the plasma conditions in a similar experiment designed to measure the 2p-3d opacity of niobium with an open M-shell in the 1-2keV photon energy region (Goldstein 1992). The 2p-3d and 2p-4d opacity of an open M-shell germanium plasma has also been measured in a similar energy region (Foster et al. 1991). Good agreement was found between the measurements and calculations of the absorption spectra using detailed configuration accounting and an unresolved transition array approximation to describe the term structure (Rose 1992; see figure 2). Finally, in this energy region, preliminary measurements have been made of the spectral profiles of absorption lines of He-like ions in a magnesium plasma for comparison with detailed lineshape code predictions (Springer et al. 1990).

A similar technique has been used to investigate K-shell absorption spectra of partially ionized CH plasmas (Davidson et al. 1991; Hammel et al. 1992) and more recently boron nitride plasmas (Hammel et al. 1992a) in the sub-keV photon energy region. Initial measurements of the L-shell absorption spectra of open L-shell aluminium plasmas have been made (Davidson et al. 1991) and the absorption spectra of open M-shell iron plasmas have been recorded in the sub-100eV photon energy region (Lee et al. 1992).

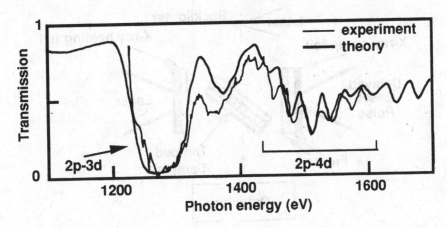

Figure 2. Comparison of the experimental transmission of germanium with the model of Rose (Rose 1992; Foster et al. 1991).

3.3   Photopumping

Well defined, LTE aluminium plasmas produced and characterised in the way discussed above in section 3.2 have been used to perform preliminary photopumping experiments designed to study non-LTE ionization/excitation kinetics (Smith et al. 1990). In these experiments, a separate laser irradiated high Z foil produces a source of keV X-rays which are used to perturb the state of excitation and ionization of the aluminium plasma. A schematic of the experimental set-up is shown in figure 3. By a suitable choice of high Z material and appropriate filtering, the emission from the second X-ray target can be tailored to photoionize a 1s electron or selectively pump some or all 1s-2p transitions of Al ions with a partially filled L-shell (which dominate the ionization balance before the X-ray pump is turned on). The 1s-2p fluorescent emission subsequently observed depends on the coupling of the excited states and ion stages to one another and the rate at which collisional and radiative transitions between them occur. This can be calculated by constructing atomic models including all the relevant excited states for all the relevant ion stages and by describing the coupling between them by a set of rate equations including the relevant collisional and radiative rates. Of course, the pump intensity and spectrum must also be known. Comparison of the predicted and measured spectra gives an indication of the accuracy of the modelling and where improvements must be made.

This method is presently complicated by the fact that the broad-band nature of the X-ray pump results in several levels being pumped simultaneously. In future experiments, it is planned to pump only the 1s-2p transition of the He-like Al ion ground state of a radiatively heated plasma with the filtered emission from a separate, directly laser irradiated Al target (Smith et al. 1990).

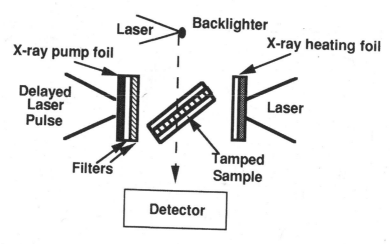

Figure 3. Schematic of experimental set-up for photopumping.

3.4 Radiative Heating and Hydrodynamics Experiments

All of the experiments described above investigate the behaviour of radiatively heated targets to some extent. The expansion of the foils can be measured with X-radiography techniques and the temperatures can be determined very accurately from ionization balance measurements. This experimental information can then be compared to hydrocode predictions. However, these targets are a special case because they have been designed so that X-rays are deposited approximately uniformly throughout the sample. This requires the sample to be fairly optically thin to most of the heating X-rays (at least after the initial heating phase) because of the exponential variation of the absorption of the radiation with optical depth. On the other hand, in, for example, the application of X-ray drive to ICF targets, it is desirable to couple as much of the incident X-ray energy as possible into inward motion of the shell surrounding the fuel, without preheating the fuel itself. Therefore, experimental investigations of how comparatively large optically thick targets behave under the impact of X-radiation are of interest.

Recently, experiments have been performed to investigate the heating of planar plastic targets by X-rays (Edwards et al. 1991). The temperature of thin chlorinated tracer layers were diagnosed as a function of time and depth in the CH targets as they were irradiated by intense, approximately Planckian soft-X-ray pulses. These pulses were generated using burnthrough foils similar to those described above (Edwards et al. 1990, Back et al. 1990). A schematic of the experimental set-up is shown in figure 4. The temperatures of the tracer layers were inferred from chlorine L-shell absorption spectra recorded with a time-resolving XUV spectrometer in the 200eV photon energy region. In this case, the burnthrough foil also provided the backlighter X-rays to form the absorption spectra. Examples of time-resolved XUV spectra obtained when the tracer layer was placed at two different depths within the plastic target are shown in figure 5. The absorption structure is largely due to 2p-3d transitions in neutral to F-like chlorine ions.

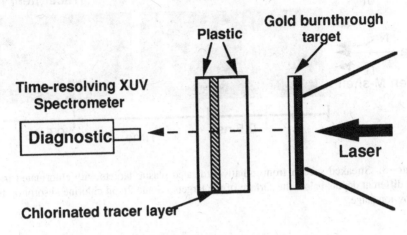

Figure 4. Schematic of experimental set-up for X-ray heating.

The two layers are heated at different rates and to different temperatures because they are buried at different depths in the plastic target. At the peak of the pulse, (approximately the centre of the streak), the absorption features are consistent with temperatures of around 40eV in the layer near the front of the target and 25eV in that near the rear. Hydrocode predictions of the temperature <u>histories</u> in the layers were found to be in reasonably good agreement with the experimental measurements. This indicates that the code was able to model with reasonable accuracy both the effect of the different thicknesses of plastic through which the X-rays passed and the subsequent heating of the tracer layers.

This study of radiative heating has been extended recently so that the hydrodynamics of the foils can be directly examined experimentally (Edwards et al. 1992). These experiments have been carried out using the high magnification (~×50) imaging technique described above (section 3.1) and shown in figure 6. The backlit images of the foils were streaked with an X-ray streak camera to obtain time-dependent records of the target motion. Preliminary radiation-hydrocode calculations are very encouraging and show good agreement with the experimental measurements. This is further evidence that the hydrocode is able to model the interaction of the X-rays with the foil targets reasonably well. These measurements also demonstrate the possibility of using this technique as an independent plasma density diagnostic by measuring the spatial extent of the plasma.

Finally, several other experiments using X-ray heating have recently been reported which are not discussed here. These include: the generation of a radiation heat-wave in a gold foil induced by an intense Planckian radiation field (T_R~200eV) generated in a cavity

(Sigel 1990); investigation of hydrodynamic (specifically, Rayleigh-Taylor) instability (Remmington et al. 1991); and investigation of hydrodynamic mix (Hansom et al. 1990).

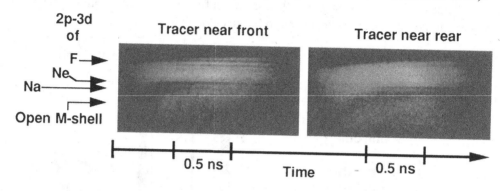

Figure 5. Streaked spectra from radiatively heated plastic targets with chlorinated tracer layers buried at different depths below the surface of the targets. Some 2p-3d chlorine absorption features are marked by ion stage.

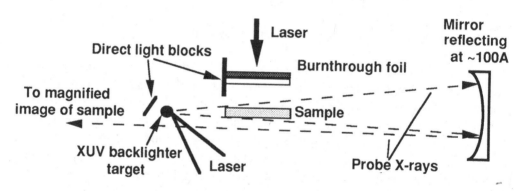

Figure 6. Schematic of set-up for hydrodynamics experiments.

4.   SUMMARY

Some experiments using soft-X-ray heating of targets have been identified and briefly discussed. Recent developments in experimental techniques have presented the possibility of studying the properties of hot, dense matter in well defined, gradient-free, controllable and reproducible plasmas for the first time in the laboratory; several such experiments have already been performed. While the plasma conditions encountered in the laboratory are generally very different to those observed in astrophysics, some of the physics problems that need to be addressed for laboratory plasmas are of a very similar nature to those which occur in astrophysics.

ACKNOWLEDGEMENT

The author is indebted to S. Rose for his comments, careful reading of the paper and for supplying figure 2. He is also indebted to O. Willi for his aid and advice in preparing the material presented and to those who supplied it: S. Davidson; R. Lee; D. Kania; E. Jannitti; J. Bruneau.

REFERENCES

Back, C. K., et al. 1990, in *Radiative Properties of Hot Dense Matter* (*Proceedings of the 4th International Workshop*), ed. W. Goldstein, C. Hooper, J Gauthier, J. Seely, and R. Lee (Singapore: World Scientific), p. 207.

Bruneau, J., et al. 1991, Phys. Rev. A , **44**, R832.

Davidson, S. J., Foster, J. M., Rose, S. J., Smith, C. C., and Warburton, K. A. 1988, Appl. Phys. Lett., **52**, 847.

Davidson, S. J. 1988a, private communication.

Davidson, S. J., et al. 1991, in Rutherford Appleton Laboratory Annual Report , RAL-91-025, p. 18.

Desselberger, M. Afshar-rad, T. Khattak, F. Vianna, S. and Willi, O. 1991, Appl. Opt., **30** 2285.

Desselberger, M. Afshar-rad, T. Khattak, F. Vianna, S. and Willi, O. 1992, Phys. Rev. Lett., **68**, 1539.

Edwards, J., et al. 1991, Phys. Rev. Lett. ,**67**, 3780.

Edwards, J., et al. 1990, in *Radiative Properties of Hot Dense Matter* (*Proceedings of the 4th International Workshop*), ed. W. Goldstein, C. Hooper, J Gauthier, J. Seely, and R. Lee (Singapore: World Scientific), p. 3.

Edwards, J., et al. 1992, in preparation.

Foster, J. M., et al. 1991, Phys. Rev. Lett., **67**, 3255.

Hammel, B. A., et al. 1992, Europhys. Lett., submitted.

Hammel, B. A., et al. 1992a, in preparation.

Hauer, A. A., Delamater, N. D., and Koenig, Z. M. 1991, Laser and Particle Beams, **9**, 3.

Hansom, J. C. V., et al. 1990, Laser and Particle Beams, **8**, 51.

Herbst, M., et al. 1982, Rev. Sci. Instrum., **56**, 803.

Kodama, R., et al. 1986, J. Appl. Phys., **59**, 3050.

Lee, R. W. 1992, in preparation.

Lee, R. W. 1992a, private communication.

Lewis, C. L. S. and McGlinchey, J. 1985, Opt. Commun., **53**, 179.

Perry, T. S., et al. 1991, Phys. Rev. Lett., **67**, 3784.

Remmington, B. A. et al. 1991, Phys. Rev. Lett., **67**, 3259.

Rose, S. J. 1991, Laser and Particle Beams, **9,** 869.

Rose, S. J. 1992, J. Phys. B., **25**, in press.

Sigel, R. et al. 1990, Phys. Rev. Lett., **65**, 587.

Smith, C. C., Davidson, S. J., Hoarty, D. J., and Foster, J. M. 1990, in *Radiative Properties of Hot Dense Matter* (*Proceedings of the 4th International Workshop*), ed. W. Goldstein, C. Hooper, J Gauthier, J. Seely, and R. Lee (Singapore: World Scientific), p. 242.

Springer, P. T., et al. 1990, in *Radiative Properties of Hot Dense Matter* (*Proceedings of the 4th International Workshop*), ed. W. Goldstein, C. Hooper, J Gauthier, J. Seely, and R. Lee (Singapore: World Scientific), p. 42.

DIELECTRONIC RECOMBINATION VERSUS CHANGE EXCHANGE : ELECTRON CAPTURE BY METASTABLE Ne–LIKE ARGON

S. Bliman[1], M. Cornille[2]

1. S. Bliman, URA 775 – CNRS, L.S.A.I – Université Paris Sud– 91405 ORSAY (FRANCE)

2. M. Cornille, UPR 176 – CNRS, DARC – Observatoire de Paris – 92195 MEUDON CEDEX (FRANCE)

Abstract

Dielectronic Recombination (DR) in the case of Ne–like Argon ions ends in the population of Na–like core excited states :

$$Ar^{8+}(2p^6)^1So + e \rightarrow Ar^{7+}(2p^5 3ln'l') \text{ with } n' \rightarrow \infty$$

The charge exchange collision process (CX) involving the long lived metastable Ne–like argon ion :

$$Ar^{8+}(2p^5 3s)^3 P_{0,2} + H \rightarrow Ar^{7+}[(2p^5 3l)^3 L \, n'l']^{2,4}L + H^+ \text{ with } n'_{max} \approx 5$$

The observation of satellites $2p^5 3ln'l'$ leaves open a certain number of questions : the decay of these excited Na–like states share between autoionization and fluorescence.

In CX since n_{max} is limited for many impurities to n_{max} of order five there is generally available one autoionization continuum $(2p^6)^1 S_0$ The case of $Ar^{8+}(2p^5 3s)^3 P_0 + H_2$ is considered. The satellites lines are seen unpolluted by the resonance lines on the one hand ; the resonance lines are oberved separately. The autoionization spectrum of the core excited Na–like Argon states is analyzed.

Introduction : In plasma modelling when studying ionization equilibria, the coronal model is of frequent use. However, given the existence of metastable ions it is now understood that charge exchange (CX) might end up in populating states which are similar to those seen in Dielectronic Recombination (DR). The comparative importance of CX Versus DR comes from the fact that the rate coefficients of CX might be 10^4 up to 10^8 times as large as DR rate coefficients. The problem of the stabilization of core excited Na–like ions is addressed.

Experimental Approach And Results :

A Ne – like argon ion beam of energy $\approx 0.38v_0 (v_0 = 1$ au of velocity $= 2.2$ 10^8 cm/s is passed in a collision chamber (inner pressure $\approx 5.10^{-5}$ Torr). The CX process is observed both by photon spectroscopic technics and Auger spectroscopy. These approaches are mandatory because it is well known that Ne–like ions have a long lived metastable state : $Ar^{8+}(2p^5 3s)^3 P_{0,2}$. The decay of $^3P_2$ to $^1S_0$ is via a M2 transition and in the case of Argon, this is a very slow transition ($\approx 10^{-4} s$)

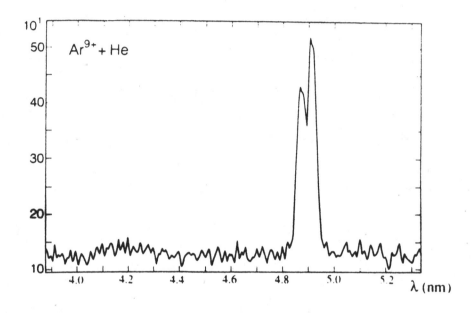

figure 1

Two further experiments have been performed. On the one hand to obtain the position of the mother lines an X–Ray spectroscopic observation of $Ar^{9+} + He \rightarrow Ar^{8+}(2p^5 nl)^{1,3} L + He^+$ has been performed ; the spectrum is shown in fig.(1) where unambiguously are identified the final decay of $(2p^5 3s)^1 P_1^0 - (2p^6)^1 S_0 (\lambda = 48,73\text{Å})$ and $(2p^5 3s)^3 P_1^0 - (2p^6)^1 S_o (\lambda = 49,18\text{Å})$. On the other hand the observation of $Ar^{8+} + He \rightarrow Ar^{7+} + He^+$, observed in the same wavelength range as the previous process, shows unambiguously that Ar^{8+} beam contains a metastable fraction and the single capture ends up at populating core excited Na–like Argon states. This is shown in fig. 2 where different groups of satellite lines are indicated with upper levels configuration. The capture level in the case of H_e as a collision partner is n=4. From these levels the stabilization shares between the autoionization and radiative channels. This is clearly seen in the Auger spectrum (Boudjema 1990). The most intense peak originates directly from the capture level and is identified as originating from $(2p^5 3s4l)^{2,4} L$ to $(2p^6)^1 S_0 + \varepsilon l$ where εl corresponds to the Auger electron (fig. 3). All other peaks in this spectrum correspond to states populated by cascades starting from $(2p^5 3s4l)^{2,4} L$. In this case the total CX cross section is of the order $2.1\ 10^{-15} \text{cm}^2$. With H as collision partner, the capture CX cross section is $1.3\ 10^{-14} \text{cm}^2$ and the capture level is n=5.

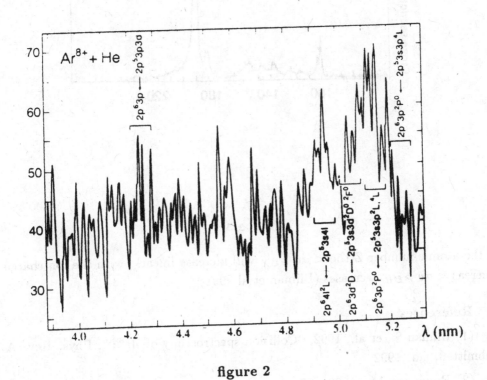

figure 2

Conclusion : The single capture (SC) by the metastable ion $Ar^{8+*}(2p^5 3s)^3 P^0_{0,2}$ populates core excited Na–like Ar levels not well known before this work. These levels are also populated in Dielectronic Recombination (DR). A close inspection of the level characteristics shows that the $^4 L$ states are not systematically metastable against autoionization as in Li–like core excited systems. It is important to differentiate between levels with $^1 L$ and $^3 L$ cores : they are different in energy and in their stabilisation (radiation and autoionization). Since rate coefficients for DR are generally much smaller than those for charge transfer and since it is likely that DR may populate preferentially the $^1 L$ core configurations, the necessity of knowing both appears. Both collision processes may well be cooperative in plasmas, e.g. for X-Ray laser schemas or in Tokamaks when neutral hydrogen beams are used for heating and diagnostics purposes. Along an isoelectronic as a function

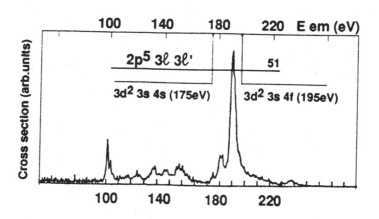

figure 3

of the atomic number Z the $< \sigma v >_{CX}$ will increase linearly with the ion charge whereas $< \sigma v >_{DR}$ decreases (Bliman et al 1992)

References

(1) Bliman S., et al., 1992, "Collision spectroscopy of Ar^{8+}", Phys. Rev. A Submitted, jan. 1992

(2) Boudjema M., 1990, "Capture électronique sur des états autoionisants d'ions multichargés en collision avec des cibles gazeuses", Thesis, Université d'Alger

Figure captions :

Figure 1 : Normalized intensity soft X-ray spectrum for $Ar^{9+}(2p^5) + He \rightarrow Ar^{8+}(2p^5nl)^{1,3}L + He^+$; near 4,9nm, the peaks correspond to $(2p^s3s)^1P_1^0 - (2p^6)^1S_0(4,873\ nm)$ and $(2p^53s)^3P_1^0 \rightarrow (2P^6)^1S_0(4,918\ nm)$

Figure 2 : Normalized intensity survey spectrum for $Ar^{8+} + $ He at 2 KeV /amu.

Figure 3 : Auger Spectrum for $Ar^8 + He$ (Boudjema Thesis)

SOME REMARKS ON THE USE OF THE TWO STATE MODEL IN HEAVY IONS COLLISIONS

BOUGOUFFA S.[1], FAUCHER P.[2]

[1]Institut des Sciences Exactes, Université de Batna, Algerie.
[2]URA 1362 du CNRS, Observatoire de la Côte d'Azur, BP 229, 06304 NICE Cedex, France.

ABSTRACT

Some recent results concerning a method on the separation of coupled differential equations are applied to illustrate some properties of the schematic model in heavy ion fusion. The theory is now extented to the three dimensional case and allows to enhance the investigation to states with $l \neq 0$

1. INTRODUCTION

Fusion of two heavy ions has customarily been described in terms of simple models which use one-dimensional potentials. Fusion is then decided by the ability of the system to penetrate through or pass over the potential energy barrier, and the model is often referred as the Barrier Potential Model (Vaz et al. 1981). However a number of experiments (Beckerman et al. 1980,1981, Jahnke et al. 1982) have clearly shown that this standard model is inadequate.

Dasso et al. (1983) considered a more general framework taking into account for the effects of inelastic collisions and transfer reactions. Their approach is based on the coupled channel formalism and is formulated in the frame of the schematic model leading to a system of two coupled differential equations for s states ($l = 0$) in the one-dimensional case.

In this work, a method on the separation of coupled differential equations developed by Cao (1988) is applied to study the channel coupling effect in the fusion theory for heavy ions. In the framework of the schematic model this method allows to enhance the investigation to states with $l \neq 0$. Results are presented for simple models.

2. THEORY

The coupled channel formalism for direct reaction process is taken from Austern (1970). Substituting the total wave function Ψ expanded in terms of channel states Φ_α in the Schrödinger equation, we obtain :

$$(H_0 + V)\Psi = E\,\Psi \tag{1}$$

$$\Psi = \Sigma_\alpha \frac{1}{r} G_\alpha(r)\,\Phi_\alpha \tag{2}$$

$$\frac{d^2 G_\alpha}{dr^2} + \frac{2\mu_\alpha}{\hbar^2}[E_\alpha - V_\alpha^{eff}(r)]\,G_\alpha(r) = \frac{2\mu_\alpha}{\hbar^2}\Sigma_{\beta \neq \alpha} V_{\beta\alpha}^{cpl}(r)G_\beta(r) \tag{3}$$

with

$$V_\alpha^{eff}(r) = \frac{\hbar^2}{2\mu_\alpha} \frac{l_\alpha(l_\alpha+1)}{r^2} + <\Phi_\alpha|V|\Phi_\alpha> \tag{4}$$

$$V_{\alpha\beta}^{cpl}(r) = <\Phi_\alpha|V|\Phi_\beta> \tag{5}$$

H_0 is the hamiltonien for intrinsic and relative motion, V is the interaction energy, and for a given channel α, μ_α is the reduced mass, l_α is the angular momentum, E_α is the relative energy

$$E_\alpha = E + \Delta E_\alpha \tag{6}$$

ΔE_α is the reaction ΔE value. The coefficients of the outgoing waves determine the various fusion cross- sections.

In order to illustrate various effects of channel coupling within a simple model we use the two state approximation. The infinite set of coupled equations (3) reduces to :

$$[\frac{d^2}{dr^2} + f_1(r)]G_1 = B_{12}G_2 \tag{7}$$

$$[\frac{d^2}{dr^2} + f_2(r)]G_2 = B_{12}G_1 \tag{8}$$

where

$$f_\alpha(r) = \frac{2\mu_\alpha E_\alpha}{\hbar^2} - \frac{l_\alpha(l_\alpha+1)}{r^2} - \frac{2\mu_\alpha}{\hbar^2} <\Phi_\alpha|V|\Phi_\alpha> \tag{9}$$

$$B_{\alpha\beta} = \frac{2\mu_\alpha}{\hbar^2} V_{\alpha\beta}^{cpl}(r) \tag{10}$$

Using the technique developed by Cao(1988) and Bougouffa(1991) the two coupled equations (7,8) can be separated. At a first order approximation, valid for weak coupling, we have to solve the two following separated differential equations :

$$\{\frac{d^2}{dr^2} + \frac{1}{2}(f_1 + f_2) \pm \frac{1}{2}\sqrt{(f_1 - f_2)^2 + 4B_{12}^2}\} Y^\pm(r) = 0 \tag{11}$$

The functions $Y^\pm$ are related to the functions G_α by a transformation matrix (Cao 1988). The solution of this uncoupled system allows to determine the two eigenphase shifts $\delta_l^\pm$ from which the inelastic cross-sections Q_l can be obtained :

$$Q_l = C \ sin^2(\delta_l^+ - \delta_l^-) \tag{12}$$

C is a quantity depending on E_1, E_2, l and can be derived analytically (Cao 1988). However, in this work, we are only interested by the position of the extrema of Q_l and we simply set : $C = 1$.

3. RESULTS

Defining

$$U_\alpha(r) = \frac{2\mu_\alpha}{\hbar^2} < \Phi_\alpha|V|\Phi_\alpha > \qquad (13)$$

$$k_\alpha^2 = \frac{2\mu_\alpha E_\alpha}{\hbar^2} \qquad (14)$$

we calculated the variation of inelastic cross-sections Q_l for the simple model :

$$U_\alpha(r) = V_\alpha \, exp(-r^2/2\sigma_\alpha^2) \qquad (15)$$

$$B_{\alpha\beta}(r) = F \, exp(-r^2/2\sigma_\alpha^f) \qquad (16)$$

In this way, the inelastic cross-section is a function of the amplitude F of the coupling strength. Figure 1 shows the variation of Q_l in the exact resonance case $\Delta k^2 = 0$ for $l = 1,3,5$. Results are obtained for the following parameters : $\mu = 1$, $V_1 = V_2 = 0$, $\sigma_1 = \sigma_2 = \sigma_f = \sigma$, $k_1^2 = 20$. The occurence of "resonance" is quite explicit. In figure 2, we display the inelastic cross-section Q_l in the non resonance case $k_2 \neq 0$ and it appears clearly a shift of the extrema.

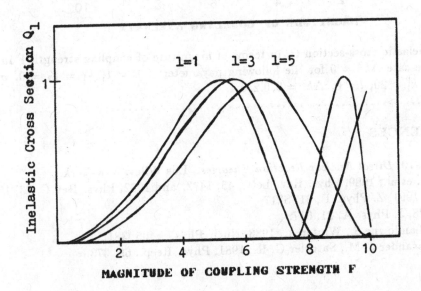

Fig 1. Inelastic cross-section Q_l in terms of the magnitude of coupling strength F in the exact resonance case $\Delta k^2 = 0$ for the following parameters : $\mu = 1$, $V_1 = V_2 = 0$, $\sigma_1 = \sigma_2 = \sigma_f = \sigma$, $k_1^2 = 20$, $l = 1,3,5$.

This first order approximation, suitable for $\Delta k^2 << k^2$, allows to obtain an enhancement and a attenuation of the inelastic cross-section in a two state model. Results obtained from this method in the case of fusion of heavy ions are quite similar to those obtained in electron-atom inelastic collisions (Bougouffa 1991)

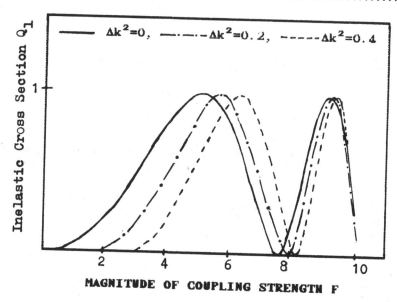

Fig 2. Inelastic cross-section Q_l in terms of magnitute of coupling strength F in the non resonance case $\Delta k^2 \neq 0$ for the following parameters : $\mu = 1$, $V_1 = V_2 = 0$, $\sigma_1 = \sigma_2 = \sigma_f = \sigma$, $k_1^2 = 20$, $l = 0$, $\Delta k^2 = 0, 0.2, 0.4$.

REFERENCES

Austern N., 1970, *Direct Nuclear Reaction Theories.*, Eds Wiley, New-York.
Beckerman M. et al., 1980, Phys. Rev. Lett., 45, 1472. and 1982, Phys. Rev C, 23, 1581.
Bougouffa S., 1991, Z. Phys. D, 21, S217.
Cao X. C.,1988, J. Phys. A, 21, 617.
Dasso C. H., Landowne S., Winter A., 1983, Nucl. Phys. A405, 381.
Vaz L. C., Alexander J. M., Satchler G. R., 1981, Phys. Reep., 69, 373.

Atomic Data for the SOHO Mission

A. Burgess[1], H.E. Mason[1] and J.A Tully[2]

[1]Department of Applied Mathematics and Theoretical Physics, Silver Street, Cambridge CB3 9EW, U.K.
[2]Observatoire de la Cote d'Azur, Centre National de la Recherche Scientifique (URA 1362), B.P. 139, 06003 Nice Cedex, France.

ABSTRACT

An interactive computer program has been developed by Burgess and Tully (1992) to assess and store electron excitation collision rates. The program is being used to prepare atomic data which is required to analyse solar and astrophysical spectra. In this paper, we present work in progress on several solar ions with spectral lines in the UV wavelength region. These will be observed by the CDS and SUMER instruments to be flown on the SOHO mission.

1 SOHO ATOMIC DATA PROJECT

The Solar Heliospheric Observatory (SOHO) mission will be flown in 1995. This will include different instruments to study solar oscillations, the solar wind and the structure of the solar atmosphere. Of particular interest are the two spectroscopic instruments, the Coronal Diagnostic Spectrometer (CDS) and the Solar Ultraviolet Measurements of Emitted Radiation instrument (SUMER), which cover the wavelength range 160-1600Å. The object of the atomic data project is to compile and assess all the electron excitation rates required for the spectral lines which will be observed by the CDS and SUMER instruments. Fifteen reviewers have been selected to write reports on the available atomic data for groups of ions. A meeting will be held in March 1992 to discuss these reports and a publication of recommended data will be produced. J. Lang has been responsible for much of the organisational work involved. The work of the reviewers is to prepare a bibliography, to assemble the electron excitation data for each transition (between J levels) and along iso-electronic sequences and to plot graphs of effective collision strength against temperature. They have also been asked to compare different results, identifying disparities and to write a preliminary assessment of the data, indicating where further work is required for SOHO.

2 OMEUPS: ASSESSMENT OF ELECTRON EXCITATION DATA

OMEUPS is an interactive program written by one of us (A.B.) which is based on a new method for analysing atomic collision data. It is easy to use, even by non-specialists in atomic physics. OMEUPS has two branches labelled OMEGA, for analysing energy-dependent collision strengths $\Omega(E)$, and UPSILON, for analysing the thermally averaged collision strengths $\Upsilon(T)$. The method is designed to analyse data for the following four types of transition in positive ions:

Type 1   Electric dipole (optically allowed).
Type 2   Electric multipole other than dipole (optically forbidden).
Type 3   Exchange (spin change, i.e. intersystem or intercombination).
Type 4   Electric dipole with abnormally small oscillator strength.

The originality of the method arises from the use of scaling techniques which remove the main energy or temperature dependence from the data and map the entire range of E or T onto the interval [0,1]. Denoting the appropriately scaled or reduced variables (E_{red}, Ω_{red}) as (x,y), they are defined in OMEGA as follows for the four types of transition:

Type 1	$x = 1 - lnC/ln(E_j/E_{ij} + C)$,	$y = \Omega_{ij}/ln(E_j/E_{ij} + e)$,	$(C > 1)$
Type 2	$x = (E_j/E_{ij})/(E_j/E_{ij} + C)$,	$y = \Omega_{ij}$,	$(C > 0)$
Type 3	$x = $ as for type 2,	$y = (E_j/E_{ij})^2\Omega_{ij}$,	$(C > 0)$
Type 4	$x = $ as for type 1,	$y = \Omega_{ij}/ln(E_j/E_{ij} + C)$.	$(C > 1)$

The indices i and j label the lower and upper energy levels. E_j is the energy of the colliding electron after excitation and E_{ij} is the transition energy. The parameter C is optimised to suit the ion and transition in question; e is the base of Napier's logarithm. After introducing data to the OMEGA branch of the program and providing an initial estimate for C, Ω_{ij} is transformed to Ω_{red} and displayed as a function of E_{red}. The program calculates a least-squares spline fit to the data and draws the corresponding curve on the screen. The values of Ω_{red} are given at the 5 equally distributed knots (i.e. $E_{red} = 0, 0.25, 0.50, 0.75, 1$). These knots can be changed interactively.

The corresponding $\Upsilon_{ij}(T)$ are evaluated by performing the Maxwell average of Ω_{ij}

$$\Upsilon_{ij}(T) = \int_0^\infty \Omega_{ij} \, exp(-E_j/kT) \, d(E_j/kT).$$

An N-point Gauss-Laguerre approximation is used to calculate this integral. The Υ can be reduced and splined in the UPSILON branch of OMEUPS.

$\Omega_{red}(1)$ is the limit to which Ω_{red} tends as $E_j/E_{ij} \to \infty$; similarly $\Upsilon_{red}(1)$ is the limit to which Υ_{red} tends as $T \to \infty$. Numerical values for the limits can be obtained fairly, easily except in the case of $\Upsilon_{red}(1)$ for a type 3 transition. They are extremely useful but have rarely been used in past extrapolation procedures.

3 ASSESSMENT OF FE IX - FE XIV

One of us (H.E.M) has been assigned the task of reviewing and assessing the ions Fe IX - Fe XIV. A lengthy report has been produced. The recommendations are summarised below:

Fe IX. The recent calculations of Fawcett and Mason (1991) are accurate with a good energy spread. No further work is required. The early results by Flower (1977a), frequently used in solar analyses, had an error in the indexing of some of the levels. Figure 1 shows a plot of Ω_{red} versus E_{red} for one of the Fe IX transitions, from the calculations by Fawcett and Mason. Figure 2 shows the corresponding plot of Υ versus $Log_{10}T$.

Fe X. For the $n = 3$ configurations, the calculations of Mason (1975) remain the most comprehensive available. Fe X needs URGENT attention, new calculations are required with an improved target.

Fe XI. The calculations of Mason (1975) remain the most comprehensive available. Fe XI needs URGENT attention, new calculations are required with an improved target.

Fe XII. The calculations of Tayal and Henry (1986) are available for the $3s^23p^3$ and $3s3p^4$ configurations. These need checking. Calculations by Flower (1987) are available for the $3s^23p^23d$ configuration also, but these could be substantially improved, with a better target and more energy values.

Fe XIII. The calculations by Fawcett and Mason (1989) are accurate with a good energy spread. The collision strengths for transitions within the ground configuration should be imroved by including the resonance cotributions.

Fe XIV. The results published by Dufton and Kingston (1991) need checking. There are some inconsistencies between their Ωs and their Υs. New calculations for this ion should be carried out with a more accurate target.

REFERENCES

Burgess, A. and Tully, J.A. 1992, *Astron. Astrophys.*, **254**, 436.

Dufton, P.L. and Kingston, A.E. 1991, *Physica Scripta*, **43**, 386.

Fawcett, B.C. and Mason, H.E. 1989, *A.D.N.D.T.*, **43**, 245.

Fawcett, B.C. and Mason, H.E. 1991, *A.D.N.D.T.*, **47**, 17.

Flower, D.R. 1987a, *Astron. Astrophys.*, **56**, 451.

Flower, D.R. 1987b, *Astron. Astrophys.*, **54**, 163.

Mason, H.E. 1975, *Mon. Not. R. astr. Soc.*, **179**, 651.

Tayal, S.S. and Henry, R.J.W. 1986, *Astrophys. J.*, **302**, 200.

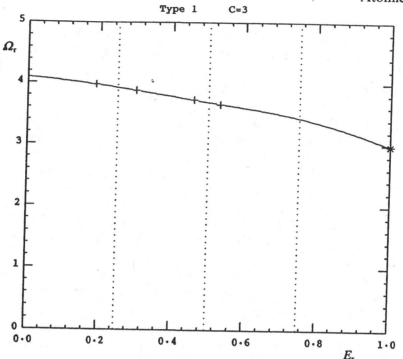

Figure 1 Reduced Collision Strength for the Fe IX transition $3p^6 \, {}^1S_0$
$\cdot \, 3p^5 3d \, {}^1P_1$

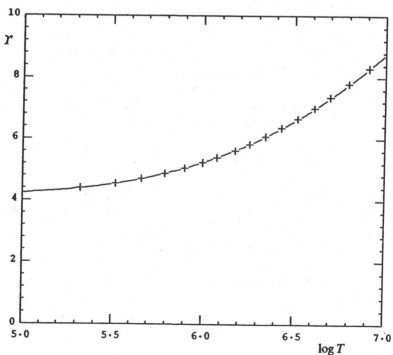

Figure 2 Averaged Collision Strength for the Fe IX transition $3p^6 \, {}^1S_0$
$\cdot \, 3p^5 3d \, {}^1P_1$

UV and Soft X-ray lines from Fe XVI Observed in Solar and Stellar Spectra

M. Cornille[1], J.A. Dubau[1], H.E. Mason[2], C. Blancard[3] and W.A. Brown[4]

[1] Observatoire de Paris, UPR 176 CNRS, DARC, 92195 Meudon-Cedex, France.
[2] Department of Applied Mathematics and Theoretical Physics, Silver Street, Cambridge CB3 9EW, U.K.
[3] Centre d'Etude de Limeil-Valeuton, 94915 Villeueuve Saint Georges Cedex, France.
[4] Lockheed Palo Alto Research Laboratory, Palo Alto, California 94304, U.S.A.

ABSTRACT

The ion Fe XVI is abundant in solar active regions and flares. Strong spectral lines from this ion have been observed over a wide wavelength range $(30 - 365 \text{Å})$ both in astrophysical and laboratory plasmas. In this paper, we present new electron scattering collision rates and compare our theoretical intensity ratios with solar observations. We note that the approximations often used to simulate Fe XVI soft X-ray lines in solar and stellar spectra are inaccurate.

1 ELECTRON EXCITATION COLLISION STRENGTHS

Accurate electron scattering calculations are now available for many of the ions which are abundant in the solar corona and transition region. The emphasis in the past decade has been on solar flare spectra (Mason, 1991). In this paper, we concentrate on the wavelength region $10 - 100 \text{Å}$, which was observed with the grazing incidence spectrograph (XSST) flown on a rocket (Acton et al, 1985). In particular, we are interested by the strong Fe XVI lines which are prominent in these spectra, corresponding to transitions between the configurations 3p - 4s, 3p - 4d, 3d - 4f and 3s - 4p. The electron scattering problem was solved using the distorted wave approximation (Eissner and Seaton, 1972) for the low partial wave values $(l \leq 25)$. The high partial wave contributions for the dipole transitions are calculated using the Coulomb Bethe Approximation. The high partial wave contributions for other transitions were calculated using a new program (distorted wave without exchange). We find that even for the s - d transitions, there are significant contributions to the total collision strengths from $l > 25$. In table 1, we give the Fe XVI theoretical level energies and the collision strengths (Ω) at one energy value.

TABLE 1 FE XVI THEORETICAL ENERGY LEVELS AND $\Omega(1, k)$

k	config.	Level	E(cm^{-1})	$\Omega(1, k)$ (at 100 Ryd.)
1	3s	$^2S_{0.5}$	0	
2	3p	$^2P_{0.5}$	277644	2.132
3		$^2P_{1.5}$	298071	4.268
4	3d	$^2D_{1.5}$	678044	0.139
5		$^2D_{2.5}$	681357	0.209
6	4s	$^2S_{0.5}$	1867439	0.112
7	4p	$^2P_{0.5}$	1977565	0.0273
8		$^2P_{1.5}$	1985477	0.0508
9	4d	$^2D_{1.5}$	2124926	0.0292
10		$^2D_{2.5}$	2126356	0.0441
11	4f	$^2F_{2.5}$	2185355	0.0377
12		$^2F_{3.5}$	2185859	0.0503
13	5s	$^2S_{0.5}$	2662852	0.0208
14	5p	$^2P_{0.5}$	2717144	0.0061
15		$^2P_{1.5}$	2721033	0.0120
16	5d	$^2D_{1.5}$	2788310	0.0085
17		$^2D_{2.5}$	2789043	0.0130
18	5f	$^2F_{2.5}$	2818761	0.0068
19		$^2F_{3.5}$	2819020	0.0091

TABLE 2 RELATIVE INTENSITIES OF THE FE XVI LINES

λ(Å)	Transition		$I/I(50.55\text{Å})$ XSST	Theory (10^7K)
50.35	3s $^2S_{0.5}$	4p $^2P_{1.5}$	2.0	1.9
50.56	3s $^2S_{0.5}$	4p $^2P_{0.5}$	1.0	1.0
54.13	3p $^2P_{0.5}$	4d $^2D_{1.5}$	1.6	1.4
54.72	3p $^2P_{1.5}$	4d $^2D_{2.5}$	2.3	2.5
54.77	3p $^2P_{1.5}$	4d $^2D_{1.5}$	0.4	0.3
62.88	3p $^2P_{0.5}$	4s $^2S_{0.5}$	2.1	2.3
63.72	3p $^2P_{1.5}$	4s $^2S_{0.5}$	3.8	4.8
66.25	3d $^2D_{1.5}$	4f $^2F_{2.5}$	2.4	2.1
66.36	3d $^2D_{2.5}$	4f $^2F_{3.5}$	3.0	3.0

2 XSST SPECTRA

In table 2, we list intensity ratios of the spectral lines corresponding to the n=3 to n=4 transitions which were observed by the XSST instrument during a solar flare (Acton *et al*,1985). It can be seen that the lines corresponding to the transitions 3p - 4s, 3p - 4d, 3d - 4f are of comparible intensity to the dipole transition 3s - 4p. The temperature which corresponds to these intensity ratios is significantly higher than that corresponding to the peak abundance of Fe XVI (2.5×10^6K).

3 SIMULATED SOLAR AND STELLAR SPECTRA

The X-ray and UV solar spectrum has been simulated by several authors using purely theoretical approximations and semi-empirical methods (see for example Mewe, 1972). The approximation which is commonly used to estimate the intensity for the solar lines is based on the $\bar{g}$ formula. The averaged collision strength Υ is related to $\bar{g}$ by

$$\Upsilon_{ij} = \frac{8\pi}{\sqrt{3}} \frac{\omega_i f_{ij}}{E_{ij}} \bar{g} \tag{1}$$

where $\omega_i f_{ij}$ is the radiative oscillator strength and E_{ij} is the energy difference between levels i and j. Υ is defined as

$$\Upsilon_{ij} = \int_0^\infty \Omega_{ij} exp(\frac{-E_j}{kT}) d(E_j/kT) \tag{2}$$

where Ω_{ij} is the collision strength for the transition i to j, E_j is the electron energy relative to the final state j, k is the Boltzmann constant and T is the electron temperature of the plasma. The electron collision rate is then given by the well known formula

$$q_{ij} = 8.63x10^6 T^{0.5} exp(-E_{ij}/kT) \Upsilon_{ij}/\omega_i \tag{3}$$

It should be emphasised that the $\bar{g}$ formulation was developed for dipole transitions. Mewe (1972) extended its use to forbidden transitions, but equn (1) then has no meaning, since f_{ij} is undefined. Mewe uses a parametric expression for $\bar{g}$.

A. Burgess has developed an interactive graphical package for assessing and storing electron excitation data (Burgess and Tully, 1992). In figure 1 we plot the reduced collision strength (Ω_r) versus the reduced energy (E_r, threshold energy is 0, infinite energy limit is 1) for the 3s - 4p transistion of the Fe XVI and in figure 2 we plot the corresponding Υ versus $Log_{10}T$. The complete set of results for Fe XVI is published in Cornille *et al*, 1992. Our Υs can be directly compared to previous estimates. We find that the type of approximation used by Mewe and other to simulate solar and astrophysical spectra is not very good for the Fe XVI lines.

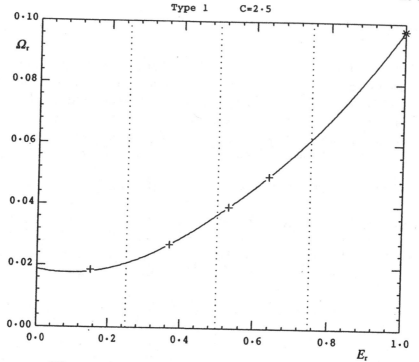

Figure 1 Reduced Collision Strength for 3s - 4p.

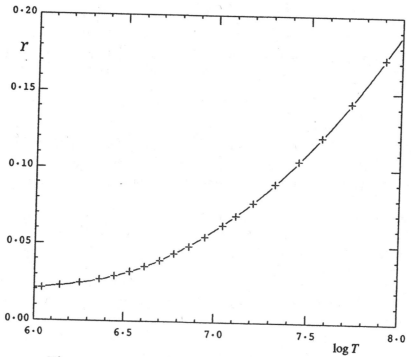

Figure 2 Averaged Collision Strength for 3s - 4p.

REFERENCES

Acton, L.W., Bruner, M.E., Brown, W.A., Fawcett, B.C., Schweizer, W. and Speer, R.J. 1985, *Astrophys. J*, **291**, 865.

Burgess, A and Tully, J.A. 1992, *Astron. Astrophys.*, **254**, 436.

Cornille, M., Dubau, J., Loulergue, M., Mason, H.E., Blancharde, C and Brown, W.A. 1992 *draft paper*

Eissner, W. and Seaton, M.J. 1972, *J. Phys. B*, **5**, 2187.

Mason, H.E. 1991, *Phil. Trans. R. Soc. Lond. A*, **336**, 471.

Mewe, R. 1972, *Sol. Phys.*, **22**, 459.

Electron Correlation in Ionization-Excitation of Helium by Electron and Proton Impact

S. Fülling[1], R. Bruch[1], P. A. Neill[1], E. A. Rauscher[1], M. Bailey[1], J. S. Thompson[1], E. Träbert[2], P. H. Heckmann[2], and J. H. McGuire[3]

[1]Physics Department, University of Nevada, Reno, NV 89557, USA
[2]Institut für Experimentalphysik III, Ruhr-Universität, D-4630, Bochum 1, Germany
[3]Physics Department, Tulane University, New Orleans, LA 70118, USA

ABSTRACT

We report experimental results on the state selective cross section ratio $\sigma^{+*}/\sigma*$ of helium for electron and proton impact. Our results show consistently higher yields for electrons than for protons at higher velocities. A comparison is made between the cross section ratios σ^{2+}/σ^{+} of helium for electron and proton impact.

1 INTRODUCTION

Recently there has been much interest in theoretical calculations (McGuire et al. 1989; Reading and Ford 1987) and in atomic collision measurements of cross sections involving many electron transitions, in particular double ionization (σ^{2+}), double excitation (σ^{**}) and ionization plus excitation (σ^{+*}) (Träbert et al. 1987; Fuelling et al. 1991; Fuelling 1991) of helium. The emphasis of this paper is to present for the first time extensive experimental absolute cross sections of ionization-excitation of He by electron and proton impact. The primary reason for this interest is the need to understand the simplest dynamical few electron effects. At high collision velocities the dominant mechanisms for two electron transitions require few electron dynamics. Correlation, produced in the interaction between two or more electrons, occurs when the full scattering wavefunction cannot be described by the independent particle model (IPM), corresponding to the time independent Hartree Fock approximation. Helium is the simplest many-electron atom, and therefore ideally suited for achieving a better theoretical understanding of many electron processes. Although some work has been conducted in this field (McGuire et al. 1989), the basic physical mechanisms producing two electron transitions at higher collision velocities, are currently not well understood.

Ionization-excitation represents an intermediate energy process when compared to double ionization and double excitation. We note that for measurements of σ^{**} we have well defined initial and final states, whereas in the case of σ^{+*} and σ^{2+} we have one or two electrons in the continuum, respectively. To shed more light on this interesting process we have performed extensive series of scattering experiments to study simultaneous ionization and excitation of He using various atomic and molecular projectiles over a large energy range. In the present work we report absolute extreme ultraviolet (EUV) cross section measurements for He^{+} (np) Rydberg states (n=2 to 4) following electron and proton impact. Such two electron processes like ionization plus excitation may be also of importance for α particle diagnostics (Frieling et al. 1991) and astrophysical plasmas and cosmological models. For example some of the species found in interstellar media contain He, He^+, H_2, H_3, H^+ and e- (Rauscher 1968, 1972 and 1973).

2 EXPERIMENTAL SET-UP

The experimental set-up has been described in detail by Fuelling (1991). In brief H^+ ions have been accelerated, focussed, mass and charge analyzed and passed, after tight collimation, into a differentially pumped gas cell and finally collected in a Faraday cup for charge normalization. Typical ion beam currents in the target area are 3-25 μA in the energy range of 50 keV to 1.8 MeV. A 1.5 m grazing incidence monochromator (Acton Research) equipped with a 600 grooves/mm grating has been used for wavelength dispersion. All measurements have been performed under single collision conditions. The gas pressure in the cell has been accurately monitored with a capacitance manometer and kept constant with a feed back control system. The Lyman transitions up to n = 5 are completely resolved. Data acquisition and control of the experiment have been accomplished by a versatile CAMAC-PC/AT system. Moreover, the relative detection efficiency of the monochromator has been determined to high accuracy over a large wavelength range. The present data have been placed on an absolute scale by additional electron impact measurements using identical excitation and detection geometries for electron and proton impact (Risley et al. 1989).

3 CROSS SECTIONS

Absolute EUV emission cross sections for ionization plus excitation (σ^{+*}) of helium atoms have been measured (Fuelling et al. 1991; Fuelling 1991). Here we report the first experimental results on state selective cross section ratios $\sigma^{+*}(np)/\sigma^*(1snp)$ of helium for electron and proton impact. These ratios are displayed in Fig. 1(a)-(c) for n=2,3 and 4 and versus velocity. An interesting result is that the σ^{+*}/σ^* ratios have a characteristic velocity dependence similar to σ^{2+}/σ^+ (double to single ionization). We expect the ratios $\sigma^{+*}(np)/\sigma^*(1snp)$ to approach σ^{2+}/σ^+ in the limit of large n values. In order to demonstrate this effect for both electron and proton projectiles, we have plotted in Fig. 2 $\sigma^{+*}(np)/\sigma^*(1snp)$ for specific projectile velocities versus $-1/2n^2$ [a.u.]. From this figure it appears that the ratio $\sigma^{+*}(np)/\sigma^*(1snp)$ does indeed approach σ^{2+}/σ^+ for higher np Rydberg levels. Our results show consistently higher yields for electrons than for protons at higher velocities.

4 CONCLUSION

In conclusion we have demonstrated that our experimental results show similarity to the double ionization process of He using projectiles with different charge signs. Both collision mechanisms show a factor of two difference when the sign of the projectile is reversed at higher impact velocities. The difference between the two-electron cross-sections for e- + He and p + He may be explained by interference of first and second orders of perturbation theory. We have shown that this explanation is satisfactory for double ionization and ionization plus excitation of He. In our case the cross-section is proportional to

$$\Sigma \,|f(\ell)|^2 = \Sigma \,|f1(\ell) + f2(\ell)|^2 \qquad (1)$$

where f1 is proportional to Z/v, and f2 proportional to $(Z/v)^2$, (Beigman 1992) where Z and v are the projectile charge and velocity respectively. Moreover we have demonstrated that the ratio $\sigma^{+*}(np)/\sigma^*$ (1snp) approaches σ^{2+}/σ^+ in the limit for large n.

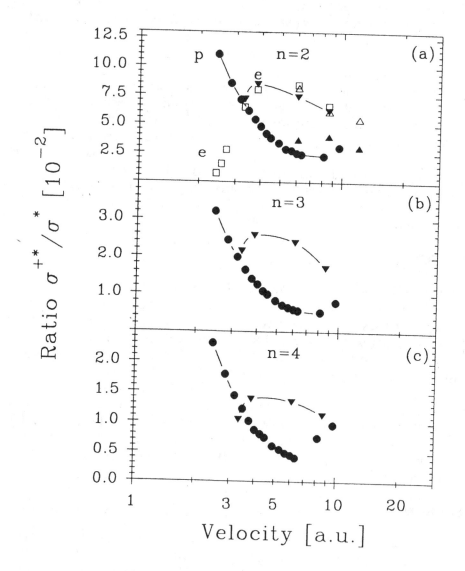

Fig. 1 -Ratio $\sigma^{+*}(np)/\sigma^*$(1snp) for (a) n=2, (b) n=3, (c) n=4. ● protons and, ▼ electrons: this work, □ electrons: (Forand et al. 1985); ▲ protons and, △ electrons: (Pedersen and Folkmann 1990).

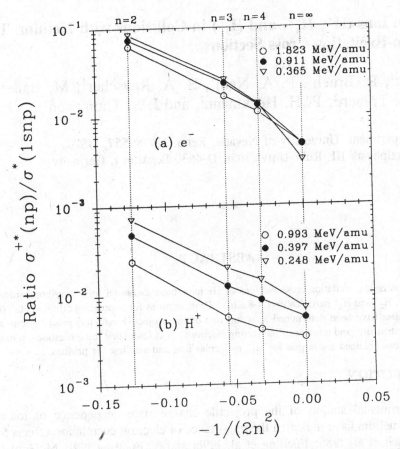

Fig. 2 - Ratio $\sigma^{+*}(np)/\sigma^*(1snp)$ for n=2,3,4 and σ^{2+}/σ^+ plotted against $-1/2n^2$ (a) for electron and (b) for protons at three different impact energies. Note that the change in the slope in the electron data for lower energies is due to threshold effects.

REFERENCES

I. Beigman 1992, private communication.
Forand, J. L., et al. 1985, J. Phys., B **18**, 1409.
Frieling, G. J., et al. 1991, in Atomic Physics of Highly Charged Ions, eds. E. Salzborn, P. H. Mokler, A. Müller (Berlin: Springer-Verlag).
Fuelling, S., et al. 1991, Nucl. Instr. Meth., B**56/57**, 275.
Fuelling, S. 1991, Doctoral Thesis, University of Nevada, Reno (unpublished).
McGuire, J. H., et al. 1989, Nucl. Instr. Meth., B**40/41**, 340.
Pederson, J. O., et al. 1990, J. Phys., B**23**, 441.
Rauscher, E. A. 1972, Lett. Nuovo Cimento, **3**, 661.
Rauscher, E. A. 1973, Lett. Nuovo Cimento, **7**, 361.
Rauscher, E. A. 1968, J. Plasma Physics, **2**, 217.
Reading, J. F. and Ford, A. L., 1987, Phys. Rev. Lett., **58**, 543.
Risley, J. S., et al. 1989, Appl. Opt. **28**, 389.
Träbert. E.. et al. 1987. Nucl. Instr. Meth.. B23. 151.

Hydrogen Ions (H^+, H_2^+, and H_3^+) in Collisions with Helium: Target Ionization-Excitation Cross Sections

S. Fülling[1], R. Bruch[1], P. A. Neill[1], E. A. Rauscher[1], M. Bailey[1], H. Wang[1], E. Träbert[2], P. H. Heckmann[2], and J. S. Thompson[1]

[1]Physics Department, University of Nevada, Reno, NV 89557, USA
[2]Experimentalphysik III, Ruhr-Universität, D-4630 Bochum 1, Germany

ABSTRACT

Ionization-excitation cross sections of He have been measured following bombardment by fast H^+, H_2^+ and H_3^+ particles($v=1.4$-8 a.u.). Cross sections for populating different He^+ (np, $n=2$-5) states have been determined as a function of the projectile velocity, mass, atomic and molecular structure, and number of interacting electrons. It is found that the ionization-excitation cross sections obtained are largest for H_3^+ molecular ions and smallest for protons.

1 INTRODUCTION

Experimental studies of the projectile charge state dependence of ionization-excitation of helium have indicated the importance of electron correlation effects for this process (Bruch et al. 1985; Fuelling et al. 1991a and b; Fuelling 1991; McGuire 1992). Experiments with diatomic and polyatomic molecules as projectiles can further elucidate the effects of correlation in many-body collision dynamics. In this paper we report data for high-energy collisions between the molecular ions, H_2^+ and H_3^+, and helium atoms, to study the effects of multiscattering centers on the ionization-excitation reaction. The H_2^+ projectile is the prototype of a covalent bond found in numerous molecules and solids. For H_2^+ the separation between the two protons, at equilibrium, is $R_0 = 1.06$ Å.

The main purpose of this study has been to measure absolute state selective ionization-excitation cross sections of helium over a broad velocity range (v:1.4 to 8 a.u.) and a large distribution of He^+(np) Rydberg states for H^+, H_2^+ and H_3^+ projectiles. These projectile characteristics differ in number of scattering centers, mass, geometry (i.e. point, linear, and triangular), number of orbital electrons, degrees of freedom, and excitation modes, including molecular fragmentation. We further note that such absolute cross section data are relevant for fusion and interstellar plasma diagnostics and modeling. For instance, the composition of interstellar plasmas has been shown to contain H^+, H_2^+, H_3^+ and He^+ ions as well as neutral He, H_2 and H_3 gases (Rauscher 1968; 1972 and 1973).

2 EXPERIMENTAL METHOD

In this work H^+, H_2^+ and H_3^+ particles have been accelerated with a 2 MV Van de Graaff and passed through a differentially pumped target cell in which the pressure can be continuously varied in the range of 0.1-300 mTorr. The pressure inside the target chamber was well below 5×10^{-6} Torr. A sophisticated electronic feedback control system has been developed which keeps the pressure in the gas cell constant. EUV radiation generated in the gas cell is observed at 90° to the particle beam by a 1.5 m grazing incidence monochromator which is absolutely wavelength and intensity calibrated. A versatile PC-CAMAC system designed for high resolution EUV spectroscopy is used for data acquisition and control (Fuelling 1991).

The flux of projectile particles incident to the gas cell is determined by monitoring the ion beam current in a post-collision Faraday cup. The ion beam current in the Faraday cup is measured as a function of target gas cell pressure and ion impact energy, so charge-changing and molecular disintegration effects can be removed from the cross section measurements. The intensity dependence on target gas cell pressure for all relevant EUV transitions has been investigated to insure that cross section measurements were obtained in a linear regime. Nonlinear effects, such as resonant absorption and collisional deexcitation, can therefore be neglected. Absolute cross sections have been measured by using the same excitation and detection geometry for electron and proton impact. In addition absolute EUV emission cross sections have been used (Risley et al. 1989) to place our data on an absolute scale.

3 RESULTS AND DISCUSSION

The experimental cross sections for ionization-excitation of He by H^+, H_2^+ and H_3^+ impact are displayed in Fig. 1 as a function of equal ion velocity. To our knowledge these measurements represent the most complete set of cross section data for $H^+ + He$, $H_2^+ + He$ and $H_3^+ + He$ at higher impact energy. We note that Schartner et al. (1991) have reported cross section measurements for H_n^+ (n=1,2,3) projectiles on He at lower energies (75-800 keV) for $He^+(2p)$ states. These data agree reasonably well with our present results. The Schartner group has also proposed a scaling relation of the type

$$\sigma(H^+) - 2\sigma(H_2^+) + \sigma(H_3^+) = 0 \qquad (1)$$

for electronic excitation processes induced by high energy H^+, H_2^+ and H_3^+ ions. This model is based on the assumption that the molecular projectiles can be treated as the sum of their constituent particles for projectile velocities $v \geq 2$ a.u. In this study we have used a more convenient scaling procedure for comparison of the measured cross section results, namely

$$\sigma(H_2^+) = 2\,\sigma(H^+) \qquad (2a)$$
$$\sigma(H_3^+) = 3\,\sigma(H^+) \qquad (2b)$$

It is evident from Fig. 1 that at equal projectile velocities, the H_3^+ ionization-excitation cross sections are larger than the corresponding H_2^+ and H^+ values for all $He^+(np)$ target states. In order to study this dependence in more detail in terms of molecular structure, number of electrons carried into the collision by the projectile ions, as well as electron correlation and interference effects due to multiple collisions, we show in Fig. 2 the cross section ratios $\sigma(H_2^+)/(2\sigma(H^+))$ and $\sigma(H_3^+)/(3\sigma(H^+))$ versus velocity. This figure clearly indicates that strong deviations from $\sigma(H_2^+)/(2\sigma(H^+)) = \sigma(H_3^+)/(3\sigma(H^+)) = 1$ occur for two electron processes, in particular for H_3^+ impact on He at higher velocities. Such striking deviations from cross section ratios close to one may be caused by geometrical effects, electron correlation and interference effects. However, further experimental and more refined theoretical models are needed to unravel the details of the collision dynamics associated with H^+, H_2^+ and H_3^+ impact on He.

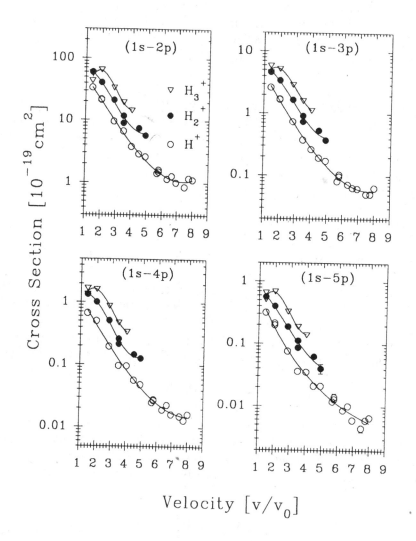

Fig. 1 - Absolute state selective cross sections for ionization plus excitation of Helium by H_n^+ (n=1,2,3) impact as a function of the projectile velocity [a.u.].

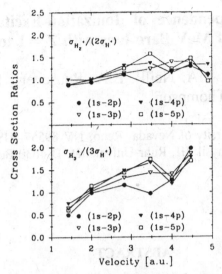

Fig. 2 - Cross section ratios (a) $\sigma(H_2^+)/(2\sigma(H^+))$ and (b) $\sigma(H_3^+)/(3\sigma(H^+))$ versus velocity [a.u.] for ionization-excitation of He by H^+, H_2^+ and H_3^+.

4 CONCLUSIONS

In summary, absolute cross sections for the most prominent $He^+(1s-np)$ transitions for n=2-5 have been obtained for high-energy $H^+ + He$, $H_2^+ + He$ and $H_3^+ + He$ single collisions. These data are of fundamental importance to study multiple scattering dynamics of complex collision systems. In particular we have found that simple scaling relations as proposed by Hasselkamp et al. (1987) break down for H_3^+ projectiles at higher impact energies.

5 REFERENCES

Bruch, R., et al. 1985, Nucl. Instr. Meth. **B9**, 438.

Fuelling. S., et al., 1991a, Z. Phys. D, **21**, 309.

Fuelling, S., et al, 1991b, Z. Phys. D, **21**, 313.

Fuelling, S., et al., 1991, Nucl. Instr. Meth. **B56/57**, 279.

Fuelling, S., 1991, Doctoral Thesis, University of Nevada, Reno (unpublished).

Hasselkamp, D., et al., 1987, Z. Phys. **D6**, 269.

McGuire, J. H., 1992, Adv. Atom. Mol. Opt. Phys., in press.

Rauscher, E. A. 1972, Lett. Nuovo Cimento, **3**, 661.

Rauscher, E. A. 1973, Lett. Nuovo Cimento, **7**, 361.

Rauscher, E. A. 1968, J. Plasma Physics, **2**, 217.

Reading, J. F., et al., 1987, Phys. Rev. Lett., **58**, 543.

Risley, J. S., et al., 1989, Appl. Opt. **28**, 389.

Schartner, K. H., et al., 1991, J. Phys. **B24**, Ll3.

Träbert. E.. et al.. 1987. Nucl. Instr. Meth. **B23**. 151.

Projectile-Charge Dependence of Ionization-Excitation of Helium Following Collisions of MeV Bare Ions with $Z_p=1$ to 6.

S. Fülling[1], R. Bruch[1], P.A. Neill[1], E.A. Rauscher[1], E. Träbert[2], P.H. Heckmann[2], and J. S. Thompson[1]

[1]Physics Department, University of Nevada, Reno, NV 89557, USA
[2]Institute für Experimentalphysik III, Ruhr-Universität, D-4630 Bochum 1, Germany

ABSTRACT

State-selective cross sections $\sigma^{+*}(nl)$ have been measured for ionization-excitation of helium in collisions with fully stripped ions. The energies of the H^+, Li^{3+}, B^{5+}, and C^{6+} were .83 MeV/u and 1.60 MeV/u. These relate to collision velocities of 5.8 and 8.0 a.u. Data for the Lyman series, $He^+(np \rightarrow 1s)$, have been measured by observation of the extreme ultra-violet spectra in the 22 to 31 nm wavelength range. The 2p-1s data is presented and analyzed in terms of a simple power law dependence on the projectile charge state, Z_p^n. For the two equivelocity series of cross sections the powers, n, derived from least squares fits to the data were 2.89 ± 0.03 and 2.83 ± 0.41.

1 INTRODUCTION

Collisions between structureless ions and simple atomic targets provide an opportunity to probe the subtleties of electron-electron, projectile-electron and projectile-nucleus interactions. Recent experiments investigating double ionization (Andersen 1988), double excitation (Giese et al. 1991) and excitation-ionization (Pedersen and Folkmann 1990) have reported data to test quantum mechanical descriptions of electron correlations in ion-helium atom collisions (Ford and Reading 1990). As the charge of projectile ions increase the nature of the collision processes may change significantly due to the larger interaction potentials (McGuire 1992).

The data reported here are state selective ionization-excitation cross sections for collisions between fully stripped ions and helium atoms. The purpose of the work is to investigate the charge state, Z_p, dependence of the cross sections and range of validity of perturbative, semiclassical or even classical models, (Bruch et al. 1985).

The ionization-excitation process is described by the reaction

$$X^{q+} + He(1s)^2 \rightarrow X^{q+} + He^+(nl) + e$$
$$\downarrow$$
$$He^+(n'l') + h\nu$$

The state selective nature of the cross sections is achieved by observing the (n,l) to (n',l') decay spectra from the residual excited ionic target. The specific reactions of interest here are for fully stripped projectiles, H^+, Li^{3+}, B^{5+} and C^{6+} producing the excited ionic $He^+(np)$ states for n=2,3,4 and 5. The Lyman series resulting from transitions to the $He^+(1s)$ ground ionic state were observed.

2 EXPERIMENTAL METHOD

The data are the result of independent investigations carried out at Bochum, Germany and Reno, Nevada (Fuelling 1991). Singly charged ions were accelerated by a tandem accelerator and subsequently stripped in a gas cell. After charge and mass selection, the fully stripped ions passed through the helium gas cell and were collected in a Faraday cup. The total charge collected during a measurement was determined by current integration and used for data normalization. Typical ion currents through the gas cell were in the range 1-30 μA.

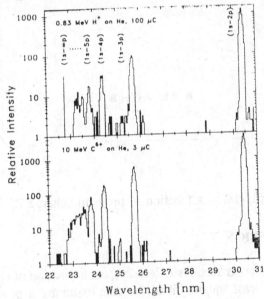

Fig. 1-Typical $He^+(np-1s)$ Lyman series spectra for H^+ and C^{6+} projectiles.

The differentially pumped gas cell was held at a constant target pressure of 10 mTorr to ensure single collision conditions. Photons emitted along an axis perpendicular to the projectile beam direction left the target cell through a 1mm exit aperture and entered the spectrometer via a 100 μm wide rectangular slit. The 2.2 m grazing incidence spectrometer (McPherson 247, 600 grooves/mm grating) was stepped through the wavelength region of interest and the spectrum was accumulated. Typical spectra are shown in figure 1. The spectra demonstrate that the wavelength resolution is sufficiently good to resolve transitions up to and including $He^+(5p-1s)$.

The projectile energies were chosen to provide two series of cross section measurements for collision velocities of 5.8. and 9.6 a.u. Relative cross sections derived from the spectra were put on an absolute scale by reference to absolute data for excitation-ionization of helium by electron impact (Risley and Westerveld 1989; Forand, Becker, and McConkey 1985). An additional systematic error of approximately 30% is introduced by this normalization process.

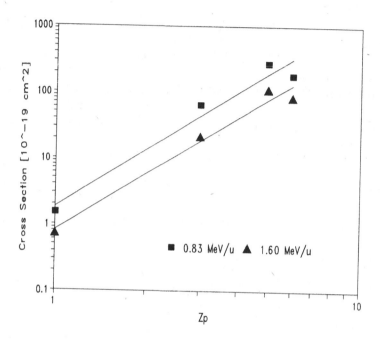

Fig. 2-Cross sections $\sigma^{+*}(2p\text{-}1s)$ as a function of projectile charge Z_p.

3 RESULTS AND DISCUSSION

Figure 2 shows the measured cross sections $\sigma^{+*}(1s\text{-}2p)$, as a function of charge state Z_p. The lines are the results of least squares fit to the data assuming a power law. They correspond to dependences of $Z_p^{2.8\pm0.4}$ and $Z_p^{2.9\pm0.3}$ for the 5.8 and 8.0 a.u. projectile velocities respectively.

The cross section values, tabulated in Table 1, range in value from 0.73×10^{19} cm^2 for H$^+$(v=8.0 a.u.) projectiles to 274.9×10^{19} cm^2 for B^{5+}(v=5.8 a.u.). The data obtained for C^{6+} projectiles lie below the corresponding data with B^{5+} projectiles. They are inconsistent with the general cross section dependence indicated by the other data, and may be an anomalous and will need further investigation. The quoted uncertainties represent one standard deviation and were derived from the counting statistics.

Although no data are available for direct comparison, $\sigma^{+*}(1s\text{-}2p)$ data have been reported, (Pedersen and Folkmann 1990), for 0.92 and 1.84 MeV/u C^{q+}(q=3 to 6) projectiles. The general trends of their results are similar to the present data and add weight to the apparent power dependence.

Projectile	Energy [MeV]	Velocity [a.u.]	σ^{+*}(1s-2p) [10^{-19} cm^2]
H$^+$	0.83	5.8	1.50 ± 0.02
H$^+$	1.60	8.0	0.73 ± 0.03
Li^{3+}	5.0	5.8	64.6 ± 0.2
Li^{3+}	9.6	8.0	21.5 ± 0.1
B^{5+}	8.33	5.8	274.9 ± 0.9
B^{5+}	16.0	8.0	110.9 ± 0.2
C^{6+}	10.0	5.8	182.4 ± 1.4
C^{6+}	19.2	8.0	84.1 ± 0.4

TABLE 1 Cross sections, σ^{+*}(1s-2p), for ionization-excitation of helium by H$^+$, Li^{3+}, B^{5+} and C^{6+} projectiles.

4 CONCLUSIONS

State-selective ionization-excitation cross sections have been measured for equivelocity fully stripped projectiles with charge states in the range $Z_p = 1$ to 6 in collision with helium atoms. The observed dependence of the cross sections on the projectile charge is consistent with the predominance of a Z_p^3 component. In the future, measurements with higher Z_p projectiles will be necessary to determine the range of projectile charge over which this behavior continues.

5 REFERENCES

Andersen L. H. 1988, in *Electronic and Atomic Collisions*, eds H. B. Gilbody, W. R. Newell, F. H. Read and A. C. H. Smith (Amsterdam: North-Holland), p. 451
Bruch R., et al. 1985, Nucl. Inst. and Meth., B9, 438
Forand J. L., Becker K., and McConkey J. W. 1985, J. Phys. B18, 1409
Ford A. L., and Reading J. F. 1990, J. Phys. B23, 2567
Fuelling S. 1991, PhD Dissertation, University of Nevada, Reno (unpublished)
Giese J. P. et al. 1990, Phys. Rev. A42, 1231
McGuire J. H. 1992, in *Advances in Atomic, Molecular and Optical Physics*, in press
Pedersen J. O. and Folkmann F. 1990, J. Phys. B23, 441
Risley J. S. and Westerveld W. B. 1989, Appl. Opt. 28, 389

Measurement of 1s2s³S₁-1s2p³P₂,₀ Intervals in Helium-Like Neon

William A. Hallett[1], Daniel D. Dietrich[2], and Joshua D. Silver[1]

[1]Dept. of Physics, Oxford University, Clarendon Laboratory, Parks Road, Oxford OX1 3PU, U.K.
[2]High Temperature Physics Division, LLNL, Livermore CA 94550

ABSTRACT

We have measured the vacuum ultra-violet 1s2s³S₁-1s2p³P₂,₀ transition wavelengths in helium-like neon by photographic spectroscopy of a recoil ion source. The results are 1248.01 ± 0.01 Å and 1277.74 ± 0.04 Å respectively and are a sensitive test of present relativistic and quantum electrodynamic calculations.

1 INTRODUCTION

The 1s2s³S₁-1s2p³P transitions in helium-like ions have been studied over a range of Z in recent years, motivated by the large contributions to the intervals of relativistic and quantum electrodynamic effects. Because the latter scale approximately as Z^4, less precise measurements at higher Z can yield comparable information to more precise experiments at low Z.

This paper briefly describes photographic spectroscopy of neon ($Z = 10$) and argon ($Z = 18$) recoil ions, yielding measurements of the Ne⁸⁺ 1s2s³S₁-1s2p³P₂,₀ wavelengths. Agreement with the recent calculation by Drake (1988) is good.

2 THE EXPERIMENT

Highly charged ions with velocities of order 10^4ms⁻¹ are produced by bombardment of a target gas by a highly charged heavy ion beam (Cocke & Olson 1991). Many of these ions are produced in excited states. Following earlier work in Oxford (Stamp 1983; Brown 1985) we used photographic detection to take advantage of Doppler widths of order $\delta\tilde{\nu}/\tilde{\nu} \sim 10^{-5}$. A 1m normal incidence concave grating spectrometer fitted with a camera and Kodak 101-01 film was used to observe the radiation from a gas target excited by an intense lanthanum beam from the Lawrence Berkeley Laboratory's SuperHILAC. Three exposures lasting about a day were taken.

TABLE 1. EXPOSURE CONDITIONS

Target	Pressure (torr)	Beam Energy (MeV/amu)	Mean Charge State	Charge Delivered (mC)
Ne/Ar	0.6	6	45+	384
Ne/Ar	0.6	3.5	37+	337
Ar	0.1	6	45+	700

Well-known transitions from low charge states of neon, argon, and air gases were excited. Some 20 of these were used as calibration lines. The films were analyzed using a modified (Brown 1985) Joyce-Loebl densitometer. This instrument can measure film densities to within a few per cent with a positional accuracy of the order of a micron. Exposure levels were estimated using a characterization of the film (Burton 1973). The lineshapes were fitted to gaussians with widths of order 0.1 Å. Because of this high resolution the lineshapes are not complicated by satellite transitions in contrast to X-ray spectra of recoil ion sources (Beyer 1985).

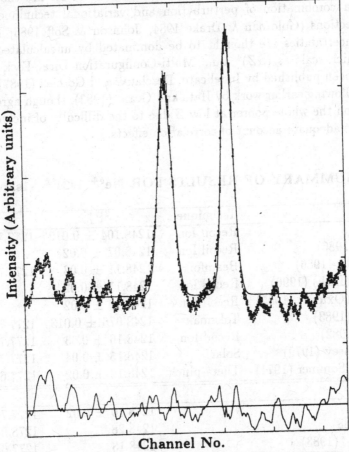

FIGURE 1. Densitometer trace of the Ne^{8+} $1s2s\,^3S_1$-$1s2p\,^3P_2$ line. To the right is a Ne^{4+} transition. Gaussian lineshape fits are also shown.

Neon/argon exposures taken at different energies yield the same wavelength for the Ne^{8+} $^3P_2$ transition suggesting that the line is not blended with a transition from a lower charge state. Comparison with the argon exposure shows that there is no blending with a line in the argon spectrum. The $^3P_0$ measurement is taken from the higher energy exposure, where both helium-like lines are above the development threshold and in the expected statistical ratio.

A re-analysis of earlier recoil ion films taken at Oxford and LBL has been made for comparison with the present experiment. The provisional value published by Brown et al. (1985) may be unreliable due to an unfortunate blend with a krypton line. The later experiment at LBL (Brown 1990) yields a value close to that from the present experiment.

3 COMPARISON WITH THEORY

Probably the most complete calculations of these intervals are those by Drake (1988) using a combination of perturbation and variational techniques, including radiative corrections (Goldman & Drake 1984; Johnson & Soff 1986; Mohr 1982). Theoretical uncertainties are thought to be dominated by uncalculated relativistic corrections which scale as $(\alpha Z)^4$ a.u.. Multi-Configuration Dirac-Fock calculations have recently been published by Indelicato, Desclaux, and Gorciex (1987), and Indelicato (1988) following earlier work by Hata and Grant (1983), though agreement with experiment is on the whole poorer at low Z due to the difficulty of including enough orbitals to take adequate account of correlation effects.

TABLE 2. SUMMARY OF RESULTS FOR Ne^{8+} $1s2s\,^3S_1$-$1s2p\,^3P_{2,0}$ (Å)

Experiment	Technique	$^3P_2$	$^3P_0$
This work	Recoil Ion	1248.104 ± 0.010	1277.74 ± 0.04
Beyer et al. (1986)	Recoil Ion	1248.07 ± 0.02	–
Berry & Hardis (1986)	Beamfoil	1248.11 ± 0.03	1277.75 ± 0.05
Brown et al. (LBL) (1990)	Recoil Ion	1248.11 ± 0.02	–
Brown et al. (Oxford) (1985)	Recoil Ion	1248.15 ± 0.02	–
Stamp et al. (1983)	Tokamak	1248.076 ± 0.013	1277.71 ± 0.02
Klein et al. (1982)	Recoil Ion	1248.16 ± 0.03	1277.79 ± 0.08
Sandlin & Tousey (1979)	Solar	1248.15 ± 0.04	1277.70 ± 0.04
Engelhardt & Sommer (1971)	Theta-pinch	1248.12 ± 0.02	1277.68 ± 0.04
Theory			
Drake (1988)		1248.10 ± 0.02	1277.70 ± 0.02
Indelicato (1988)		1248.28	1278.04
Hata & Grant* (1983)		1248.13	1277.64

* Corrected according to Hata & Grant (1984)

This work was partially performed under the auspices of the U.S. Department of Energy by the Lawrence Livermore National Laboratory under contract No. W-7405-ENG-48 and the U.K. Science and Engineering Research Council.

The authors would like to acknowledge the staff of the SuperHILAC, and Mr Paul Piperski (LBL) for technical support during the experiment.

REFERENCES

Berry H.G., Hardis J.E., 1986, *Phys. Rev. A* **33**, p. 2778

Beyer H.F., Deslattes R.D., Folkmann F., LaVilla R.E., 1985, *J. Phys. B* **18**, p. 207

Beyer H.F., Folkmann F., Schartner K.-H., 1986, *Z. Phys. D* **1**, p. 65

Brown J.S. *et al.*, 1985, *Nucl. Instr. and Meth. B***9**, p. 682

Brown J.S., 1990, private communication.

Burton W.M., Hatter A.T., Ridgeley A., 1973, *Appl. Opt.* **12**, p. 1851

Cocke C.L., Olson R.E., 1991, *Phys. Rep.* **205**, p. 155

Drake G.W.F., 1988, *Can. J. Phys.* **66**, p. 586

Engelhardt W., Sommer J., 1971, *Ap. J.* **167**, p. 201

Goldman S.P., Drake G.W.F., 1984, *J. Phys. B* **17**, p. L197

Hata J., Grant I.P., 1983, *J. Phys. B* **16**, p. 523

Hata J., Grant I.P., 1984, *J. Phys. B* **17**, p. 931

Indelicato P., Gorceix O., Desclaux J.P., 1987, *J.Phys. B* **20**, p. 639

Indelicato P., 1988, *Nucl. Instr. Meth. B* **31**, p. 14

Johnson W.R., Soff G., 1986, *At. Nucl. Dat. Tab.* **33**, p. 405

Klein H.A. *et al.*, 1982, *J. Phys. B* **15**, p. 4507

Mohr P.J., 1982, *Phys. Rev. A* **26**, p. 2338

Sandlin G.D., Tousey R., 1979, *Ap. J.* **227**, p. L107

Peacock N.J., Stamp M.F., Silver J.D., 1984, *Phys. Scr. T***8**

Energy Levels And Oscillator Strengths For Transitions In Helium-Like Fe XXV And Ni XXVII

L. K. Harra[1], A. W. Boone[2], P. H. Norrington[2], F. P. Keenan[1] and A. E. Kingston[2]

[1]Department of Pure and Applied Physics, The Queen's University of Belfast, Belfast BT7 1NN, Northern Ireland.
[2]Department of Applied Mathematics and Theoretical Physics, The Queen's University of Belfast, Belfast BT7 1NN, Northern Ireland.

ABSTRACT

Configuration interaction (CI) wavefunctions are used to calculate energy levels and oscillator strengths for all significant electric dipole (E1), electric quadrupole (E2), magnetic dipole (M1) and magnetic quadrupole (M2) transitions among the $1s^2$, $1s2l$ and $1s3l$ states of He-like Fe XXV and Ni XXVII. Accurate wavefunctions are also obtained using the fully relativistic MCDF method and similarly employed to calculate these same energy levels and oscillator strengths. Derived energy levels are compared to each other and with previous results, and indicate that the MCDF method gives data which are closer to the experimental energies. The calculated CI and MCDF A-values are found to be in good agreement, but differ significantly in some cases from these and previous authors.

1. INTRODUCTION

Energy levels and oscillator strengths for transitions in He-like ions are intrinsically important for the interpretation of observational data obtained from laboratory and astrophysical sources.

Here we calculate the energy levels and oscillator strengths for all significant electric dipole (E1), electric quadrupole (E2), magnetic dipole (M1), and magnetic quadrupole (M2) transitions among the $1s^2$, $1s2l$ and $1s3l$ states of He-like Fe XXV and Ni XXVII.

2. CONFIGURATION INTERACTION (CI) CALCULATIONS

This calculation was performed using the general configuration code CIV3 (Hibbert 1975, Glass and Hibbert 1976, Hibbert and Scott 1982). This is an analytic calculation in which the radial functions are dependent on a number of variational parameters, which are varied to make the energy eigenvalue, $E^{(j)}$, a minimum for an appropiate choice of j. For the Fe XXV and Ni XXVII calculations it was found that the 1s2s $^1S_0$ and the 1s2p $^3P_1$ states are nearly degenerate, and a non-standard approach was used to obtain the correct ordering. The energies were eventually calculated in intermediate coupling including all the relativistic terms ie spin-orbit, Darwin, mass correction, spin-spin and spin-other orbit.

3. MULTI-CONFIGURATIONAL DIRAC FOCK (MCDF) CALCULA-TIONS

This code (Grant et al 1980, Dyall et al 1989) follows a numerical method and rather than optimising one orbital at a time as in CIV3, all the orbitals are optimised silmutaneously on an average energy of all the configurations. It is fully relativistic and includes Breit and QED corrections.

4. RESULTS

Tables 1 and 2 show the energy levels compared to previous calculations and experimental results for Fe XXV and Ni XXVII. Tables 3 (the data is taken from the following sources;a CIV3, b GRASP, c Shirai *et al* (1990), d Pradhan (1983) , g Dalgarno, Johnson and Lin (1977), i Dalgarno (1971)) and 4 (the data is taken from the following sources; a GRASP, b CIV3, c Corliss and Sugar (1981)) show selected oscillator strengths and transition probabilities. The energies for both ions compare well with previous results, with the MCDF result now being the most accurate available. In general, there is a good agreement for most electric dipole transitions. The large variations in the transition probability (which depends on ΔE^3) are due mainly to the differences in energy separation for the different calculations.

REFERENCES

Corliss and Sugar 1981, J.Phys.Chem.Ref.Data, 10,197.

Dalgarno,A., 1971, N.B.S. Special Publication 353,47.

Dalgarno,A., Johnson,W.R., and Lin,C.D., 1977, Phys.Rev.A15,154.

Doschek,G.A., 1972, Space Science Reviews,13,765.

Dyall, K.G.,et al, 1989. Comput.Phys.Commun. 55,425.

Fuhr,J.R., Martin,G.A. and Wiese,W.L., 1988, J.Phys.Chem.Ref.Data 10,197.

Glass,R., and Hibbert,A.,1976.Comput.Phys.Commun.11,125.

Grant,I.P.,et al,1980.Comput.Phys.Commun.21,207 and 233.

Hibbert,A.,1975. Comput.Phys.Commun. 9,141.

Hibbert,A.,and Scott,N.S.,1982. Commut.Phys.Commun.28,189.

Pradhan,A.K., 1983, Phys.Rev.A28,2113.

Shirai,T. et al, 1990, J.Phys and Chem.Ref.Data 19,127.

Table 1. Selected energy levels in Rydbergs relative to the ground state (first column) calculated by CIV3 and GRASP and compared with Shirai et al (1990), Pradhan (1983) and Ermoleav & Jones (given in Pradhan 1983).

	CIV3	GRASP	Shirai	Ermolaev & Jones	Pradhan
1s2s $^3S_1$	488.2919	487.8297	487.7799	487.8453	488.5291
1s2p $^3P_0$	490.5359	489.9916	489.9086	489.9809	490.3875
1s2p $^3P_1$	490.6865	490.1417	490.0559	490.1299	490.6609
1s2s $^1S_0$	490.6928	490.1610	490.0898	490.1502	490.5808
1s2p $^3P_2$	491.7775	491.2221	491.1400	491.1875	491.7766
1s2p $^1P_1$	493.2262	492.5702	492.4708	492.5206	493.1257

Table 2. Selected Energy Levels in Atomic Units for Ni XXVII Relative to the Ground State Calculated by CIV3 and GRASP compared with compilations of theoretical calculations obtained by Corliss and Sugar (1981), Kelly (1987) and Fuhr, Martin and Wiese (1988), and from solar spectra (Doschek 1972).

	CIV3	GRASP	Corliss& Sugar	Doschek	Kelly	Fuhr, Martin & Wiese
1s2s $^3S_1$	284.74080	284.02919	284.14907		284.17166	
1s2p $^3P_0$	285.71469	285.19378	285.34420			
1s2p $^3P_1$	285.80253	285.28043	285.43077		285.43068	285.43077
1s2s $^1S_0$	285.84143	285.30428	285.43214			
1s2p $^3P_2$	286.57004	286.04090	286.17028			
1s2p $^1P_1$	287.34694	286.75415	286.87651	287.10646	286.87651	

Table 3. Selected energy differences, transition type, transition probabilities, absorption oscillator strengths and line strengths for a sample of transitions calculated between the 17 target states for Fe XXV.

Transition	Type	ΔE(a.u.)	f_{ij}(abs)	$A_{ji}(s^{-1})$	S_{ij}(a.u)
$1^1S_0 - 2^3P_1$	E1	a $2.4534E+02$	a $4.21E+13$	a $6.53E\text{-}02$	a $3.99E\text{-}04$
		b $2.4507E+02$	b $4.40E+13$	b $6.83E\text{-}02$	b $4.18E\text{-}04$
		c $2.4503E+02$	c $4.42E+13$	c $6.87E\text{-}02$	c $4.21E\text{-}04$
		d $2.4533E+02$	d $3.77E+13$	d $5.85E\text{-}02$	d $3.58E\text{-}04$
$1^1S_0 - 2^1P_1$	E1	a $2.4661E+02$	a $4.70E+14$	a $7.22E\text{-}01$	a $4.39E\text{-}03$
		b $2.4629E+02$	b $4.75E+14$	b $7.31E\text{-}01$	b $4.45E\text{-}03$
		c $2.4624E+02$	c $4.57E+14$	c $7.03E\text{-}01$	c $4.28E\text{-}03$
		d $2.4656E+02$	d $4.63E+14$	d $7.11E\text{-}01$	d $4.33E\text{-}03$
$1^1S_0 - 2^3S_1$	M1	b $2.4391E+02$	b $2.13E+08$	b $3.35E\text{-}07$	b $1.55E\text{-}04$
		g $2.08E+08$			
		h $2.00E+08$			
$1^1S_0 - 2^3P_2$	M2	a $2.4589E+02$	a $6.60E+09$	a $1.70E\text{-}05$	a $4.83E\text{-}02$
		b $2.4561E+02$	b $6.58E+09$	b $1.70E\text{-}05$	b $4.85E\text{-}02$
		g $6.55E+09$			

Table 4. Energy differences, transition type, transition probabilities, absorption oscillator strengths and line strengths for a sample of the transitions between the 17 target states for Ni XXVII.

Transition	Type	ΔE(a.u.)	f_{ij}(abs)	$A_{ji}(s^{-1})$	S_{ij}(a.u)
$1\,^1S_0 - 2\,^3P_1$	E1	a $2.8528E+02$	a $8.8183E\text{-}02$	a $7.6864E+13$	a $4.636E\text{-}04$
		b $2.8581E+02$	b $8.4302E\text{-}02$	b $7.3751E+13$	b $4.425E\text{-}04$
		c $2.8534E+02$	c $8.83E\text{-}02$	c $7.70E+13$	c $4.64E\text{-}04$
$1\,^1S_0 - 2\,^1P_1$	E1	a $2.8675E+02$	a $7.0834E\text{-}01$	a $6.2381E+14$	a $3.71E\text{-}03$
		b $2.8735E+02$	b $7.1112E\text{-}01$	b $6.2887E+14$	b $3.71E\text{-}03$
		c $2.8688E+02$	c $6.83E\text{-}01$	c $6.02E+14$	c $3.57E\text{-}03$
$1\,^1S_0 - 2\,^3S_1$	M1	b $2.8403E+02$	b $5.2593E\text{-}07$	b $4.5440E+08$	b $2.80E\text{-}09$
$1\,^1S_0 - 2\,^3P_2$	M2	a $2.8657E+02$	a $2.3272E\text{-}05$	a $1.2281E+10$	a $1.22E\text{-}07$
		b $2.8604E+02$	b $2.2947E\text{-}05$	b $1.2065E+10$	b $1.20E\text{-}07$

The K- and L-Shell Absorption Spectra of CIV

E. Jannitti[1], M. Gaye[2], P. Nicolosi[3], G. Tondello[3], P. Villoresi[3], and
F. Xianping[3]

[1]Istituto Gas Ionizzati, CNR, Padova, Italy
[2]ITNA Université de Dakar, Senegal
[3]Dipartimento di Elettronica ed Informatica, Università di Padova, 35100, Padova, Italy

ABSTRACT

The absorption spectra of the Li-like CIV ion have been studied in the soft X-ray spectral range. Both the K-shell and the L-shell spectra have been obtained with the two laser-produced plasmas technique. The K-shell spectrum has been observed between 25-45 Å. The discrete transitions, corresponding to jumps of the 1s inner electron, as well as the photoionization continuous spectrum have been observed. The L-shell spectrum appears at longer wavelengths and the transitions of the optical 2s electron have been observed between 200-250 Å.

1 INTRODUCTION

The method of two laser-produced plasma can be used for obtaining absorption spectra of multiply charged ions. In fact the laser produced plasmas appear very suitable for generating both a relatively dense and reproducible column of the absorbing ions and also a reproducible source emitting a strong pulse of EUV continuous radiation.

In a laser produced plasma the various ionization stages follow a different time evolution; while the highest ionization stages have essentially the duration of the laser pulse, the lowest ones appear at very later time during the cooling phase of the target. In addition, due to the expansion of the plasma, these species have also steep density gradients versus the distance from the target surface and tend to be mostly concentrated in different spatial regions (Jannitti *et al.* 1988). In the case of high Z material target the soft X-ray emission from the plasma has essentially a continuous distribution with the wavelength, a duration comparable to the the laser pulse ($\sim$ 20 ns) and in addition, when looking at the laser-target interaction region where the density approaches the critical value, it results of very high brightness and small physical size.

Consequently by using a suitable time delay between the absorbing plasma and the pulse of the background continuum, it is possible to resolve the spectra of different ions. Furthermore it is necessary to take into account the expansion properties of the absorbing plasma; so, by probing the plasma with the background continuum at different distances from the target, it is also possible to optimize the absorption of a single ion species.

2 THE EXPERIMENT

The scheme of the experiment is reported in Fig.1. The output pulse of a Q-switched ruby laser (6 J,15 ns) is splitted into two parts. One beam (Beam 1), with about 70 % of the total laser energy, is sharply focused on a plane tungsten target (T_1) and generates a plasma emitting a strong and uniform continuous XUV radiation. The second one (Beam 2), with the remaining energy, is focused on a graphite target (T_2) with a sphero-cylindrical lens producing the absorbing plasma. The grazing incidence toroidal mirror (M, 87°) focuses the XUV radiation in one direction on the entrance slit of a grazing incidence spectrograph and in the other direction on the focal plane of the spectrograph.

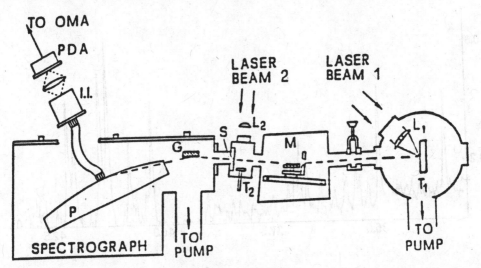

Fig. 1 Schematic of the experimental setup

The absorbing plasma is generated close to the entrance slit of the spectrograph. In this way by using the toroidal mirror is possible to probe the absorbing plasma in a very narrow region and to fill the aperture of the spectrograph, increasing considerably the S/N ratio of the observed spectra. By moving the target T_2 up and down, the absorbing plasma can be probed at various distances from the target. Finally the delay between the generation of the two plasmas can be continuously varied. A detailed description of the experiment has been already reported elsewhere (Jannitti et al. 1988). The spectrum of the K-shell of the CIV has been observed with a 2m 1200 l/mm grating, while that one of the L-shell has been recorded with 2m radius, 576 l/mm grating.

3 OBSERVATIONS

The CIV ground level in only 83.5 eV over the ground state of the neutral CI and its ionization energy is only 64.5 eV. Consequently it is a very difficult task to obtain a spectrum without contributions from the adjacent CIII and CV ions. The K-shell CIV transitions appear between 35 and 43 Å very near the CV $1s^2$-1snp n=2,3 resonance lines while the L-shell transitions correspond to smaller energies. Both these spectra have common initials levels: the ground $1s^2$2s and the $1s^2$2p excited level, and indeed both these states are populated. The CIV K-shell spectrum has been already observed and it has been reported elsewhere in a previous paper(Jannitti et al. 1990). Here we have observed in some slightly different experimental conditions again this spectrum between about 33 and 43 Å, that is reported in Fig. 2.

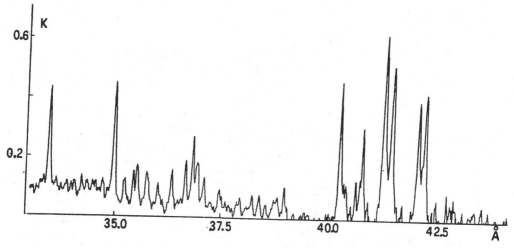

Fig. 2 CIV K-shell spectrum

The experimental parameters have been: laser power density on the C target about 4.5×10^9 W/cm^2, focal spot size 0.7×2.8 mm^2, delay 15 ns, distance from the target surface 0.6 mm. In the spectrum are clear both the discrete transitions and the photoionization jump at about 34 Å. The discrete transitions correspond to transitions of the 1s electron with configuration $1s^2$2s-1s2snp and $1s^2$2p-1s2pnp, transitions with n up to 6 have been already observed(Jannitti et al. 1990) .Some contribution of CIII and CV ions is also appearing; CIII lines appear at 41.8, 42.2, 42.3 Å while the resonance lines $1s^2$-1snp, n=2,3,4 of CV are also present. In the same experimental conditions, the L-shell spectrum of CIV has been observed between 200-250 Å. In this case a less dispersive grating has been used. The absorption coefficient is reported in Fig. 3 where the lines $1s^2 2s - 1s^2 np$, n=5,6,7 are clearly shown. The lines appear broadened mostly via Stark effect. Indeed these broadening results decreasing with the distance from the target by recording the spectrum at larger distances. Before deriving the absorption coefficient, the continuous as well as the absorption spectra have been corrected for the contributions due to the higher order diffration. This contribution has been evaluted by observing discrete lines of Be, B, C. In addition for reducing these contributions, an 1000 Å thick Al filter has been installed in the spectrograph.

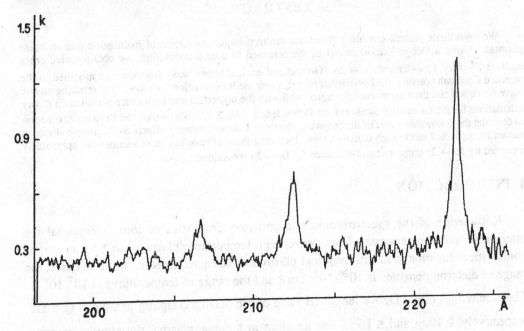

Fig. 3 CIV L-shell spectrum

From these measurements we plan to derive the column density of the ions on the $1s^2 2s$ and $1s^2 2p$ levels (Jannitti *et al.* 1988) and to derive a measurement for the photoionization cross-section.

4 REFERENCES

Jannitti, E., Nicolosi, P., Tondello, G. 1988, in *X-ray and VUV Interaction Data-Bases, Calculations and Measurements*, ed. N. K. Del Grande (Proc. SPIE 911), p.157.

Jannitti, E., Nicolosi, P., Tondello, G. 1990, *Phys. Scr.* **41**, p. 458.

Electron-Impact Excitation of Hydrogenic Ions in Dense Plasmas

Young-Dae Jung[1]

Space Science Laboratory, ES-65, NASA/Marshall Space Flight Center, Huntsville, AL 35812, U. S. A.

ABSTRACT

We investigate plasma-screening effects on electron-impact excitation of hydrogenic ions in dense plasmas. Using a Debye-Hückel model of the screened Coulomb interaction, we obtain scaled cross sections $Z^4 \cdot \sigma$ for $1s \rightarrow 2s$ and $1s \rightarrow 2p$. Ground and excited bound wave functions are modified in the screened Coulomb potential (Debye-Hückel model) using the Ritz variational method. The resulting atomic wave functions and their eigen-energies agree well with the numerical and high-order perturbation theory calculations for the interesting domain of the Debye length $\Lambda/a_Z \geq 10$. We employ the Born approximation to describe the continuum states of the projectile electron. Plasma screening effects on the atomic electrons cannot be neglected in the high density cases. Including these effects, the cross sections are appreciably increased for $1s \rightarrow 2s$ transitions and decreased for $1s \rightarrow 2p$ transitions.

1 INTRODUCTION

Knowledge of the spectroscopic and collision properties of ions is essential for interpretation of line emission from dense, high-temperature plasmas, such as in inertial confinement fusion and in astrophysical plasmas of compact objects. In these cases, the range of electron densities is 10^{20}-10^{23} cm^{-3} and the range of temperatures is 10^7-10^8 K. Thus the ranges of the Debye length (Λ) and of the plasma coupling parameter (Γ) are, respectively, $\geq 10 a_Z$ and $\leq 10^{-2}$. For an atom in a dense plasma, the atomic electrons experience substantially screened Coulomb potentials. In this case, screening effects on the wave functions cannot be completely neglected, although this is done in much of the literature. This neglect of screening effects causes serious errors for the cross sections as well as for the excitation thresholds. Thus in this paper we calculate the electron-impact excitation cross sections of hydrogenic atoms using our modified wave functions and energies. We use the Debye-Hückel model for the interaction potential for the electron-impact excitation process. And the Born approximation is used to describe the continuum states of the projectile electron.

A Coulomb correction near excitation threshold will be taken into account in the future work using Coulomb focussing [1] with modified effective charges due to plasma-screeing effects.

---

[1]NAS/NRC Resident Research Associate

2 ATOMIC WAVE FUNCTIONS

For the domain of interest ($\Lambda/a_Z \geq 10$, $\Gamma \leq 10^{-2}$), the Debye-Hückel model reliably describes the potentials of the $1s$, $2s$, and $2p$ states of a hydrogenic ion (with nuclear charge Z) embedded in a dense plasma. Using the Ritz variational method with hydrogenic trial wave functions, we obtain effective Bohr radii and eigen-energies for the ground ($1s$) and excited ($2s$ and $2p$) states:

$$\alpha_{nl} = a_Z/\eta_{nl}, \tag{1}$$

$$\langle E_{nl} \rangle = -(Z/n)^2 Ry \cdot (1 - \delta_{nl}) \tag{2}$$

where $\eta_{1s} \approx 1 - \frac{3}{4}(a_Z/\Lambda)^2 + (a_Z/\Lambda)^3$, $\eta_{2s} \approx 1 - 12(a_Z/\Lambda)^2 + 56(a_Z/\Lambda)^3$,

$\eta_{2p} \approx 1 - 10(a_Z/\Lambda)^2 + 40(a_Z/\Lambda)^3$, $\delta_{1s} \approx 2(a_Z/\Lambda) - \frac{3}{2}(a_Z/\Lambda)^2 + (a_Z/\Lambda)^3$,

$\delta_{2s} \approx 8(a_Z/\Lambda) - 24(a_Z/\Lambda)^2 + 56(a_Z/\Lambda)^3$, $\delta_{2p} \approx 8(a_Z/\Lambda) - 20(a_Z/\Lambda)^2 + 40(a_Z/\Lambda)^3$.

For (Λ/a_Z) ≥ 10, the above results are in good agreement with numerical calculations [2] and high-order perturbation theory calculations [3]. If we neglect the plasma-screening effects on the atomic wave functions, $\eta_{nl} = 1$ and $\delta_{nl} = 0$.

3 ELECTRON-IMPACT EXCITATIONS

When a hydrogenic ion with nuclear charge Z is embedded in dense plasma, the interaction potential for electron-impact excitation is simply the screened Coulomb interaction (see [4] — [7]):

$$V'(\mathbf{r}, \mathbf{r}_1) = \left(-\frac{Ze^2}{r} + \frac{e^2}{|\mathbf{r} - \mathbf{r}_1|} \right) \exp(-r/\Lambda) \tag{3}$$

where $\mathbf{r}$ and $\mathbf{r}_1$ are the positions of the incident and bound electrons, respectively. With this interaction potential and the wave functions in §2, we obtain the scaled cross sections for the $1s \rightarrow 2s$ and $1s \rightarrow 2p$ excitation processes, respectively:

$$Z^4 \cdot \sigma_{2s,1s} = 576 \frac{(\eta_{1s}\eta_{2s})^3}{(\eta_{1s} + \eta_{2s}/2)^2} \frac{\pi a_0^2}{\bar{\varepsilon}_i} D_{2s,1s} \tag{4}$$

where $D_{2s,1s} = \int_{Q_{min}}^{Q_{max}} \left[\frac{A_1}{(b_1^2 + Q^2)} - \frac{A_1}{(b_2^2 + Q^2)} + \frac{A_2}{(b_2^2 + Q^2)^2} + \frac{A_3}{(b_2^2 + Q^2)^3} \right]^2 Q \, dQ,$

$Q_{max} = \bar{\varepsilon}_i^{1/2} + (\bar{\varepsilon}_i - 3\xi_{2s,1s}/4)^{1/2}$, $Q_{min} = \bar{\varepsilon}_i^{1/2} - (\bar{\varepsilon}_i - 3\xi_{2s,1s}/4)^{1/2}$,

$\xi_{2s,1s} = 1 - 6(a_Z/\Lambda)^2 + 20(a_Z/\Lambda)^3$, $\bar{\varepsilon}_i = (\hbar^2 k_i^2/2m)/Z^2 Ry$,

$b_1 = a_Z/\Lambda$, $b_2 = a_Z/\Lambda + (\eta_{1s} + \eta_{2s}/2)$, $A_1 = \frac{15}{4}(\eta_{1s} + \eta_{2s}/2)^{-3}(a_Z/\Lambda)^2 \left[1 - \frac{44}{9}(a_Z/\Lambda) \right]$,

$$A_2 = \frac{1}{9}(\eta_{1s} + \eta_{2s}/2)^{-1}(a_Z/\Lambda)\left\{1 - \frac{135}{4}(a_Z/\Lambda)\left[1 - \frac{44}{9}(a_Z/\Lambda)\right]\right.$$
$$\left. - 45(\eta_{1s} + \eta_{2s}/2)^{-1}(a_Z/\Lambda)^2\left[1 - \frac{44}{9}(a_Z/\Lambda)\right]\right\},$$

$$A_3 = \frac{4}{9}\left[a_Z/\Lambda + (\eta_{1s} + \eta_{2s}/2)\right]^2\left\{1 - \frac{45}{4}(\eta_{1s} + \eta_{2s}/2)^{-1}(a_Z/\Lambda)^2\left[1 - \frac{44}{9}(a_Z/\Lambda)\right]\right\},$$

and

$$Z^4 \cdot \sigma_{2p,1s} = 256 \cdot \left(\frac{3}{2}\right)^{12} \frac{\eta_{1s}^3 \eta_{2p}^5}{(\eta_{1s} + \eta_{2p}/2)^{10}} \frac{\pi a_0^2}{\bar{\varepsilon}_i} D_{2p,1s} \tag{5}$$

where

$$D_{2p,1s} = \int_{Q_{min}}^{Q_{max}} \left\{\frac{c_1}{Q}\left[\arctan\left(\frac{Q}{c_2}\right) - \arctan\left(\frac{Q}{b_1}\right)\right] + \frac{B_1}{(c_2^2 + Q^2)} + \frac{B_2}{(c_2^2 + Q^2)^2} + \frac{B_3}{(c_2^2 + Q^2)^3}\right\}^2 \frac{dQ}{Q},$$

$$Q_{max} = \bar{\varepsilon}_i^{1/2} + (\bar{\varepsilon}_i - 3\xi_{2p,1s}/4)^{1/2}, \quad Q_{min} = \bar{\varepsilon}_i^{1/2} - (\bar{\varepsilon}_i - 3\xi_{2p,1s}/4)^{1/2},$$

$$\xi_{2p,1s} = 1 - \frac{14}{3}(a_Z/\Lambda)^2 + 12(a_Z/\Lambda)^3, \quad c_1 = (2/3)^6(a_Z/\Lambda), \quad c_2 = a_Z/\Lambda + (\eta_{1s} + \eta_{2p}/2),$$

$$B_1 = (2/3)^6(a_Z/\Lambda)(\eta_{1s} + \eta_{2p}/2), \quad B_2 = (2/3)^6(a_Z/\Lambda)\left[a_Z/\Lambda + (\eta_{1s} + \eta_{2p}/2)\right],$$

$$B_3 = (2/3)^6\left[a_Z/\Lambda + (\eta_{1s} + \eta_{2p}/2)\right]^3(\eta_{1s} + \eta_{2p}/2)^3.$$

4 DISCUSSION

Fig. 1 and Fig. 2 show the plasma-screening effects on the wave functions for the $1s \rightarrow 2s$ and $1s \rightarrow 2p$ excitations of hydrogenic ions. Plasma screening effects on the atomic electrons cannot be neglected. Including these effects, the cross sections are appreciably increased for $1s \rightarrow 2s$ transitions and decreased for $1s \rightarrow 2p$ transitions.

Further details for these results will be published. I am grateful to Dr. S. L. O'Dell for useful discussions and comments. This work was done while the author held a National Research Council Research Associateship at NASA/Marshall Space Flight Center.

REFERENCES

[1] Jung, Y.-D., 1992, *Ap. J.*, submitted
[2] Rogers, F. J., Graboske, H. C., & Harwood, D. J., 1970, *Phys. Rev. A*, **1**, 1577
[3] Iafrate, G. J., & Mendelsohn, L. B., 1969, *Phys. Rev.*, **182**, 24
[4] Hatton, G. J., Lane, N. F., & Weisheit, J. C., 1981, *J. Phys. B*, **14**, 4879
[5] Whitten, B. L., Lane, N. F., & Weisheit, J. C., 1984, *Phys. Rev. A*, **29**, 945
[6] Deb, N. C., & Sil, N. C., 1984, *J. Phys. B*, **17**, 3587
[7] Pundir, R. S., & Mathur, K. C., 1984, *J. Phys. B*, **17**, 4245

Fig. 1—Scaled cross sections $Z^4 \cdot \sigma_{2s,1s}$ (πa_0^2) ($1s \rightarrow 2s$) for hydrogenic ions. The solid lines represent the cross sections given by eq. (4) (including plasma-screening effects on the wave functions); the dashed lines represent the cross sections without plasma-screening effects.

Fig. 2—Scaled cross sections $Z^4 \cdot \sigma_{2p,1s}$ (πa_0^2) $(1s \rightarrow 2p)$ for hydrogenic ions. The solid lines represent the cross sections given by eq. (5) (including plasma-screening effects on the wave functions); the dashed lines represent the cross sections without plasma-screening effects.

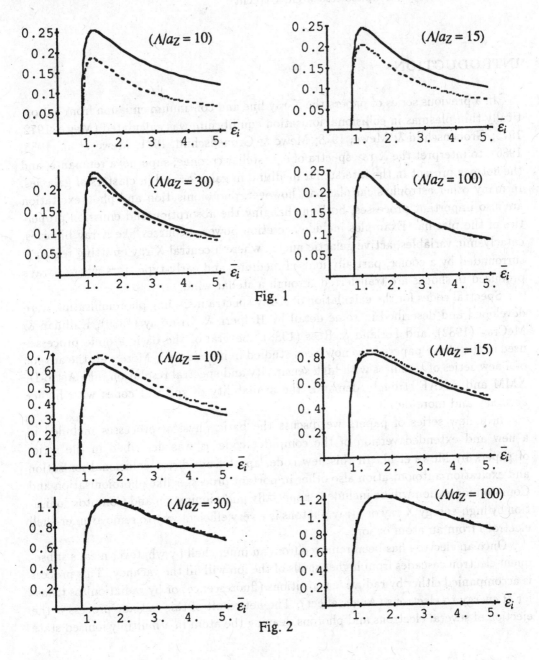

Fig. 1

Fig. 2

Multiple Auger ionisation and fluorescence processes for Be to Zn

J.S. Kaastra[1], R. Mewe[2]

[1]SRON Laboratory for Space Research, Leiden
[2]SRON Laboratory for Space Research, Utrecht

INTRODUCTION

In a previous series of papers the X-ray line and continuum emission from hot optically thin plasmas in collisional ionisation equilibrium was calculated (Mewe, 1972, 1975; Gronenschild & Mewe, 1978; Mewe & Gronenschild, 1981; Mewe et al., 1985, 1986), to interpret the X-ray spectra of e.g. stellar coronae, supernova remnants, and the hot gas present in the interstellar medium, in galaxies and in clusters of galaxies. In many other astrophysical plasmas, however, photoionisation and photoexcitation are also important processes, both in changing the absorption and emission properties of the plasma. Examples include accretion-powered sources like X-ray binaries, cataclysmic variables, active galactic nuclei, where a central X-ray emitting region is surrounded by a cooler, partially ionised medium, and early-type stars where X-rays produced in shocks are transferred through a stellar wind.

Spectral codes for the calculation of X-ray spectra including photoionisation were developed and described in some detail by Halpern & Grindlay (1980), Kallman & McCray (1982), and Ferland & Rees (1988). Several of the basic atomic processes used in the above papers have now been studied in more detail. Moreover, the advent of a new series of satellites with high sensitivity and spectral resolution, like Astro-D, XMM and AXAF strongly demands the availability of spectral codes with higher accuracy and more detail.

In a new series of papers we discuss the basic physical processes included in a new and extended version of the computer code as was described in the series of papers by Mewe et al. In this new code, apart from direct collisional ionisation and excitation-autoionisation also other important processes like photoionisation and Compton ionisation will be included. Especially photoionisation and Compton ionisation by high-energy X-ray or γ-ray photons is a very efficient way to remove inner-shell electrons from an atom or ion.

Once an electron has been removed from an inner shell by whatever means subsequent electron cascades from higher shells of the ion will fill the vacancy. This process is accompanied either by radiative transitions (fluorescence) or by radiationless transfer of energy to electrons (Auger effect). The net result of this cascade process is the ejection of several electrons and photons, leaving the atom in a multiply ionised state

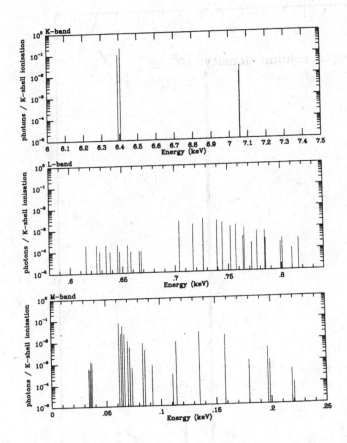

Fig. 1—Fluorescent spectrum of Fe I in the K, L and M band for K-shell ionisation.

(e.g. the removal of one K-shell electron in neutral iron may lead to the creation of Fe X by such a series of cascades). This multiple ionisation has an important effect in calculations of the ionisation balance in photo-ionised plasmas. Since many different transitions are possible, a probability distribution for the number of electrons and fluorescent photons emitted during the cascade results.

We have calculated the probability distribution for the number of emitted photons and ejected electrons for inner-shell ionisation of all atoms and ions from H–Zn. Also the spectrum of the ejected Auger electrons as well as the fluorescent photon spectrum are calculated. The calculations are based upon scaling of the Auger and radiative transition rates for neutral atoms as obtained by McGuire using a non-relativistic Hartree-Fock-Slater treatment, and evaluation of the individual probabilities of all possible cascade sequences.

Details of the calculation and tables containing the fluorescence yields for all individual lines as well as the probability distribution for the number of electrons emitted during the cascade are given by Kaastra & Mewe (1992). Here we present graphically some of our results.

The Laboratories for Space Research Leiden and Utrecht are supported financially

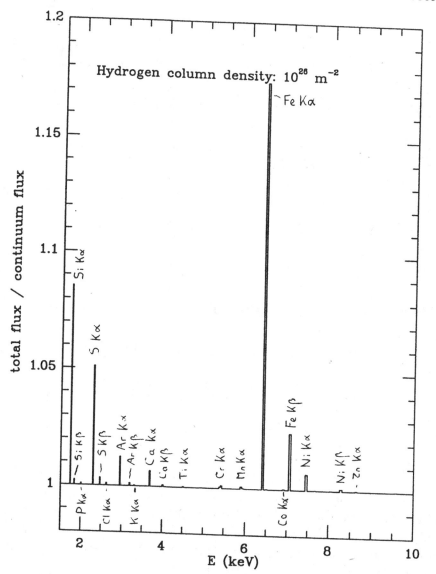

Fig. 2—Fluorescent spectrum of a neutral gas with cosmic abundances. Only the optical thin region of the spectrum is shown. The calculation is done for an E^{-2} input photon spectrum.

by NWO, the Netherlands Organization for Scientific Research.

REFERENCES

Ferland, G.F., and Rees, M.J., 1988, *Ap. J.*, **332**, 141.
Gronenschild, E.H.B.M., and Mewe, R., 1978, *Astr. Ap. Suppl.*, **32**, 283.
Halpern, J.P., and Grindlay, J.E., 1980, *Ap. J.*, **242**, 1041.

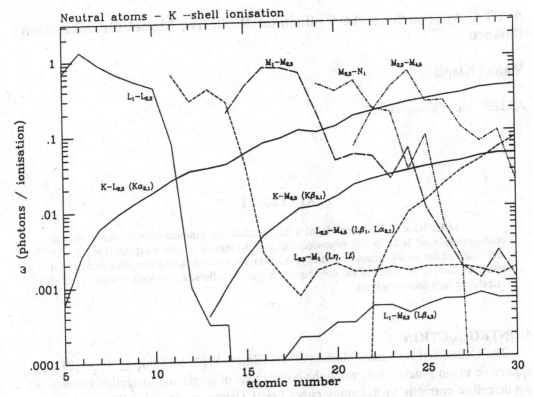

Fig. 3—Fluorescence yield ω (photons / ionisation) for K-shell ionisation of all neutral atoms from B-Zn. The lines are identified by the lower and upper level, respectively, with the common line name in ().

Jacobs, V.L., Rozsnyai, B.F., 1986, *Phys. Rev. A*, **34**, 216.

Kaastra, J.S., and Mewe, R., 1992, submitted to *Astr. Ap. Suppl.*

Kallman, T.R., and McCray, R., 1982, *Ap. J. Suppl.*, 50, 263.

McGuire, E.J., 1969, *Phys. Rev.*, **185**, 1.

McGuire, E.J., 1970a, *Phys. Rev. A*, **2**, 273.

McGuire, E.J., 1970b, *Sandia laboratories report* SC-RR-70-50.

McGuire, E.J., 1971a, *Phys. Rev. A*, **3**, 587.

McGuire, E.J., 1971b, *Phys. Rev. A*, **3**, 1801.

McGuire, E.J., 1971c, *Sandia laboratories report* SC-RR-710075.

McGuire, E.J., 1972a, *Phys. Rev. A*, **5**, 1052.

McGuire, E.J., 1972b, *Sandia laboratories report* SC-RR-710835.

Mewe, R., 1972, *Sol. Phys.*, **22**, 459.

Mewe, R., 1975, *Sol. Phys.*, **44**, 383.

Mewe, R., and Gronenschild, E.H.B.M., 1981, *Astr. Ap. Suppl.*, **45**, 11.

Mewe, R., Gronenschild, E.H.B.M., and van den Oord, G.H.J., 1985, *Astr. Ap. Suppl.*, **62**, 197.

Mewe, R., Lemen, J.R., and van den Oord, G.H.J., 1986, *Astr. Ap. Suppl.*, **65**, 511.

A UTA Approach to the Collisional Radiative Model for Ionization Balance

Marcel Klapisch

ARTEP* Inc., Columbia, Md.

ABSTRACT

The Collisional Radiative Model rate equations are summed over the states of each configurrations, while the energy dependent factor of the transition rates is expended in a Taylor series around the average transition energy. We obtain corrections to the transition rates which take into account the width of the configurations and the influence of selection rules, without calculating each and every level.

1. INTRODUCTION

The Collisional Radiative Model (CRM) (Bates, 1962, McWhirter, 1963 ,1978) is applicable in the general case, when the populations of levels and ionization balance are not described correctly by assuming either Local Thermodynamic Equilibrium (LTE) or the low electron density limit known as the "corona model". It consists, in principle, in writing the rate equations for the populations of all levels of all ions of an atom in a plasma – a finite number M due to "continuum lowering"– through all possible mechanisms. In matrix form, it reads

$$\frac{d\mathbf{N}}{dt} = -\mathbf{N} \bullet \mathbf{R}(T_e, n_e),$$

where $\mathbf{N}$ is the vector of M unknown populations and $\mathbf{R}$ is the MxM matrix of rates, accounting for all possible transition mechanisms, which are functions of electron temperature and density (T_e, n_e).

M being of the order of millions for heavy atoms in hot plasmas, an accurate ionization balance (IB) computation based on the CRM is presently out of reach. However, its need is urgently felt for many important applications. To name just a few, simulation of X-ray lasers, evaluation of radiative energy losses in fusion plasmas, and accurate diagnostics analysis using Unresoved Transition Arrays (UTA) (Bauche, Bauche-Arnoult and Klapisch, 1987)(Zigler, *et al.*, 1987) would all be much more precise with a reliable IB model.

Several codes were compared by Stone and Weisheit (1986). Either these are specific to a few ions, or they employ drastic approximations, like hydrogenic orbitals, that would not be adequate for our purpose. In any case, the comparison of different results show such large variations that they would appear unreliable.

In this work, we propose an alternative approach that would retain the characteristics of complex spectra, i.e. broad configurations with thousands of levels, without the need of computing all the levels. Thus the computing time would scale as the square of the number of configurations only.

2. BASIC FORMULA

We propose to sum the rate equation over all states of each configuration, assuming the levels are populated according to their statistical weight, but *without ignoring the fact that rate coefficients generally depend on the transition energy.*

Let us consider a generic transition "strength", i.e. probability or cross section. Since any transition operator can be expanded in tensorial – possibly multi-particle – operators, any transition strength connecting levels a and b, defined by quantum numbers (α_a, J_a) and (α_b, J_b), can be written as:

$$T(a,b) = \sum_k t_k(\alpha_a J_a, \alpha_b J_b) \Theta_k(E_{a,b})$$

where t_k is the square of a purely angular, energy independent matrix element of a tensorial operator of rank k composing T, $E_{a,b}$ is the transition energy $E_a - E_b$, and all the energy dependence is relegated in the "radial" factor Θ_k.

The transition strength between configurations A and B in the summed CRM equations is obtained as usual by averaging over initial states and summing over final states:

$$T_{A,B} = \frac{1}{g_A} \sum_{a,b} T(a,b),$$

where g_A is the statistical weight of A. Let us expand $\Theta_k(E_{a,b})$ in a Taylor series around the energy E_{av} corresponding to the difference between averages of final and initial configurations:

$$\Theta_k(E_{a,b}) = \Theta_k(E_{av}) + (E_{a,b} - E_{av}) \frac{\partial \Theta_k}{\partial E} + \tfrac{1}{2}(E_{a,b} - E_{av})^2 \frac{\partial^2 \Theta_k}{\partial E^2} + \cdots$$

we obtain:

$$T_{A,B} = \frac{1}{g_A} \sum_k \left\{ S(t_k) \Theta_k(E_{av}) \left[1 + \mu_1^{(k)} \frac{1}{\Theta_k} \frac{\partial \Theta_k}{\partial E}\Big|_{E_{av}} + \tfrac{1}{2} \mu_2^{(k)} \frac{1}{\Theta_k} \frac{\partial^2 \Theta_k}{\partial E^2}\Big|_{E_{av}} \right] \right\}, \quad (1)$$

where $S(t_k)$ is obtained by the usual sum rules and the moments

$$\mu_n^{(k)} = \frac{\sum\limits_{a,b} (E_a - E_b - E_{av})^n t_k(a,b)}{\sum\limits_k t_k(a,b)}$$

are a generalization of the UTA moments (Bauche, Bauche-Arnoult and Klapisch; 1987) involving other operators than the electric dipole. Analytical formulas can be derived for these operators with the same methods that were used for the electric dipole UTA

(Bauche-Arnoult, Bauche and Klapisch; 1979). The derivatives of Θ_k can be evaluated numerically. It is a simple matter to multiply by a velocity distribution of the free electrons (i.e. Maxwellian) and integrate to obtain the transition rates.

3.RADIATIVE TRANSITION

In this case, the rate coefficient is Einstein's A transition probability. Applying the above to electric dipole radiative transition probabilities yields

$$T_{A,B} = \frac{1}{g_A} CSE_{av}^3 I^2 \left(nl_a, nl_b\right) \left[1 + 3\frac{\delta}{E_{av}} + 3\frac{v}{E_{av}^2}\right]. \quad (2)$$

Here C is a constant depending on the units, S is the sum rule for the line strengths, and I is the radial dipole integral. The correcting bracket takes into account the shift δ of the transition array due to selection rules and the variance v or "width" of the UTA.

These corrections will be important in $\Delta n=0$ transitions. For instance, for ionized rare earths with a $4d^n$ ground configuration, the ratio δ/E_{av} can be more than 10% (Mandelbaum, $et\ al.$ 1987). The second order correction will be important for high–n overlapping configurations. These corrections will influence the population of the first excited configuration $4d^{n-1}4f$ and therefore the ionization balance.

4.COLLISIONAL EXCITATION

It was shown (Bar-Shalom, Klapisch and Oreg; 1988) that collisional excitations can be represented by a sum of multipole, one-electron operators acting only on the bound electrons of the atom, which are multiplied by radial integrals Q_k that include the sum over all partial waves and exchange. Let us write the collision strength, as in equation (1),

$$\Omega_{A,B} = \frac{1}{g_A} \sum_k S(\Omega_k) \Theta_k (E_{av}) \left\{ 1 + \mu_1^{(k)} \frac{1}{Q_k} \left.\frac{\partial Q_k}{\partial E}\right|_{E_{av}} + \tfrac{1}{2} \mu_2^{(k)} \frac{1}{Q_k} \left.\frac{\partial^2 Q_k}{\partial E^2}\right|_{E_{av}} \right\},$$

where $S\ (\Omega_k)$ refers to the angular sum rule for collision strength. We obtain now corrections for each multipole of Ω. The formulas for the moments of the multipoles can be worked out easily (Bauche, Bauche-Arnoult and Klapisch; 1987). It was also shown by (Bar-Shalom, Klapisch and Oreg; 1988) that the Q_k integrals were nearly linear with transition energy over a wide range, except for $k=1$, for which the second order correction will not be negligible. Consequently, it is expected that the largest corrections will occur in the dipole allowed component of the transitions, particularly for $\Delta n=0$.

For the dipole allowed component, the moments are identical to the electric dipole ordinary UTA. Corrections of the order of 50% can be expected.

5.DISCUSSION

In order to implement this project, we have to establish the formulas for the moments corresponding to other operators, i.e. auto-ionization, radiative recombination, collisional ionization, three body recombination. This work is in progress .

For dielectronic recombination, this model would have to treat separately the electron capture and the radiative parts of this two step process. Any influence of metastable levels would be ignored. The only way to avoid this would be to employ the full CRM.

The main time gains with respect to the full level-by-level model occur in handling matrices: building the Hamiltonian, and the various transition matrices, as well as solving the CRM rate equations. These gains might be several orders of magnitude. However the computing time of Q_k integrals scales already like the number of orbitals because of the factorization-linearization method (Bar-Shalom, Klapisch and Oreg, 1988). In this calculation, there would be little gain, if any, only in the case configuration interaction is neglected. Unfortunately, this is the most time consuming part of building the rate equations.

Both the development of the model and the running of the computer code would involve substantial effort. However, if one develops such a code for ionization balance, the implementation of the corrections described here would involve a small increment in computing time, with the great advantage of being consistent with the UTA model, and giving more reliable results than obtained by supposing degenerate configurations.

* Mailing address: Naval Research Laboratory, Code 4694, Washington,D.C. 20375.

REFERENCES

Bar-Shalom, A., Klapisch, M. and Oreg, J., 1988, *Phys. Rev.*, **38**, 1773.

Bates, D. R., Kingston, A. E. and McWhirter, R. W. P., 1962, *Proc. Roy. Soc.*, **A267**, 297.

Bauche-Arnoult, C., Bauche, J. and Klapisch, M., 1979, *Phys. Rev.*, **A20**, 2424.

Bauche, J., Bauche-Arnoult, C. and Klapisch, M., 1987, *Advances At. Mol. Phys.*, **23**, 131.

Mandelbaum, P., Finkenthal, M., Schwob, J. L. and Klapisch, M., 1987, *Phys. Rev.*, **A35**, 5051.

McWhirter, R. W. P. and Hearn, A. G., 1963, *Proc. Phys. Soc.*, **82**, 641.

McWhirter, R. W. P., 1978, *Phys. Reports*, **37**, 165.

Stone, S. R. and Weisheit, J. C., 1986, *J. Q. R. S. T.*, **35**, 67.

Zigler, A., et al., 1987, *Phys. Rev.*, **A35**, 280.

Artificial Neural Networks for Plasma X-ray Spectroscopic Analysis

J. T. Larsen[1], W. L. Morgan[2], and W. H. Goldstein[3]

[1]Cascade Applied Sciences, Inc., P.O. Box 4477, Boulder, CO 80306
[2]Kinema Research, 18720 Autumn Way, Monument, CO 80132
[3]Lawrence Livermore National Laboratory, Livermore, CA 94550

ABSTRACT

Artificial neural networks have been applied to a variety of signal processing and image recognition problems. The feed-forward, back-propagation technique is well suited for the analysis of scientific laboratory data, which is viewed as a pattern-matching problem. We summarize the concepts and algorithms as implemented on a personal computer, and illustrate the method using a nonLTE theoretical atomic physics model for k-shell x-ray spectroscopy of a high density, high temperature aluminum plasma.

1 BASIC FEATURES OF ARTIFICIAL NEURAL SYSTEMS

Compared to the most advanced artificial intelligence systems running on the fastest digital computers, the human brain is far more capable of solving pattern recognition/analysis problems. It has many features desirable for automated data analysis systems: it is robust and fault tolerant; it is flexible and easily adapts to a new environment by "learning;" it can deal with fuzzy, probabilistic, noisy or inconsistent information; it is highly parallel; and it is small, compact and dissipates very little power.

Artificial neural systems were first developed in the 1940's, saw a brief revival of interest in the 1960's, and assumed significant status as a computational tool in the 1980's. However, nearly all of the artificial neural network development has sought to model neuro-psychological processes or has found application in image processing or process control. Very little attention has been directed towards applying the methods to other laboratory research problems.

There are many types of artificial neural systems, but all are modelled along the lines of biological neural networks, often somewhat simplistic in sophistication. The brain is composed of about 10^{11} neurons of many different types. Tree-like networks of nerve fibers called dendrites are connected to the cell body, or soma. Extending from the cell body is a single long fiber called the axon, which branches into strands and substrands. At the ends of these are the transmitting ends of the synaptic junctions to other neurons. The axon of a typical neuron makes a few thousand synapses with other neurons.

A simple mathematical model of a neuron treats it is a binary threshold unit in which the weighted sum of its inputs from other units generates an output signal according to whether this sum is above or below a prescribed threshold. The weight representing the strength of the synapse connecting the two neurons can be positive or negative corresponding to an excitatory or inhibitory response respectively. Although simple, this model neuron is a powerful computational element; a synchronous assembly of these elements is capable (in principle) of universal computation for suitably chosen weights.

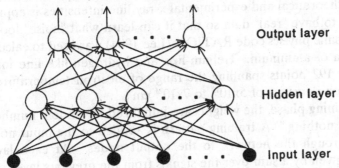

Fig. 1 A multilayer perceptron transfers an input pattern to an output pattern.

One of the earliest implementations of this model neuron was in a network architecture called a perceptron, in which the neural units were organized into layers with feed-forward connections between one layer and the next. A schematic of a simple multilayer perceptron is shown in Fig. 1. Such a network has (in general) a highly nonlinear mapping of the input pattern to the output pattern. For the past twenty years the primary difficulty has been how to find the appropriate weights between neurons for a given computational task.

Recent attention to multilayer perceptrons has resulted in the development of a weight-determining algorithm known as back-propagation. Early perceptrons could only solve problems that were linearly separable. This limitation is overcome by having the network "learn" to recognize the important features of the input pattern. This corresponds to adjusting the weights of the connections between neurons on successive layers. Additionally, the threshold rule of the simple model has the effect of removing information that is needed for the network to learn successfully; the output from a neuron has to have continuous values, yet still be limited to the interval 0:1.

The back-propagation method (or generalized delta rule) is founded on the principle of supervised (reinforced) learning, and is not unlike that found in certain biological systems. This training phase necessitates the use of example pairs from which a "mistake" can be calculated. An example pair consists of an input stimulus together with an expected response; the error is then the difference between the desired output and the computed output. This error term, calculated on the output neurons, is then apportioned to the weights feeding the output layer. While it is impossible to calculate an error function for hidden layers, since the desired output there is not known, a correction term can be computed and propagated back to the next layer from which the corresponding weights may be changed.

2 TRAINING THE BACK-PROPAGATING NETWORK

Before a feed-forward, back-propagating neural network can be used for routine data analysis, it must "learn" the weights on the connections between the processing elements (nodes). A training database of numerous examples, consisting of tens to hundreds of input-output pairs, must be prepared. For the application of interest here, a mixture of theoretical and experimental x-ray line intensities is appropriate. (The network likes to have "real" data so that it can learn what "noise" looks like.)

The nonLTE atomic physics code RATION (Lee 1984) was used to calculate the k-shell x-ray spectra of aluminum. Helium-like and hydrogen-like line intensities were calculated for 192 points spanning the range of electron temperatures 50 to 1200 eV and electron densities of 5×10^{19} to 7×10^{21} cm^{-3}.

To begin the training phase, the weights are initialized to random numbers, and the network "knows nothing." A training pattern is applied to the input nodes and then propagated through the network to the output nodes. At each layer, the output of node i is given by a sum over the signals from the previous layer of nodes

$$x_i = g(\Sigma w_{ij} x_j) \tag{1}$$

where g is the nonlinear activation function (the sigmoid, $g(y) = [1 + \exp(-y)]^{-1}$ is commonly used), w_{ij} is the weight connecting nodes i and j, and x_j is the output of the j-th node from the previous layer. An error measure or cost function is calculated at the output layer; it is commonly taken to be the sum of the squares of the differences of the calculated and expected outputs on each node. The correction term to the output layer weights is found using the gradient descent rule, which says that the correction is proportional to the derivative of the cost function with respect to each weight. In a similar fashion, the correction terms to the more deeply imbedded weights are found.

This procedure is performed, in turn, for each of the training patterns. After cycling through the patterns once, the order of the patterns is randomized, and the procedure is repeated continually until the convergence criteria on the cost function is achieved. The shuffling of the patterns is necessary to make the path through weight-space stochastic thus allowing a wider exploration of the cost surface.

For this demonstration, a network architecture with eight input nodes, two output nodes, and two hidden layers of fifteen nodes each was found to be near optimal. A total of 375 connections (weights) were required. A network architecture is deemed optimal by studying the histograms of the weights (which should resemble a near normal distribution) and a graphical representation of the distribution of weights known as a Hinton diagram.

Once the weights have been found by the iterative training process, the weights may be saved for future production runs; it is not necessary to recompute the weights for similar plasma conditions. The vast majority of the computer time required by a neural network is in the training phase; the back-propagation step is algorithmically more complex than the feed-forward step used only in production.

3 PRELIMINARY RESULTS

The first test of neural networks for data analysis used the above network architecture and a training dataset consisting of relative line intensities for nine H-like and He-like transitions in aluminum. The network was trained to a 10% convergence criteria on 95 example pairs from the 192 pattern set mentioned above. The test "unknown" data was generated using the same atomic physics model with the temperature/density points chosen to mimic realistic laboratory data. The results are shown in Table 1.

TABLE 1. Plasma parameters determined by the neural network.

Network values		Correct values	
T_e (eV)	n_e (cm^{-3})	T_e (eV)	n_e (cm^{-3})
218	2.3e+20	230	2.5e+20
236	2.7e+20	250	3.0e+20
299	2.8e+20	310	3.0e+20
369	4.7e+20	370	5.0e+20
437	5.8e+20	430	6.0e+20

These results are within the 10% convergence criteria used for training.

4 CONCLUSIONS

Much of the recent excitement about neural networks is their inherent ability to generalize to new situations. After being trained on a number of example pairs, the network can often induce a complete relationship that interpolates and extrapolates from the examples in a sensible way.

The algorithms that allow a network to do this are not immediately obvious, and there is little in the literature that provides guidance. There remains considerable work to demonstrate the full utility of neural networks for extracting relevant physical parameters from laboratory data; these first efforts are encouraging. Research is now in progress to explore the details of the simple algorithms as well as develop more complex neural network models. The effort is directed toward reducing the time needed to train the network and to improving the minimization of the cost function (accuracy).

5 REFERENCES

Lee, R. W., Whitten, B. L., and Stout II, R. E. 1984, *J. Quant. Spectrosc. Radiat. Transfer* **32**, 91.

Effect Of Excitation-Autoionization On Fractional Abundances Of Highly Ionized KrI- To NiI-Like Heavy Elements In Coronal plasmas

D. Mitnik[1], P. Mandelbaum[1], J.L. Schwob[1], J. Oreg[2], A. Bar Shalom[2] and W.H. Goldstein[3]

[1]Racah Institute of Physics, The Hebrew University, Jerusalem 91904,Israel.
[2]Nuclear Center of The Negev, P.O. Box 9001, Beer Sheva, Israel.
[3]Lawrence Livermore National Laboratory, Livermore ,P.O. Box 808,California 94550

ABSTRACT

In the present work , improved results of detailed computations for Excitation-Autoionization (EA) processes through 3d-4d and 3d-4f electron impact inner-shell excitations for KrI-,BrI-,GaI-,ZnI- and CuI-like ions have been performed and the behaviour of the EA rate coefficients along the isoelectronic sequences has been analysed. The results were introduced in coronal equilibrium model for Praseodymium and Dysprosium . In this model, the EA effect was introduced for the other neighbouring sequences by using a mean branching ratio towards autoionization. As a result of the introducing EA processes, the ion most abundances are displaced by as much as 35% for some ions towards lower electron temperature.

1 INTRODUCTION

Excitation of inner-shell electrons into highly excited autoionizing levels followed by autoionization, is an important process that can significantly enhance the total ionization rate. EA processes have been studied mainly for ions pertaining to isoelectronic sequence having simple ground-state structure, in particular for ions of the NaI isoelectronic sequence (Cowan and Mann 1979; Griffin,Pindzola and Bottcher 1986). In this sequence, computations have shown that the EA rate coefficients can be comparable to that of direct electron impact ionization.

For ions isoelectronic to KrI and up to CuI one expects that EA could be even more important since direct ionization competes with excitation of the ten 3d M-shell electrons. Computations for GaI- and ZnI-like isoelectronic sequences have been performed recently (Mandelbaum et al. 1990; Oreg et al. 1991) and showed that EA processes through 3d-4l electron impact excitation could enhance by a factor up to four the total ionization rate coefficient at the most abundance electron temperature of the ion. But, in contradistinction with the NaI-sequence , the isoelectronic trend of the EA rate coefficient along GaI and neighbouring sequences showed an abrupt discontinuity due to closing of autoionization channels as Z of the element increases. In the present work, improved results for EA rates through 3d-4d and 3d-4f inner-shell transitions involving hundreds of levels are presented, and the systematic isoelectronic trend is analysed in details for the GaI,ZnI and CuI isoelectronic sequences. The results of these computations are used in the coronal equilibrium model for Praseodymium and Dysprosium to show the effects of EA processes on ionization balance.

2 EA COMPUTATIONS

Details of the theoretical methods used in these computations are given in our previous work (Mandelbaum *et al.* 1990). The EA rate coefficient $S^{EA}(g,k)$ from a level g of the ground configuration $3d^{10}4s^x4p^y$ of a ion of charge r to a given level k within the $3d^{10}4s^x4p^{y-1}$ or $3d^{10}4s^{x-1}4p^y$ configuration of the next ion r+1, through inner-shell excitation to any autoionizing level j $3d^94s^x4p^y4d$ or $3d^94s^x4p^y4f$ configuration of ion r, is given by:

$$S^{EA}(g,k) = \sum_j \left[Q_{gj} A_{jk}^a / \left(\sum_{k'} A_{jk'}^a + \sum_i A_{ji} \right) \right] \qquad (1)$$

where Q_{gj} is the electron impact excitation rate coefficient from the ground level to level j. A_{jk}^a is the autoionization coefficient from level j to level k, and A_{ji} is the Einstein spontaneous radiative decay coefficient to a lower level i of ion r. Energies and radiative decay coefficients were computed using the relativistic RELAC code (Klapisch *et al.* 1977); the collisional excitation rate coefficients were computed in the semirelativistic distorted-wave approximation using the factorization method (Bar-Shalom, Klapisch and Oreg 1988) ; and the autoionization coefficients have been calculated with a new code included in the HULLAC package (J. Oreg *et al.* 1991). Results of these computations are shown in Figures 1 and 2. Figure 1 shows the mean branching ratio B for inner-shell 3d-4d and 3d-4f electron impact excitations towards autoionization along the GaI, ZnI and CuI isoelectronic sequences computed at $kT_e = E_i$ (E_i is the first ionization energy of the ion). For each sequence there is a steep decreasing in B when inner-shell excited levels fall below the ionization limit as Z increases, closing autoionization

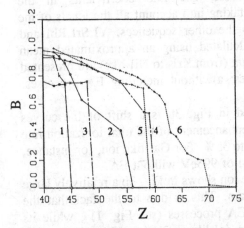

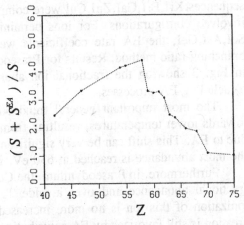

Fig. 1 Mean branching ratio for EA processes through inner-shell excited configurations $3d^94s^x4p^y4d$ (curves 1,3,5) and $3d^94s^x4p^y4f$ (curves 2,4,6) (at $kT_e = E_i$) in CuI (curves 1,2), ZnI (curves 3,4) and GaI (curves 5,6) isoelectronic sequences.

Fig. 2 Enhancement of collisional ionization rate (at $kT_e = E_i$) due to Excitation-Autoionization through 3d-4l inner-shell transitions from $4p_{1/2}$ initial level in GaI isoelectronic sequence

channels. For the GaI sequence this change is striking at $Z=59$ (for 3d-4d excitations) and $Z=70$ (for 3d-4f). For the ZnI sequence, this occurs at $Z=49$ (for 3d-4d) and $Z=60$ (for 3d-4f) and for the CuI sequence at $Z=45$ (for 3d-4d) and $Z=56$ (for 3d-4f). Figure 2 shows the enhancement of the collisional ionization rate coeficient $(S + S^{EA}) / S$ due to EA processes from the lower ground level $3d^{10}4s^24p_{1/2}$ in GaI-like sequence, at $kT_e = E_i$. S is the direct collisional ionization rate coefficient including inner-shell ionization, calculated using Lotz's formulas (Lotz 1967). The curve shows a maximum at $Z=58$, for heavier elements the effect of the closing of EA channels begins to be important.

3 CORONAL MODEL EQUILIBRIUM

In a coronal plasma at steady state, the relative densities of two consecutive ions r and r+1 of a given element is given by:

$$n_{r+1} / n_r = (S_r + S_r^{EA})/\alpha_{r+1} \qquad (2)$$

where α_{r+1} is the total rate coeffficient for radiative and dielectronic recombination. Direct and EA ionizations are assumed to involve the lower level of the ground configuration only.

The system of coupled equations (2) was computed using modified hydrogenic formulas for radiative recombination (Von Goeler et al. 1975) and approximate expressions for dielectronic recombination (Burgess 1965; Merts et al. 1976). This system was solved as a function of the electron temperature T_e for Praseodymium (Z=59) and Dysprosium (Z=66). The calculations included the successive ions isoelectronic to YI up to CrI. In this model, EA rate coefficients in the sequences: KrI, BrI, GaI, ZnI, CuI were computed taking into account all the levels of the involved configurations. For ions pertaining to the other sequences: YI, SrI, RbI and SeI, AsI, GeI, the EA rate coefficients were calculated using an approximated mean branching ratio method. Results for Praseodymium (from KrI- to NiI-like) are presented in Fig. 3 showing the fractional ion abundances: a)without including EA processes; b)including EA processes.

The most important general trend observed in Fig. 3b is a shift of the curves towards lower temperatures, resulting from the enhancement of the total ionization rate due to EA. This shift can be very significant, up to 35%. For GaI-like ion, for instance, the most abundance is reached at 630 eV, instead of 995eV without EA.

Furthermore, in Praseodymium, the CuI-like ion shows in Fig. 3b a relatively large fractional abundance and over a wide T_e domain. This is due to the fact that the ionization of this ion is no more increased by EA processes (see Fig. 1), while its creation is still favored by EA contribution in the ZnI-like ionization (although through 4f only). This behaviour changes with Z. For Dysprosium, computations show that the same occurs, but in this case the ZnI-like ion fractional abundance is the larger, as it can be explained from Fig. 1.

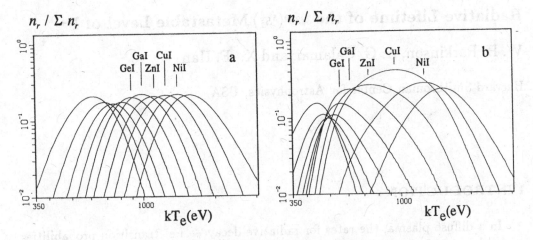

Fig. 3 Fractional abundances for Praseodymium ions Pr^{+23}-Pr^{+32}.
a-without Excitation-Autoionization processes.
b-including Excitation-Autoionization processes.

In conclusion, the present results have emphasized the dominant EA contribution in ionization processes in $3d^{10}4s^x4p^y$ ground configuration ions, leading to a drastic shift in coronal equilibrium. The dependence of the effect on Z, due to the crossing of inner-shell excited levels below the ionization limit as Z increases, represents a striking feature in these electronic sequences.

4 REFERENCES

Bar Shalom,A.,Klapisch,M., Oreg,J. 1988, *Phys. Rev. A*38,1773.

Burgess,A. 1965,*Astrophys. J.* **141**,1588.

Cowan,R.D., Mann,J.B. 1979, *Astrophys. J.* **232**,940.

Griffin,D.C.,Pindzola,M.S.,Bottcher,C. 1986,*Phys. Rev. A*36,3642.

Klapisch,M.,Schwob,J.L.,Fraenkel,B.S., Oreg,J. 1977,*J. Opt. Soc. Am.* **61**,148.

Lotz,W. 1967,*Astrophys. J. Suppl.* **14**,207.

Mandelbaum,P., *et al.* 1990, *Phys. Rev. A*42,4412.

Merts,A.L.,Cowan,R.D.,Magee,N.H. 1976,La-6220-MS,Los Alamos,New Mexico.

Oreg,J.,Goldstein,W.H.,Klapisch,M.,Bar-Shalom,A. 1991,*Phys. Rev. A*44,1750.

Von Goeler,S. *et al.* 1975. *Nucl. Fus.* **15**.301.

Radiative Lifetime of the $3s3p^3(^5S_2^o)$ Metastable Level of P+

W. H. Parkinson, A. G. Calamai, and X. -F. Han

Harvard-Smithsonian Center for Astrophysics, USA

1 INTRODUCTION

In a diffuse plasma, the rates for radiative decay — i.e. transition probabilities or A-values — of metastable atomic levels can be the same order of magnitude as the rates for excitation and de-excitation by electron collisions. Intensity ratios involving lines from metastable levels are sensitive to electron densities and temperatures, and are often the basis of the best diagnostic techniques (Doschek 1985) used to determine these plasma parameters. However, the precision of these diagnostic methods critically depends on an accurate knowledge of the A-values of the allowed and forbidden decays involved. Consequently, accurate radiative lifetime and branching ratio measurements are required.

Theoretical predictions for the radiative lifetimes of the $3s3p^3(^5S_2^o)$ metastable level of low-charge-state ions in the Si isoelectronic sequence are quite uncertain (Ellis 1989). The various theoretical methods (Ellis and Martinson 1984, Huang 1985, Biémont 1986) employed to calculate A-values for intercombination transitions involving the $3s3p^3(^5S_2^o)$ metastable level in low-charge-state ions produced results that differ by as much as an order-of-magnitude. Thus, precise measurements of radiative lifetimes and branching ratios for transitions from the $^5S_2^o$ levels of these ions are required, not only to extend the available atomic physics data base, but also as essential tests of the methods employed in theoretical atomic spectroscopy (Ellis 1989).

We have measured the radiative lifetime of the $3s3p^3(^5S_2^o)$ metastable level of P+. This is the first lifetime measurement of the $3s3p^3(^5S_2^o)$ level belonging to a low charge-state ion in the Si isoelectronic sequence. The position of the P+$(^5S_2^o)$ level relative to the fine-structure levels of the $3s^23p^2$ ground configuration has been established by Martin and his coworkers (Martin *et al.* 1985). Because the $^5S_2^o$ level is the lowest level of the first excited configuration in P+, radiative decay by a direct electric-dipole transition is forbidden by the LS selection rule $\Delta S = 0$. Consequently, intercombination transitions represent the dominant decay mode of the $3s3p^3(^5S_2^o)$ metastable level. In P+, either a 221.1 or 219.6 nm photon is emitted when the $^5S_2^o$ state undergoes an intercombination transition to either the $3s^23p^2(^3P_2$ or $^3P_1)$ level, respectively.

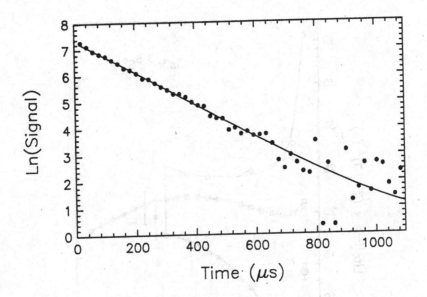

Fig. 1—Example decay curve. Natural logarithm of the first 53 channels of a 128 channel decay curve of signal counts vs time as the metastable $3s3p^3(^5S_2^o)$ population of stored P+ ions decays. Only the first 53 channels are shown to emphasize the time dependence of the decay. The solid-curve represents the natural logarithm of the nonlinear least-squares fit of $N_o e^{-\gamma t} + B$ to the decay counts.

2 EXPERIMENTAL METHOD

A thorough discussion of our experimental technique has been published elsewhere (Calamai *et al.* 1992), only a brief discussion of the method used to measure the radiative lifetime of the $3s3p^3(^5S_2^o)$ metastable level of P+ is given for completeness. A metastable ion population was produced inside a cylindrical radio frequency ion trap by electron bombardment of PH_3 vapor at pressures ranging from 5 to 40×10^{-8} Torr. The trap, operated with potential well depths ranging from 12 to 20 eV, consisted of two circular, wire-mesh, end caps and a cylindrical, wire-mesh, ring electrode with physical dimensions $r_o = z_o = 1.67$ cm. After the ions were created, the radiative decay of the metastable-state population was monitored versus time by focusing some of the light emitted by the decaying $^5S_2^o$ population onto a 19-nm-bandwidth interference filter in front of a photomultiplier tube operated in single-photon counting mode. This generated a decay curve, from which the mean lifetime was obtained from the decay rate parameter γ of a nonlinear least-squares fit of a single-exponential plus a constant background to the decay counts, as shown in Figure 1. The measured decay rates did not exhibit a dependence on the source PH_3 pressure. Thus, our result for the radiative lifetime of the $3s3p^3(^5S_2^o)$ metastable level of P+ is the inverse of the mean decay rate obtained from a total of 36 decay curves and yields a value of 167 ± 12 μs for the lifetime.

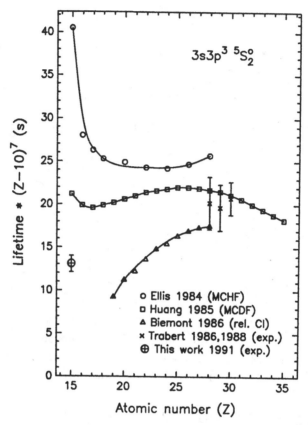

Fig. 2—Scaled mean lifetime versus atomic number. Mean lifetime of the $3s3p^3(^5S_2^o)$ level versus atomic number Z in the Si isoelectronic sequence; the lifetimes are multiplied by a factor of $(Z-10^7)$. Theoretical results correspond to semiempirical MCHF calculations (Ellis and Martinson 1984) ($\bigcirc$); *ab initio* MCDF calculations (Huang 1985) ($\square$); and semiempirical calculations using relativistic CI wavefunctions (Biémont 1986) ($\triangle$). Results of beam-foil lifetime measurements (Träbert 1986, Träbert 1988) are shown ($\times$) with the reported error bars, as is the $P^+(^5S_2^o)$ lifetime measurement reported in this work ($\oplus$).

3 DISCUSSION

Figure 2 illustrates the results of three theoretical methods used to determined the radiative lifetime of the $3s3p^3(^5S_2^o)$ level in the Si isoelectronic sequence: the semiempirical multiconfigurational Hartree-Fock MCHF (Ellis and Martinson 1984), and the *ab initio* multiconfigurational Dirac-Fock MCDF (Huang 1985) methods, and a semiempirical method using relativistic configuration interaction CI wavefunctions (Biémont 1986). Because the radiative lifetime of a level decaying by an intercombination transition with $\Delta n = 0$, $\Delta S = 1$ roughly scales as Z_c^{-7}, where Z_c is the partially screened nuclear charge, the lifetimes plotted in Fig. 2 have been multiplied by a factor of $(Z-10)^7$. Three beam-foil measurements (Träbert 1986, Träbert 1988) and the result of our work for the $3s3p^3(^5S_2^o)$ lifetime in Si-like ions are also plotted in Fig. 2, showing the reported uncertainties.

For the $3s3p^3(^5S_2^o)$ level of high-charge-state ions, where jj coupling applies, the MCDF results should provide accurate A-values and lifetimes (Ellis 1989). Prior to the lifetime measurement reported in this paper, the only experimental results for the $3s3p^3(^5S_2^o)$ lifetime corresponded to beam-foil measurements (Träbert 1986, Träbert 1988) on Ni^{14+}, Cu^{15+}, and Zn^{16+} ions, which confirmed the validity of the MCDF method for high-charge-state ions. In the case of low-charge-state species, Ellis (Ellis 1989) has discussed the intractability of theoretically determining the $3s3p^3(^5S_2^o)$ lifetime. Our measured result for the P^+ $(^5S_2^o)$ lifetime does not agree with either of the theoretical (Ellis and Martinson 1984, Huang 1985) results or their extrapolation (Biémont 1986).

Although Ellis' (Ellis and Martinson 1984) theoretical lifetime for the $^5S_2^o$ level of P^+ is approximately 100% greater than the calculation of Huang (Huang 1985), branching ratios deduced from the theoretical transition probabilities agree to about 35%. By using our measured decay rate and reported uncertainty, and adopting the average branching ratios predicted by the theoretical A-values with the difference between the theoretical branching ratios serving as the error for the adopted branching ratios, we estimate the following A-values for the intercombination transitions $A(^5S_2^o \rightarrow\ ^3P_2) = 4629 \pm 881$ s^{-1} and $A(^5S_2^o \rightarrow\ ^3P_1) = 1359 \pm 579$ s^{-1} in P^+. We hope to be able to measure the intercombination line branching ratios in the future, so that the A-values can be more accurately estimated. Ellis (Ellis 1989) has pointed out that such measured line strengths would lead to better spin-orbit parameters, and thus to a yardstick for improvement of the theoretical description of Si–like ions.

This work was supported in part by NASA Grants No. NAGW - 1596 and NAGW - 1687 to Harvard University.

REFERENCES

E. Biémont, Phys. Scripta **33**, 324 (1986); J. Opt. Soc. Am. **B3** 163, (1986).

A. G. Calamai, X.-.F Han, and W. H. Parkinson, (*scheduled for publication*) Phys. Rev. **A 45** #5, (March 1992).

G. A. Doschek, in *Progress in Atomic Spectroscopy*, ed. A. Temkin (New York: Plenum, 1985), p. 175.

D. G. Ellis and I. Martinson, Phys. Scripta **30**, 255 (1984).

D. G. Ellis, Phys. Scripta **40**, 12 (1989).

K. -N. Huang, At. Data Nucl. Data Tables **32**, 503 (1985).

W. C. Martin, R. Zalubas, and A. Musgrove, J. Phys. Chem. Ref. Data **14**, 751 (1985).

E. Träbert, Z. Phys. **D 2**, 213 (1986).

E. Träbert, P. H. Heckmann, and W. L. Wiese, Z. Phys. **D 8**, 209 (1988).

Proton and Heavy Particle Excitation
of the $2s^2 2p^5\,{}^2P_{3/2} \rightarrow 2s^2 2p^5\,{}^2P_{1/2}$ Transition
in Fluorine-like Zn XXII, Kr XXVIII and Mo XXXIV

R. H. G. Reid, V. J. Foster and F. P. Keenan

School of Mathematics and Physics, The Queen's University of Belfast, Belfast BT7 1NN, Northern Ireland.

ABSTRACT

Cross sections and rate coefficients for excitation of the $2s^2 2p^5\,{}^2P_{3/2} \rightarrow 2s^2 2p^5\,{}^2P_{1/2}$ transition in fluorine-like Zn XXII, Kr XXVIII and Mo XXXIV by proton (p), deuteron (d), triton (t) and α-particle (α) impact have been calculated by a close-coupled semi-classical method. The $2s2p^6\,{}^2S$ states were taken into account by a polarization potential. For each ion, at temperatures close to or below the temperature of maximum fractional abundance of that ion, the p-,d- and t-rates are comparable and are much larger than the α-rates. However at high temperatures the situation is reversed, with the α-rates about twice those due to the other particles.

The usefulness, for plasma diagnostic purposes, of the emission line ratios involving the $2s^2 2p^5\,{}^2P_{3/2}$, $2s^2 2p^5\,{}^2P_{1/2}$ and $2s2p^6\,{}^2S$ levels of fluorine-like ions has been discussed by several authors (Doschek and Feldman 1976, Keenan *et al.* 1989, and references cited therein). The object of the present work is to provide the reliable atomic data needed in such diagnostic analysis.

We have calculated the cross section for the process

$$\mathrm{Kr}^{27+}(2s^2 2p^5)\,{}^2P_{3/2} \;+\; \mathrm{H}^+ \;\longrightarrow\; \mathrm{Kr}^{27+}(2s^2 2p^5)\,{}^2P_{1/2} \;+\; \mathrm{H}^+ \qquad (1)$$

We have also considered the ions Zn^{21+} and Mo^{33+}, and in addition to protons (p), we have considered deuterons (d), tritons (t) and α-particles (α) as the perturber.

The cross sections have been calculated by the close-coupled semi-classical method, which has been used by several authors (*cf* Reid 1988) and is essentially the method used for Coulomb excitation of nuclei (*cf* Alder and Winther 1975). In our calculation, we have used the symmetrized formulation, and we have modified the form of the interaction to allow for penetration of the ion's electron cloud. Also, we have included the effects of the $2s2p^6\,{}^2S$ states by means of a polarization potential (*cf* Alder and Winther 1975). This is necessary for the high-Z ions under consideration, because the ratio of the $2s^2 2p^5\,{}^2P_{1/2} - 2s2p^6\,{}^2S$ separation to the $2s^2 2p^5\,{}^2P_{3/2} - 2s^2 2p^5\,{}^2P_{1/2}$ separation decreases as Z increases. The physical data have been taken from Cheng *et al.* (1979). Further details of the cross section calculation are given by Reid *et al.* (1992).

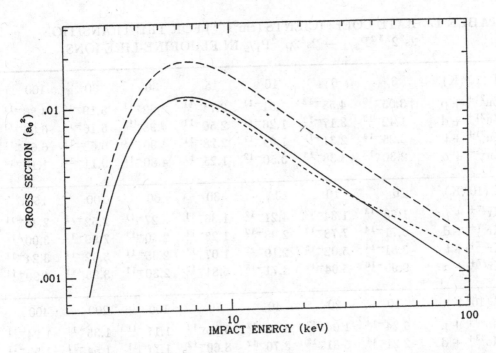

Fig. 1   Calculated cross sections for the $2s^2 2p^5 \, ^2P_{3/2} \rightarrow 2s^2 2p^5 \, ^2P_{1/2}$ transition in $Kr^{27+} + H^+$ collisions. ——— ^2P states close-coupled, ^2S states included by a polarization potential; – – – ^2P and ^2S states close-coupled; — — — ^2P states close-coupled, ^2S states absent.

Figure 1 illustrates the necessity of including the $2s2p^6 \, ^2S$ states. If only the $2s^2 2p^5 \, ^2P_{3/2}$ and $2s^2 2p^5 \, ^2P_{1/2}$ states are included, the cross section is overestimated by more than 60%. Figure 1 also shows that the effect of the $2s2p^6 \, ^2S$ states on the cross section is adequately described by the polarization potential.

The rate coefficients, obtained by averaging the cross sections over a Maxwellian distribution, are given in Table 1. At the lower temperatures, the extra Coulomb repulsion experienced by the α-particle keeps its collisions more distant, and hence the α-rates are much smaller than the rates for the singly charged perturbers. With the same energy and charge, the perturber with the smallest velocity (*i.e.* largest mass) has the smallest cross section. The temperature at which the ion has its maximum fractional abundance in ionization equilibrium lies in this low-temperature regime. At extremely high temperatures, Coulomb repulsion is not the dominant feature, and the critical parameters are the charge and velocity of the perturber. Hence heavier and more highly charged perturbers become more significant, as noted by Walling and Weisheit (1988). Table 1 shows the α-rates becoming largest at high temperature.

TABLE 1. RATE COEFFICIENTS $(cm^3 s^{-1})$ FOR THE TRANSITION $2s^2 2p^5\, {}^2P_{3/2} \rightarrow 2s^2 2p^5\, {}^2P_{1/2}$ IN FLUORINE-LIKE IONS.

T $(10^6$ K)	3	6	10	15	30	60	100
$Zn^{21+} + p$	†3.03^{-13}	4.58^{-12}	1.51^{-11}	2.74^{-11}	4.62^{-11}	5.19^{-11}	4.86^{-11}
$Zn^{21+} + d$	1.42^{-13}	3.17^{-12}	1.26^{-11}	2.56^{-11}	4.98^{-11}	6.16^{-11}	6.06^{-11}
$Zn^{21+} + t$	8.25^{-14}	2.31^{-12}	1.04^{-11}	2.28^{-11}	4.90^{-11}	6.52^{-11}	6.68^{-11}
$Zn^{21+} + \alpha$	2.30^{-15}	3.38^{-13}	3.50^{-12}	1.25^{-11}	4.80^{-11}	9.11^{-11}	1.10^{-10}

T $(10^6$ K)	6	10	15	30	60	100	150
$Kr^{27+} + p$	1.60^{-13}	1.34^{-12}	4.21^{-12}	1.36^{-11}	2.27^{-11}	2.52^{-11}	2.48^{-11}
$Kr^{27+} + d$	6.71^{-14}	7.78^{-13}	2.98^{-12}	1.23^{-11}	2.40^{-11}	2.90^{-11}	3.00^{-11}
$Kr^{27+} + t$	3.51^{-14}	5.02^{-13}	2.19^{-12}	1.07^{-11}	2.35^{-11}	3.02^{-11}	3.24^{-11}
$Kr^{27+} + \alpha$	9.95^{-16}	5.04^{-14}	4.71^{-13}	5.81^{-12}	2.30^{-11}	3.95^{-11}	4.99^{-11}

T $(10^6$ K)	10	20	40	70	100	200	400
$Mo^{33+} + p$	6.24^{-14}	1.04^{-12}	4.90^{-12}	9.36^{-12}	1.17^{-11}	1.35^{-11}	1.24^{-11}
$Mo^{33+} + d$	2.21^{-14}	5.81^{-13}	3.76^{-12}	8.60^{-12}	1.17^{-11}	1.54^{-11}	1.53^{-11}
$Mo^{33+} + t$	1.03^{-14}	3.64^{-13}	2.93^{-12}	7.61^{-12}	1.10^{-11}	1.59^{-11}	1.68^{-11}
$Mo^{33+} + \alpha$	3.46^{-16}	5.27^{-14}	1.19^{-12}	5.61^{-12}	1.08^{-11}	2.28^{-11}	3.04^{-11}

† x^{-y} stands for $x \times 10^{-y}$

Acknowledgement. VJF gratefully acknowledges support by the Department of Education for Northern Ireland.

REFERENCES

Alder, K. and Winther, A. 1975, *Electromagnetic Excitation* (Amsterdam: North Holland)

Cheng, K. T., Kim Y.-K. and Desclaux, J. P. 1979, *At. Data Nucl. Data Tables* **24**, 111.

Doschek, G. A. and Feldman, U. 1976, *J. Appl. Opt.* **43**, 3083.

Keenan, F. P., Reid, R. H. G. and McCann, S. M. 1989, *Journal de Physique* **50**, C1-565.

Reid, R. H. G. 1988, *Adv. Atom. Molec. Phys.* **25**, 251.

Reid, R. H. G., Foster, V. J. and Tully, J. A. 1992, *J. Phys. B: At. Mol. Opt. Phys.*, submitted.

Walling, R. S. and Weisheit, J. C. 1988, *Phys. Reports* **162**, 1.

Spectra of Laser-Produced Rare-Earth Ions and Theoretical Survey of Nickel-Like Resonance Lines from Br VIII to U LXV

P. Renaudin[1], C.A. Back[1], C. Chenais-Popovics[1], J.-P. Geindre[1], J.-C. Gauthier[1], C. Bauche-Arnoult[2], E. Luc-Koenig[2] and J.-F. Wyart[2]

[1]Laboratoire de Physique des Milieux Ionisés and Laboratoire d'Utilisation des Lasers Intenses, Ecole Polytechnique, Palaiseau, France

[2]Laboratoire Aimé Cotton, CNRS II, Bât. 505, Centre universitaire, F-91405 Orsay, France

ABSTRACT

The emission spectra of Sm, Gd, Tb and Ho produced by high power laser irradiation have been observed in the range 4.8-9.3Å. Nickel-like resonance lines have been identified by means of the relativistic parametric potential method and the mixing of 19 odd configurations from Br VIII to U LXV has been studied. The transition arrays $3d^N$-$3d^{N-1}4f$ emitted by ions isoelectronic with Co, Fe, Mn, Cr and V are interpreted in the UTA formalism.

Nickel-like ions are present in plasmas produced under various conditions. Their resonance lines are prominent features in the spectra of high-Z elements irradiated by Nd:YAG lasers at intensities 10^{14}-10^{15} Wxcm^{-2} (Tragin et al. 1988, Mandelbaum et al. 1991). Soft X-ray amplification has been demonstrated in Eu^{35+} and Yb^{42+} (MacGowan et al. 1987), in Ta^{45+} and W^{46+} (MacGowan et al. 1990) and it is predicted to be enhanced by photopumping in Tl^{53+} (Nilsen 1991). For such studies and for plasma modelisation, a better knowledge of the Ni I isoelectronic sequence is needed and the ionization limits have been already predicted by Tragin et al. (1989). The present work had two aims: i) extending the observations in the region of rare earths where earlier measurements were either inaccurate or scarce (Burkhalter et al. 1974, Zigler et al. 1987, von Goeler et al. 1988); ii) predicting the wavelengths and transition probabilities in an approximation which stays valid in a large range of Z-values.

1. EXPERIMENTAL ARRANGEMENT

The experiments were performed with the Nd:glass laser system of the Laboratoire d'Utilisation des Lasers Intenses (LULI). The targets were different high Z elements: samarium (Z=62), gadolinium (Z=64), terbium (Z=65) and holmium (Z=67). The plasma was created by one laser beam focused to an intensity of 5.10^{14}-10^{15} W/cm^2 with a pulse duration of 500 ps and a wavelength 0.53 μm. The focal spot, which was 70 μm in diameter, determines the spectral resolution. The time-integrated spectra were recorded on SB X-ray film using simultaneously two Bragg crystal spectrographs. One spectrograph used a flat pentaerythritol (PET) crystal (2d=8.742Å) to cover the energy range 3.2-5.9 Å with about 5 mÅ spectral resolution and the second used a flat ammonium dihydrogen phosphate (ADP) crystal (2d=10.64Å) to cover the neighboring energy range 5.5-9.2Å with 4-7 mÅ spectral resolution. The line of sight of the two spectrometers were 45° and 30° respectively measured from the target surface.

The targets consisted of a wire coated with the elements or a powder. For each different element, we added two different tracers on the targets (MgO or $AlCl_3$) to obtain the wavelength calibration. In Table 1, the wavelengths of nickel-like transitions measured in the four elements have been collected. The linewidth is usually broad for the tracer elements and this generally limits the accuracy of the measurements.

TABLE 1. Measured Ni-like resonance transition wavelengths (in Å). Code for transitions An: $3d^{10} - 3d^9nf$ (3/2,5/2), Bn: $3d^{10} - 3d^9nf$ (5/2,7/2), Cn: $3p^63d^{10} - 3p^53d^{10}nd$ (1/2,3/2) Dn: $3p^63d^{10} - 3p^53d^{10}nd$ (3/2,5/2), E4: $3s^23p^63d^{10} - 3s3p^63d^{10}4p$ (1/2,3/2).

Sm XXXV (±0.004Å)

A5	7.247	Blend	C4	7.708	
B5	7.358	Blend	D4	8.326	
E4	7.391				

Gd XXXVII (±0.003Å)

C5	5.438		
D5	5.806	B5	6.676
A6	5.841	E4	6.752
B6	5.928	C4	7.035
A5	6.573	D4	7.638
		A4	8.533
		B4	8.771

Tb XXXVIII (±0.004Å)

A5	6.279		
B5	6.387	C4	6.736
E4	6.465	D4	7.325

Ho XL (±0.005Å)

C4	6.177		
D4	6.753	A4	7.490
		B4	7.706

2. NICKEL-LIKE IONS

The isoelectronic sequence of nickel is the simplest one for studying the mixings which also affect more complex configurations of the neighbouring isoelectronic sequences. In (Tragin *et al.* 1988), the odd J=1 levels of the sequence Ta^{45+}-Pb^{54+} had been calculated in 3 separate groups located in different energy ranges. These groups overlap in the case of ions of lower charge and this led us to introduce 19 non-relativistic configurations for describing the crossings which perturb the strongest transitions when going from moderately- to highly-charged ions. These configurations are $3s^23p^63d^9np$ and -nf (n=4-8), $3s^23p^53d^{10}ns$ and -nd and $3s3p^63d^{10}np$ (n=4-6). The relativistic parametric potential method (Koenig 1972, Klapisch *et al.* 1977) was used for calculating the wavelengths and transition probabilities (length form) of 51 electric dipole transitions in 50 ions. The ground state was assumed to be pure $3d^{10}$, but the mixing of the 51 odd relativistic configurations was determined at the first order of perturbation. The results for Yb XLIII are given in Table 2 as an example. The isoelectronic consistency of the energies obtained in independent calculations has been checked. Energies given by adjustable polynomials with 6 coefficients agree with the RELAC raw data to 5.10^{-5} of the energies, in absence of close perturbers.

All $3d^{10}$-$3d^9nf$ $(j_1,j_2)1$ oscillator strengths are characterized by smooth dependences on ionic charge which can be broken dramatically when crossing $3p^5ns$,-nd and $3s^14p$ configurations. In figure 1, several destructive interferences are noticed on the 3d-6f most intense transitions. By this way, the present work is a step in the future study of enhancement or extinction of unresolved transition arrays. Configuration mixing is not only related with the crossing of isoelectronic sequences of levels. It has been shown, by applying the RELAC code to various sets of interacting configurations, that the gf-values of $3d^{10}$-$3d^94f$ are lowered (by about 20%) when upper $3d^9nf$ configurations are taken into account. This fact had been noticed also in the MCHF study of $3p^5$-$3p^4nd$ transitions in Ar II (Smid and Hansen 1981). Recent theoretical studies have been compared with our results: the MCDF method was used on a restricted basis of 4 odd configurations (Quinet and Biémont, 1991) and the neglected mixings with $3d^95f$ and -5p led to gf values (3p-4d transition) which are 2.4 times too large in Ag XX and 1.33 too small in Cs XXVIII. The present work is in better agreement with Zhang *et al.* (1991) who considered all configurations of the 3^174^1 and 3^175^1 odd complexes in a DHF approach.

TABLE 2. Resonance transitions in Yb XLIII. Theoretical wavelengths λ are in Å, transition probabilities gA are in s^{-1} (.124+14 means 0.124×10^{14}). The squared amplitude of the leading component is given under 'purity', the eigenfunction being normalized to one. If purity is lower than .99, a comment is given in notes.

λ	gA	Configuration	Purity	Note
8.38183	.124+14	3d^{9}4p (3/2,1/2)	.9902	
8.23816	.256+14	3d^{9}4p (5/2,3/2)	.9844	a
7.97895	.299+13	3d^{9}4p (3/2,3/2)	.9923	
7.08694	.471+14	3p^{5}4s (3/2,1/2)	.9961	
6.91698	.134+13	3d^{9}4f (5/2,5/2)	.8946	a
6.82458	.217+15	3d^{9}4f (5/2,7/2)	.7340	a
6.61855	.880+15	3d^{9}4f (3/2,5/2)	.8270	a
6.25757	.978+13	3p^{5}4s (1/2,1/2)	.9922	
6.05932	.708+13	3p^{5}4d (3/2,3/2)	.9608	a
6.00402	.211+15	3p^{5}4d (3/2,5/2)	.9647	a
5.48383	.722+13	3d^{9}5p (5/2,3/2)	.9952	
5.44888	.918+13	3d^{9}5p (5/2,5/2)	.7711	b
5.44356	.128+15	3p^{5}4d (1/2,3/2)	.5221	b
5.40053	.500+12	3s^{1}4p (1/2,1/2)	.6570	b
5.36423	.301+13	3d^{9}5p (3/2,3/2)	.9932	
5.23479	.497+14	3s^{1}4p (1/2,3/2)	.9938	
5.17875	.577+12	3d^{9}5f (5/2,5/2)	.9236	a
5.15035	.209+15	3d^{9}5f (5/2,7/2)	.9568	a
5.04940	.319+15	3d^{9}5f (3/2,5/2)	.8821	a
4.78786	.163+14	3p^{5}5s (3/2,1/2)	.9968	
4.68579	.298+13	3d^{9}6p (5/2,3/2)	.9969	
4.63176	.341+13	3d^{9}6p (3/2,1/2)	.9978	
4.59835	.941+12	3d^{9}6p (3/2,3/2)	.9973	
4.56001	.157+12	3d^{9}6f (5/2,5/2)	.9267	a
4.54725	.141+15	3d^{9}6f (5/2,7/2)	.9137	a
4.53845	.647+13	3p^{5}5d (3/2,3/2)	.9604	a
4.52388	.131+15	3p^{5}5d (3/2,5/2)	.9690	a
4.46817	.161+15	3d^{9}6f (3/2,5/2)	.9857	a
4.39105	.397+13	3p^{5}5s (1/2,1/2)	.9972	
4.32153	.210+13	3d^{9}7p (5/2,3/2)	.9990	
4.26396	.260+13	3d^{9}7p (3/2,1/2)	.9973	

λ	gA	Configuration	Purity	Note
4.25532	.764+11	3d^{9}7f (5/2,5/2)	.9392	a
4.24844	.102+15	3d^{9}7f (5/2,7/2)	.8976	ac
4.24654	.301+13	3d^{9}7p (3/2,3/2)	.9616	
4.17868	.376+13	3d^{9}7f (3/2,5/2)	.5822	c
4.17799	.150+15	3p^{5}5d (1/2,3/2)	.5839	
4.12159	.878+13	3p^{5}6s (3/2,1/2)	.6537	c
4.11785	.755+12	3d^{9}8p (5/2,3/2)	.6548	
4.07950	.615+11	3d^{9}8f (5/2,5/2)	.9415	a
4.07515	.791+14	3d^{9}8f (5/2,7/2)	.9368	a
4.06127	.506+12	3d^{9}8p (3/2,1/2)	.9980	
4.05096	.104+12	3d^{9}8p (3/2,3/2)	.9981	
4.02045	.168+14	3s^{1}5p (1/2,1/2)	.6996	c
4.01292	.159+13	3p^{5}6d (3/2,3/2)	.6831	
4.01029	.101+14	3d^{9}8f (3/2,5/2)	.8326	c
4.00858	.141+15	3p^{5}6d (3/2,5/2)	.8029	
3.97059	.158+14	3s^{1}5p (1/2,3/2)	.9909	
3.82254	.255+13	3p^{5}6s (1/2,1/2)	.9993	
3.73088	.391+14	3p^{5}6d (1/2,3/2)	.9994	
3.55647	.114+14	3s^{1}6p (1/2,1/2)	.9992	
3.53588	.159+14	3s^{1}6p (1/2,3/2)	.9994	

Notes

a : the breakdown of pure jj coupling is due to mixing within the same non-relativistic configuration

b : the breakdown of pure jj coupling is due to mixing with two other non-relativistic configurations.

c : the breakdown of pure jj coupling is due to mixing with the next upper level.

b4 : crossing with 3p^{5}3d^{10}4d
c4 : crossing with 3s3p^{6}3d^{10}4p
a5 : crossing with 3p^{5}3d^{10}5s
b5 : crossing with 3p^{5}3d^{10}5d
d6 : crossing with 3d^{9}6p

Fig. 1. Oscillator strengths of the strongest 3d-6f transitions determined by means of the relativistic parametric potential method from Br VIII to U LXV.

3. UNRESOLVED TRANSITION ARRAYS $3d^N$-$3d^{N-1}4f$

The average wavelength and full width at half maximum (FWHM) of unresolved transition arrays have been calculated according to (Bauche *et al.* 1987). The comparison between theory and experiment is given in Table 3. The effect of the asymmetry (third order momentum of the weighted wavenumber distribution) is larger than the experimental accuracy of the wavelength (0.02-0.03Å). Two experimental FWHM values are given. The largest one assumes that there is no background emission whereas the smallest one assumes that the UTA's are located on top of a continuum. The theoretical FWHM lies in between.

TABLE 3. Average wavelengths and widths of unresolved transition arrays $3d^N$ - $3d^{N-1}4f$. All values are in Å.

TRANSITIONS		Samarium		Gadolinium		Terbium		Holmium	
		λ	FWHM	λ	FWHM	λ	FWHM	λ	FWHM
[Co] $3d^9$ - $3d^8 4f$	Th	9.09	0.16	8.30	0.16	7.95	0.16	7.31	0.16
80 lines	Exp			8.34	0.11-0.17	7.97	0.13-0.20	7.34	0.13-0.20
[Fe] $3d^8$ - $3d^7 4f$	Th	8.85	0.18	8.10	0.17	7.76	0.17	7.14	0.16
721 lines	Exp	8.89	0.14-0.22	8.12	0.13-0.18	7.80	0.13-0.20	7.16	0.14-0.20
[Mn] $3d^7$ - $3d^6 4f$	Th	8.63	0.19	7.91	0.18	7.58	0.18	6.99	0.17
2860 lines	Exp	8.63	0.12-0.20	7.92	0.13-0.18	7.58	0.11-0.22	7.01	0.10-0.20
[Cr] $3d^6$ - $3d^5 4f$	Th	8.42	0.19	7.72	0.19	7.41	0.18	6.84	0.18
5540 lines	Exp	8.42	0.12-0.20	7.72	0.11-0.19	7.40	0.10-0.22		
[V] $3d^5$ - $3d^4 4f$	Th	8.22	0.20	7.55	0.19	7.25	0.19		
5540 lines	Exp	8.20	Blend	7.52	Blend	7.24	Blend		

REFERENCES

Bauche, J., Bauche-Arnoult, C. and Klapisch, M., 1987, Adv. At. Molec. Phys. **23** 131.

Burkhalter, P.G., Nagel, D.J. and Whitlock, R.R., 1974, Phys. Rev. **A9**, 2331.

von Goeler, S., Beiersdorfer, P., Bitter, M., Bell, R., Hill, K., LaSalle, P., Ratzan, L., Stevens, J., Timberlake, J., Maxon, S. and Scofield, 1988, J., J. Physique Coll. C1, **49**, 188.

Klapisch, M.,Schwob, J.L.,Fraenkel, B.S and Oreg, 1977, J. J. Opt. Soc. Am. **67**, 148.

Koenig, E., 1972, Physica **62**,393.

MacGowan, B.J. et al. 1987, Phys. Rev. Letters **59**, 2157.

MacGowan, B.J. et al. 1990, Phys. Rev. Letters **65**, 420.

Mandelbaum, P., Seely, J.F., Brown, C.M., Kania, D.R., Kauffman, R.L.,1991 Phys. Rev. **A44**, 5752

Nilsen, J., 1991, Physica Scripta **43**, 596.

Quinet, P. and Biémont, E., 1991, Physica Scripta **43**, 150.

Smid, H. and Hansen, J.E., 1981, J. Phys. **B 14**, L811.

Tragin, N., Geindre, J.-P., Monier, P., Gauthier, J.-C., Chenais- Popovics, C., Wyart, J.-F. and Bauche-Arnoult, C., 1988, Physica Scripta **37**, 72.

Tragin, N., Geindre, J.-P., Monier, P., Gauthier, J.-C., Chenais-Popovics, C., Luc-Koenig, E. and Wyart, J.-F., 1989, Phys. Rev **A39**, 2085.

Zhang, H.L., Sampson, D.H. and Fontes, C.J., 1991, At. Data Nucl. Data Tab. **48**, 91.

Zigler, A., Givron, M., Yarkoni, E., Kishinevsky, M., Goldberg, E., Arad, B. and Klapisch, M., 1987, Phys. Rev. **A35**, 280.

MODEL POTENTIAL METHOD FOR THE CALCULATION OF THE ATOMIC CHARACTERISTICS OF NE-LIKE SYSTEM

U.I.SAFRONOVA[1] , and J.-F.WYART[2]

[1] Institute of Spectroscopy,Russian Academy of Science,Russia
[2] Lab.Aime Cotton,C.N.R.S.11,Bat.505.,Centre Univ.,France

1.Introduction

The Ne-isoelectronic sequence has been studied rather in detail, especially in last years and this is connected with the problem of X-ray lasers. Energy levels, transitions probabilities, oscillator strengths were calculated by very different methods both for 2-2 and 3-3 transitions. In the presentwork we analyze only such contributions in which calculations were carried out for large interval of Z: Z-expansion method- MZ (Vainshtein and Safronova 1977), Hartree-Fock-Pauli method-HFP (Zhang et al.1987), Dirac-Fock method-MCDF (Cogordan and Lunell 1986), model potential method - MP (Ivanova and Glushkov 1986, Gulov and Ivanova 1991).

Energy levels $1s^2 2s^2 2p^5 3l$ and $1s^2 2s2p^6 3l$ were calculated on the basis of another variant of model potential method. Algorithm and program of the method were given in papers (Klapisch 1971, Luc-Koenig 1971, Klapisch et al.1977) The calculations of all characteristics (energy levels,transition probabilities,wavelengths, oscillator strengths) were carried out in high interval of Z=18-92. The following ions have been chosen for comparison with experimental data (Z=18,30,34,36,39,47,49,54,60,79,83,92). Transitions 2-3, 2-3,3-3 were analyzed. We present only part of the results for avoiding excessive tabulations. We pay main attention to the comparison of our results with other theoretical results and experimental data.

2. Energy Levels. The Results of Calculations

As mentioned above, the calculations of energy levels made on the basis of the RELAC program were carried out for the ions of the isoelectronic sequence in the interval Z=18-92. Here, present only a part of the results to illustrate our conclusions.

All the results are given in the intermediate coupling scheme but, as usual, designations of the levels are given in pure coupling.In the MCDF, RELAC,MP methods the jj coupling is used,in the HFP,MZ methods the LS coupling is used In this article we use designations in LS and jj couplings.

Table I gives a comparison with other calculations and experimental data (Beirsorfer 1986,1988).The main source of error in the MCDF method is the effect of correlation. The accuracy of the MP method depends on new experimental data. As seen from Table 1, the difference between calculations does not exceed 0.1%, that is 10^3-10^4 cm^{-1}. The discrepancies are similar for all levels. In our opinion, this is the estimation of accuracy for 4 theoretical methods and we hope that these data will be useful for new experiments.The agreement with experiment is better for RELAC program data.

TABLE 1.

Energy levels (cm^{-1}) Of Ne-like ions. Comparison of differentmethods:1-RELAC,2-MZ ,3-MP ,4-MCDF ,5-exp.

Level		Z=30	Z=34	Z=39	Z=47	Z=54
2p$_-$3p$_+$	1	8815340	11975727	16598223	25481147	34725219
^{3}D$_2$	2	8805162	11972123	16590894	25484039	34749666
	3	8814000	11976700	16597900	25498200	34735400
	4	8807292	11973525	16590053	25472957	34716700
	5				25478000	34724100
2p$_+$3p$_+$	1	8878030	12078975	16781126	25889077	35476419
^{3}P$_2$	2	8869213	12076926	16774540	25887871	35485349
	3	8876100	12079600	16780800	25907300	35493100
	4	8870346	12077551	16774170	25883323	35478861
	5				25889000	35478400
2p$_-$3p$_+$	1	9058442	12402659	17393235	27313328	38134416
^{1}D$_2$	2	9049728	12401377	17383862	27302100	38119724
	3	9057400	12405900	17394000	27333200	38153600
	4	9050381	12402136	17384681	27304117	38123971
	5				27314000	38137000
2p$_+$3s	1	8496242	11591181	16135351	24881748	33997625
^{1}P$_1$	2	8493137	11597146	16135533	24894797	34036225
	3	8496000	11595900	16136400	24901700	34012500
	4	8489240	11592983	16129045	24876331	33992560
	5				24884000	34000900
2p$_-$3s	1	8683110	11922148	16755133	26313887	36645691
^{3}P$_1$	2	8680765	11929830	16754105	26319478	36661350
	3	8683300	11929000	16757200	26335600	36667700
	4	8676142	11925270	16748626	26309042	36659632
	5				26314000	36645800
2p$_+$3d$_-$	1	9262549	12536574	17331007	26591822	36316264
^{3}P$_1$	2	9256979	12536477	17325504	26589465	36321592
	3	9264200	12540400	17334600	26614300	36338200
	4	9255377	12535792	17325472	26587872	36315357
	5				26597000	
2s3p$_+$	1	10264609	13825864	19067627	29435092	40679529
^{1}P$_1$	2	10232072	13786979	19030339	29387109	40614358
	3	10243700	13796700	19047800	29432300	40673600
	5				29408300	40649500
2p$_+$3s	1	8475800	11566782	16105947	24844433	33952902
^{3}P$_2$	2	8473771	11573721	16107122	24858215	33952902
	3	8476100	11572000	16107500	24864900	33968500
	4	8468371	11567823	16098620	24837519	33946206
	5				2484900	33957200

3. Transitions Probabilities. Ions with Z=30-39

The calculations of the probabilities of the transitions between excited states are more complicated than the calculations of the probabilities of the transitions from the excited states to the ground state. In the latter case we have only seven transitions for the Ne-like system (Aglitski et al.1989). The number of the transitions between excited states is higher (123 transitions of the type 3-3 and 72 transitions of the type 2-2). The systematization of these transitions was made in the work on the base of LS coupling (Pokleba and Safronova 1981,1982).

The ions with Z=30,34,36,39 are considered with more details, because it is possible to made a full comparison almost for all type of transitions(2s3d-2p3d,2s3d-2s3p,2s3p-2s3s, 2s3p-2p3p, 2p3d-2p3p, 2p3p-2p3s) for these ions on the basis of the data presented by (Pokleba and Safronova 1981).

Taking into account that the spread in values of the gW is very large from one to another transition it is considered that the agreement of the results of two calculations is rather good. Taking into account transpositions of levels the difference between the calculations do not exceed a factor 10. This difference is 20-50% in most of the cases. It is satisfactory for transitions probabilities without change of principal quantum number.

4. Transition probabilities. Ions with Z=47,49,54,60,79,83,92

The selection for these ions was driven by the present experimental investigations of transitions 3-2, but these first works will certainly be extended in future. We present transition probabilities for few transitions in order to investigate the Z-dependence of the gW.

TABLE 2. Z-dependence of gW in un. $10^8 s^{-1}$ (2pj 3pj'{1}-2p$_+$3s{2})

Z	2p$_-$3p$_+$	2p$_+$3p$_-$	2p$_-$3p$_-$	2p$_+$3p$_+$
18	11.6	81.5	1.09	25.3
30	17.2	126	1.49	1.21
34	19.8	153	2.19	1.47
36	26.3	188	3.01	6.64
39	36.7	216	4.11	22.4
47	122	292	9.90	138
49	174	312	12.5	195
54	471	366	22.7	433
60	1960	438	47.3	1050
79	28800	764	450	13600
83	18500	864	702	22700
92	12700	1140	1820	69600

Table 2 illustrates rather complicated dependence for this interval of Z. For example, the probability gW increases smoothly for transition 2p 3p {1}-2p 3s{2} in the interval Z=18-39, increases strongly in the interval Z=39-79 and decreases from Z=79 to Z=92. Z-dependencies of gW for these four transitions are very different. The transition 2p 3p {1} - 2p 3s{2} has the largest gW$^+$for small values of Z and the smallest value for Z=92.

5. Asymptotic Limit for the Line Strengths

The asymptotic limit for the line strengths of the 2-2 and 3-3 transitions was obtained and compared with the results of exact calculation by the program RELAC in the work (Safronova and Wyart 1992).The differences don't exceed 20%, except for $2s_{1/2}$ $-2p_{1/2}$ transitions which eventually disagree by factors as 2-3.Therefore, we can predict the values of the line strengths for the high Z rather successfully on the basis of analytical formula.

As we mentioned above the transitions with two electron jumps take place also, therefore all intermediate cases are possible (Safronova and Senashenko 1984).In Table 2 three variant are given. The transition 2p 3p {1} - 2p 3s{2} is a transition with changing j and display a strong dependence on Z ($W \cong Z^{10}$). The transition 2p 3p {1} - 2p 3s{2} is two electron jump but without changing of j and Z-dependence is very weak ($W \cong Z$).It is confirmed by gW presented in Table 2. There is changing of two quantum numbers in two other transitions which leads to irregular Z-dependencies of W. It should be mentioned that gW is minimum for the transitions with $W \cong Z$ and the differences between W values range from 10^3 to 10^5 for these two types of transitions for Z=92.The largest disagreements between gW and gW(asympt), calculated in pure jj coupling occur for small gW values and the importance of numerical accuracies might be the reason for this.

Acknowledgements. One of us (J.-F.Wyart) is grateful to E.Luc-Koenig and M.Klapisch for the possibility of using the unpublished computer code RELAC.

REFERENCES

Aglitskii,E.V.,*et al*. 1989, *Phys.Scripta*, **40**, 60.
Beirsdorfer,P.,*et al*.1986, *Phys.Rev.*, **A34**, 1297.
Beirsdorfer,P.,*et al*.1988, *Phys.Rev.*, **A37**, 4153.
Cogordan,J.A. and Lunell,S. 1986, *Phys.Scripta*,**33**, 406.
Gulov,A.V.,and Ivanova,E.P. 1991, *A.D.N.D.T.*, **49**, 1.
Ivanova,E.P.,and Glushkov,A.V. 1986, *J.Q.S.R.T.*, **36**, 127.
Klapisch,M. 1971, *Comp.Phys.Communications*, **2**, 239.
Klapisch,M., Schwob,J.L., Fraenkel,B.S.,and Oreg,J. 1977, *J.O.S.A.*,**67**, 148.
Luc-Koenig,E. 1971, *Physica*, **62**, 393.
Pokleba,A.K.,and Safronova,U.I. 1981, "preprint of ISAN".
Pokleba,A.K.,and Safronova,U.I. 1982,*Optic and Spect.*,**53**,12.
Safronova,U.I. and Senashenko,V.S. 1984, *Theory of spectra of multicharged ions* (Moscow: Energoatomizdat).
Safronova,U.I.,and Wyart,J.-F. 1992,*Phys.Scripta* "in press".
Vainshtein,L.A.,and Safronova,U.I. 1977, in *Spectroscopical constants for atoms*,ed.U.I.Safronova (Moscow: USSR Academy of Sciences), p.3.
Zhang,H.,Sampson,D.H.,Clark,R.E.H.,and Mann,J.B. 1987, *A.D.N.D.T.*, **37**, 17.

Laboratory Absorption Spectra of Interstellar Molecules: Measurements on CO and N_2 at ~30 and 295 K.

G. Stark[1], Peter L. Smith[2], K. Yoshino[2], W. H. Parkinson[2], K. Ito[3], and M. H. Stevens[4]

[1] Physics Department, Wellesley College, Wellesley, MA 02173
[2] Harvard-Smithsonian Center for Astrophysics, 60 Garden St., Cambridge, MA 02138
[3] Photon Factory, National Laboratory for High Energy Physics, Tsukuba, Ibaraki 305, Japan
[4] Department of Earth & Planetary Science, The Johns Hopkins University, Baltimore, MD 21218

ABSTRACT

Accurate atomic and molecular spectroscopic data are required for analysis of astronomical spectra and for modelling properties of and processes in many astronomical objects. If there are uncertainties in the spectroscopic data, there will be uncertainties in the astronomical studies. In order to improve the extreme ultraviolet spectroscopic database for CO and N_2, we have used high resolution spectrometers ($R \approx 170,000$) to measure wavelengths and absorption cross sections. In order to simplify spectral regions with overlapping bands at room temperature, we have used a supersonic expansion technique to cool these molecules to ~30 K and have studied absorption by most of the bands in the 91 - 100 nm wavelength region. These observations complement our measurements of the room temperature absorption cross sections of CO and N_2.

INTRODUCTION

Modelling of the complex photochemical processes in astronomical objects, such as interstellar clouds, circumstellar shells, cometary comae, and planetary atmospheres, requires accurate spectroscopic data – line wavelengths, line widths and profiles, and absorption cross sections or oscillator strengths (f-values) – for the atomic and molecular species involved. If there are uncertainties in the laboratory data, there will be concomitant uncertainties in atomic and molecular abundances and reaction rates, and, thus, in our knowledge of properties and processes in astronomical objects.

In the 91.2 - 100 nm spectral region, where absorption of photons by molecules often leads to dissociation, laboratory spectra for many molecules at room temperature show bands that are both overlapped and perturbed. Reliable band oscillator strengths cannot be extracted from such spectra. Consequently, synthetic absorption spectra and models of photon-induced processes for molecules at other temperatures, such as those of interstellar clouds and circumstellar envelopes, can be uncertain.

The difficulties are compounded if the measured f-values or cross sections must be determined from low-resolution spectra.

In order to provide the high quality data needed for astronomy, we have developed a laboratory program for high-resolution measurements of vacuum ultraviolet (VUV) and extreme ultraviolet (EUV) wavelengths and absorption cross sections. Some results are reported here.

MEASUREMENTS

Our wavelength measurements are made with the 6.65-m spectrometer at the Center for Astrophysics (Yoshino et al. 1980). The photoabsorption cross section meausurements were performed with the the 6.65-m spectrometer at the Photon Factory, where synchrotron radiation provides a background continuum (Ito et al. 1986). Both spectrometers have demonstrated resolving powers ($\lambda/\Delta\lambda$) of about 175,000, far greater than that used in most previous measurements of EUV absorption cross sections for molecules of astromonical interest.

In order to reduce the overlapping of the molecular bands seen in the laboratory spectra, we have used a supersonic expansion technique (Levy 1984, Vaida 1986) to cool molecules to $\sim$30 K. We have studied absorption by most of the bands of CO and N_2 in the 91 - 100 nm wavelength region. The intensity distributions within individual bands show that we have achieved a rotational temperature of about 30 K. These observations complement our measurements of the room temperature absorption cross sections of CO (Stark et al. 1991) and N_2 (Stark et al. 1992).

RESULTS AND COMPARISON WITH OTHER WORK

CO: Cross Section Measurements at 30 K

Comparison of our 295 K (Stark et al. 1991) and 30 K EUV spectra for CO (Smith et al. 1991) shows overlapping bands at 295 K; determination of reliable oscillator strengths for individual bands from such spectra is problematic. The difficulties are compounded if the line widths are not known, as is the case for absorption lines to the predissociating levels of CO considered here.

A comparison of the differences between our measured absorption spectrum for CO at 30 K and the modelled low-temperature "synthetic spectra" published by Eidelsberg & Rostas (1990) led to the realization that the latter was created with incorrect band intensities and line widths. The ratios of integrated cross sections are shown in Table 1. The significance of the low temperature cross section measurements for interstellar and circumstellar chemistry remains to be investigated.

	this work	Eidelsberg & Rostas
L(0)-X(0)	1.00 .	1.00 .
L'(1)-X(0)	0.7 ± 0.3	1.6 ± 0.2
K(0)-X(0)	2.0 ± 0.5	2.7 ± 0.3

TABLE 1: Comparison of Integrated Cross Sections at 30 Ǩ

CO: Errors in Literature Wavelength Data For CO

In the course of analyzing our room temperature spectral data, we discovered a systematic wavelength error of about 10 mÅ in the new atlas of EUV wavelengths for CO by Eidelsberg et al. (1991). Smith et al. (1992) discuss this error, much larger than the line widths, and also point out that, because of perturbation of the upper energy levels of CO, many of the calculated wavelengths in the atlas of Eidelsberg et al. (1991) are unreliable.

New Cross Section Data for N_2 at 295 K

EUV cross sections of N_2 are needed for analysis of observations of the atmospheres of Triton and Titan obtained during flybys by the Voyager II satellite , and for studies of terrestrial airglow. The literature data available prior to our measurements were obtained with resolving powers of $\sim$30,000 at most, which is far less than required to resolve the line shapes. When cross sections are measured with instrumental resolution that is less than the line widths, significant distortions will occur; in particular, peak cross sections are systematically underestimated and line wing cross sections are systematically overestimated. In order to evaluate the literature data, we remeasured the cross sections of N_2 at 295 and 30 K.

Our room temperature cross section results for N_2, which were measured with a resolution of about 6 mÅ, are very different from those of Gürtler et al. (1977), who had a resolution of about 30 mÅ. The qualitative differences can be explained by the difference between the resolving powers used. Our data (Stark et al. 1992) tend to support some of the measurements of N_2 cross sections that preceded the work of Gürtler et al.

This project was supported in part by NASA Grant NAGW-1596, by the Japan Society for the Promotion of Science, and by the Robert Balk Fund. The measurements were made with the approval of the Photon Factory Advisory Committee (Proposal No. 90-149).

REFERENCES

Eidelsberg, M., J. J. Benayoun, Y. Viala & F. Rostas 1991, *Astr. Ap. Suppl.*, **90** 231.

Eidelsberg, M. & F. Rostas 1990, *Astr. Ap.*, **235**, 472.

Gürtler, P., V. Saile & E. E. Koch 1977, *Chem. Phys. (Letters)*, **48**, 245.

Ito, K., T. Namioka, Y. Morioka, Y. Sasaki, H. Noda, K. Goto, T. Katayama M. Koike 1986, *Appl. Opt.*, **25**, *837*.

Levy, D. 1984, *Scientific American*, **250**, (Feb.), 96.

Smith, P., G. Stark, K. Yoshino, K. Ito & M. H. Stevens 1991, *Astr. Ap.*, **252**, L13

Smith, P., K. Yoshino, G. Stark & C. A. Shettle 1992, *Astr. Ap.*, submitted.

Stark, G., K. Yoshino, P. L. Smith, K. Ito & W. H. Parkinson 1991. *Ap. J.*, **369** 574.

Stark, G., P. L. Smith, K. P. Huber, K. Yoshino, M. H. Stevens, & K. Ito 1992 *J. Chem. Phys.*, submitted.

Vaida, V. 1986. *Acc. Chem. Res.*, **19**, 114.

Yoshino, K, D. E. Freeman & W. W. Parkinson 1980. *Appl. Opt.*, **19**, 66.

The Effect of Fast Electrons, Penning and CX Processes on the XUV to VUV Emission of Al, Y and Zn Ions from a PID Plasma

D. Stutman[1], M. Finkenthal[1], A.K. Bathia[2], J.L. Schwob[1], S. Regan[3], M. May[3] and H.W. Moos[3]

[1]Racah Institute of Physics, The Hebrew University of Jerusalem, Israel

[2]Goddard Space Flight Center, Greenbelt, MD 20771

[3]Department of Physics and Astronomy, The Johns Hopkins University, Baltimore, MD 21218

ABSTRACT

Spectra of one to three times ionized Al, Zn and Y have been recorded in the 100-3000 Å range from Penning Ionization Discharges (PID) with open and closed anode configuration. By varying at constant current the gas pressure from a few mtorr to several torr we observed the fast electron produced Y IV and Al III spectra at 100-350 Å, the influence of Penning ionization on the spectra of Al II and the charge exchange produced Zn II emission between 800-2200 Å.

1 EXPERIMENT

The plasma sources used in the experiments have been described elsewhere (Finkenthal et al. 1991).

The spectra have been recorded simultaneously by two types of instruments: A 2 m grazing incidence Schwob-Fraenkel spectrograph (Filler, Schwob and Fraenkel 1977) equipped with interchangeable 600 and 133 lines/mm gratings, which covered the ranges 50-400Å, and 100-1800Å respectively. Also, a photometrically calibrated, 1 m time resolved-multispectral grazing incidence spectrometer, GRITS (Hodge, Stratton and Moos 1984) has been used in the 100-300Å range. At longer wavelengths, 0.2m VUV Minuteman scanning monochromators, equipped with 3600 lines/mm, MgF_2 coated and aberration corrected holograpic gratings, covered the range 300-4000Å. The detectors were p-terphenyl or sodium salicylate coated (1 mg/cm^2), high sensitivity photomultipliers.

The sensitivity curve of the monochromator used on the experiments at Hebrew University was determined in the 300-3000Å range using branching ratio lines.

2 RESULTS

2.1 The Low Pressure Regime

In our previous work, on Al at low pressure (1-10mtorr), we pointed out the role of fast electrons in producing large populations of the auto-ionizing states, $2p^53131'$ (1=s,p) of Al III

(M.Finkenthal et al. 1989). Here we present evidence of fast electrons (E~100eV) obtained from the enhancement of the $4p^6 \, {}^1S_0$-$4p^5 \, 4d \, {}^1P_1$ transition in the Y IV spectrum.

The space resolved measurements of Y and Ne ions emission showed that the Y IV 4d levels are populated by excitation from the ground state $4p^6 \, {}^1S_0$ of the ion and not by inner-shell ionization of either Y II or Y III.

The temperature and density of these slow electrons have been estimated at about 3eV and respectively 2×10^{13} cm^{-3}, using the predictions of collisional-radiative models for the line intensity ratios of Al II (Finkenthal et al. 1990) and Al III (Finkenthal et al. 1989) ions.

In Fig. 1, we present the predictions for the $4p^6 \, {}^1S_0$ - $4p^5 \, 4d \, {}^1P_1$ and $4p^6 \, {}^1S_0$ - $4p^5 \, 4d \, {}^3P_1$ line ratio, as obtained from a collisional radiative model for the Y IV ion, including the ground $4p^6$ and the excited $4p^5 4d$ levels. The level populations have been computed at low temperatures (1-5 eV) and assuming a variable fraction of 40 eV 'fast' electrons. At the temperature and density above estimated for our PID, the ratio predicted in the absence of fast electrons is by a factor of more than 10 smaller than the experimental one. The observed ratio is consistent with a fast (above 40 eV) electron fraction of at least 1%.

This is consistent with the measurements of the absolute intensity of the emission from Al III autoionizing levels (Finkenthal et al. 1991); together with the population mechanism we proposed earlier for these levels (Finkenthal et al. 1989), we arrive at a fast electron density above 10^{12} cm^{-3}, i.e. at least 5% of the total electron density.

2.2 The Intermediate Pressure Regime

In the intermediate pressure range (several hundred mtorrs), significant enhancements of the Al II ion transitions with upper levels 1-2 eV below the threshold for Penning reactions with metastable He were observed: for instance, the $3s3p \, {}^3P_2$-$3s3d \, {}^3D_3$/$3s^2 \, {}^1S_0$ - $3s3p \, {}^1P_1$ ratio approaches unity (Fig. 2). Our previous computations for the Al II ion (Finkenthal et al. 1990) predicted at the electron temperature of less than 1 eV, which was diagnosed in the discharge in this range of pressures, a ratio of maximum 1/10. We explain the observed ratio, mainly by the reaction:

$$Al \, I + (He)^* \rightarrow (Al \, II \, 3s3d \, {}^3D) + He + e + 2 \, eV$$

and partly by increased self-absorption of the singlet transition. The anomalous intensity of Al III emission at this low temperature is partially attributed also to the Penning ionization of Al II by He metastables (energy defect 1 eV).

2.3 The High Pressure Regime

In the high pressure regime (several torr), the dominant ionization mechanism observed for the

Zn atom is resonant charge exchange mainly into the 5d levels of Zn II, but also into $3d^9 4s^2$ levels. Using the efficiency calibration of the monochromator, we could measure that the Zn II spectrum is in proportion of about 80-90% excited by CX into the above levels, evidencing a strong population inversion on the 4p-5d (VUV) and 4d-4f (visible) transitions of the ion.

At the same time, a drastic reduction in the intensity of the transitions of Al II populated by Penning processes can be observed in the spectrum of the high pressure regime; this reduction cannot be entirely accounted for only by self-absorbtion The spectrum of a Zn-Al-He discharge in this regime is shown in Fig. 3.

We advance the explanation that, in our discharge - as well as in the active medium of CX based rare gas/metal vapor visible lasers - the rare gas metastables mediate the formation of the CX dominated spectrum. A population of rare gas ions, in excess of that reflecting the electronic temperature, is created through non-thermal electron ionization of the He metastables, which otherwise would tend to accumulate at high pressures. This process is favored in comparison to direction ionization: it has been shown that in He glow discharges in this pressure range, the most probable energy of the non-thermal electrons drops below the energy required to ionize the He atom from its ground state (Gill and Webb 1979).

3 CONCLUSIONS

We found that from the point of view of the emitted spectra and of the relevant excitation/ionization mechanisms, the pressure range we spanned can be divided roughly in the following domains:

a)   from a few mtorr to a few tens of mtorr: the spectra of certain ions evidence a significant fast electron component in the electron energy distribution.

b)   the range from tens to hundreds of mtorr: the bulk electron temperature and density decrease and Penning processes become an important part of the ionization/excitation mechanisms.

c)   above a few torr: the dominant ionization mechanism is resonant charge-exchange; thus, population inversion may occur on certain VUV transitions.

4 ACKNOWLEDGEMENTS

This work was funded by a grant of the Basic Research Fund of the R & D Authority of the Hebrew University and partly supported by a subcontract with the Johns Hopkins University - DOE Grant No. DEFGO 1-86ER53214.

5 REFERENCES

Filler, A., Schwob, J.L. and Fraenkel, B.S. 1977, in Proc. 5th Intl. Conf. on VUV Rad. Phys., eds. M.C. Castex, M. Pouey and N. Pouey (Paris: Centre National de la Recherche

Scientifique), p. 86.

Finkenthal, M., Littman, A., Stutman, D. and Bathia A.K., 1989, *J. Phys. B: At. Mol. Opt. Phys.*,
   22, p. L115

Finkenthal, M. *et al.* 1990, *Phys. Scr.*, **41**, 502.

Finkenthal, M. *et al.* 1991, in Proc. 8th APS Conf. on At. Processes in Plasmas.

Gill, P. and Webb, C.E. 1979, *J. Phys. D: Appl. Phys.*, **10**, 299.

Hodge, W.L., Stratton, B.C. and Moos, H.W. 1984, *Rev. Sci. Instrum.* **55**,16.

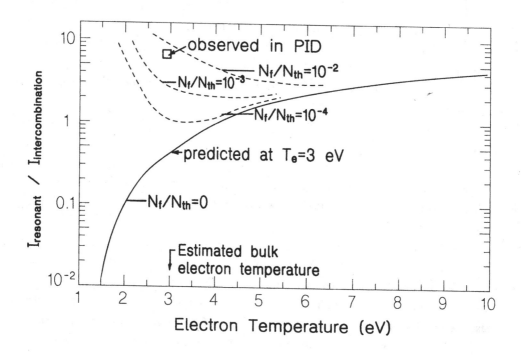

Fig. 1   Computed 4p-4d resonant intercombination line intensity ratio for the Kr-like YIV ion at an electron density of 2×10^{13} cm^{-3}.

Solid line: maxwellian distribution at T$_e$.

Dashed lines: maxwellian at T$_e$ + variable contribution (Nf/Nth) of fast (40 eV) electrons.

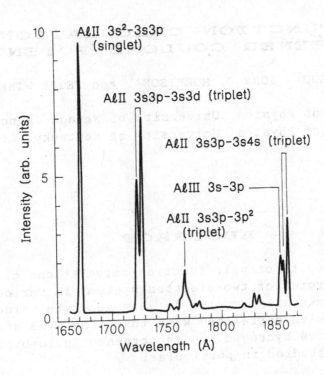

Fig. 2   Al-He spectrum between 1600 and 1900 Å in the intermediate pressure regime.

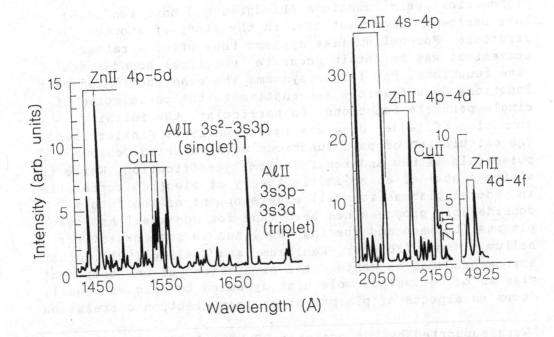

Fig. 3   Zn(Cu)-Al-He spectrum between 1400 and 2200 Å in the high pressure regime.

PAIR FUNCTION CALCULATIONS WITH SCREENED COULOMB POTENTIALS[']

ZHENGMING WANG[1], JOHN C. MORRISON[2], and PETER WINKLER[1]

[1] Department of Physics, University of Nevada, Reno.
[2] Department of Physics, University of Kentucky, Louisville.

ABSTRACT

The results of pair function calculations of ground state energies of two-electron systems in various plasma environments are compared to very accurate values obtained with correlated wave functions. The stability of the negative hydrogen ion in screened Coulomb potentials has been studied in particular.

1 INTRODUCTION

Numerical pair functions (Lindgren and Morrison 1986) have become an important tool in the study of atomic structure. For helium-like systems they offer a rather convenient way to obtain accurate, numerical bound-state wave functions. For larger systems the construction of pair functions is a feasible alternative to the correlation of single particle functions, in particular, the infinite summation of ladder diagrams (Morrison 1973; Winkler 1986). The calculation of pair functions with screened Coulomb potentials is new and requires some justification. While the ultimate goal of an *ab initio* theory of electron correlation in general plasmas is still a development of the future considerable progress has been made for non-ideal hydrogen plasmas (Lehmann and Ebeling 1991) and to some extent for helium plasmas (Förster, Kahlbaum, and Ebeling 1991). Other approaches such as Monte Carlo simulations of one component plasmas or equivalent molecular dynamics techniques usually focus on aspects of plasmas other than electron correlation

['] Work supported by DOE contract DE-FG08-90ER14160

so that their results are of limited use for the application
to atomic or ionic processes in plasmas. One valuable
exception are the very few existing treatments of two
component plasmas (Dharma-wardana and Perrot 1982). Such
calculations evaluate distribution functions for ion-ion
pairs as well as ion-electron pairs. These quantities are
related to averaged one-particle densities with exchange and
correlation effects included to some extent. Static
potentials derived from such quantities provide valuable
tools to include important plasma effects into atomic
structure and scattering calculations in zeroth order.

Here we have studied, in particular, the stability of the
negative hydrogen ion in various screening environments. We
proceed from the simple Debye screening model to more
realistic potentials obtained from pair distribution
functions of two-component plasma simulations. Reliability
and accuracy of the present approach are monitored by
comparison to highly accurate results obtained from
correlated wave functions.

2   THE ATOMIC PAIR FUNCTION

The theory of atomic pair functions will be outlined only
briefly. More relevant details can be found in a pioneering
publication by Mårtensson (1979). Although we developed our
computer code independently because of the modifications
required by the inclusion of plasma screening the
description of the finite difference procedure provided
there can serve as a guide also here.

For two-electron systems the equation for the pair
function ρ_{ab} is given by

$$[\epsilon_a + \epsilon_b - h_0(1) - h_0(2)]|\rho_{ab}> = \sum_{rs \notin D} |rs><rs|V|ab+\rho_{ab}> -$$

$$\sum_{cd \in D} \rho_{cd}<cd|V|ab+\rho_{ab}> \quad (1)$$

Here D is the model space spanned by a subset of
eigenfunctions of an approximate Hamiltonian consisting of a
sum of single-particle operators $h_0(i)$. We notice that the
pair function ρ_{ab} complements the product functions $|ab>$.
On the right is a functional of ρ but starting with a guess
we use the function obtained in the (n-1)st step as input
for the n-th iteration. The one-electron problem connected

with h_0 has to be solved first:

$$h_0(i) = -\nabla_i^2/2 - Z/r_i + u(r_i). \tag{2}$$

In order to include Debye screening u is chosen as

$$u(r) = -Z \, [\exp(-r/D) - 1]/r. \tag{3}$$

More realistic choices for u(r) have been studied also.

3 RESULTS

In Table 1 we present results of the present calculation for inreasing Debye screening (decreasing D) and compare the ground state energy of the negative hydrogen ion to the binding energy for theneutral atom in the same plasma environment. Both values decrease with increasing screening. As long, however, as the value in the last column is more negative than the corresponding entry in column 2 the negative ion is bound. We notice that at values of D = 34 or less the two-electron system ceases to be bound. What happens when we go beyond the bound regime? Does the system emit the second electron immediately into the continuum or do some traits of the combined system persist even though the energy coexists with the one electron continuum? Our studies are not yet at the stage to provide a quantitative answer. The fact that our calculations yield states that have distinct energies and electron densities not much different from the bound situations may be introduced by the chosen computational approaches. A similar question arises for heavily screened one-electron systems for which it could be answered in the affirmative (Wang, Winkler, Pickup, and Elander 1989). The results in the last column have been obtained from correlated wave functions in the generator coordinate representation described elsewhere (Winkler and Porter 1974) and have been included here only to assess the accuracy of the present calculations given in columns 3 and 4. One further point is worthwhile mentioning: While the results in the last column are obtained from a variational calculation the same is not true for the numerical pair function results which may happen to lie below the true energy values. In view of the demonstrated accuracy of the results this distinction is of little relevance for all practical purposes.

D	H-atom	H⁻-ion (N=45)	H⁻-ion (extrapolated)	H⁻-ion (GCM)
∞	-0.500000	-0.527613	-0.527322	-0.527750
1000	-0.499001	-0.525871	-0.525696	-0.525754
200	-0.495019	-0.517935	-0.517883	-0.517818
100	-0.490074	-0.508135	-0.508118	-0.508018
50	-0.480296	-0.488846	-0.488662	-0.488808
35	-0.472049	-0.472854	-0.472540	-0.472745
34.25	-0.471430	-0.471702	-0.471535	-0.471588
34	-0.471225	-0.471305	-0.471249	-0.471191
33.5	-0.470848	-0.470543	-0.470343	-0.470343
32	-0.469668	-0.468149	-0.467981	-0.468699
20	-0.451816	-0.434280	-0.434033	-0.442189

TABLE 1 The ground-state energy of the H⁻-ion (last column) for some values of the Debye parameter D (column 1) is compared to the ground state energy of the neutral atom (column 2). Columns 3 and 4 show the results of the present pair function calculation obtained on a grid of 45 by 45 points and the values of the Richardson extrapolation, respectively. In the pair function results s,p and d partial waves have been consistently included.

4 REFERENCES

Dharma-wardana M. W. C. and Perrot F. 1982, Phys. Rev. A 26, 2096.

Förster A., Kahlbaum T., and Ebeling W. 1991, in Proc. XX. Int. Conference on Phenomena in Ionized Gases (in press).

Lehmann H. and Ebeling W. 1991, Z. Naturforsch. 46a, 583.

Lindgren I. and Morrison J. 1986, Atomic Many-Body Theory (Heidelberg: Springer Verlag, 2nd edition).

Mårtensson A.-M. 1979, J. Phys. B.: Atom. Molec. Phys. 12, 3995.

Morrison J. 1973, J. Phys. B.: Atom. Molec. Phys. 6, 2205.

Wang Z., Winkler P., Pickup B. T., and Elander N. 1989, Chem. Phys. 135, 247.

Winkler P. 1986, Int. J. Quant. Chem. S 19, 201.

Winkler P. and Porter R. N. 1974, J. Chem. Phys. 62, 2038.

Study of Heliumlike Neon Using an Electron Beam Ion Trap

B. J. Wargelin[1], P. Beiersdorfer[2], and S. M. Kahn[1]

[1]Department of Physics and Space Sciences Laboratory, University of California, Berkeley, CA 94720

[2]Lawrence Livermore National Laboratory, University of California, Livermore, CA 94550

ABSTRACT

The 2-to-1 spectra of several astrophysically abundant He-like ions are being studied using the Electron Beam Ion Trap (EBIT) at Lawrence Livermore National Laboratory. Spectra are recorded for a broad range of plasma parameters, including electron density, energy, and ionization balance. We describe the experimental equipment and procedure and present some typical data.

1 INTRODUCTION

The 2-to-1 spectra of heliumlike ions (specifically the forbidden [z], intercombination [x and y], and resonance [w] lines, and also the lithiumlike satellite line [q]) are among the most useful diagnostics available for the study of X-ray emitting astrophysical and laboratory plasmas. As first pointed out by Gabriel and Jordan (1969), these lines may be used as sensitive indicators of electron density (using the line ratio $R=z/[x+y]$) and electron temperature (using $G=[x+y+z]/w$).

These diagnostics will be particularly important for interpreting the high resolution spectral data coming from future soft X-ray satellite missions such as the Advanced X-ray Astrophysics Facility (AXAF) and the X-ray Multi-Mirror Mission (XMM). Reliable interpretation of such data, however, will require a correspondingly complete and accurate understanding of the atomic processes involved. Several authors have constructed detailed computer models of He-like systems (e.g. Mewe and Schrijver 1978ab; Pradhan and Shull 1981, McKenzie and Landecker 1982), but the many cross sections and transition rates involved are typically accurate to only 10 or 20 percent, and significant errors may accumulate. Indeed, experimental studies on tokamaks (e.g. Bitter et al. 1979, Källne et al. 1983, Keenan et al. 1989) have pointed out some significant deficiencies in theoretical models. Further laboratory measurements, with detailed control over plasma parameters, are therefore necessary to compare with predictions and guide the development of theory.

We are using an Electron Beam Ion Trap (EBIT) to study X-ray emission from several He-like ions of astrophysical interest. Neon is the centerpiece of this investigation, since its critical density (where the line intensity ratio R is most sensitive to electron density) coincides with the practical density range of EBIT, roughly 5×10^{11} to 10^{13} cm^{-3}. In addition to Ne^{8+}, we are also studying O^{6+}, S^{14+}, and Fe^{24+}, measuring relative line intensities as functions of electron density, electron energy, and ionization balance.

2 EBIT

The Electron Beam Ion Trap (EBIT) at Lawrence Livermore National Lab is an electron beam ion source specifically designed to permit study of X-ray emission from highly charged ions. It is described in detail by Levine, *et al.* (1988), and many of its capabilities are discussed by Beiersdorfer *et al.* in these proceedings ("X-ray Spectroscopy with EBIT"). Ions are confined in an electrostatic "trap" where they may be further ionized and excited by a 70-μm-diameter, nearly monoenergetic electron beam. The beam energy can be quickly raised or lowered anywhere between 500 eV and 30 keV, allowing study of the plasma while ionizing, recombining, or in equilibrium. As previously mentioned, the electron density can also be varied, with a typical value of 5×10^{12} cm^{-3}. Six viewing ports ring the electron-ion interaction region, permitting several detectors and spectrometers to study the X-rays produced. Figure 1 shows the layout of the three instruments we use for our investigation of He-like neon.

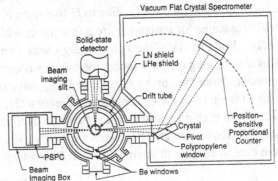

Figure 1. A schematic of the instruments used for study of He-like neon on EBIT, shown in cross section. Three of the six observation ports are used.

The primary instrument is the Flat Crystal Spectrometer (FCS), which was designed to work under vacuum because of the low energy of the X-rays we are studying (about 900 eV for He-like neon). Bragg angles up to 60° may be used, allowing the study of 2-to-1 lines in He-like oxygen (around 570 eV) with TlAP crystals ($2d = 25.76$ Å). Diffracted X-rays are detected by a one-dimensional position-sensitive proportional counter, which uses a thin window made of 4-μm-thick aluminized polypropylene. Both the crystal and detector may be repositioned while under vacuum. Spectral resolution depends upon which of the interchangeable diffraction crystals is being used, but typical values of $\lambda/\Delta\lambda$ are 500 for TlAP and 1300 for mica.

To permit the determination of electron density in EBIT, we have constructed a beam imaging system which uses the pinhole camera effect. A very narrow slit (7.8 μm) images the electron-ion interaction region onto a position-sensitive proportional counter with a magnification factor of 40. This provides the electron beam diameter, typically 70 μm. The electron density is then computed using the formula $n_e = I/(v_e A e)$, where I is the electron beam current, v_e is the electron velocity, A is the beam area, and e is the electron charge. The beam area changes very little with energy, so a constant electron density may be maintained by varying the beam current in proportion to electron velocity.

3 EXPERIMENTAL PROCEDURE

The power of EBIT lies in its ability to obtain X-ray spectra from highly charged ions under widely varying but precisely controlled conditions. A salient feature of EBIT is that spectra are accumulated as a function of electron energy rather than temperature. By making measurements of a cross section or transition rate at several electron energies, the net rate for that atomic process at any temperature can then be computed by weighting the individual rates according to the appropriate Maxwellian distribution.

To take full advantage of this capability, an "Event Mode" data acquisition system has recently been developed that records the instantaneous electron beam energy for every detected photon, permitting spectra to be collected as the beam energy is varied continuously over a desired range. A spectrum of He-like neon obtained in this manner, with the beam energy sweeping between 600 and 1150 eV, is shown in Figure 2. For comparison, the direct excitation thresholds for w, y, and z are 922, 916, and 905 eV, while the ionization potential for He-like to H-like neon is 1196 eV.

Typical timescales for ionization are a few msec, while recombination proceeds much more slowly. For this data, the beam energy was swept up and down with a 1 msec period, thus maintaining a steady ionization balance dominated by He-like ions, with no H-like component. Note the curved tails on w and q that extend below threshold. These are satellite lines arising from dielectronic recombination (He- to Li-like). The tails just below threshold are from high n satellites, while the more distinct KLL (n=2) satellite band occurs around 680 eV. Excitation cross sections are easily extracted from such data, and may be seen for the case of He-like titanium (Z=22) in Beiersdorfer *et al.* (these proceedings).

Figure 2. He-like neon spectrum taken with a TlAP crystal in the Flat Crystal Spectrometer, plotted in "event mode" fashion. The horizontal axis is the usual dispersion axis, while the vertical axis records electron energy.

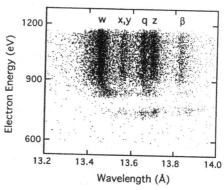

In our study of He-like neon, we use the Event Mode system to map out all the resonances below and near threshold using very narrow energy bins of 20 eV. Above the ionization potential for He-like to H-like ions, the excitation cross sections of the 2-to-1 lines become smooth and slowly varying, so we use larger energy bins, collecting data up to several times the direct excitation threshold energy. The temperature sensitive line ratio $G=[x+y+z]/w$ is then determined for all temperatures or nonthermal energy distributions.

In addition to recording spectra as a function of energy, we also measure the line ratio $R=z/[x+y]$ as a function of electron density. As described previously, the electron density can be varied by a factor of ten or more, while the density is measured using the beam imaging system.

In any astrophysical plasma there is always a mix of ion states, so we also examine the effect of Li-like and H-like ions on He-like spectra. Innershell ionization of Li-like ions enhances z emission, while recombination of H-like ions produces relatively more x, y, and z emission than w. We use EBIT's ability to establish widely varying ionization balances to separate out and quantize these different contributions. For example, setting the baseline electron beam energy at several times the ionization potential for He-like to H-like ions establishes an ion balance dominated by H-like (and bare) ions. We then lower the beam energy below the direct excitation threshold for He-like 2-to-1 lines and collect a spectrum arising entirely from the recombination of H-like into He-like ions. After a few msec (much less than the time for a significant fraction of H-like ions to recombine) the beam energy is raised again to maintain the ionization balance, and the process is repeated.

4 SUMMARY

The Electron Beam Ion Trap is an extremely versatile tool for studying atomic processes in highly charged ions, and is well suited to investigate the many mechanisms involved in the emission of 2-to-1 lines in heliumlike ions. EBIT offers unprecedented control over many plasma parameters, and spectra may be obtained under widely varying conditions of energy, density, and ionization balance. With the intention of developing and improving He-like ion diagnostics for use in analyzing astrophysical spectra, we have constructed a soft X-ray crystal spectrometer and developed experimental techniques which fully utilize the capabilities of EBIT, and have begun a systematic investigation of neon and several other astrophysically important elements.

We wish to thank D. Nelson and D. Carter for their expertise in assembling the spectrometer. This work is supported by the California Space Institute and by NASA under grant number NAGW-2688. Work performed at LLNL is under the auspices of the U.S. Department of Energy under contract number W-7405-ENG-48.

REFERENCES

Beiersdorfer, P., et al., these proceedings.

Levine, Morton A., et al. 1988, Phys. Scripta, T22, 157.

Bitter, M., et al. 1979, Phys. Rev. Letters, 43, 129.

Gabriel, A.H., and Jordan C. 1969, M.N.R.A.S., 145, 241.

Källne, E., Källne, J., and Pradhan, A.K. 1983, Phys. Rev. A, 28, 467.

Keenan, F.P., et al. 1989, Phys. Rev. A, 39, 4092.

McKenzie, D.L., and Landecker, P.B. 1982, Ap. J., 259, 372.

Mewe, R., and Schrijver, J. 1978a, Astr. Ap., 65, 99.

Mewe, R., and Schrijver, J. 1978b, Astr. Ap., 87, 55.

Pradhan, A.K., and Shull, J. Michael. 1981, Ap. J., 249, 821.

Electron Impact Ionization of Lithiumlike Ions: Ti^{19+}, V^{20+}, Cr^{21+}, Mn^{22+}, and Fe^{23+}

K. L. Wong, P. Beiersdorfer, M. H. Chen, R. E. Marrs, K. J. Reed, J. H. Scofield, D. A. Vogel, and R. Zasadzinski

Lawrence Livermore National Laboratory, University of California, Livermore, California 94550

ABSTRACT

The Livermore electron beam ion trap (EBIT) has been used to measure the cross section for electron impact ionization of lithiumlike Ti, V, Cr, Mn, and Fe. The measurement was made using a novel x-ray technique. The ionization cross sections were determined at one energy corresponding to the KLL dielectronic resonance of heliumlike ions and are measured relative to the direct excitation cross section of the lithiumlike line q. The uncertainties in the ionization cross sections are estimated to be about 13%. Good agreement is found with results from a relativistic distorted-wave code.

1 INTRODUCTION

Accurate knowledge of ionization cross sections and rates is important for determining the charge-state balance and is required for proper interpretation of spectra observed in high temperature plasmas. It is important, for example, for inferring the radial transport of impurity ions such as Ti, Cr, and Fe in tokamaks from spectral observations and for determining the power balance. Previous ionization measurements on the transition metals have been performed using crossed beams (Fe^{15+}) [1], theta-pinches (Ti^{8+}) [2], and tokamaks (Fe^{21+}) [3].

In the following we describe measurements of ionization of lithiumlike ions from the transition metals (Ti-Fe), i.e., of the process $1s^2 2s + e^- \rightarrow 1s^2 + 2e^-$. The measurements were made on the Livermore electron beam ion trap (EBIT) using a technique that relies on x-ray observations [4]. In the current set of experiments the electron beam is a factor of two to three times larger than the crossed beam experiments (≤ 1.5 keV). Therefore, we can investigate the high energy dependence of the cross sections.

2 DESCRIPTION OF TECHNIQUE

EBIT was designed as an x-ray source [5, 6]. Therefore, we have devised a novel technique for measuring the electron impact ionization cross section of lithiumlike ions based on studying the x rays emitted from EBIT at 90° with respect to the electron beam. The measurement scheme uses the fact that for electron beam energies equal to the KLL dielectronic recombination resonance energies for heliumlike ions, dielectronic recombination of heliumlike ions is to a good approximation balanced by ionization of lithiumlike ions. In this approximation, where we have neglected radiative recombination and charge exchange recombination onto heliumlike ions, the abundance ratio of the helium- and lithiumlike charge states is approximately equal to the ratio of the cross sections for dielectronic recombination and ionization. As a result, an x ray emitted following dielectronic recombination is a good indicator that an ionization event has occurred. Taking into account radiative and charge exchange recombination, we can determine an ionization cross section by counting the n=2→1 x rays. The measurement is normalized to the excitation cross section of the lithiumlike line q ($1s2s2p\ ^2P_{3/2} - 1s^22s\ ^2S_{1/2}$).

The general equation for the density of heliumlike ions in steady-state is given by:

$$\frac{dn_{He}}{dt} = \frac{j_e}{e}(\sigma_i n_{Li} - \sigma_{DR} n_{He} - \sigma_{RR} n_{He}) - n_0 v_0 \sigma_{cx} n_{He} = 0, \tag{1}$$

where j_e is the effective current density, e is the charge of the electron, σ_i is the ionization cross section of lithiumlike ions, σ_{DR} is the cross section of dielectronic recombination onto heliumlike ions, σ_{RR} is the radiative recombination cross section onto heliumlike ions, σ_{cx} is the cross section of charge exchange recombination of neutral background gases with heliumlike ions, n_{He} and n_{Li} are the number densities of ground-state heliumlike and lithiumlike ions respectively, n_0 is the density of neutral background atoms, and v_0 is the relative velocity of the neutrals and ions. Solving for σ_i:

$$\sigma_i = \frac{n_{He}}{n_{Li}}\left(\sigma_{DR} + \sigma_{RR} + \frac{en_0 v_0}{j_e}\sigma_{cx}\right), \tag{2}$$

The x-ray intensities at 90° produced by dielectronic recombination onto heliumlike ions and direct excitation of the lithiumlike line q can be expressed as:

$$I_{DR} = \frac{j_e}{e}\sigma_{DR} n_{He} W_{DR} G, \tag{3}$$

$$I_q = \frac{j_e}{e}\beta_r \sigma_q n_{Li} W_q G, \tag{4}$$

where σ_q is the excitation cross section of the lithiumlike line q, β_r is the radiative branching ratio of line q. W accounts for the angular distribution of the x rays, the linear polarization of the x rays, and the reflectivity of the analyzing crystal. G is the solid angle subtended by the detector. Substituting these into Eq. (2), we get:

$$\sigma_i = \frac{I_{DR}}{I_q}\frac{\beta_r \sigma_q W_q}{W_{DR}}\left(1 + \frac{\sigma_{RR}}{\sigma_{DR}} + \frac{en_0 v_0}{j_e}\frac{\sigma_{cx}}{\sigma_{DR}}\right). \tag{5}$$

The last two terms in Eq. (5) are small compared to one. We use radiative branching ratios calculated using the methods of Chen [7], the excitation cross section of q is calculated using the distorted-wave code of Zhang, Sampson, and Clark [8], the radiative recombination cross sections are calculated with the methods described by

Scofield [9], dielectronic recombination are calculated by the method of Chen [10], and the charge exchange cross section is determined from Janev et al. [11].

3 RESULTS

The results for lithiumlike Ti, V, Cr, Mn, and Fe (Z=22-26) are given in Fig. 1. Because of the technique (measurement at the KLL dielectronic recombination resonance) all measurements were made at an energy corresponding to 2.3 times the threshold energy for ionization of lithiumlike ions. We estimate the uncertainty in the measurements to be about 13%. The results are compared to theoretical values calculated by using the relativistic distorted-wave code of Zhang and Sampson [12]. The measurements agree to within 5-18% with the theoretical results.

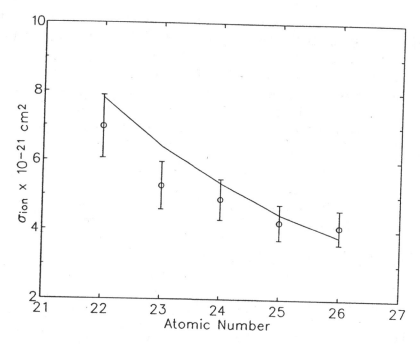

Fig. 1—Measurement of the electron impact ionization cross section of lithiumlike ions versus atomic number. The measurements were made at an energy corresponding to 2.3 times the threshold energy for ionization of lithiumlike ions. The solid curve represents results from the relativistic distorted-wave calculation.

We wish to thank E. Magee and D. Nelson for their technical assistance in these measurements. This work was performed under the auspices of the U. S. Department of Energy by Lawrence Livermore National Laboratory under contract No. W-7405-ENG-48.

REFERENCES

[1] D. C. Gregory, L. J. Wang, F. W. Meyer, and K. Rinn, Phys. Rev. A, **35**, 3256 (1987).

[2] R. U. Datla, and J. R. Roberts, Phys. Rev. A, **28**, 2201 (1983).

[3] J. S. Wang, H. R. Griem, R. Hess, and W. L. Rowan, Phys. Rev. A, **38**, 4761 (1988).

[4] K. L. Wong, P. Beiersdorfer, D. Vogel, R. Marrs, and M. Levine, Z. Phys. D (Supplement) **21**, S197 (1991).

[5] M. A. Levine, R. E. Marrs, J. R. Henderson, D. A. Knapp, and M. B. Schneider, Phys. Scr. **T22**, 157 (1988).

[6] R. E. Marrs, M. A. Levine, D. A. Knapp, and J. R. Henderson, Phys. Rev. Lett. **60**, 1715 (1988).

[7] M. H. Chen, Phys. Rev. A **31**, 1449 (1985).

[8] H. L. Zhang, D. H. Sampson, and R. E. H. Clark, Phys. Rev. A **41**, 198 (1990).

[9] J. H. Scofield, Phys. Rev. A **40**, 3054 (1989).

[10] M. H. Chen, Phys. Rev. A **33**, 994 (1986).

[11] R. K. Janev, D. S. Belić, and B. H. Bransden, Phys. Rev. A**28**, 1293 (1983).

[12] H. L. Zhang and D. H. Sampson, Phys. Rev. A **42**, 5378 (1990).

TOWARD REALISTIC ELECTRON-ION POTENTIALS IN PLASMAS[']

YONG YAN, HONGBIN ZHAN, YINCHUN ZHANG and PETER WINKLER

Department of Physics, University of Nevada, Reno NV 89557

ABSTRACT

Realistic electron-ion potentials have been derived from plasma distribution functions and subsequently used in atomic pair function calculations.

1 INTRODUCTION

The potential between electrons and ions plays a key role in plasma studies as well as in studies of atomic processes in plasmas. The ultimate goal of an *ab initio* description of plasmas is a formidable task of many-body theory. In particular for non-ideal plasmas in which the kinetic energy average is much smaller than the potential energy the difficulties are compounded by the existence of bound ionic or atomic states. This requires an accurate treatment of both the global and local properties of the plasma. Considerable progress has been made recently in the formulation of a fundamental theory of non-ideal hydrogen plasmas (Lehmann and Ebeling 1991) and applications to helium plasmas have been reported (Förster, Kahlbaum and Ebeling 1991) but the *ab initio* treatment of more general plasma situations is still mostly a matter of the future. Since the calculation of atomic processes in plasmas on the other hand requires the detailed and accurate knowledge of the local environment of the ions it appears timely to study alternative approaches to extract more realistic electron-ion potentials from detailed plasma calculations plasmas than provided by Debye theory. At present the majority of plasma calculations are based on the one component plasma model (OCP) which takes the electron

[']Work supported by DOE contract DE-FG08-90ER14160

component as a background providing a uniform charge density throughout all space mainly to ensure charge neutrality. While this approach may be adequate for high-energy plasmas it is inadequate for non-ideal plasmas. It is particularly inadequate for the study of atomic processes in plasmas because the artificial immobility of the electron background introduces stronger spatial fluctuations of the calculated ion densities and the associated potentials than are to be expected in reality. Fortunately there are a few calculations in the two-component plasma model (TCP) available (Dharma-wardana and Perrot 1982) the results of which have enabled us to design strategies for the refinement of potential functions to be used in future calculations of atomic processes in various plasma environments. Unfortunately such TCP computations are still all too rare and the existing ones do not directly cover the region of interest for the study atomic processes.

2 DENSITY-FUNCTIONAL FORMULATION

Here we summarize the main relations from density functional theory applied to hydrogen plasmas. The equations for the electron subsystem are given by

$$[-\tfrac{1}{2} \nabla^2 + V_e(r)] \Phi_v(r) = \epsilon_v \Phi_v(r)$$

where v designates a quantum state (v,l,m) if ϵ is negative and (k,l,m) if $\epsilon = \tfrac{1}{2}k^2$ referring to a scattering state.

$$V_e(r) = - [1/r + V_P(r)] + V_e^{xc}(r) - V_e^{xc}(R)$$

where

$$V_P(r) = \int \frac{\rho(r')-n(r')}{|\underline{r} - \underline{r}'|} d\underline{r}'$$

and the exchange potentials are taken in local density approximation. The key quantity here is the distribution function for an ion-electron pair

$$g_{ie}(r) = n(r)/n_{av}$$

which is related to the following densities

$$n(r) = n^b(r) + \Delta n^f(r) + n_{av}$$

with

$$n^b(r) = \sum_{\nu,l} (2l+1) |\Phi_{\nu,l}|^2 \, f(\nu,\bar{\mu}_e)$$

$$\Delta n^f(r) = \pi^{-2} \int_0^\infty k^2 dk \{ f(k,\bar{\mu}_e) \Sigma (2l+1) [R_{kl}^2(r) - j_l^2(rk)] \}$$

$$f(\nu,\bar{\mu}_e) = 1/[1+\exp(\epsilon_\nu - \bar{\mu}_e)\beta] \qquad (\beta = 1/k_B T).$$

For the ion subsystem the most relevant relations are the pair function equation

$$g_{ii}(r) = \exp[-\beta V_I(r)]$$

with

$$V_i(r) = [1/r + V_r(r)] + V_i^c(r)$$

where the average correlation potential is given by

$$V_i^c(r) = -(1/\beta) \rho_{av} \int [h(\underline{r}') + \beta V_i(\underline{r}')] h(\underline{r}-\underline{r}') \, d\underline{r}'.$$

Finally we define

$$\ln g(r) = -\beta[1/r + V_P(r)] + \rho_{av} \int [h(\underline{r}') - \ln g(\underline{r}')] h(\underline{r}-\underline{r}') \, d\underline{r}'$$

with $h(r) = g(r) - 1$.

3 REALISTIC POTENTIALS

Once the pair functions g_{ie} and g_{ii} have been obtained by iteration we use the densities $n(r) = n_{av} g_{ie}(r)$ and $\rho(r) = \rho_{av} g_{ii}(r)$ in Poisson's equation $\nabla^2 V(r) = -4\pi(\rho(r) - n(r))$ for the evaluation of a 'realistic' potential. This term is used here to express the fact that important features of the plasma environment have been included albeit in an approximate way. Subsequent computations can be considerably accelerated if we use potential functions which are fitted to the realistic potentials. In Figure 1 we present a fit employing the analytical form of a Debye-Laughton potential

$$V_{DL}(r) = A \exp(-r/D)[1/r - B \, r^\mu \exp(-Cr/D)].$$

This form reproduces the potential curve obtained from one of the few existing TCP calculations (Dharma-wardana and Perrot 1982) so well that in the graph the curves cannot be distinguished. We have used potentials of this form in resonance calculations previously (Wang, Winkler, Pickup, and Elander 1989) and more recently in pair function calculations which are included in a companion paper in the present volume.

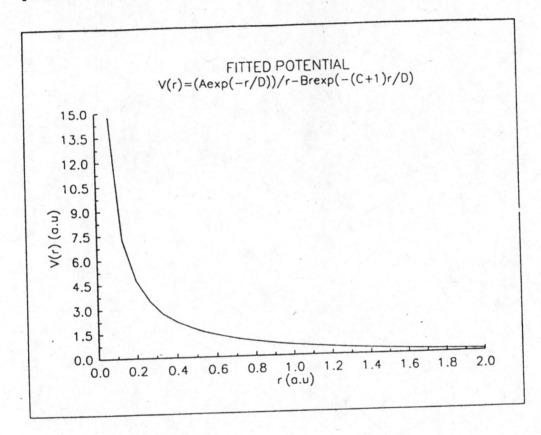

FIGURE 1 Fit of an analytic function to a realistic potential for hydrogen plasma ($\Gamma=10$). The radius of the correlation sphere is 2 a.u. The parameters of the Debye-Laughton potential are $A=1, B=.223, C=.596, D=4.737, \mu=1$.

4 REFERENCES

Dharma-wardana M. W. C. and Perrot F. 1982, Phys. Rev. A 26, 2096.

Förster A., Kahlbaum T., and Ebeling W. 1991, in Proc. XX. Int. Conference on Phenomena in Ionized Gases (in press).

Lehmann H. and Ebeling W. 1991, Z. Naturforsch. 46a, 383.

Wang Z., Winkler P., Pickup B. T., and Elander N. 1989, Chem. Phys. 135, 247.

UV Spectroscopy

Astronomical Observations with Normal Incidence Multilayer Optics II: Images of the Solar Corona and Chromosphere

Arthur B.C. Walker, Jr.[1] , Troy W. Barbee Jr.[2] and Richard B. Hoover[3]

1 Departments of Physics and of Applied Physics and Center for Space Science and Astrophysics, Stanford University, Stanford, CA 94305
2 Lawrence Livermore National Laboratory, Livermore, CA 94558
3 Space Science Laboratory Marshall Space Flight Center, Huntsville, AL 35812

ABSTRACT

The first high resolution x-ray images of an astronomical object (the solar corona) formed with normal incidence multilayer optics, were obtained in late 1987. We briefly describe some of the developments which have occurred in multilayer optics technology since 1987, and discuss capabilities and limitations that these developments imply for solar observations. We then describe the development of a new multilayer instrument for solar observations which uses several different optical configurations, and operates at a variety of soft x-ray, EUV and FUV wavelengths. We briefly discuss the results which have been achieved with this new instrument.

1. INTRODUCTION

Multilayer coatings for the efficient reflection of soft x-ray, EUV and FUV radiation at normal incidence were developed in the period from 1971-1981, primarily as a result of the independent work of E. Spiller and T.W. Barbee, Jr. and their collaborators. Spiller first described the development of multilayer interference coating for the VUV and FUV in 1971 (Spiller, 1974), and fabricated the first structure for use at EUV wavelengths in 1972 (Spiller, 1992). Haelbich and Kunz (1976) reported the first experimental results on EUV multilayers in 1976. Techniques for the fabrication of multilayer structures with the perfection on atomic scales necessary for efficient reflection of EUV and soft x-ray radiation were first developed by Barbee and Keith (1978), and by Haelbich, Segmuller and Spiller (1979). In 1981, Spiller (1981) and Barbee (1981) demonstrated that multilayer structures were a viable tool for soft x-ray optics. The first successful x-ray/EUV imaging experiments with multilayer optics were performed in the laboratory in 1981 (Underwood and Barbee, 1981; Henry, Spiller and Weisskopf, 1982) and several groups began the development of multilayer astronomical telescopes in the early 1980's (see Walker, Barbee, Hoover and Lindblom, 1988a and Catura and Golub, 1988 for a review of this early work). Underwood *et al* (1987) made the first astronomical observation with a multilayer telescope, obtaining an image of an active region in the corona with modest resolution (~ 10 - 15 arc seconds) at 44 Å, corresponding to Si XII (T ~ 2,000,000 K) emission. The first high resolution multilayer images of the sun were obtained by Walker, Barbee, Hoover and Lindblom (1988a,b), and are shown in Figures 1a and 1b. Figure 1a, was obtained with a pseudo-Cassegrain (*ie.* spherical rather than conic mirrors) telescope of 62.5 mm aperture which was coated to reflect radiation in a 12 Å bandpass centered at 173 Å; the solar image has a resolution of ~ 1.5 arc seconds. This result, dramatically demonstrated the power of multilayer optical systems for astronomical

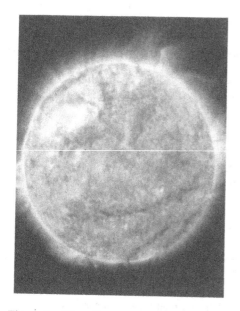

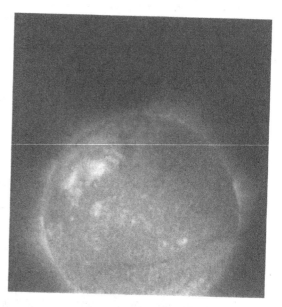

Figure 1a. The solar corona photographed in the emission of the resonance lines of Fe IX at 171 Å and Fe X at 174.5 Å on 23 October, 1981 at 1809 UT. The length of the exposure was 200 seconds and the images were recorded on Kodak XUV 100 film. The image is dominated by material at ~ 1,000,000 K.

Figure 1b. The solar corona as photographed on 23 October, 1987 at 1809 UT in a bandpass centered at ~ 256 Å, which contain lines of He II and Fe XIV. The telescope used to record this image was a single reflection Herschellian. The resolution of the image is limited to ~ 4 arc seconds by the resolving power of the film.

research, and spurred the application of multilayer optics to other disciplines, such as biology and lithography. In 1989, Golub *et al* obtained an image of the solar corona in a bandpass centered at 63.5 Å which contains strong lines of Fe XVI and Mg X, using a large Herschellian mirror of 254 mm aperture, demonstrating the effectiveness of the multilayer technology at soft x-ray wavelengths. At the time of this publication, multilayer optical systems have obtained high resolution images of the solar corona in nine band passes in the wavelength range $44 \text{ Å} \leq \lambda \leq 304 \text{ Å}$.

The advantages of multilayer optics for solar observations can be summarized as follows: *(i)* Compared to grazing incidence optics, normal incidence multilayer optics display lower scatter, suffer less from geometrical aberrations, are less expensive to fabricate and easier to mount and align, and can be made more compact (by folding the optical path). Consequently, it is possible to approach the resolution limit imposed by diffraction for such systems. The combination of filled apertures, and high reflection efficiencies provide high sensitivity optical systems in a compact package. *(ii)* The compact nature and low cost of multilayer telescopes capable of high resolution observations permits the use of multiple telescopes, and therefore the observation of plasma over a broad range of temperatures. *(iii)* The spectral resolving capability of multilayer mirrors (typically $\lambda/\Delta\lambda \sim 10$-$100$), sufficient in most cases to isolate a single line multiplet, is a powerful analytical tool, permitting observations with temperature specificity over the entire range of plasma temperatures in the solar atmosphere, $7{,}000 \text{ K} < T < 30{,}000{,}000 \text{ K}$. *(iv)* Multilayer gratings offer the prospect of attaining high spectral resolution ($\lambda/\Delta\lambda \sim 10{,}000$) stigmatic images of the solar atmosphere in the spectral range $10 \text{ Å} < \lambda < 400 \text{ Å}$.

2. MULTILAYER OPTICS

The concept of multilayer and multilayer interference coated optics is quite simple; materials with contrasting optical properties are stacked, with Angstrom scale precision, on a suitably figured substrate so that the reflections from the successive interfaces are coherently added to form a strong reflected beam. The wavelength of the reflected light is determined by the period of the multilayer structure (typically 20 Å - 500 Å, permitting radiation of wavelength 40 Å $< \lambda <$ 1000 Å to be reflected). The properties of the reflected light such as intensity, bandpass and image quality are, however, strongly dependent on a number of factors, including the relative thicknesses and absorption coefficients of the materials, the uniformity of the layers and the nature of the interfaces, and the smoothness of the substrate. Furthermore, the performance of a multilayer optical system is strongly dependent on the use of filters to exclude long wavelength light which is specularly reflected by the multilayer, and high resolution detectors to record the image. Consequently, fundamental advancements in four areas of technology underlie the results that have been achieved and that are anticipated in the near future in imaging XUV [we refer to the soft x-ray and EUV spectral regions ($\sim$ 1A to 1000 Å) as the XUV] and FUV radiation; (i) optical coatings for the XUV and FUV, (ii) ultra smooth mirror substrates, (iii) XUV and FUV filters, (iv) high resolution XUV/FUV sensitive photographic emulsions.

Ultrasmooth Mirror Substrates: The surface quality of mirror substrates for XUV and FUV mirrors is of critical importance for two reasons: (i) the efficiency of multilayer coatings is dependant on substrate surface quality [this dependance becomes critical for wavelengths below $\sim$ 80 Å, since reflectivity ε decreases as $\varepsilon = \varepsilon_t \exp[-(2\pi\sigma/d)^2]$ where σ is the RMS roughness of the substrate, d is the multilayer lattice constant, and ε_t is the theoretical efficiency for an ideal multilayer, and (ii) scattering from surface inhomogeneities decreases image contrast and degrades image quality. P. Baker has shown that figured optical surfaces of very high quality [with figures accurate to $\lambda/100$ when tested at $\sim$ 6300 Å (Hoover et al, 1990a)] can be polished to have a surface smoothness of less that 2 Å RMS in Zerodur substrates and less that 0.5 Å in sapphire substrates, by using a flow polishing technique (Baker, 1989). These substrates are of sufficient quality to permit images to be formed at the level of 0.1 arc second resolution.

Multilayer Coatings for the XUV/EUV: The development of multilayer coating technology (Barbee, 1990) for the XUV has now progressed to the point that the reflection efficiency of optical coatings for the wavelength regions $\sim$ 100 $< \lambda <$ 350 Å are approaching the levels predicted for "perfect" multilayer structures. We have recently completed the measurement of reflectivity for Ritchey Chrétien and Cassegrain Telescopes developed for the *Multispectral Solar Telescope Array (MSSTA)* rocket payload (Hoover et al, 1990b; Barbee et al, 1991). The measured reflectivities, typically $\sim$ 50% (at $\sim$ 150 Å) to $\sim$ 25% (at $\sim$ 300 Å), are $\sim$ 70% - 80% of the reflectivity for ideal multilayer structures. A measure of the progress that has been achieved is demonstrated by the comparison of the reflectivity of the mirrors of the Cassegrain Telescope which obtained the image of Figure 1a ($\sim$ 25%), with the reflectivity of our recently completed 171 Å - 175 Å Ritchey Chrétien Telescope mirrors ($\sim$ 43%). For two reflections, this results in an almost 300% increase in throughput! The reflectivity of contemporary multilayers at wavelengths below $\sim$ 70 Å, typically 10%, is too low to permit the use of double reflection telescopes for the weak solar lines in this spectral range. The single reflection Herschellian configuration has been used for the multilayer telescopes in this wavelength interval. However, significant progress has been made in perfecting the interfaces between layers to improve reflection efficiency (Barbee, 1990) for these short period multilayers. The fractional bandpass ($\delta\lambda/\lambda$) of a multilayer mirror can range between $\sim$ 1% and 20%, allowing, within limits, a mirror to be specifically designed for a particular application (*i.e.* narrow band imaging, providing an broadband image for a spectrometer, etc.) An important development for astronomical applications is the fabrication of multilayer structures with low contrast between the refractive indices of the materials forming a layer pair, permitting many layers to contribute to the reflected beam, resulting in improved resolution (Barbee, 1990).

Filters and Films: For stellar sources of soft x-ray, EUV and FUV radiation such as the sun, the intensity of ultraviolet, visible, and infrared radiation exceeds the intensity of the short wavelength radiations by factors between 100 and 1,000,000. For multilayer optical systems, two multilayer reflections are able to discriminate against nearby offband radiation by factors of ~ 100 - 300, consequently if the objective is to obtain an image of structures as seen in a strong solar line such as He II λ 304 Å , Fe IX λ 171 Å, etc., then we only need to be concerned with "leakage" from distant XUV lines, and from UV and visible radiation. Unfortunately, multilayer coatings are reasonably efficient (~ 50%) reflectors for UV and visible light. If a polychromatic detector such as film is used, it is necessary to attenuate off-band radiation by the use of filters. Rejection of visible light by a factor of ~ 10^{10} must be achieved. This is generally accomplished by the use of thin (typically ~ 1000 Å - 2000 Å thick) metallic (*i.e.* aluminum or beryllium) or metal coated plastic films supported by nickel mesh; other materials are used for additional attenuation at specific wavelengths. This is a challenging aspect of multilayer imaging technology; Powell (1989) and Spiller *et al* (1990) discuss filters in detail.

All of the high resolution multilayer images obtained so far have been recorded on XUV sensitive photographic emulsions prepared by Kodak. Hoover *et al* (1990c) have carried out detailed measurements on the 2 emulsions which we have used, XUV 100, which has a resolving power of 200 *lines/mm* (i.e. 2.5 micron "pixels") and the XUV 649 emulsion, which has a resolution of 2000 *lines/mm* (i.e. 0.25 micron "pixels"). With the XUV 100 film, the resolution of the telescopes flown so far (3500 mm focal length) is limited to 0.3 arc seconds, with the XUV 649 emulsion, resolution exceeding 0.1 arc second is possible. While less sensitive than the XUV 100 film, the XUV 649 emulsion was sufficiently sensitive to allow images to be obtained with the *MSSTA* telescopes for the strong lines. We present images obtained with both films in this paper.

Multilayer Gratings: Spectroscopic analysis in the XUV and EUV is now possible using stigmatic instrument configurations previously confined to the visible, ultraviolet and far-ultraviolet, as a result of the development of multilayer gratings. Barbee (1989) successfully fabricated and tested the first multilayer grating; he has shown that these combined microstructure optical elements disperse wavelengths Bragg-diffracted by the multilayer structure so that all constructive interference occurs at essentially constant angles relative to the zero order Bragg-diffracted beam. Bixler, Barbee and Dietrich (1989) have tested multilayer gratings operating at ~ 200 Å, and have demonstrated resolution as high as λ/Δλ ~ 2000 at normal incidence. These very exciting results demonstrate that *it is now possible to fabricate multilayer gratings with properties which are tailored to the needs of high resolution solar XUV spectroscopic observations.*

Optical Configurations: In a previous review [Walker · *et al*, 1990a (Paper I)] we discussed optical configurations for multilayer telescopes; more recently we analyzed the factors governing the resolution of multilayer telescopes in depth (Walker, *et al*, 1991). The effective use of conventional optical configurations, such as the Ritchey Chrétien, the Herschellian, and the off-axis parabola has already been demonstrated by the sub-arc second quality images presented in Section 3. We have also demonstrated the use of a multilayer tertiary mirror in conjunction with a Wolter I primary (Walker *et al*, 1988b). The use of a multilayer mirror as an imaging Bragg crystal in a Johann curved crystal spectrograph has been demonstrated in the laboratory (Walker *et al*, 1990a). Walker *et* *al* (1990b) have discussed the design of multilayer coronagraphs. Keski-Kuha, Thomas and DaVilla (1992) have obtained a high resolution solar spectra using a multilayer coated grating in the Rowland circle geometry, and Walker *et al* (1992) are planning to fly an objective grating to obtain spectrally dispersed stigmatic solar images. *In summary, all of the optical configurations discussed by Walker et al (1990a) have been successfully used in solar observations, or have been studied in depth.*

Walker *et al*, (1991) pointed out that the use of aplanatic mirror configurations such as the Ritchey Chrétien, can permit resolutions of 0.05 - 0.1 arc-second to be achieved over fields of several arc minutes. For the sun, this resolving power corresponds to ~ 35-70 kilometers, which approximates the photon meanfree path in the lower chromosphere. To achieve this resolving power at 5000 Å, apertures of 1.25 - 2.50 meters are required, at H Lyman α (1215.6 Å), apertures of 30-60 cm are sufficient, while at He II Lyman α (304 Å) apertures of 7.5 - 15 cm will suffice. Because they permit observations at FUV, EUV and soft x-ray wavelengths, multilayer optical systems offer a distinct advantage for high resolution observations.

3.0 SOLAR OBSERVATIONS WITH MULTILAYER OPTICS

Progress on basic problems relating to the structure and dynamics of the solar chromosphere, corona, and corona/solar wind interface has been limited by the quality of the available soft x-ray/extreme ultraviolet (XUV) data. In the past, instruments designed to obtain such data have been forced to compromise on three or four of the five primary goals of the observational astronomer [spectral resolution, spatial resolution, temporal resolution, field of view, and temperature determination (achieved by simultaneous observations in several lines or wavelength bands)] to concentrate on the remaining one or two goals. This limitation is due in large measure to the fact that the reflection efficiency of conventional mirrors at normal incidence for EUV and soft x-ray radiation is low, therefore precluding the use of the powerful techniques developed for visible and ultraviolet spectroscopy. The techniques that were available, such as grazing-incidence optics and mechanically collimated Bragg spectrometers, have achieved important but limited success. The images in Figure 1 have demonstrated the power of normal incidence multilayer optics to achieve high angular resolution astronomical images with moderate spectral resolution ($\lambda/\Delta\lambda \sim 30$ to 100) in the wavelength region $30 < \lambda < 400$ Å.

A Comprehensive Solar Chromospheric/Coronal Observatory: Based on these capabilities we decided to develop a comprehensive rocket borne solar observatory designed to achieve high angular resolution temperature specific images which would permit us to delineate the structure and dynamics of the solar atmosphere, from the chromosphere (7,000 K), through the hottest structures present in the corona (30,000,000 K). This instrument (Walker *et al*, 1990c), the *Multi Spectral Solar Telescope Array (MSSTA)*, utilizes five 127 mm aperture multilayer Ritchey Chrétien telescopes and two 63.5 mm Cassegrain telescopes covering the spectral range from ~ 150 to ~ 350 Å and as many as six multilayer Herschelian telescopes covering the spectral range $40 < \lambda < 150$ Å. Each telescope is able to isolate line multiplets excited over a narrow temperature range, providing full disk images of diagnostic quality arising from structures in the solar atmosphere ranging in temperature (T) from T ~ 50,000 K (He II) to 10,000,000 K (Fe XX). The soft x-ray and EUV images are supplemented by full disk high resolution far-ultraviolet (FUV) images in H I Ly-α (λ ~ 1216 Å) and C IV (λ ~ 1548/1550 Å) that are obtained by two Ritchey Chrétien telescopes of the same design as the XUV telescopes. The first flight of the *MSSTA* payload occurred on 13 May, 1991 The resulting data sets are intended to address several fundamental problems related to the following solar phenomena: *(i)* the morphology and energetics of the fine structure of the solar chromosphere/corona interface, including the "chromospheric network," spicules, prominences, cool loops, and the magnetic field. *(ii)* the structure, energetics, and evolution of high temperature coronal loops. *(iii)* The large scale structure and dynamics of the corona, including the solar wind interface (represented by phenomena such as polar plumes). *(iv)* The structure and evolution of hot flare generated coronal loops.

The selection of individual telescopes for a particular flight of the *MSSTA* is expected to vary. The thermal responses of a typical set of telescopes which are under consideration for the second flight of the *MSSTA* is shown in Figure 2. The set of thermal response curves presented in Figure 2 provides a very powerful diagnostic tool for the solar atmosphere.

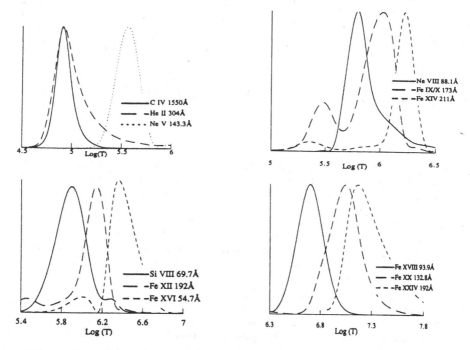

Figure 2. Calculated temperature response of a set of multilayer telescopes to the solar atmosphere. The response of the H Lyman α telescope, which is peaked at 7,000 - 10,000 K, is not shown.

Recent Multilayer Observations of the Sun: In Figures 3b-d we present some of the results of the 13 May, 1991 flight of the *MSSTA* payload. Figure 3a represents the configuration of the solar magnetic field. Figure 3b is an image of the chromosphere obtained with the *MSSTA* H Lyman α telescope, and Figure 3c and 3d are images of the corona at ~ 1,500,000 K (Fe XII) and ~ 2,500,000 K (Fe XIV) obtained with two of the *MSSTA* XUV telescopes. Figure 3b was obtained with a 127 mm aperture Ritchey Chrétien telescope, Figure 3c was obtained with a 75 mm aperture Herschellian telescope, and Figure 3d was obtained with a 62.5 mm aperture pseudo-Cassegrain telescope. Both 3b and 3c display sub-arc-second resolution (~ 0.7 arc seconds); Figure 3d is limited to ~ 1.5 arc seconds resolution. A number of coronal features are visible in the XUV images, including bright points, coronal holes, polar plumes, active region loops, and filaments. The chromospheric network is clearly seen in Figure 3b. The dominant influence of the magnetic field (Figure 3a) is apparent in both the chromospheric and coronal structures.

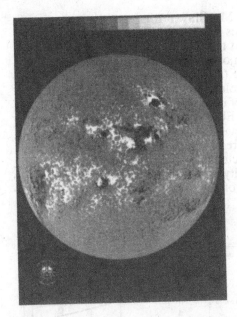

Figure 3a. National Solar Observatory magnetogram for 13 May, 1991. (Courtesy of Jack Harvey.)

Figure 3b. Image of the Chromosphere in H Lyman α at 1907 UT on 13 May, 1991. The image is recorded on 649 film.

Figure 3c. Image of the corona at λ 193 Å (Fe XII) at 1907 UT on 13 May, 1991. The material imaged is at ~ 1,500,000 K. The image is recorded on 649 film.

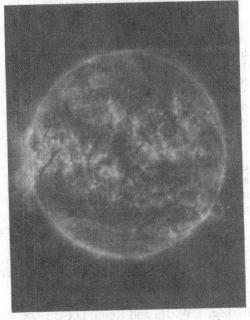

Figure 3d. Image of the corona at λ 211 Å (Fe XIV) at 1907 UT on 13 May, 1991. The material imaged is at ~ 2,500,000 K. The image is recorded on XUV 100 film.

The Analysis of the Coronal Plasma with Multilayer Images: To demonstrate the power of high resolution multilayer XUV images for the analysis of the coronal plasma, we have analyzed the density and temperature in two of the polar plumes observed in Figure 1a. Polar plumes are believed to be the major source of the high-speed solar wind streams associated with coronal holes. Withbroe (1986) has reviewed models of polar plumes and pointed out the importance of high-resolution (~ 1 arc second) observations of polar plumes to an understanding of the physics of coronal holes. Previous observations of polar plumes (on Skylab) in the soft x-ray and EUV have been limited to distances of 0.4 solar radii or less above the limb and resolution ~ 3 - 5 arc seconds. Our observations extend to 0.6 solar radii and have ~ 1.5 arc second resolution.

Figure 4a represents a comparison of the predicted and observed emission profile of 2 plumes, and Figure 4b represents the derived density of these plumes. Clearly, multilayer images provide a powerful technique for the study of large scale solar structure, as Walker *et al* (1990b) point out.

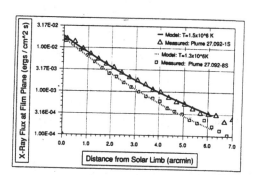

Figure 4a. The predicted and observed emission from typical polar plumes observed in Figure 1a, based on an isothermal model with solar wind flow included.

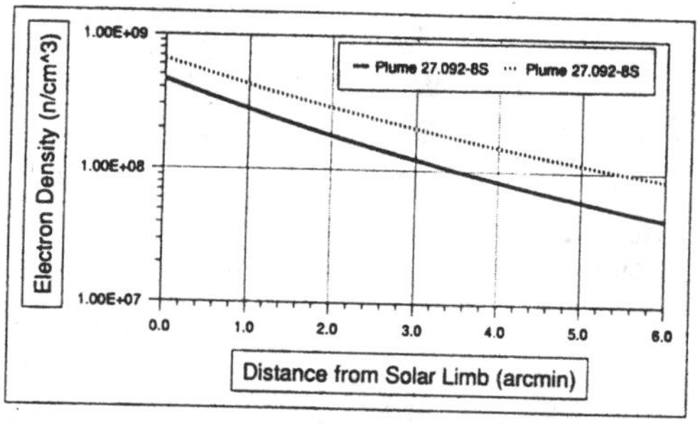

Figure 4b. Density versus height for typical polar plumes observed in Figure 1a, derived from an isothermal model with solar wind flow included

4.0 CONCLUSIONS

Multilayer optics have been shown to be highly effective for the study of both the large scale structure and the fine scale structure of the solar atmosphere. The *best* soft x-ray and EUV images of the sun obtained to date, with 0.7 arc second resolution, have been achieved using these relatively new techniques. Several satellite borne multilayer instruments for the long term study of the sun are under development, or have been selected for study (Walker *et al*, 1990d). A multilayer coated grating has already been successfully flown, and additional high resolution instruments containing gratings are under development.

References

Baker, P.C., 1989, *Proc. SPIE* **1160**, 263.

Barbee, T.W. Jr., and Keith, D.C., 1978, *X-ray Instrumentation for Synchrotron Radiation Research*, eds. H. Winick and G. Brown, SSRL Report **78/04**, III-26.

Barbee, T.W. Jr., 1981, in *Low Energy X-Ray Diagnostics*, D.T. Atwood and B.L. Henke. eds. *Proc AIP* **75**. 124.

Barbee, T.W. Jr., 1990, *Optical Engineering* **29**, 711.
Barbee, T.W. Jr., Weed, J.W., Hoover, R.B., Allen, M.J., Lindblom, J.F., O'Neal, R.J.,Kankelborg, C.C., DeForest, C.E., Paris, E.S., Walker, A.B.C. Jr., Willis, T.D., Gluskin, E., P. Pianetta, P. and P. C. Baker, 1991, *Optical Eng.*, **30**, 1061.
Barbee, T.W. Jr., 1989, *Rev. Sci. Instr.*, **60**, 1588.
Bixler, J.V., Barbee T.W. Jr. and Dietrich, D.D., 1989, *Proc SPIE*, **1160**, 648.
Catura, R.C. and Golub, L., 1988, *Rev. Phys. Appl*, **23**, 1741.
Golub, L., Herant, M., Kalata, K., Lovos, I., Nystrom, G., Prado, F., Spiller, E. and Wilczynski, J., 1990, *Nature* **344**, 842.
Haélbich, R.P. and Kunz, C., 1976, *Optics Comm.*, **17**, 287.
Haélbich, R.P., Segmuller, A. and Spiller, E., 1979, *Appl, Phys. Lett.*, **34**, 184.
Henry, J.P., Spiller, E. and Weisskopf, M., 1982, *Appl Phys. Lett.*, **40**, 25.
Hoover, R.B., Baker, P.C., Hadaway, J.B., Johnson, R.B., Gabardi, D.R., Walker, A.B.C. Jr., Lindblom, J.F., DeForest, C.E. and O'Neal, R.H., 1990a, *Proc. SPIE* **1343**, 189.
Hoover, R.B., Barbee, T.W. Jr., Baker, P.C., Lindblom, J.F., M. J. Allen, DeForest, C., Kankelborg, C.C., O'Neal, R.H., Paris, E. and Walker, A.B.C. Jr., 1990b, *Optical Eng.* **29**, 1281.
Hoover, R.B., Walker, A.B.C. Jr., De Forest, C.E., Allen M.J. and Lindblom, J.F., 1990c, *Proc, SPIE* **1343**, 189.
Keski-Kuha, R.A., Thomas, R.J. and Davilla, J.M., 1992, *Proc. SPIE*, **1546**, 614.
Powell, F., 1989, *Proc SPIE* **1160**, 37.
Spiller, E., 1974 in *Space Optics*, B.J. Thompson and R. R. Shannon, eds., pp. 570-581, (Washington, D.C.: National Academy of Sciences).
Spiller, E., 1992, *Proc. SPIE*, **1546**, 489.
Spiller, E., 1981, In Low Energy X-Ray Diagnostics, in D.T. Atwood and B.L. Henke eds., *Proc AIP*, **75**, 131.
Spiller, E., Grebe K. and Golub, L., 1990, *Optical Engineering* **29**, 625.
Underwood, J.H. and Barbee, T.W. Jr., 1981, *Nature*, **294**, 429.
Underwood, J.H., Bruner, M.E., Haisch, B.M., Brown W.A. and Acton, L.W., 1987, *Science* **238**, 61.
Walker, A.B.C. Jr., Barbee, T.W. Jr., Hoover, R.B. and Lindblom, J.F., 1988a, *Science.* **241**, 1781.
Walker, A.B.C. Jr., Barbee, T.W. Jr., Hoover, R.B.and Lindblom, J.F., 1988b, *J. de Physique Coliques*, **C1 49**, C1-175.
Walker, A.B.C. Jr., Lindblom, J.F.,O'Neal, R.H. Jr., Hoover, R.B., Barbee, T.W. Jr., 1990a, *Phys Scripta.* **41**, 1053.
Walker, A.B.C., Jr., Allen, M.J., Barbee, T.W. Jr. and Hoover, R.B., 1990b, *Proc. SPIE*, **1343**, 415.
Walker, A.B.C. Jr., Lindblom, J.F., O'Neal, R.H., Allen, M.J., Barbee, T.W. Jr., Hoover, R.B., 1990c, *Optical Eng.* **29**, 581.
Walker, A.B.C. Jr., Lindblom, J.F., Timothy, J.G., Barbee, T.W. Jr., Hoover, R.B. and Tandberg-Hanssen, E., 1990d, *Optical Eng.* **29**, 698
Walker, A.B.C. Jr., Lindblom, J.F.,Timothy, J.G., Hoover, R.B., Barbee, T.W. Jr., Baker, P.C. and Powell, F.R., 1991, *Proc SPIE*, **1494**, 320.
Walker, A.B.C. Jr., Barbee, T.W. Jr. and Hoover, R.B., 1992, private communication.
Withbroe, G.L., 1986, in *Solar Flares and Coronal Physics using P/OF as a Research Tool*, E. Tandberg-Hanssen, R.M. Wilson, H.S. Hudson, eds., *NASA Conf. Proc.* **2421**. 221.

Question from Giovanni Peres, Osservatorio Astrofisico di Catania, Italy: What is the spatial resolution of the films you have used?

Answer: It is 2.5 microns for XUV 100, and 0.25 microns for the XUV 649 emulsion.

Question from Suzanne Hawley, IGPP, LLNL: Can you use your EUV data to derive a differential emission measure distribution for the solar corona?

Answer: Yes. For example, for the polar plumes analyzed in Figures 4a and 4b, the analysis predicts an isothermal plasma at 1,300,000 K in one case, and 1,500,000 K in the second case. For the *MSSTA* observations, we can use simultaneous observations in 2 or more bandpasses to derive emission measure.

Question from Loren Acton, Lockheed Palo Alto Research Laboratory: Can you describe the differences in coronal structure which are observed in your different spectral channels?

Answer: For the Fe IX/X, Fe XII, and Fe XIV images presented the differences are subtle. The plumes are brightest in the Fe XII images, indicating a temperature near 1,500,000 K, the coronal loops are brightest in the Fe XIV images, indicating temperatures above 2,500,000 K. Emission not associated with active regions is most strongly observed in the Fe IX/X images, and the coronal holes are more distinct in the Fe XII and Fe XIV images than in the Fe IX/X images. For the next *MSSTA* flight, we hope to include bandpasses dominated by Fe XVI and Fe XVIII which would be sensitive to material at 4,000,000 K and 6,000,000 K respectively.

Issues in Solar EUV Observations

John C. Raymond

Harvard-Smithsonian Center for Astrophysics

ABSTRACT

EUV observations of the Sun have enormous potential for testing models of the structure, density, dynamics and energetics of the chromosphere, transition region, and corona. These tests are seriously hampered by uncertainties in the currently available measurements (especially intensity calibration) and in the atomic rate coefficients.

1. INTRODUCTION

The EUV spectral range presents enormous opportunities. It provides strong spectral lines formed over the full temperature range from the chromosphere to the corona. It permits a direct measurement of the total radiative losses over much of this range, the basic observable for studies of transition region energetics. Dozens of intensity ratios provide diagnostics of electron density and temperature, as well as measurements of the abundances of a dozen or so elements.

The EUV must have some drawbacks as well, or there would be no solar X-ray or optical astronomers. These drawbacks are technical and perhaps historical, in that x-ray and near UV techniques were relatively easily adapted to solar observations. The basic difficulties in the EUV were the difficulty of making low noise, high sensitivity windowless detectors, and the low reflectivity of common optical materials at short wavelengths, requiring grazing incidence optics below about 400Å. Both technologies have improved rapidly in recent years. A less technical problem is absorption by neutral hydrogen, which can greatly complicate the interpretation of emission line intensities. We will return to the H^0 opacity later.

The early EUV experiments were carried out from rockets and OSO satellites. The earliest extensive EUV observations with high spatial resolution came from the SO55 and SO82 experiments aboard Skylab in 1973, and those data are still being used. More recently the CHASE (Lang *et al.* 1991) and SERTS (Davila 1992; Neupert *et al.* 1992) experiments have measured large numbers of line intensities, while rocket flights (*e.g.* Hassler, Rottman and Orrall 1991) have measured flow and turbulent velocities of hot material. The recent development of multilayer optics has brought about imaging experiments with high spatial and temporal resolution in individual spectral lines (A. Walker, this volume; Herant *et al.* 1991; Sobel'man *et al.* 1990). The SOHO satellite, due for launch in 1996, includes 4 EUV experiments: **CDS** (Patchett *et al.* 1988), **SUMER** (Wilhelm *et al.* 1988), **EIT** (Delaboudiniere *et al* 1988) and **UVCS** (Kohl *et al.* 1988) which will vastly improve the quality and quantity of the observations.

This paper will focus not on the values of emission measure, density, temperature, or abundance derived from EUV measurements, but on the accuracy with which we can derive them. The following section summarizes the observables and atomic rates needed for such derivations, and the final section estimates the level of accuracy now available.

2. DERIVED QUANTITIES

2.1 Emission Measure

This is the basic measurement of the amount of hot gas as a function of temperature, and it is the primary characteristic which a model of coronal heating must explain. Figure 1 shows the emission measures derived from Skylab SO55 observations at two times during a solar flare (Doyle and Raymond 1983). Uncertainties in atomic rates or in measured fluxes should show up as scatter of the points about a smooth curve. The scatter in this plot appears modest, but notice the discrepant points near Log T = 5.4. Also, there are not often many points near any one temperature, so that the apparent smoothness is not a good test of the accuracy of the points. Attempts to construct emission measure curves with 50-100 line intensities typically yield RMS deviations around 50% (Raymond 1988; 1989; Lang, Mason and McWhirter 1990). These errors are attributed in roughly equal measure to observational error (calibration uncertainty, line blends, or misidentification) and to errors in the atomic rates and ionization state (excitation, ionization and recombination rate coefficients).

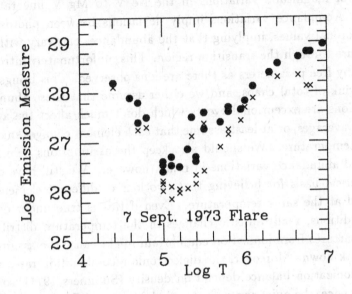

Figure 1. Emission Measures of the 7 Sept. 1973 flare from Doyle and Raymond (1983) at 12:55 UT (solid circle) and 14:03 UT (X).

Lyman continuum absorption and errors in the assumed abundances may also contribute to the scatter. It is also possible that the basic physical assumptions of the model are not fulfilled, and that the scatter in ratios of observed to predicted line intensities results from departures from ionization equilibrium (rapid heating or cooling), non-Maxwellian electron velocity distributions, or photoionization. We will not be able to test these exciting possibilities until we are confident that atomic rates and the observed emission line intensities are accurate at the 10-15% level. Relative emission measure distributions of different solar features, such as coronal holes, quiet sun, active regions and flares, can be derived without worry about the atomic rates or absolute instrumental calibration. Even the relative emission measures could be seriously in error, though, if Lyman continuum absorption is important and no lines longward of the Lyman limit are available.

2.2 Elemental Abundances

Elemental abundance determinations can really be considered part of the emission measure determination unless one has so little data that one or the other must be assumed. Therefore the same uncertainties enter. One can avoid many of the problems by comparing lines of different elements which are formed at the same temperature and which lie at the same wavelength. Widing and Feldman (1989) have found order of magnitude variations in the Ne V to Mg V line ratios at 400Å. Abundance variations imply deviations between photospheric and coronal values, implying that the abundances can vary with temperature through the transition region. This, unfortunately introduces as many free parameters as there are ions observed. This brings us to the brink of total chaos, and we either assume that these abundance variations are exceptional events which don't much affect the average global averages, or at least assume that each element changes smoothly with temperature. We should also keep the assumptions behind the derived abundance variations in mind, however. We still have only a theoretical basis for believing that two ions of different elements are formed at the same temperature. Even if this is true under one set of conditions, really drastic changes in the temperature distribution or departures from ionization equilibrium might cause the assumption to break down. Moreover, the dielectronic recombination rates which enter ionization balance depend on density (Summers 1974), so that, for instance, the quiet sun ionization balance may differ from the active region ionization balance by $\approx 30\%$. More recently, Reisenfeld (1992) and Reisenfeld et al. (1992) have shown that the enhancement of the dielectronic recombination rate by Stark mixing in an electric field (the Holtzmark field or a DC electric field) can significantly change the C III and C IV emissivities at transition region densities.

2.3 Electron Density

An up-to-date review of density diagnostic line ratios is given in Keenan's paper in this volume. From the point of view of the attainable accuracy, the important point is that each density sensitive line ratio involves 2 to 4 collision strengths and a radiative transition probability of at least one intercombination or forbidden transition. In many cases the excitation among fine structure levels by protons must also be taken into account. Even if all the relevant cross sections are known to $\approx 10\%$. The accumulated uncertainty is substantial, particularly if the lines are separated in wavelength so that a calibration uncertainty must be added. Worse still, most of the ratios of strong lines change by only a factor of 3 to 5 over a factor of 10-100 density range, so that the error in derived density is larger than the atomic rate or calibration errors.

Fortunately, there are several possible checks. The most basic is that most line ratios have a high or low density limit, or both. Any measured ratio outside the predicted limits shows that there is trouble somewhere, and C III and S V and Fe XII are perhaps good examples of improvements in atomic rates driven by observed line ratios (e.g. Dupree, Foukal and Jordan 1976; Dufton et al. 1986; Tayal et al. 1991). On the other hand one cannot help but think that there must be cases where erroneous observations have driven us to choose incorrect sets of atomic rates.

A second test of density diagnostics would be a plot of n_e vs. T derived from a large number of line ratios. This is somewhat difficult because any one ratio is density sensitive only over a limited density range. Typical analyses consider no more than half a dozen density measurements from any one spectrum, and it seems that a factor of 2 to 3 uncertainty in derived density is about as well as one can do (e.g Doyle et al. 1985; Keenan 19xx). It should also be kept in mind that most density diagnostic ratios are somewhat sensitive to temperature as well, and that some ratios are sensitive to Lyman continuum absorption.

2.4 Temperature Diagnostics

A temperature-sensitive line ratio is the intensity ratio of two lines of a single ion which show different temperature dependences due to differing excitation thresholds (via the Boltzmann factor) or differing recombination contributions or cross section energy dependences. Temperature sensitive line ratios have received far less attention than density diagnostics for several reasons. First, there is a touching faith that the solar plasma should be in ionization equilibrium except under exceptional circumstances, so that a temperature diagnostic would only tell you something you already know from ionization balance calculations. Second, a really good temperature diagnostic would involve lines whose excitation potentials differ by at least kT_m where T_m is the temperature where the ion is formed. In most cases this would involve a greater wavelength range than is available from a single instrument, and it would require an accurate excitation rate for a $\Delta n \geq 1$ quantum number changing transition.

Finally, if one is to derive a departure from ionization equilibrium, one requires a reliable equilibrium calculation, while the scatter in T_m among different calculations is typically $\Delta \log T_m \approx 0.1$ (et al. Arnaud and Rothenflug 1985; Arnaud and Raymond 1991). Most of the attempts thus far to model the solar EUV spectrum neglect the density dependence of dielectronic recombination, though this can be quite important for transition region ions (Summers 1974). Still worse, recent work on the Stark mixing of high n doubly excited levels (Riesenfeld 1992) shows that the Holtzmark field at transition region densities affects the C IV dielectronic recombination rate, and stronger fields may well be present.

REFERENCES

Arnaud, M., and Rothenflug, R. 1985, *Astr. Ap.*, **60** 425.

Davila, J. 1992, this volume

Delaboudiniere, J.P. *et al.* 1988, in *The SOHO Mission*, ESA SP-1104, p. 43

Doyle, J.G., and Raymond, J.C. 1983, *Solar Physics*, **90**, 97

Doyle, J.G., Raymond, J.C., Noyes, R.W., and Kingston, A.E. 1985, *Astrophys. J.*, **297**, 816.

Dufton, P.L., Hibbert, A., Keenan, F.P., Kingston, A.E., and Doschek, G.A. 1986, *Astrophys. J.*, **300**, 448.

Dupree, A.K., Foukal, P.V., and Jordan, C. 1976, *Astrophys. J.*, **209**, 621.

Herant, M., Pardo, F., Spiller, E., and Golub, L. 1991, *Astrophys. J.*, **376**, 797.

Hassler, D.M., Rottman, G.J., and Orrall, F.Q. 1991, *Astrophys. J.*, **372**, 710.

Kohl, J.L. *et al.* 1988, in *The SOHO Mission*, ESA SP-1104, p. 49

Lang, J., Mason, H.E., and McWhirter, R.W.P. 1991, *Solar Physics*, **129**, 31

Neupert, W.M., Brosius, J.W., Thomas, R.J., and Thompson, W.T. 199, this volume

Patchett, B.E. *et al.* 1988, in *The SOHO Mission*, ESA SP-1104, p. 39

Radiation from Hot, Thin Plasmas, J.C. Raymond 1988, in *Hot Thin Plasmas in Astrophysics*, R. Pallivicini, ed. (Kluwer: London), p. 3.

Highly Charged Ions in Astrophysics, in *Physics of Highly-Ionized Atoms*, R. Marrus, ed. (Plenum: New York) p. 189.

Reisenfeld, D.B. 1992, submitted to *Ap. J.*

Reisenfeld, D.B., Raymond, J.C., Young, A.R., and Kohl, J.L. 1991, *Ap. J.*, in press

Sobel'man, I.I., *et al.* 1990 *Sov. Astron. Lett.*.**16**. 137.

Summers, H.P. 1974, Appleton Laboratories Internal Memo AL-R-5

Tayal, S.S., Henry, R.J.W., Keenan, S.P., McCann, S.M., and Widing, K.G. 1991, *Astrophys. J.*, **369**, 567.

Walker, A.B.C., Jr. 1992, this volume

Widing, K.G., and Feldman, U. 1989, *Astrophys. J.*, **344**, 1046.

Wilhelm, K., *et al.* 1988, in *The SOHO Mission*, ESA SP-1104, p. 31

Tokamak Spectroscopy in the UV - Is This The Golden Age?

Alan T. Ramsey

Princeton Plasma Physics Laboratory,
Princeton University

ABSTRACT

Ultraviolet spectroscopy on tokamak plasmas, with its ability to measure resonance lines of both light impurities (O, C) and of the usual heavy impurities (Cr, Fe, Ni) has been the workhorse instrumentation in impurity monitoring. It also has been a productive technique in the study of particle transport by allowing measurement of the radial distribution of charge states of intrinsic impurities, and the temporal evolution of line radiation from injected impurities. The ion temperatures of the current tokamaks are large enough that Doppler broadening can be used to give accurate values of T_i, and Doppler shifts yield bulk plasma rotation velocities.

The most fruitful modern technique in both UV and visible tokamak spectroscopy is the study of spectra from neutral beam induced charge exchange. The rapid radiative decay of charge exchange excited atoms means that the light contains velocity and temperature information (via shift and broadening) and spatial information (the radiation comes from the intersection of the spectrometer and neutral beam sightlines). The next generation of tokamaks, however, will not be so easily probed by beams. The plasma will be bigger and denser so that beam attenuation (with current methods) will limit charge exchange work to about the outer 20% of the plasma radius.

Thus, our success may mean that our best modern technique is played out. Although we may be able to find solutions to allow limited charge exchange recombination spectroscopy to be done in such plasmas, it is likely that earlier approaches may find renewed application. Moreover, with radiation from fusion reactions reaching very high levels, not only must the spectrometer detector be well shielded, but near grazing incidence systems will suffer high levels of radiation streaming down the sightline. This will encourage less grazing, near-UV systems, perhaps looking again at M1 transitions and intercombination lines which occur at longer wavelengths than the usual E1 resonance lines for these plasma ions.

1 INTRODUCTION

Plasma spectroscopy predates research on controlled thermonuclear fusion, and continues with a vigorous life of its own outside the field. In the fusion effort, however, it has such an important role that it seems worthwhile to review it here separately. I shall outline what the tasks of tokamak plasma spectroscopy are and how they are performed. Although fusion research is quite directed, I shall mention briefly how some real spectroscopy is done on the side. A discussion of the future of fusion research and what that means to spectroscopy will suggest that we may be living now in a golden age, where easy access to the light source and the (relatively) friendly environment for the experiments will soon be ended. The next generation of fusion devices will be armored, radiation emitting behemoths who will yield their secrets only grudgingly, and at great cost.

This review will focus on fusion research done on tokamaks. For reasons having less to do with scientific merit than the constraints of funding, the world is concentrating its attention and money on the most promising path to the near exclusion of all else. While this lack of breadth in research is undoubtedly bad from a purely scientific point of view, such a limitation does not seriously limit the scope of this review. The techniques described here are also used on, for example, stellerators and reversed field pinches, with changes only to allow for the different time scales involved.

In the next section, we shall review the classical task of spectroscopy - impurity monitoring - and what information it can yield. In Section 3, we shall look at the uses of spectroscopy in direct measurement of plasma parameters such as temperature and motion. Section 4 will address briefly the question, "Does anyone do real spectroscopy on tokamaks?" Section 5 contains an outline of the future role of spectroscopy, and will show why the part becomes more difficult to play. Possible solutions to the problems will be presented.

The examples shown in this paper will be from spectroscopy in the ultraviolet. "Ultraviolet" here means wavelengths too short for viewing through a window - about 2000 Å - and too long for convenient use of crystals as Bragg diffractors - about 18 Å. The reader must remain aware, however, that what plasma spectroscopists do involves the full range of visible through UV wavelengths. Indeed, I will suggest later that the problems of UV spectroscopy may soon outweigh its advantages, and plasma spectroscopy may return to the visible where it started.

2 IMPURITY MONITORING

A tokamak is a magnetic confinement device whose plasma has the shape of a torus. The major confining field is around the long dimension of the torus. It is generated by a set of toroidal field coils, typically 10 to 20 in number. The plasma threads through these coils. The plasma carries a sizable current circulating in the toroidal direction (this is what makes it a tokamak) and the magnetic field associated with this current together with the toroidal field results in magnetic field lines that rotate in poloidal angle as they go around the torus. This rotational transform of the field gives the tokamak plasma its unique stability.

In the current generation of large machines (JET in England, JT-60 in Japan, TFTR and DIII-D in the US, and Tore-Supra in France) the toroidal fields are 1 - 5 Tesla, the plasma currents are 0.5 - 5 MA, the plasmas 0.5 - 2 m in diameter, and the major radius of the toroid is 1 - 3 m. In all cases, there is something which forms the primary plasma-machine interface for plasma particles which are not magnetically confined. This primary plasma interface is called the limiter, and in a well-constructed machine it generates most of the impurity content of the plasma.

What *is* an impurity? The fusion reaction for the foreseeable future is the D-T one; that is, $^2H + ^3H \rightarrow \alpha + n + 18$ MeV. Although JET has recently made several plasmas using a small amount of tritium as a trial of their handling abilities, its radioactivity ($^3H \xrightarrow{t_{1/2}=12\,yrs} {}^3He + e^+ + \nu + 18$ keV) makes it a nuisance to deal with. For that reason, magnetically confined fusion research today uses H or D as sole fuel gas. Everything else is an impurity. The most common limiter material is graphite, although refractory metals have been and will be used, and JET has had success with a Be limiter. The vacuum vessel itself is always made of stainless steel. Water vapor adsorbs on all surfaces in a vacuum vessel and slowly desorbs during operation. The common impurities in all tokamaks, therefore, are C, O, Fe, Ni, and Co. Add to these elements used as getters, as well as structural and electrical elements, and you can find Be, B, Al, Si, K, Cu, W, and Ta. Air leaks, surprisingly uncommon, add N. Helium is often used as a temporary fuel gas, or as a plasma source to clean the limiter.

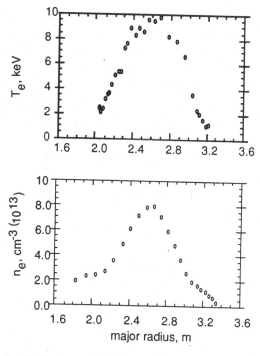

Fig. 1. The temperaturea and density profiles for TFTR during a high-power beam heated shot

What the tokamak operator wants to know is usually simple. "What fell in during that shot?" "Why is the radiated power so high?" "Is the iron higher than yesterday?" To answer these questions, both qualitatively and quantitatively, all machines have routine impurity monitors. What should this monitor be able to see?

Figure 1 shows typical electron temperature and density profiles of TFTR during auxiliary heating. The heating (by the injection 25 MW of neutral particles beams into the plasma) has resulted in a 10 keV central electron temperature. The electron temperature is what determines what ionic states and excitation levels we will see in the tokamak; the time scale of the plasmas (1 to 10 seconds) and the average confinement time of the particles (0.1 to 1 second) means that the plasma is usually in some kind of gross equilibrium. The density is low enough that the plasma is optically thin for all lines. In fact, this rather low density would limit brightness were not the size of the plasma so large.

To view the range of impurities listed above over the temperature range shown in Fig. 1, Fonck and co-workers at Princeton's Plasma Physics Lab developed the survey spectrometer SPRED(Fonck, Ramsey and Yelle 1982). An optical schematic is shown in Fig. 2. The 100 Å - 1100 Å range may be read out every 20 ms if desired. The resolution (~1.8 Å) is sufficient to isolate most lines. Fig. 3 shows the SPRED spectral range, and the location of the longer wavelength member of the Li-like doublet for elements of atomic number 8 through 35. The brightness of the Li doublet and its (usual) 2:1 intensity ratio makes these transitions an excellent diagnostic for the presence of an impurity. Elements with Z<8, Be, B, and C, have bright lines that fall into SPRED's range, too, so that virtually any element present in a tokamak can be uniquely and quickly identified. SPRED is usually located viewing radially across the plasma toward the tokamak's center. This integrated view includes radiation from all temperature and density zones. Figure 4 shows the radial location where the Li-like state becomes most prominent for the TFTR discharge of Fig. 1. You can see that, for this rather hot plasma, even Ge (used as a trace element to study transport) is ionized to this high state.

Finally, for completeness, I feel obliged to show a spectrum from TFTR taken by our SPRED. Figure 5 shows the spectrum from a shot during auxiliary heating (20 MW of neutral injection, 4 MW of ion-cyclotron frequency rf heating) with a helium puff for density control. The spectrum contains the stainless steel lines (best seen between 100 Å and 150 Å), He (very bright at 304 Å) and carbon everywhere. What is most notable here is the near-total absence of oxygen; several locations where O lines are easily seen are marked (the O VI doublet at 1032-1036 Å is one), although there is nothing there. This is a dull, although common, spectrum on TFTR - it is mostly carbon. The tiny unmarked lines in the spectrum are mostly C, too, although a few are Fe, Ni, or Co. There is nothing more exotic in the plasma visible even to a very high throughput spectrometer like SPRED.

So much for qualitative detection. An equally important task is to determine how much of a particular impurity is in the plasma. SPRED can be absolutely calibrated. One method is to use an absolutely calibrated visible spectrometer (itself calibrated by use of an absolutely calibrated tungsten ribbon lamp) and the simultaneous measurement of a pair of lines, one visible and one UV, from the same level. The known brightness ratio of these

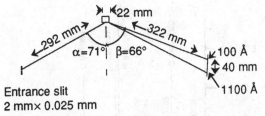

Entrance slit
2 mm × 0.025 mm

Fig. 2 The SPRED grating is a Jobin-Yvon Type IV holographically produced grating especially designed for this instrument. The grating blank is fused silica with a toroidal figure. There is little loss of light due to astigmatism with a 2.5 mm entrance slit, which gives SPRED its very high throughput. The grating has a flat focal field over the 40 mm detector surface. The grooved area in operation is usually masked to 21 × 3 mm. The grooves are ion-etched, and the grating is coated with gold.

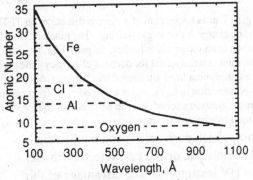

Fig. 3 We see here the wavelength of the upper member of the Li-like doublet for those elements which fall in the SPRED range of 100 Å to 1100 Å.

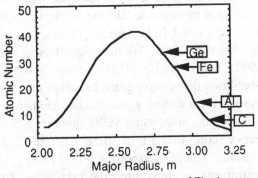

Fig. 4 Using the electron temperature of Fig. 1, we display here the radius where the 3-electron ion reaches its brightest point and starts to ionize away. The flags on the curve show the locations of several common elements.

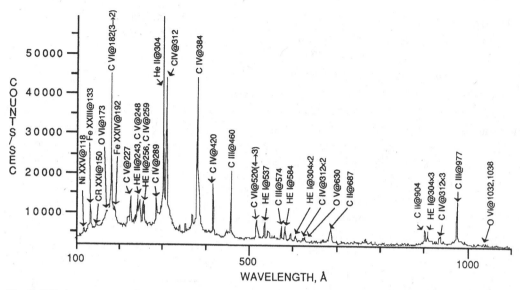

Fig. 5 This is a spectrum of a plasma discharge on TFTR with auxiliary heating. The presence of He in this discharge is due to gas puffing. The machine is well conditioned, so that the only major impurity is carbon from the graphite limiter. In particular, note the near total absence of oxygen from water. The spectrum is uncorrected for detector efficiency - the peak of the efficiency curve is at about 250 Å. The most prominent lines are caused by charge exchange from the injected D neutrals. The line at 182 Å is the $3\rightarrow2$ transition in C VI, and is a cascade line from capture into a higher n level. The He II line at 304 Å is largely electron excited, but the existence of this much He II is due to a shift in the ionization balance caused by charge exchange.

lines, which does not depend on the details of the plasma or the excitation, calibrates the UV instrument. The advantage of this technique is that it is done *in situ*. The disadvantage is that there is a shortage of available line pairs in most plasmas. An alternative method is to take the instrument to a calibrated light source, such as a synchrotron source. We have done this at PPPL, using NIST's SURF facility, and found that the results agreed to within the measurement accuracy with the branching ratio points. Since using SURF involves a heavy commitment of time and the attendant risk of transporting a calibrated instrument, we have used the branching ratio technique to track the calibration, using the SURF data to give the shape of the curve between the scattered calibration points.

With your spectrometer calibrated, what

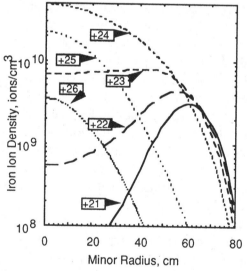

Fig. 6 The iron ion density distribution in a plasma during the beam-heated phase. The central electron temperature is 8.0 keV; the central electron density is 8.2×10^{13} cm^{-3}. This distribution is calculated by the impurity transport code MIST, using measured density and temperature profiles for a high confinement TFTR plasma.

can you see? Figure 6 shows the result of a model calculation of the spatial distribution of the high ionization states of iron. The calculation was done by the *Multi-Ionic Species Transport* code, named MIST(Hulse 1983). With SPRED viewing through the plasma, all of the charge states are seen, and the observed brightness of a line is an integral over the emission region of the parent ion, with the ionic density and electron density and temperature as parameters. MIST is used self-consistently, in the following way. Lines can be observed that arise at the edge of the plasma (near to the iron source) and at the center, where the iron has been transported by the plasma. MIST tells you what the brightness ratio of these lines is for an assumption about the transport of the iron in the plasma. Once you have found transport parameters (usually expressed as an effective diffusion constant and a convective velocity) that give a satisfactory match of the relative brightnesses across the plasma, the code tells you what the absolute impurity density must be to yield the observed brightnesses. Table 1 shows the results of such an exercise for Fe at two times in a discharge. By doing such a fit to all the major impurities, one can obtain an estimate of the total radiated power from the plasma. When this is compared with the measured value from the wide angle bolometer, the results can be quite close. This gives us considerable confidence in the procedure.

While such analyses give some information about plasma transport mechanisms, the integrated nature of the data smooths out too many details which we would like to know. Another way to obtain transport parameters is to watch transport as it occurs. Using a laser to blow a trace impurity off the back of a glass slide and into the plasma, we have used SPRED to study the temporal evolution of the impurity as it penetrates into the core of the plasma and then is lost as it random-walks back to the edge and is swept off by the limiter. Figure 7 shows such temporal behavior for a core state of Ge, injected as described above(Stratton et al. 1989). This technique has been fully exploited, and has been quite useful.

As useful as impurity injection with integrated views is, you must understand that anything which measures transport attracts a crowd in fusion physics. Transport (of heat, particles, momentum) determines confinement, and confinement is all. It is easy, relatively speaking of course, to put energy into a plasma.

Fe in the Ohmic Phase			
Ionic State	λ, Å	Brightness	B_{obs}/B_{fit}
+15	335	2.3×10^{13}	1.09
+15	360	1.0×10^{13}	0.93
+22	133	1.9×10^{13}	1.09
+23	192	2.0×10^{13}	1.09
+23	255	7.5×10^{12}	0.80
$P_{rad} = 17$ kW $P_{in} = 1$ MW			

Fe in the Beam Phase			
Ionic State	λ, Å	Brightness	B_{obs}/B_{fit}
+15	335	7.3×10^{13}	1.14
+15	360	2.7×10^{13}	0.84
+22	133	1.6×10^{14}	0.48
+23	192	5.2×10^{14}	0.94
+23	255	3.0×10^{14}	1.08
$P_{rad} = 241$ kW $P_{in} = 13$ MW			

Table 1 This is the result of MIST modeling of the brightnesses of 5 Fe lines from the edge and the center of the TFTR plasma. The fit was done with a diffusion constant of D = 1.0 M^2/s, and the assumption that the impurity density has the same shape as the electron density (this is the same as assuming a flat Z_{eff} profile).

Ohmic heating in a tokamak, driven by the transformer action used to maintain the plasma current, can furnish a few megawatts (a 1 volt drop around the torus plasma loop multiplied by a few megamperes of plasma current). Anything composed largely of ions in a magnetic field obviously will absorb rf power at a number of frequencies, so rf heating is a widely used technique. The injection of highly energetic neutral particles (neutral to penetrate the confining magnetic fields) to heat the plasma is a use of technology not dissimilar to that used in the internal combustion engine: It appears impractical and crude at first encounter, but has been driven by necessity to unlikely levels of utility and sophistication. The reduction of energy loss from a plasma, however, has yielded

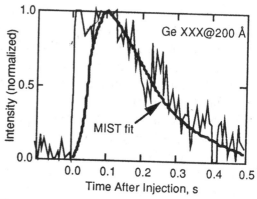

Fig. 7 Using SPRED with its radial sightline, impurity injection has been used to study transport. The temporal evolution of the Ge XXX line has been fit by MIST. In this ohmic discharge, Ge XXX peaks in the center of the plasma column. To obtain the fit, MIST used a diffusion constant of D = 1.3 m^2/s along with some convective velocity. In the experiment, both core and edge lines are fit similtaneously. The rise just after Äction is due to an unresolved line from a lower ionization state.

less to our efforts. Transport in tokamaks is called "anomalous" because we understand it poorly at best. Anything which enables a closer look is pursued.

If radial, integrated, sightlines give only crude spatial information, why not look with several tangential sightlines, each of which sees a little deeper into the plasma? A fan array of sightlines, each absolutely calibrated, can yield a radially resolved profile of the observed line. Such systems, like the one developed by Moos and co-workers at Johns Hopkins(Richards, Moos and Allen 1980), have been developed. They are limited, however, by the need to do what amounts to an Abel inversion, with that implicit differentiation of the data which increases noise substantially. What was needed was true spatial resolution in measurements.

Such a technique is charge exchange recombination spectroscopy (CXRS), and is the premier method for obtaining spatially resolved data on optical emission in a tokamak, both in the UV and the visible(Isler and Murray 1983)(Fonck, Darrow and Jaehnig 1984). This technique utilized the reaction between an energetic neutral atom (injected along a narrow, well known path by a neutral injector) and a thermal ion in the bulk plasma. The reaction, for a case of usual interest in tokamaks, may be denoted: H(energetic)+C^{+6}(thermal)→H^{+1}(still energetic)+C^{+5}(still thermal)+γ(182 Å)+other γ's. The excited lifetime of the product ion C^{+5*} is so short that it will have moved only a centimeter or so by the time the photon is emitted. Thus, the radiation is localized by the intersection of the sightline of the spectrometer and the neutral atom beam. Figure 8 shows the SPRED sightline on TFTR and its intersection with the neutral injectors. One intersection is near the center of the plasma, and was used to directly measure the density of carbon in the plasma in a region where it is completely stripped(Stratton et al. 1990). This is not a weak signal: Fig. 5 shows the C^{+5} $3p \rightarrow 2s$ line at 182 Å that was used for this work. It is one of the brightest lines in the spectrum. The cross sections are as big as kinetic cross sections for many cases; that is, about the size of the atom, 1 Å^2. Figure 9 shows Stratton's results compared with the usual measure of C density using visible bremsstrahlung. The value of Z_{eff} ($Z_{eff} = \Sigma Z_i \times (Z_i n_i)/n_e$, the average charge of an ion in the plasma weighted by the fraction of electrons it contributes to the total electron density) is used as a global measure of plasma purity. A pure hydrogen plasma has $Z_{eff}=1$; a pure carbon plasma has $Z_{eff}=6$. It can be seen in Fig. 9 that the agreement with the visible bremsstrahlung value of Z_{eff} is very good when the Z_{eff}'s of the C from CXRS and the metal contributions are added.

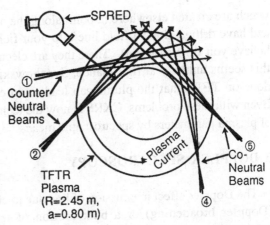

Fig. 8 The UV spectrometer SPRED looks across 2 neutral beams, each of which has 3 sources. In most plasmas no signal can be seen from beamline 2 by the time that the beams have penetrated to the intersection of the SPRED sightline. For precise measurements, two of the three beamline 1 sources can be turned off, giving radial resolution of only a few cm.

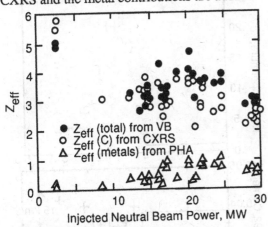

Fig. 9 Using CX radiation from the C VI line at 182 Å, and calculating the beam attenuation before it reaches the viewline intersection with SPRED, the absolute C density was calculated. That C density is here compared with the global measure of impurities, the Z_{eff}. The Z_{eff} of the metals, shown separately, was measured by the PHA system. The sum of the C Zeff and the metals Zeff is very close to the global value from visible bremsstrahlung, as we expect.

The benchmarking of a technique as shown in Fig. 9 is essential. All methods have serious potential problems, and CXRS is no exception. As Figure 8 shows, the beam must traverse substantial plasma before it reaches the sightline of the spectrometer. This attenuation must be known exactly, because the brightness of the CXRS line depends linearly on the excitation rate, which is the beam intensity. The beam may traverse several mean free paths before the intersection with the spectrometers sightline is reached, and thus the overall attenuation is a very strong function of the plasma density. This is one of the reasons that JET's von Hellerman has remarked that one of the limiting parameters for extracting data from plasma spectroscopy is the plasma density(von Hellermann 1992). Another critical problem is the effect of what is called the plasma plume. The plume is composed of ions in the same charge state you are observing, which are created elsewhere (a point along the neutral beam which you are not observing) and have drifted along a field line into your field of view (although not at the radius you believe you are measuring). There they are electron-impact excited, and radiate. Although this seems unlikely at first glance, Synakowski has shown in He transport work he has done on TFTR that the plume can lead to an error of nearly 100%(Synakowski 1992). Even with these problems, CXRS is now the technique of choice for the local measurement of plasma parameters by spectroscopic means.

3 DOPPLER SPECTROSCOPY

The Doppler effect in emission of line due to either a random, thermal motion of the ions (Doppler broadening) or a bulk motion of the ions (Doppler shift) contains useful information about the plasma ions which is hard to measure otherwise. As ion temperatures reach values of 30 keV, as they do on TFTR during neutral injection heating, the fraction of the stored energy in the hot ions becomes crucial to the energy balance. Most tokamaks have a high resolution, grazing incidence spectrometer like SOXMOS which can reliably measure the 0.2 Å - 0.5 Å linewidths of core ions at these temperatures(Schwob et al. 1987). For example, SOXMOS (a 2 m, 1/2 ° grazing incidence Rowland circle instrument) has been used at TFTR to measure the core ion temperature during neutral injection heating by observing the 192 Å line of Fe XXIV(Strachan et al. 1987). With a good knowledge of the instrumental width (which we measured *in situ* on low ionization states during the plasma breakdown before any heating has taken place) it was possible to measure temperatures as low as several keV. (The instrumental width of our SOXMOS is about 8 keV for Fe at 192 Å.)

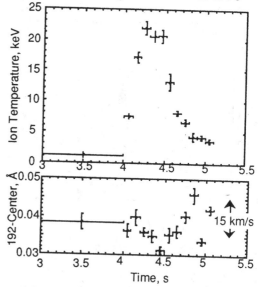

Figure 10 shows the temperature history of a TFTR discharge. The center of a line can be determined more easily than its width; accuracy of 10% of the line width is easy, and 1% can be achieved if the lineshape is well known. The lower part of Fig. 10 shows the line center during the heating injection. SOXMOS was installed nominally radial to the torus of TFTR, but with a velocity resolution of 5 km/s, we can see the effect of a small off-normal angle in the installation. A tangentially viewing instrument can measure toroidal rotation with good accuracy.

Fig. 10 The Fe XXIV line at 192Å is a good choice for Doppler width measurements. Here, heating is applied from 4 to 4.5 s. The instrumental width of SOXMOS at this line is about 8 keV. However, the instrumental width is so well characterized that reliable values for T_i were obtained down to ohmic discharge values of a few keV.

4 BUT CAN WE DO REAL SPECTROSCOPY?

There have been real spectroscopists looking at fusion devices from the beginning of the field, identifying unknown lines, making precision measurements of line positions, and generally improving our knowledge of atomic structure. As the magnetic fusion energy effort has become more mission-oriented in the past 10 years or so, this work has lost much of its support at the same time that the environment of the tokamak has become more hostile. Spectroscopists still persist, however, because the rewards can be substantial. Sugar and his coworkers at NIST are a good example. Working with Rowan at the University of Texas's TEXT tokamak, they have used photographic plates(!) to determine lines centers to 0.005 Å in the Cu I and Zn I isoelectronic series to elucidate certain features in the level structures(Sugar et al. 1991). At these high values of effective charge, QED corrections (particularly the self-energy) become quite large. Because of its size, the time scale of its discharge, and the high electron temperature, a tokamak is a very attractive spectral source. And although the environment, both physical and bureaucratic, will become even less attractive in the next generation of machines, the much higher electron temperature, density, and plasma duration will create a situation where real spectroscopists will continue to insinuate themselves into the programs.

5 A LOOK AT THE FUTURE

The current generation of large tokamaks will be replaced by a device or devices burning tritium. The best defined machine currently is the International Tokamak Experimental Reactor (ITER), an effort supported by all the major players in fusion research. There may be a smaller machine as an intermediate step, but this is uncertain now. We will use ITER as a reference point in this discussion.

Whatever the details, the next tokamak will use D and T as fuel, and have $Q \gg 1$. Perhaps it will be ignited. It will probably have a higher confining field and a larger plasma current than present tokamaks. Because of the high neutron flux and the attendant damage to structure (accumulated damage will be several displacements per atom for ITER), as well as the larger current and fields, the machine will be much less accessible for diagnostics. Indeed, the present ITER design has diagnosticians scrambling to find access for their instruments. Some of the problems can be overcome by brute force, some will yield to new ideas, and some may force us to fall back to older methods.

Figure 11 shows some projected ITER profiles taken from simulations by Redi,(Redi, Cohen and Synakowski 1991) with the TFTR profiles of Fig. 1 overlaid on it. In both axes TFTR (and, indeed, even the larger JET) is dwarfed by ITER. First, the good news. The central electron temperature (for this particular ITER scenario) is 30 keV. Heavy ion fanciers will note that this will allow observation of the L_α line of ^{92}U. The lower graph is where the trouble lies, however. Although the central density is only twice that of a peaked TFTR profile, the radial extent of this high density region is 9 times greater.

To see what that means for charge exchange measurements, we look at some calculations by Schilling in Fig. 12(Schilling 1992). They show the brightness of the C^{+5} n=8→7 transition at 5292 Å, a commonly used CXRS line, for a probe beam of 1 A. The radius is normalized so that ITER and TFTR can be compared on the same axis. We can see that the outer 20% of the ITER plasma is as accessible as TFTR (and what goes on in the outer edge is crucial for tokamak confinement). The signal from the core will be attenuated by a factor of 100, however.

By increasing the neutral beam voltage we improve penetration, but at the cost of a lower cross section for charge exchange. Roughly speaking, we cannot improve the signal by increasing beam voltage beyond about 120 keV. Beam intensity is our only knob to turn up, and that is near its limit with current technology.

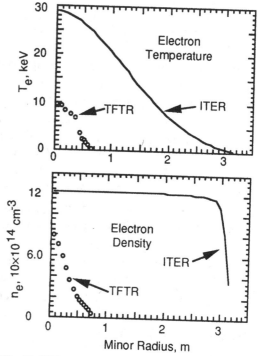

Fig. 11 ITER temperature and density profiles from a discharge simulation, with the TFTR profiles from Fig. 1 overlaid.

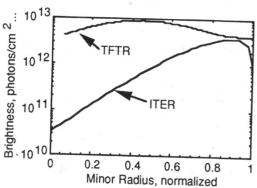

Fig. 12 The brightness of the C^{+5} n=8→7 CX line for a 1 A beam in a simulation for ITER and TFTR, showing the effects of the ITER density and size. The radius is normalized separately for each machine.

Although some are optimistic that subtlety and ingenuity can still recover data with present techniques, Fonck and coworkers are looking into pulsed beam techniques(Fonck 1992). Extremely intense neutral beams of very short duration an be produced. Using gated detectors, the data accumulation time can be made very short, with an attendant reduction in background and noise. The correspondence with laser pulse measurements of Tompson scattering, universally used to determine T_e and n_e in plasmas, is very close. Whether we can overcome factors of 100 is yet to be seen.

The higher density may cause other problems. Certain very low signal diagnostics, such as attempts to see He line radiation from fusion produced α-particles, are limited by photon statistics on the background. The bremsstrahlung continuum depends on n_e^2, while the He signal will increase more slowly.

Radiation levels will rise by at least a factor of 200 for any D-T machine; for ITER, the factor will be much greater. Omnidirectional radiation from (n,γ) reactions induced by thermalized neutrons in the machine enclosure causes both noise and damage in detectors. Solid state array detectors, now almost universally used in UV spectroscopy, are more susceptible than photomultiplier tubes. Charge-coupled devices, the detector of choice in current systems, have a lower damage threshold than photodiode arrays, and will need extensive shielding. This may be possible with a spectrometer with a large angle between the entrance and exit beam lines (as in the case of SPRED - see Fig. 2). Grazing incidence systems like SOXMOS, however, are almost unshieldable, since radiation streams down the entrance arm and directly onto the detector. Add to that the intrinsic lower throughput of a grazing incident spectrometer, and the signal to noise ratio rapidly becomes unusable. Already on TFTR, the background levels on SOXMOS during high neutron shots saturates the detector in 100 ms accumulation times. (SPRED, with an identical detector system, is only mildly affected.)

Solutions to the streaming problem are to get the exit beam from the disperser well away from the entrance beam. An important region of UV spectroscopy is 19 Å to 40 Å, which contains the resonance lines of oxygen and carbon, the main light impurities in tokamaks. It will be crucial to monitor this radiation in ITER.

A technique for doing this may be the use of multiple layer mirrors in the UV. Essentially Bragg reflectors, they can produce high reflectivity at non-grazing angles with sufficient spectral resolutions to monitor isolated lines (like oxygen L_α at 19 Å). Finkenthal's group at Johns Hopkins is aggressively pursuing this path.

Figure 13 shows the reflectivity of a mirror use by Zwicker (of Finkenthal's group) and coworkers at General Atomic to look at plasmas from DIII-D(Zwicker 1992). The 8% reflectivity and a 0.6 Å fwhm at this wavelength make a powerful combination, and suggest such elements as plasma facing dispersers on an ITER-class tokamak. A possible problem: the enhanced sensitivity of such an exacting structure (the mirror in Fig. 13 has 100 layers) to neutron damage. We should note here that this spectral region overlaps that exploited by X-ray crystal spectroscopy, and Schumacher and coworkers at Garching have developed an elegant system using crystals and X-ray detectors to monitor these lines. The large angles of diffraction effectively remove the detectors from streaming and allow a great amount of space for shielding(Schumacher 1991).

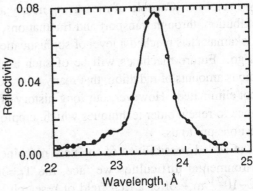

Fig. 13 The reflectivity of a 100 layer Monel/B$_4$C multilayer mirror. A Mirror like this has been used to monitor impurities on DIII-D by performing an angular scan. This resolution is sufficient to separate the O and C resonance lines in this wavelength region.

Another solution is to use light guide techniques. UV can be transmitted, although with some loss, by total internal reflection down many centimeters of hollow capillary tube. Shielding can be placed around the capillaries, and streaming greatly reduced. A high throughput 7-chord spectrometer with a light guide input, operating from 60 Å to 250 Å, has been built by Krieger and coworkers and is at Princeton waiting testing on a tokamak.

Even with some success in these techniques, we still may find the most prudent course is to retreat to earlier techniques. Even if we lose the precise spatial resolutions offered by CXRS because of beam penetration failure, the use of temporal evolution of injected impurities to study transport, as well as modeling with chordally integrated data, is still a powerful technique.

Spectrometers with near-normal reflections and working at longer wavelengths can be used to study magnetic dipole and intercombination lines, which usually occur at longer wavelengths than the resonance lines from a particular ionic species. These lines grow brighter as Z increases, and as T_e increases, we reach higher effective Z's. Another possibility is to retreat to the visible, where fiber optic cable allows the light to be transported to a shielded room. However, the use of fiber optics will be difficult, too, because of radiation induced luminescence and opacity, already seen at present radiation levels on TFTR(Ramsey and Hill 1992). Perhaps remote visible spectrometers using relay mirrors and shielded lenses can overcome this(Ramsey 1986)(Morsi et al. 1991).

6 CONCLUSIONS

The contribution that spectroscopy makes to tokamak research is perhaps the largest of any discipline. Capable of measuring nearly any phenomenon from the plasma current distribution, through transport and fluctuations, to impurity content, plasma spectroscopy on tokamaks has reached a level of sophistication and power which probably is as far as it can go. Future machines will be of such a physical configuration, and produce such copious amounts of radiation, that many of our current techniques will be severely limited if not eliminated. However, our long history of spectroscopic research now offers us the chance to renew older techniques which, employed skillfully, will probably answer most questions put to us.

The added possibilities of the new machines are exciting. If we can overcome the environmental difficulties we face, the T_e=30 keV, 3 m long plasma columns with $n_e \ell \sim 10^{21}$ m^{-2} offer a rich field of research. With cleverness, and the persistence to overcome a certain amount of bureaucratic inertia, real spectroscopists may find the next generation of tokamaks the best platform ever for UV spectroscopy.

ACKNOWLEDGEMENTS

My colleagues have been very generous with their help during the preparation of this review. In particular, conversations with M. von Hellerman (JET), R. Fonck (U. Wisc.), and B. Stratton (Princeton) have been particularly useful. Data from and conversations with the following also are gratefully acknowledged: B. Denne (JET), K. Hill,M. Redi,

G. Schilling, and E. Synakowski (Princeton), and E. Källne (KTH, Stockholm).
This work was supported the United States Department of Energy under contract DE-AC02-76-CHO-3073.

References

Fonck, R. J., 1992, *private communication.*

Fonck, R. J., Darrow, D. S., and Jaehnig, K. P., 1984, *Phys. Rev. A* , **29**, 3288.

Fonck, R. J., Ramsey, A. T., and Yelle, R. V., 1982, *Appl. Op.* , **21**, 2115.

Hulse, R. A., 1983, *Nucl. Technol./Fusion* , **3**, 259.

Isler, R. C., and Murray, L. E., 1983, *Appl. Phys. Lett.* , **42**, 355.

Morsi, H. W., et al. in Controlled Fusion and Plasma Physics (Proc. Eighteenth European Conference on Controlled Fusion and Plasma Physics, Berlin, 1991) (1991) **IV**: 2611.

Ramsey, A. T., 1986, *Rev. Sci. Instrum.* , **57**, 2017.

Ramsey, A. T., and Hill, K. W., 1992, *Rev. Sci. Instrum.* , submitted),

Redi, M. H., Cohen, S. A., and Synakowski, E. J., 1991, *Nuc. Fusion* , **31**, 1689.

Richards, R. K., Moos, H. W., and Allen, S. L., 1980, *Rev. Sci. Instrum.* , **51**, 1.

Schilling, G., 1992, *private communication.*

Schumacher, U., 1991, *private communication.*

Schwob, J. L., Wouters, A. L., Suckewer, S., and Finkenthal, M., 1987, *Rev. Sci. Instrum.* , **58**, 1601.

Strachan, J. D., et al., 1987, *Phys. Rev. Lett.* , **55**, 1004.

Stratton, B. C., et al., 1989, *Nucl. Fusion* , **29**, 437.

Stratton, B. C., et al., 1990, *Nucl. Fusion* , **30**, 675.

Sugar, J., et al., 1991, *JOSA B* , **8**, 1795.

Synakowski, E., 1992, *private communication.*

von Hellermann, M., 1992, *private communication.*

Zwicker, A. P., 1992, *private communication.*

FUV Plasma Diagnostics Available to the Hopkins Ultraviolet Telescope

William P. Blair[1], Knox S. Long[2], Charles W. Bowers[1],
Gerard A. Kriss[1], and Arthur F. Davidsen[1]

[1]Center for Astrophysical Sciences, The Johns Hopkins University
[2]Space Telescope Science Institute

ABSTRACT

The Hopkins Ultraviolet Telescope (HUT) flew aboard the Astro-1 space shuttle mission in December 1990, observing a wide range of astrophysical objects in the FUV (912 - 1860 Å; $\Delta\lambda = 3$–6 Å) range and a handful of objects in the EUV (415 - 912 Å; $\Delta\lambda = 1.5$–3 Å) region. Approximately 40 hours of spectrophotometric observations were obtained for 77 astronomical sources including quasars, galaxies and galaxy clusters, globular clusters, cataclysmic variable stars, supernova remnants, planetary nebulae, white dwarf stars, Be stars, cool coronal stars, and solar system objects (plus a wealth of information on the airglow produced in the earth's upper atmosphere). The sub-Lyman α region was of particular interest because of the many potential plasma diagnostics in this spectral region, including O VI $\lambda\lambda$ 1032,1038 (both in emission and in absorption), the H_2 Lyman and Werner bands of hydrogen, as well as transitions of ions of intermediate mass elements. In addition, the extended continuum coverage to the Lyman limit provides considerable leverage to continuum fits in low-redshift active galaxies, cataclysmic variables, and hot stars. We discuss some of these diagnostics and how they are being used to interpret the HUT spectra of various objects.

1 INTRODUCTION

The Hopkins Ultraviolet Telescope (HUT) flew on the space shuttle Columbia in December 1990 for 9 days as part of the Astro-1 space shuttle mission (STS-35), the first Spacelab mission to be dedicated to a single discipline. The Astro Observatory was an attached payload that used the shuttle for power, pointing, and communications. Three ultraviolet telescopes (including HUT) were mounted and aligned on the Instrument Pointing System, provided to NASA by ESA and used previously on the Spacelab-2 mission in 1985. An X-ray telescope and spectrograph (the Broad Band X-ray Telescope – see article by R. Petre from this meeting) was mounted on a separate pointing system and could co-point or offset from the pointing position of the ultraviolet instruments. The ultraviolet telescopes were operated by the astronauts from the aft flight deck. The instruments could also be operated from the ground during periods of TDRS contact, and were run in this manner for more than half the mission after on-board computer hardware problems made it impossible for the crew to operate the instruments directly.

HUT consists of an f/2 – 0.9 m iridium-coated primary mirror that feeds a prime focus spectrograph with a holographically-ruled grating, a microchannel-plate inten-

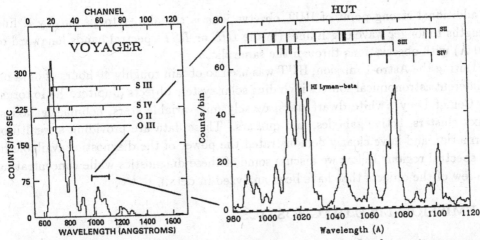

Figure 1—A comparison of *Voyager* and HUT data for the Io plasma torus around Jupiter. Even with HUT's moderate resolution the two broad blends in the *Voyager* spectrum resolve into many separate features belonging to S^+, S^{+2}, and S^{+3}.

sifier, and photon-counting detector. The grating is coated with osmium for good FUV reflectivity and the microchannel-plate is coated with cesium iodide to provide ultraviolet sensitivity. In first order the spectrograph covers the region from 830–1860 Å at 0.51 Å pixel^{-1} with a resolution of $\sim$ 3 Å. Objects with sufficient EUV flux can be observed in second order (415–930 Å; resolution about 1.5 Å pixel^{-1}), using an aluminum filter to block the first order spectrum. (EUV measurements with HUT are described by R. A. Kimble *et al.*, this conference.) Various circular or rectangular apertures could be selected by rotating a focal plane slit wheel, which also included a sealed position to maintain the spectrograph under vacuum during ground operations and testing. A TV camera viewed the focal plane for target acquisition and guiding. Further details of the telescope, spectrograph, and calibration can be found in Davidsen *et al.* (1992a).

The primary purpose of HUT was to open the spectral band from 912–1200 Å to general scrutiny for the first time. This band contains many important plasma diagnostics, including O VI $\lambda\lambda$ 1032,1038, the H$_2$ Lyman and Werner bands of hydrogen, and transitions of many ions of intermediate mass elements. The continuum coverage to the Lyman limit also provides sensitivity to higher temperatures in continuum fits of low-redshift active galaxies, cataclysmic variables, white dwarfs, and other hot stars. While *Copernicus* was used to observe many objects at high spectral resolution in this band, it could only observe the brightest stars. The Ultraviolet Spectrometers (UVSs) on the *Voyager* spacecraft have been used to successfully observe a number of astronomical sources in this band, but at much lower spectral and spatial resolution than HUT. Figure 1 shows a comparison of the *Voyager* UVS and HUT data for the Io plasma torus around Jupiter (cf. Moos *et al.* 1991). The difference in resolving power of the HUT observation is immediately obvious as the two large blends seen in the *Voyager* data are resolved into many separate features.

An additional strong point of HUT observations is the ability to tie sub-Ly α line strengths to longer wavelength lines (in the *IUE* or *HST* spectral bands longward of 1200 Å) with observations through the same slit.

During the Astro-1 mission, HUT was used to obtain roughly 40 hours of data on 77 different astronomical sources including solar system objects (Jupiter, the Io torus, and Comet Levy), white dwarf stars, cataclysmic variable stars, nebulae, galaxies, galaxy clusters, active galaxies, and quasars. These data are providing a wealth of information and have clearly demonstrated the power of the diagnostics available in this spectral region. Below we discuss some of these diagnostics while concentrating on a few of the objects that have been analyzed in detail.

2   EMISSION LINE DIAGNOSTICS

2.1   SPECTRA OF SUPERNOVA REMNANT SHOCK WAVES

Observations of supernova remnant (SNR) shock waves below 1200 Å are of great interest because of the new diagnostics available in this wavelength range. Of particular importance is the resonance doublet of O VI at 1032,1038 Å, which samples a hotter portion of the post-shock flow than other UV emission lines and can be used as a diagnostic of faster shock waves. Shock model calculations such as those of Hartigan *et al.* (1987) and Raymond *et al.* (1988) are fairly sophisticated, but are virtually untested for the lines expected below 1200 Å. O VI $\lambda\lambda$ 1032,1038 was seen in absorption when *Copernicus* observed stars behind the Vela SNR (Jenkins *et al.* 1976), and Shemansky *et al.* (1979) and Blair *et al.* (1991a) discuss data for the Cygnus Loop obtained with the *Voyager* 2 UVS. However, the large *Voyager* field of view blended the spectra of many filaments together and the ~ 33 Å effective resolution of the *Voyager* observations made even line identifications difficult.

Because of its proximity, well-resolved filamentary structure, and relatively low reddening (E(B - V) = 0.08; Fesen *et al.* 1982), the Cygnus Loop is an important object for the study of shock waves and their interaction with the interstellar medium. The Cygnus Loop is the prototypical "middle-aged" SNR. At a distance of 770 pc (Fesen *et al.* 1982; Hester *et al.* 1986), its $\sim 3 \times 4°$ angular extent corresponds to a linear dimension of $\sim 40 \times 54$ pc. HUT was used to observe two filaments in the Cygnus Loop during Astro-1. The first was an optically bright "radiative" filament near the eastern limb (Blair *et al.* 1991b). The second filament was on the northeastern limb of the remnant on the very edge of the X-ray emitting region (i.e. presumably at the position of the primary shock front – see Long *et al.* 1992). The filament at this position is faint optically, emitting almost nothing but hydrogen Balmer line emission, although the ultraviolet emission lines are in general within a factor of two of the bright optical filaments (Raymond *et al.* 1983). A 9.4″ × 116″ aperture was placed on each filament, and reduced flux-calibrated spectra from these two observations are shown in Figure 2a and 2b. These spectra are from roughly 650 s

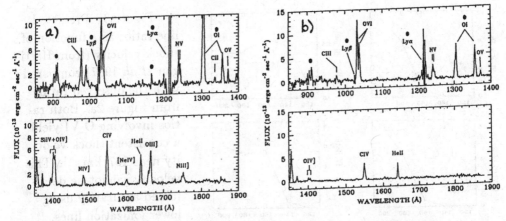

Figure 2—The HUT spectra for two filaments in the Cygnus Loop supernova remnant. a) The spectrum of a bright radiative filament on the eastern limb, and b) the spectrum of a nonradiative filament at the position of the primary shock front in the northeast. Note the strong O VI emission lines at both positions and the almost total lack of intermediate and low ionization lines at the position of the primary shock front.

night portions of the observations when contamination due to terrestrial airglow was minimized.

The HUT spectra in Figure 2 are the first such data to fully characterize the entire sub-Lyman α region of interstellar shock waves. Looking first at Figure 2a (i.e. the radiative shock position), the dominant emission lines below Ly α are C III $\lambda977$, N III $\lambda991$, and a very strong O VI doublet seen in the 2:1 ratio expected for optically thin conditions. The C III and N III lines can both be used with longer wavelength features to determine the electron temperature in the regions emitting these ions (see below). A number of fainter features are seen throughout the spectrum, many of them for the first time in SNR spectra (Blair *et al.* 1991b). A rough comparison with published shock models confirms that the lines expected to be strong in the sub-Ly α region are indeed those which we observe.

Restricting ourselves to the highest ionization stages, the comparison of O VI:N V: C IV to shock model predictions is very interesting. Figure 3 shows the predicted ratios as a function of shock velocity for the models of Hartigan *et al.* (1987) along with the values inferred from the data of Figure 2a. Both the O VI:N V and O VI:C IV ratios are consistent with a peak shock velocity of 167 km s^{-1} at the observed position. The ratios of N V:C IV and C IV:C III are consistent with this value, but have "saturated" and cannot be used to determine the velocity for shocks this fast (see Figure 3). Hence, above about 160 km s^{-1}, O VI is a crucial diagnostic of shock velocity. The consistency of the shock velocities determined using lines of different elements argues for the correctness of the assumed abundances, which in the case of the Hartigan *et al.* models were solar.

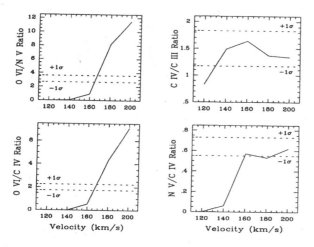

Figure 3—Predicted line ratios as a function of shock velocity (from Hartigan *et al.* 1987) compared with the observed ratios from Figure 2a. Both ratios involving O VI yield a consistent shock velocity near 167 $km\,s^{-1}$. This velocity cannot be determined accurately using lower ionization lines.

Optically, the radiative filament observed with HUT would be classified as an "incomplete" shock; the optical [O III] $\lambda5007$ emission is very strong relative to Hα, indicating that only the hotter portion of the recombination and cooling zone is present. The strength of the optical [O III] lines suggests that the observed filament represents a fairly recent encounter between an interstellar cloud and the shock wave. Many of the intermediate and low ionization UV lines are actually too strong relative to a steady flow shock model for $\sim 167\,km\,s^{-1}$. This suggests either that slower velocity shocks are mixed within the HUT aperture, or possibly that this fast shock is thermally unstable, similar to the picture described by Innes *et al.* (1987). Further high dispersion optical spectra with spatial resolution will be necessary to distinguish between these two pictures.

Although the optical properties of the second (nonradiative) filament observed with HUT are quite different, it may simply represent a more extreme version of the incomplete radiative shock described above. The HUT spectrum in Figure 2b shows some similarities to Figure 2a when only the high ionization lines are considered: strong O VI, N V, C IV, and He II $\lambda1640$ are present in both spectra. However, this filament shows very little emission from the cooler ions. The implication is that the nonradiative filament represents an even more recent shock/cloud encounter than the radiative filament in Figure 2a, so that only the hottest portion of the cooling zone is present.

The O VI lines in Figure 2b deviate somewhat from the 2:1 ratio expected for the optically thin case. This is due to resonance line scattering in the nonradiative filament, which is thought to be a sheet of gas being viewed in an edge-on geometry (Long *et al.* 1992). We have used the observed ratio to correct the spectrum for resonance line scattering and determine intrinsic line intensities at this position. Comparison of these line intensities with model calculations indicates a shock speed near 180 $km\,s^{-1}$ for this filament. We also find it necessary to decrease the carbon abundance to a value near that expected for diffuse interstellar clouds in order to match the line intensities at this position. In conjunction with the observation at the other position observed with HUT, a consistent picture is derived if interstellar dust grains have not yet had time to be destroyed since the shock passage at the non-radiative position, while the dust grains have been sputtered away at the radiative position, enhancing the carbon abundance over typical interstellar values.

2.2 EMISSION LINE RATIOS IN ACTIVE GALAXIES

Active galaxies were also an important class of objects for HUT. Low redshift objects such as NGC 4151 (Seyfert 1) and NGC 1068 (Seyfert 2) had never been observed below Ly α even though they are the nearest and brightest members of their respective classes. The HUT observation of NGC 4151 is described by Kriss et al. (1992a). Below we discuss the HUT observation of NGC 1068 (cf. Kriss et al. 1992b).

NGC 1068 is the prototype Seyfert 2 galaxy, and the discovery of polarized emission from broad permitted lines with widths of a few thousand $km\,s^{-1}$ (the defining characteristic of Seyfert 1 galaxies) by Antonucci and Miller (1985) is important evidence of a possible link between the two classes. Lawrence and Elvis (1982) have suggested that Seyfert 1's and 2's are similar phenomena related by varying degrees of obscuration and orientation. The nuclear continuum and broad line region are obscured from direct view by an opaque, molecular torus, and a hot electron plasma scatters the radiation into our line of sight. With the additional diagnostic emission lines at wavelengths shortward of Lyα made accessible by the FUV sensitivity of HUT, our observations of NGC 1068 were planned to help unravel the complex physical conditions in the nuclear region.

In two separate pointings centered on the optical nucleus of NGC 1068, we used circular apertures of 18″ and 30″. Only the 18″ aperture data will be discussed in detail here. The highest quality data were obtained during orbital night when the airglow is at a minimum; this portion of the integration totalled 2060 s. Figure 4 shows the flux-calibrated HUT spectrum of NGC 1068 from this observation. The spectrum obtained through the 30″ aperture is qualitatively similar, but the continuum is $\sim$35% brighter. Since the larger aperture includes portions of the 1 kpc starburst ring, it is plausible that the additional light is due to young stars and H II regions.

The moderately strong emission lines C III λ977 and N III λ991 in our spectrum of NGC 1068 were unexpected. These emission lines have never been seen in spectra of other AGN, including others observed with HUT during Astro-1. These lines are not expected to be strong in photoionized gas because they require high excitation temperatures. The emission line spectrum of NGC 1068 throughout the HUT wavelength range bears a qualitative resemblance to the HUT spectrum of the radiative shock in the Cygnus Loop described above, which could indicate an important contribution from shock heated material.

To make a more quantitative assessment of whether collisional processes in addition to photoionization are necessary to explain the NGC 1068 emission line spectrum,

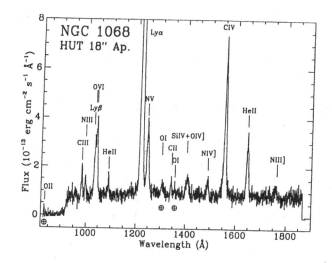

Figure 4 – The flux-calibrated HUT spectrum of NGC 1068. Prominent emission lines are marked and airglow emission is indicated with an earth symbol. Lyα from NGC 1068 is blended with geocoronal Lyα. Note in the region below Lyα the strong emission from C III λ977, N III λ991, Lyβ, O VI $\lambda\lambda$1032,1038, and He II λ1085.

we use the C III line ratio $(I(\lambda1907) + I(\lambda1909))/I(\lambda977)$ and the N III line ratio $I(\lambda1750)/I(\lambda991)$ as temperature diagnostics (cf. Kriss *et al.* 1992b for details). Since the C III] $\lambda\lambda$1907,1909 lines are outside the HUT wavelength range, we use the flux from IUE archival spectra (Snijders *et al.* 1986). We assume purely collisional excitation in the low density limit ($n_e < 10^5$ cm^{-3}). This gives a temperature of 26,700 K for the C III line ratio, and a temperature of 24,000 K using the N III line ratio. We note that higher extinction or corrections for resonance line scattering will lead to higher temperatures. These are very difficult temperatures to support with photoionization (Kwan and Krolik 1981) and we therefore suggest that collisional heating may be a significant mechanism in the nucleus of NGC 1068. Possible sources of this heating include thermal emission from the hot nuclear wind proposed by Krolik and Begelman (1986) as the scattering medium, or shocks from the interaction of this wind with the interstellar medium. Other alternatives include shocks from radio plasma interacting with clouds in the narrow line region (Evans *et al.* 1991), or supernova shocks from starburst regions in or near the nucleus.

3 ABSORPTION LINE DIAGNOSTICS

In addition to emission line diagnostics, sources with continuum spectra below Lyα permit us to search for absorption by intervening material (either interstellar or associated with the source directly). Accretion disks in cataclysmic variables, for instance, are strong sources of ultraviolet continuum. HUT observations of the cataclysmic variable Z Cam in outburst showed many absorption lines in the sub-Lyα region, including strong O VI, N III, C III, Lyman lines of atomic hydrogen, and others (Long *et al.* 1991). Likewise, we have been able to detect O VI in absorption against the quasar 3C 273, implying a substantial hot halo surrounding our Galaxy (Davidsen *et al.* 1992b). Below we discuss in more detail an observation that detected the H$_2$ Lyman and Werner bands in absorption.

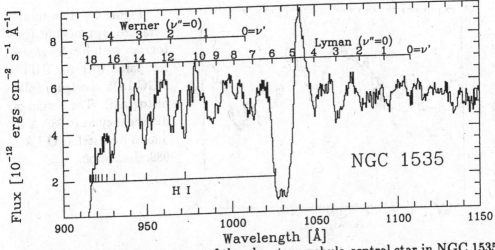

Figure 5—The HUT spectrum of the planetary nebula central star in NGC 1535.
Note the strong P Cygni feature at O VI, and the strong FUV continuum from the
central star. Expected positions of absorptions due to atomic hydrogen (below) and
the Lyman and Werner bands of molecular hydrogen (above) are indicated.

A bright, high-excitation planetary nebula, NGC 1535, was observed during the
Astro-1 mission. The central star of this PN is thought to have a temperature near
70,000 K (Mendez et al. 1988) and has relatively little extinction (E(B - V) ≤ 0.1;
Sabbadin et al. 1984, Mendez et al. 1988). Hence, it is a strong source of FUV
continuum emission. This nebula has a galactic latitude of −40° and lies at a distance
of ~1.6 kpc, well out of the galactic plane. HUT was used to observe the central star
for 700 s before offsetting to obtain a nebular spectrum. Figure 5 shows the resulting
flux-calibrated spectrum below 1150 Å. No reddening correction has been applied to
the data in this Figure. (Note: a value of E(B - V) = 0.1 appears necessary to match
the 70,000 K temperature of the central star with our observations.)

The spectrum consists of the central star continuum, a strong P Cygni profile of
O VI (and N V and C IV at longer wavelengths), and a series of absorption features
with depths ranging from ~10–50 percent of the continuum. The P Cygni lines have
a terminal velocity of ~ 3000 km s^{-1}, indicative of a very strong wind emanating from
the central star. In Figure 5 we have indicated the positions of the Lyman series of
atomic hydrogen and the Lyman and Werner bands of molecular hydrogen. The line
positions shown in Figure 5 are those of the strong R(0) ($J'' = 0$, $J' = 1$) rotational
transitions between the ground vibrational state ($v'' = 0$) and the upper vibrational
states indicated. The coincidence between the absorption features and the molecular
transitions clearly shows a substantial column of H$_2$ is present along the line of sight
to NGC 1535.

To analyze this spectrum, we have synthesized H$_2$ spectra, smoothed to the HUT
resolution, and compared these model results to the observed spectrum of NGC 1535
(Bowers et al. 1992). In our models, we have constrained the relative J'' populations

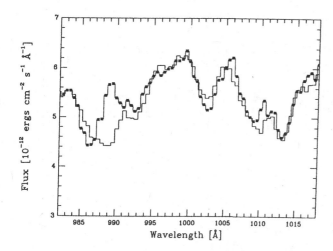

Figure 6 – Two temperature H_2 model fit (histogram with black squares) compared with the HUT NGC 1535 spectrum (plain histogram). The region of discrepancy near 989 Å is due to interstellar O I λ 989 absorption.

to follow a distribution characterized by one or two temperatures. This permits the number of models to be kept manageable and is descriptive of many of the configurations which might be expected. We have concentrated on a 40 Å region of the spectrum near 1000 Å which contains both Lyman and Werner bands but very little contamination from atomic or interstellar transitions; the single exception is O I λ989. Also, our models show a good deal of temperature sensitivity for the transitions in this spectral region.

Figure 6 shows our best two temperature model fit to the data. The two temperatures are 144 K and 1570 K. (A single temperature model with T = 400 K fits reasonably well also, but is significantly worse for the feature near 987 Å.) From *Copernicus* observations, the half thickness of the H_2 layer associated with the galactic plane is only 65 pc (Scheffler and Elsasser 1988). Because of the location of NGC 1535 far off the plane and the high temperatures inferred for the H_2 population structure, we believe the H_2 is associated with the PN. Depending on the assumed distribution of this material in the PN (and the assumed Doppler b value), we find that the mass of H_2 involved is very small, $\leq 0.1\ M_\odot$. Such a small amount of H_2 would not have been detectable in the infrared. This demonstrates the power of the ground-state-connected FUV transitions for detecting very small amounts of H_2.

4 SUMMARY

It is clear that during Astro-1 we have barely scratched the surface in utilizing the many plasma diagnostics available to HUT. In many cases only the nearest or brightest examples of a class of objects have been observed and in some cases whole classes of objects were missed entirely. The ultraviolet telescopes are scheduled to fly again in mid-1994 on the Astro-2 mission. We look forward to expanding on the work begun with Astro-1.

The building of HUT and the successful flight of Astro-1 involved the best creative and technical efforts of many people; to properly thank them all in this space would be impossible. We especially acknowledge the work of many people at NASA Marshall Space Flight Center and at the Johns Hopkins University and Applied Physics Laboratory for their efforts to make this flight possible. The Hopkins Ultraviolet Telescope project is supported by NASA grant NAS 5-27000 to the Johns Hopkins University.

REFERENCES

Antonucci, R. R. J., and Miller, J. S. 1985, *Ap. J.*, **297**, 621.

Blair, W. P., *et al.* 1991a, *Ap. J.*, **374**, 202.

Blair, W. P., *et al.* 1991b, *Ap. J.*, **379**, L33.

Bowers, C. W., *et al.* 1992, *Ap. J.*, submitted.

Davidsen, A. F., *et al.* 1992a, *Ap. J.*, **391**, in press.

Davidsen, A. F., *et al.* 1992b, *Ap. J.*, submitted.

Evans, I. N., Ford, H. C., Kinney, A. L., Antonucci, R. R. J., Armus, L., and Caganoff, S. 1991, *Ap. J.*, **369**, L27.

Fesen, R. A., Blair, W. P., and Kirshner, R. P. 1982, *Ap. J.*, **262**, 171.

Hartigan, P., Raymond, J. C., and Hartmann, L. H. 1987, *Ap. J.*, **316**, 333.

Hester, J. J., Raymond, J. C., and Danielson, G. 1986, *Ap. J.*, **303**, L17.

Innes, D. Giddings, J. R., and Falle, S. A. E. G. 1987, *M.N.R.A.S.*, **226**, 67.

Jenkins, E. B., Silk, J., and Wallerstein, G. 1976, *Ap. J. Suppl.*, **32**, 681.

Kriss, G. A., *et al.* 1992a, *Ap. J.*, **392**, in press.

Kriss, G. A., *et al.* 1992b, *Ap. J.*, submitted.

Krolik, J. H., and Begelman, M. C. 1986, *Ap. J.*, **308**, L55.

Kwan, J., and Krolik, J. H. 1981, *Ap. J.*, **250**, 478.

Lawrence, A., and Elvis, M. 1982, *Ap. J.*, **256**, 410.

Long, K. S., *et al.* 1991, *Ap. J.*, **381**, L25.

Long, K. S., *et al.* 1992, *Ap. J.*, submitted.

Mendez, R. H., Kudritzki, R. P., Herrero, A., Husfeld, D., and Groth, H. G. 1988 *Astr. Ap.*, **136**, 113.

Moos, H. W., *et al.* 1991, *Ap. J.*, **382**, L105.

Raymond, J. C., *et al.* 1983, *Ap. J.*, **275**, 636.

Raymond, J. C., *et al.* 1988, *Ap. J.*, **324**, 896.

Sabbadin, F., Bianchini, A., and Hamzaoglu, E. 1984, *Astr. Ap.*, **136**, 183.

Scheffler, H., and Elsasser, H. 1988, in *Physics of the Galaxy and Interstellar Matter* (Springer-Verlag), 361.

Shemansky, D. E., Sandel, B. R., and Broadfoot, L. 1979, *Ap. J.*, **239**, 35.

Snijders, M. A. J., Netzer, H., and Boksenberg, A. 1986, *M.N.R.A.S.*, **222**, 549.

Stellar Chromospheric and Transition Region Studies Using The Goddard High Resolution Spectrograph

Alexander Brown[1]

[1]Joint Institute for Laboratory Astrophysics, University of Colorado and National Institute of Standards and Technology

ABSTRACT

The Goddard High Resolution Spectrograph (GHRS) on the Hubble Space Telescope (HST) is providing ultraviolet spectra of cool star chromospheres, transition regions, and coronae of a quality previously unobtainable for stars other than the Sun. This instrument offers greatly improved sensitivity, spectral resolution, and temporal resolution over previous satellites. These capabilities are illustrated with examples of stellar spectra obtained during the Science Verification and Cycle 0 phases of the HST mission. These spectra allow investigation of the ranges of plasma temperature and density and dynamic effects such as flows, turbulence, and flares occurring in cool star outer atmopsheres. Stars for which GHRS spectra have been obtained include Capella (α Aur, G9 III + G0 III), γ Dra (K5 III), Aldebaran (α Tau, K0 III), Betelguese (α Ori, M2 I), AR Lac (RS CVn binary), and AD Leo (M4 flare star).

1 INTRODUCTION

The capabilities of the GHRS represent a major advance for the study of the ultraviolet (UV) spectra emitted by cool star chromospheres, transition regions (TR), coronae and winds. The processes that heat these atmospheric regions and provide the driving mechanisms for stellar winds are presently poorly understood, and new observations with high spectral resolution and signal-to-noise can hopefully provide important advances in our knowledge concerning cool star outer atmospheres.

The GHRS is a spectrograph that offers a range of dispersion which are choosen by rotating a grating carrousel. Full details of the instrument are given in the GHRS Instrument Handbook and SV Report available from the Space Telecope Science Institute. The low dispersion mode has a nominal resolution of order 2,000, while the intermediate mode provides a nominal resolution of 20,000. An echelle mode allows observation of small intervals of spectrum at a resolution of 90,000. Spectra obtained at these three resolutions are illustrated in Fig. 1. A target can be observed through one of two entrance apertures; with the small 0.25 x 0.25 arcsec aperture the full design spectral resolution is achieved, while use of the larger 2 x 2 arcsec aperture results in a spectral resolution that is about a factor of two lower than the design values due to the large HST point spread function. The technical advantages of GHRS over previous UV astronomical satellites that were capable of observing the spectra of cool stars, such as IUE and *Copernicus*, include

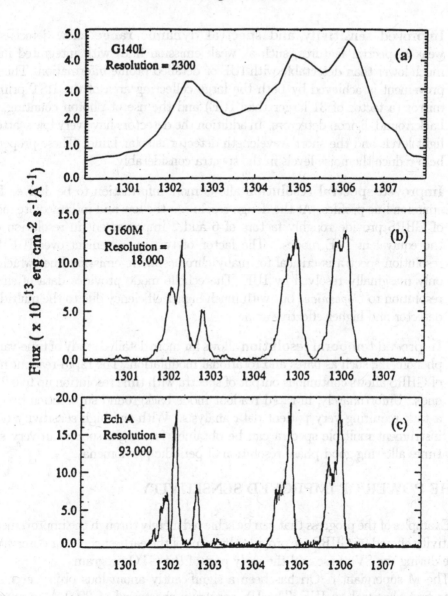

Figure 1: Three GHRS spectra of the same spectral region, the O I triplet of γ Dra near 1300Å. All these spectra were obtained using the small science aperture and illustrate the design resolution of the spectrograph. The increase in spectral information with increasing resolution is dramatically illustrated. The O I profiles show the influence of absorption due to a cool stellar wind on optically thick resonance lines of neutral atoms.

- **Improved sensitivity and spectral dynamic range** allow detection of weaker spectral features, such as weak emission lines with integrated fluxes much lower than detectable with IUE or detailed profile information. This improvement is achieved by both the large collecting area of the HST primary mirror (a factor of 31 larger than IUE) and the use of photon counting, low background digicon detectors. In addition the detectors have very low scattered light levels and the short wavelength detector is solar blind. These properties help reduce the noise levels in the spectra considerably.

- **Improved spectral resolution** allows more information to be derived from emission line profiles. At the design resolution the low and intermediate modes of GHRS provide roughly factors of 6 and 2 improvement in resolution over the equivalent IUE modes. The factor of two improvement over IUE high resolution spectra is critical for many chromospheric emission lines which are only marginally resolved by IUE. The echelle mode provides data of similar resolution to *Copernicus* but with much higher efficiency due to the multidiode detector and higher effective area.

- **Improved temporal resolution** allows far more detailed study of time-variable phenomena, such as flares and rotational modulation. The rapid-readout mode of GHRS allows continuous output of spectra with time resolution up to 0.2 seconds. Unfortunately, in rapid readout mode background subtraction becomes a task requiring very patient data analysis. With the high sensitivity of the instrument multiple spectra can be obtained for bright sources in very short times allowing good phase resolution of periodic phenomena.

2 THE POWER OF IMPROVED SENSITIVITY

Examples of the progress that can be achieved merely through the improvement in sensitivity offered by GHRS are provided by some of the earliest cool star observations made during the SV phase and the early part of the GTO program.

The M supergiant α Ori has been a significantly anomalous object ever since it was first observed by IUE. The UV spectrum shortward of 2000 Å was unlike that shown by other cool, low-gravity stars. Many of the emission lines that were strong in other stars were absent in the spectrum of α Ori and the features that were present were not readily identifiable. The low resolution G140L spectrum obtained by Carpenter clearly shows the reason for this previous confusion and opens significant new potential modelling techniques for M supergiant outer atmospheres.

The spectrum of α Ori is dominated by a continuum originating from the chromosphere surrounding the star that provides a background radiation field against which CO molecules in the shell far from the star absorb. (See Fig 2.) Many of the strongest emission lines seen from other stars are so self absorbed due to the large amounts of circumstellar material that they are barely visible

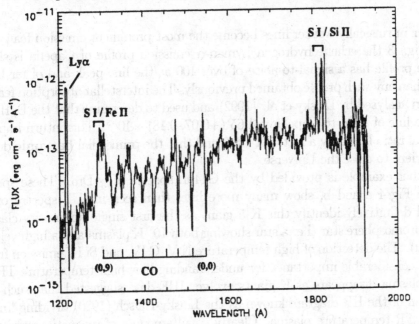

Figure 2: Low resolution spectrum of α Ori showing a chromospheric continuum formed at 8000 K and with a turbulent velocity of 20 km s^{-1}, based on Fe II absorption profiles seen near 1650 Å in higher resolution spectra. Strong CO absorption lines are seen due to molecules in the circumstellar shell. These molecules are at a temperature of $\sim$ 500 K and indicate a turbulent velocity of 5 km s^{-1} in this region.

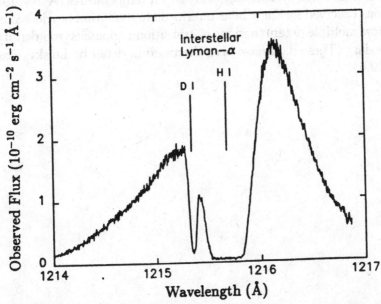

Figure 3: GHRS Echelle spectrum of the H Lyman-α line of Capella. Superposed on the stellar emission line are interstellar absorption lines dur to hydrogen and deuterium.

and other intrinsically weaker lines become the most prominent emission features.

In Fig. 3 the echelle hydrogen Lyman-α emission profile of Capella is shown. This line profile has a signal-to-noise of over 100 at the line peak and of far higher quality than any such profile obtained previously. The interstellar absorption features have been analysed by Linsky et al. (1992) and used to determine that the D/H ratio along the line of sight to Capella is 1.65 (+0.07,-0.18) $\times 10^{-5}$. This inturn implies a primordial ratio larger by a factor of 1.5-3 and that the primordial **baryonic** density is insufficient to close the Universe.

Another example is provided by the G140L spectra of γ Dra. These spectra, shown in Fig 4 a and b, show many more lines than seen in IUE spectra of this star, and definitively identify this K 5 giant as the first single luminosity class III hybrid-chromosphere star (i.e. a star showing both 10^5 K plasma and a high velocity, cool wind). The detection of high temperature N V, C IV, and Si IV emission from γ Dra is of considerable importance for understanding how high temperature TR and coronal plasma disappears as K giants evolve. IUE data suggested that such stars in a region of the HR diagram known as the Linsky-Haisch (1980) dividing line did not have TR temperature plasma. Clearly small amounts of magnetic activity can persist among the late K giants.

As a final example, the 1400 Å spectral region of the binary Capella (Fig. 5) shows weak intersystem lines of O IV and S IV at a signal-to-noise not seen before in stellar spectra. Capella is a 104 day spectroscopic binary consisting of a slowly rotating G9 III primary and a more rapidly rotating G0 III secondary. The G0 star dominates the ultraviolet spectrum of the system, especially at TR temperatures (Ayres, 1988). The spectral regions observed for Capella at intermediate resolution provide considerable data that allow multiple determinations of transition region electron density for the Capella secondary. These data have been discussed in detail by Linsky, Brown, and Carpenter (1991).

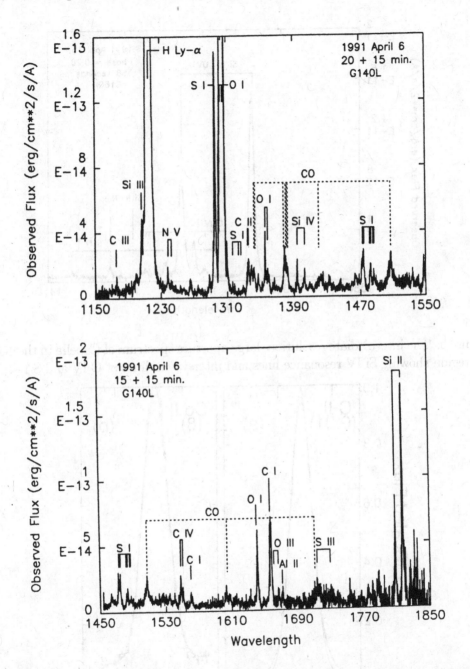

Figure 4: The low resolution GHRS spectrum of γ Dra contains over 80 emission lines formed in the chromosphere and transition region of this star. The presence of weak C IV emission, indicative of 10^5 K plasma, is unique for a normal late K giant and suggests that many giants previously thought to have only cool chromospheres may still retain small amounts of magnetic activity.

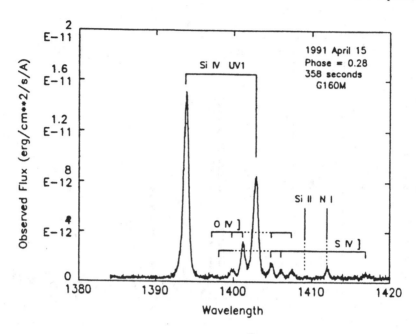

Figure 5: The intermediate resolution, large aperture spectrum of Capella in the 1400 Å region showing Si IV resonance lines and intersystem lines of O IV and S IV.

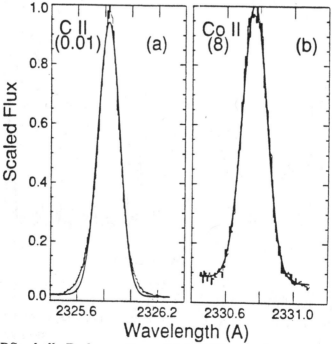

Figure 6: GHRS echelle B observations of (a) C II] and b) Co II lines with best fit Gaussians overplotted in heavy lines (Carpenter et al., 1991)

3 THE POWER OF IMPROVED SPECTRAL RESOLUTION

High quality emission line profiles can be obtained using GHRS with sufficient resolution to resolve the profiles and allow detailed study of tubulence and flows in cool star outer atmospheres with unprecidented precision. Fig. 6, from Carpenter et al.(1991), shows the C II λ 2325 and Co II λ 2330 lines of α Tau at echelle resolution. There is clearly additional broadening of the C II due to some form of turbulence or other mass motion. Also with accurate line profiles and wavelength scales it is possible to measure systematic flows in stellar outer atmospheres. For α Tau Carpenter et al.found that the C II intersystem lines indicated a downflow of 4 km s^{-1} in their region of formation.

Intermediate resolution observations of the RS CVn binary, AR Lac, have been made by Walter, which are being used to attempt active region Doppler imaging in both chromospheric (Mg II h and k) and TR (C IV) emission lines. While the emission lines of the rapidly rotating components of an RS CVn binary system are very broad, the emission emitted from individual active region can be very narrow and the additional resolution provided by GHRS is important for investigations such as these. This is the first time that such observations have been possible for TR plasma.

The γ Dra O I spectra shown in Fig. 1 show how additional information is gained with increasing spectral resolution. The λ1302 line has the largest optical depth and the λ1306Åthe least. The wind absortion feature becomes progressively stronger with increasing line optical depth. These profiles in combination with Mg II profiles obtained at the same time can be used to make a detailed model of the wind density, velocity and acceleration.

4 THE POWER OF IMPROVED TIME RESOLUTION

Rapid readout observations have been obtained for two dM flare stars, AD Leo and AU Mic. The AD Leo observations of Bookbinder contained an extraordinary flare, which showed extremely red-shifted C IV and Si IV emission associated with the flare peak. The motions implied by these red shifts is over 1,800 km s^{-1}, an extraordinary velocity never seen previously in a stellar TR flare. Fig. 7 shows the changes in the TR emission line profiles. In the upper panel, a contour plot is presented that shows both how the flare peaked very rapidly and decayed slowly. The flare response is seen most strongly in the TR plasma. In the lower panel, the C IV emission line profile is plotted for various stages during the flare. The extreme red-shifted emission is restricted to the 35 second period corresponding to the flare peak.

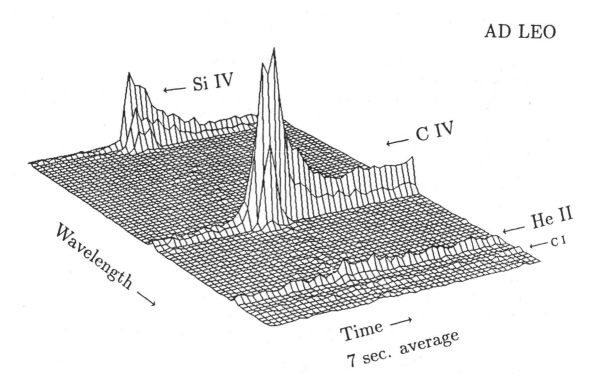

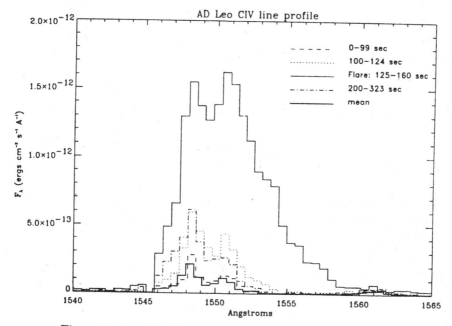

Figure 7: Rapid readout GHRS data of the AD Leo flare.

The work described in this paper results primarily from the efforts of the GHRS Instrument Definition Team (IDT) during Science Verification and Guaranteed Time Observations. Those who have provided material for this review include J. L. Linsky (JILA), K. Carpenter, R. Robinson, G. Wahlgren (Goddard Space Flight Center), F. Walter (SUNY Stony Brook), J. Bookbinder (Center for Astrophysics), B. Savage, A. Diplas (U. of Wisconsin), and J. Brandt (LASP, U. of Colorado). This research was supported by NASA grant S-56500-D to the National Institute of Standards and Technology.

REFERENCES

Ayres, T. R. 1988, *Ap. J.*, **331**, 467.

Carpenter, K. G., Robinson, R. D., Wahlgren, G. M., Ake, T. B., Ebbetts, D. C., Linsky, J. L., Brown, A., and Walter, F. M. 1991, *Ap. J. (Lett.)*, **377**, L45.

Linsky, J. L., Brown, A., and Carpenter, K. G., 1991, in *The First Year of HST Observations*, ed. A. L. Kinney and J. C. Blades (Baltimore: STScI), p. 70.

Linsky, J. L., Brown, A., Gayley, K., Diplas, A., Savage, B. D., Ayres, T. R., Landsman, W., Shore, S. N., and Heap, S. R. 1992, *Ap. J.*, submitted.

Linsky, J. L., and Haisch, B. M. 1979, *Ap. J. (Lett.)*, **229**, L27.

LYMAN THE FAR ULTRAVIOLET SPECTROSCOPIC EXPLORER
WARREN MOOS

Center for Astrophysical Sciences
The Johns Hopkins University
Baltimore, Maryland 21218

1 ABSTRACT

Lyman FUSE is an Explorer class mission which will perform astronomical spectroscopic observations in the 100 to 1550 Å wavelength range with a high spectral resolution capability from 912 to 1250 Å. The instrumentation will have a long slit capability with an angular resolution of order one arcsec. The instrumentation will have high sensitivity with an effective area ~50 cm^2 and a dark rate < 1 count/pixel/hr. A CCD fine error sensor will permit acquisition and tracking of objects with V>16. With these capabilities, the mission is envisioned both as a follow on to the Copernicus mission - but with vastly increased capability - and as bridge to the gap between the HST cutoff near 1200 Å and the much shorter wavelengths addressed by x-ray missions.

2 INTRODUCTION - MISSION OVERVIEW

The Lyman FUSE mission has been described in detail in several previous publications (See the Phase A Report by Moos et al. 1989, also Moos 1990, 1991). Recently, the mission plan has been modified with a decision by NASA to assign a dedicated spacecraft instead of utilizing the Explorer Platform constructed for EUVE. This paper briefly reviews the scientific basis for the mission and the status of the technical preparation.

The mission is presently in Phase B with Phases C/D expected in FY 95. Technical and budgetary constraints lead to a launch date of 2000. The mission will be launched on a Delta rocket which can place a ~1300 kg package in a highly elliptical high- earth 24 hour orbit with an apogee ~ 70,000 km. Typically ~ 16 hrs of viewing above the radiation belts would be available each day. There is a clear advantage in operational simplification in using such an orbit. However, both weight and radiation damage to the components are potential areas of risk. NASA Goddard Space Flight Center is presently conducting a detailed study to evaluate the technical feasibility of a high-earth orbit.

Most of the observing time (of order 90%) will be available to the general astronomical community through a Guest Observer Program. The mission has received broad community support which as demonstrated by endorsements in two successive reports by the Astronomy and Astrophysics Survey Committees (Field et al. 1982, Bahcall et al. 1991).

The Canadian Space Agency is participating with NASA in the definition and development of the FUSE mission. Canada will supply both fine error sensor and the baffle system for the telescope. Additional contributions are being made by UK and French scientists.

A mission of this size requires the efforts of many more scientists and engineers than could be listed here. These efforts are gratefully acknowledged. Table 1 lists the members of the Phase B Science Team.

TABLE 1 - Phase B Science Team

The Johns Hopkins University
H. Warren Moos, Principal Investigator
A. Davidsen
P. D. Feldman

Goddard Space Flight Center
A. Boggess
A. Michalitianos
J. Osantowski
G. Sonneborn, Project Scientist
B. Woodgate

University of California
C. S. Bowyer
R. Malina
O. Siegmund

University of Colorado
W. C. Cash, Jr.
J. L. Linsky
M. Shull

Dominion Astrophysical Obs.
J. B. Hutchings

University College London
A. Willis

University of Hawaii
L. L. Cowie

Smithsonian Astrophysical Obs.
A. Dupree
L. P. Van Speybroeck

Kitt Peak National Obs.
R. Green

Princeton University
E. Jenkins

Univ. of Wisconsin, Madison
B. Savage

Stanford University
J. G. Timothy

University of Chicago
D. G. York

Institut d'Astrophysique
A. Vidal-Madjar

3 SCIENTIFIC GOALS

It is no surprise to participants at this conference that the spectral region covered by the Lyman FUSE mission provides a unique set of diagnostics for addressing astrophysical problems. A large number of astrophysically important species have their most important - in many cases their only - transitions in this spectral region e.g. deuterium, molecular hydrogen, helium, O^+ to O^{5+}, Ne to Ne^{5+}, S^+ to S^{12+}, A and A^+. The number of interstellar lines increases dramatically below 1200 Å and access to this spectral region is essential for studying the interstellar medium. Emission and absorption transitions in this spectral region correspond to species which exist at very different temperatures ranging from molecules at very low temperatures to highly ionized species such as Fe^{23+} near 10^7 K. In addition ratios of emission intensities provide electron densities over the range of 10^3 to 10^{14} electrons per cubic centimeter. The high spectral resolution and high signal to noise will permit the determination of gas velocities as small as a kilometer per second.

The HUT mission has given us a tantalizing glimpse of what can be accomplished in this spectral region. (For example, see the paper by Blair et al. in this volume). The limiting spectral resolution on HUT is about 3 Å. Even with this value, HUT has indicated new information about the interstellar medium. However, for detailed studies of the gas between stars and galaxies, much higher spectral resolution is required. Copernicus (Rogerson et al. 1973), launched in 1972, was the only mission covering the spectral range down to 912 Å with sufficient spectral resolution to accomplish this goal. Unfortunately, the sensitivity of Copernicus limited observations to objects closer than ~10^3 pc and primarily to wavelengths longer than 1000Å. Figure 1 shows the spatial coverage of the Copernicus mission compared to a schematic of the galaxy. FUSE will not only have a much higher sensitivity (a factor close to 10^5) which automatically extends the range, but will have access to new classes of objects as "candles" for absorption studies: active galactic nuclei to measure the gas along lines of sight to far outside the galaxy and white dwarfs for the local interstellar medium. In addition, the use of a glancing incidence telescope extends the sensitivity to well below the H photoionization edge at 912 Å.

The glancing incidence telescope design under consideration will transmit to below 100 Å. As a consequence, with little increase in complexity, it has been possible to incorporate an EUV capability into the spectrograph as a follow on to the EUVE mission. (See the article by Malina in this volume.)

It has been clear from the earliest feasibility studies that the Lyman FUSE mission will make significant advances in three major areas. (For a detailed discussion, see the Phase A Report, Moos et al. 1989.) It will improve our knowledge of the early universe by making accurate determinations of the light nuclei abundances, including deuterium. Second, the mission will advance our understanding of the evolution and fundamental processes in galaxies. Measurements will range from the hot gas in discs and halos of galaxies to examining supernovae and their remnants. Third, the many diagnostic lines in this spectral region will provide fundamental insight into a number of problems associated with the evolution of stars and planetary systems. Examples are accretion processes, winds and magnetic activity in cool stars, and studies of planetary atmospheric excitation processes. Because of limited space, we discuss (and only briefly) only two topics as examples: deuterium and the Io torus. Also, any discussion of a mission with broad observational capabilities such as FUSE must contain the caveat that often the most important discoveries which result from such a mission are quite different from those predicted by its developers.

Deuterium. The wavelength region covered by FUSE provides unique access to the absorption lines of atomic deuterium. (The resonance line at 1216 provides access in the HST/IUE region, but for all but the shortest lines of sight it is necessary to go to the higher Lyman transitions to prevent the broad absorption line due to the vastly more abundant H from swamping the D absorption.) Because of the high sensitivity of FUSE, the "candles" used for absorption line spectroscopy of the intervening gas will include local hot white dwarfs, more distant hot stars both in our galaxy and in other galaxies, and QSOs for distant lines of sight.

It is difficult to find astrophysical processes which create rather than destroy the weakly bound deuterium nucleus. There appears to be no known process which would produce the observed abundance of deuterium. As a consequence, it is believed that the almost all deuterium nuclei were formed at $\sim 10^2$ s in the early expansion of the universe. The number of nuclei depends on the baryonic density as the temperature falls below 100 keV. As a consequence, the abundance of this fossil nucleus is a sensitive indicator of the baryonic contribution to the mass density of the universe. (Strictly speaking the baryon to photon ratio, but the latter is determined by the background radiation.) A critical issue is deducing the primordial abundance of deuterium from present day measurements.

A number of papers have provided comprehensive discussions of primordial nucleosynthesis using both the standard big bang nucleosynthesis (SBBN) model and various models which assume baryonic inhomogeneity. Boesgaard and Steigman (1985) have provided a detailed review More recently, Walker et al. (1991) have used the latest nuclear crosss sections (including the neutron lifetime) to recalculate the primordial abundances of the light nuclei. Typically measurements of the interstellar D/H ratio have produced values near one and a half parts per hundred thousand. Most recently Linsky et al. have use the HRS on HST to determine with unprecedented accuracy that the value

along the 14 pc line of sight to Capella is between 1.42 and 1.76 x 10^{-5}. (Linsky et al.
1992) Note that all of these measurements are for lines of sight less than 1100 pc. As
Fig.1 dramatically demonstrates, we have sampled only a small part of our galactic disc -
our knowledge of the D/H ratio is truly restricted to the local swimming hole. Walker et al.
have used the ^{3}He/^{4}He ratio in meteorites to determine the deuterium abundance relative to
hydrogen in the solar nebula as greater than 1.8 x 10^{-5}, setting a lower limit for the
primordial value. As the deuterium abundance decreases with baryonic density this lower
limit sets an upper limit to the baryonic density. However, they find that within the
framework of the SBBN model a more restrictive upper limit to the baryonic density (about
40% lower) is set by the ^{7}Li abundance in halo (population II) stars. The same meteoritic
data was also used to determine an upper limit of 1 x 10^{-4} to the D+^{3}He abundance, which
provides a restrictive lower limit to the baryonic density. Walker et al. also reviewed
inhomogeneous models of the early universe briefly, concluding that in this case nucleon to
photon ratios are increased by at most a factor of two over the standard model.

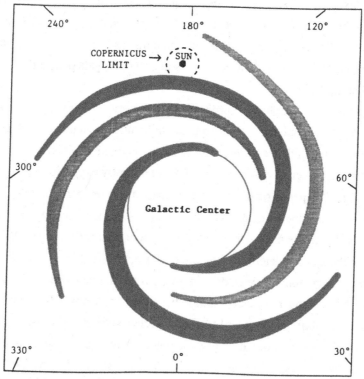

Fig. 1.-Range of the Copernicus mission compared to size of the Galaxy. Properties
such as the D to H ratio have been examined only in environments with a very limited range
of evolutionary histories. Schematic of the Galaxy is after Audouze and Israël, 1985.

The power of FUSE will be its ability to study the D abundances at many different types of locations both in the Milky Way Galaxy, in other galaxies and in the medium between. Thus by sampling regions with very different evolutionary histories, it will be possible to address such issues as a quantitative determination of the degree of deuterium destruction by astration. A related speculative question is the possibility of more recent sources of this nucleus. Finally, a goal of all studies of light nuclei abundances must be an examination of the fundamental paradigm and reassurance that the SBBN model is in fact the proper description.

Io Plasma Torus. Atoms and molecules ejected by volcanic activity on this satellite are ionized and trapped on Jovian magnetic field lines. The lines have a much shorter rotational period than the Io and as a consequence the gas is swept out about the planet. This gas is the dominant source of the magnetospheric plasma. The solar wind contribution is much less. The heating of the plasma is poorly understood. A number of S II and S III emissions in this spectral range are excellent diagnostics of the temperature and electron density.

Even the injection mechanisms whereby the neutrals find their way from the surface and volcano vents to the torus are poorly understood. (Schneider et al. 1989) In addition to evaporation and direct injection by volcanic events, more exotic mechanisms such as sputtering of the residual atmosphere and even of the surface by large plasma currents in the vicinity of Io may be significant. A related question is the elemental abundances of the torus plasma ions. The best evidence is that the torus is made up of sulfur and oxygen with a dash of sodium. (Brown et al. 1983). Potassium emission has also been reported. (Trafton 1975)

Recently, two new spectra of the Io torus were obtained by the HUT instrument on the space shuttle ASTRO-1 mission in December 1991. The spectra coverage was 830-1864 Å in first order and 415-932 Å in second order. (See Blair et al. in this volume and references therein for a more detailed description of the instrument.) As a consequence these spectra provide advance insight into the results expected from FUSE. The spectral resolution in first order was 3 Å for one spectrum and 6 Å for the other. This is a significant improvement over the ~30Å resolution of the Voyager spectra (Broadfoot et al. 1979) A preliminary report has been given by Moos et al. 1991. The spectra contain a large number of features (~100) These are due to the ions of sulfur and oxygen in agreement with the analysis of the Voyager data. A number of line in this spectrum are sensitive to variations in electron temperature and density. In particular, the S III multiplet at $\lambda\lambda$ 1190, 1194, 1201 shows intensity ratios of 1/2.3/1.6 whereas for collisional equilibrium at the nominal 60,000 K temperature of the torus electrons, the expected ratios are 1/3/5. As discussed by Moos et al., the relative strength of the 1201 feature depends on the population of the J=2 sublevel at 833 cm^{-1} in the ground ^{3}P state and hence this level must have a population much smaller than that expected for collisional equilibrium. Scans of the torus with density and temperature sensitive lines can be used to measure both spatial and temporal variations in the electron density and temperature. A puzzling feature of this spectra is the fact that they show no features other than those due to the elements oxygen and sulfur. What other species are present? Detection of these other species will provide insight in to the poorly understood mechanisms whereby material on Io is injected into the torus. A preliminary analysis of the HUT spectra and previously published IUE spectra (Moos and Clarke. 1981) has permitted the determination of a set of preliminary upper limits for potential torus species but no detections of new species. Measurements by FUSE with a higher sensitivity and wavelength resolution are expected to lead to either detections or much lower upper limits for these species.

4 INSTRUMENTATION

Figure 2 shows the FUSE optical system. The primary problem facing the designers of the FUSE observatory is the classic one facing all those associated with constructing astronomical instrumentation. How does one achieve the highest sensitivity and spectral sensitivity within the limits set for size, weight and (by implication) budget. FUSE is an Explorer mission and as such is limited to the capabilities of the Delta launch vehicle. This has led to the instrumental dimensions discussed below. The angular resolution requirement is on the order of a arcsecond; even though it has been obtained in a number of previous space observatories, it is still a significant requirement.

Below 1200 Å, optical transmission becomes a significant problem leading to windowless detectors and the requirement that the number of reflections be kept to a minimum. The instrumentation utilizes both normal incidence and grazing incidence optics to accomplish this goal. The telescope will be a 64 cm diameter F/8 Wolter II telescope. With a large grazing angle of ~ 10° compared to x-ray telescopes, it transmits to 100 Å. As a consequence it is possible to extend the spectrograph into the EUV, bridging the range from 1200 Å down to 100 Å.

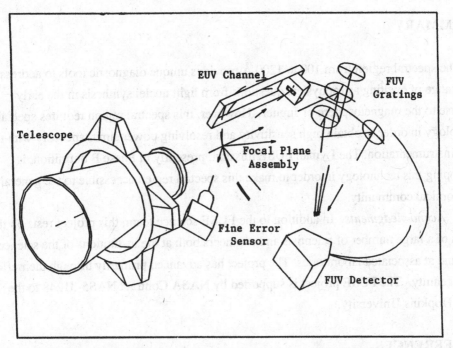

Fig. 2.-FUSE optical system. The ~10 degree grazing incidence telescope forms an image at the focal plane. Location in the focal plane is maintained by the fine error sensor. Light is diverted by the focal plane assembly to one of the FUV gratings or the EUV channel, depending on which spectrograph aperture is illuminated. The high resolution FUV grating on the Rowland circle plus the two others above and below span 912 - 1250 Å. The survey grating to the right spans 400 - 1550 Å. All of the FUV gratings use the same detector; the EUV channel uses a separate detector.

High resolution measurements from 1250 down to 912 Å will be made by Rowland mount gratings with a resolution of R~30,000. The sensitivity of this system will be very high. The effective area will be ~50 cm^2 and the dark count will be less than one per hour per pixel. A survey capability from 400 to 1550 Å at R~300 will also be included. These gratings will not be simple spheres, but will utilize non-spherical surfaces (Cash 1984) and possibly non-classical ruling techniques. The FUV detector will utilize a microchannel plate; both the MAMA (See Timothy et al. in this volume.) and Delay Line (See Siegmund et al. in this volume.) readout technologies are under evaluation for the FUSE missions. For the region from 100 to 350 Å, variable-line glancing incidence gratings are planned (analogous to those carried by EUVE; see Malina et al in this volume) with Delay Line microchannel detectors. A Fine Error Sensor with a CCD will monitor the telescope focal plane both for target acquisition and in order to maintain the stellar target on the ~1 arcsec wide slit.

5 SUMMARY

The spectral region from 100 to 1200 Å provides unique diagnostic tools to address a wide range of exciting astrophysical questions from light nuclei synthesis in the early universe to the magnetosphere of Jupiter. However, this spectral region requires special technology in order to obtain high sensitivity and resolving power simultaneously with the same instrumentation. The Lyman FUSE mission, presently in Phase B definition, is developing this technology in order to make this spectral region accessible to the general astronomical community.

Acknowledgments. In addition to the FUSE science team, this project rests on the efforts of a large number of scientists and engineers both at the institutions of the science team and at associated laboratories. The project has advanced primarily through their efforts and ingenuity. The FUSE project is supported by NASA Contract NAS5-31248 to the Johns Hopkins University.

6 REFERENCES

Audouze, J., and Israël, G. 1985, *The Cambridge Atlas of Astronomy* , (Cambridge: Cambridge Univ. Press), 308

Bahcall, et al., 1991, *The Decade of Discovery in Astronomy and Astrophysics* (Washington, D.C.: National Academy Press), 23

Boesgaard, A. M. & Steigman, G. 1985, *Ann. Rev. Astr. Ast.*, **23**, 319

Broadfoot, A. L., et al. 1979, *Science*, **204**, 979

Brown, R. A., Pilcher, C. B., & Strobel, D. F. 1983, in *Physics of the Jovian Magnetosphere*, ed. A. Dessler (Cambridge: Cambridge Univ. Press), 197

Cash, W. C. 1984, *Apl. Optics*, **23**, 4518

Field, G. B., et al. 1982 *Astronomy and Astrophysics for the 1980's* (Washington, D.C.: National Academy Press), 144

Linsky, J. L., et al. 1992, *Ap. J.* Submitted.

Moos, H. W. & Clarke, J. T. 1981, *Ap. J.*, **247**, 354

Moos, H. W., et al. 1991, *Ap. J.*, **382**, L105

Moos, H. W., et al. 1989, *Lyman The Far Ultraviolet Spectroscopic Explorer: Phase A Study Final Report*, Johns Hopkins University, Baltimore, Maryland.

Moos, H. W. 1991, *Adv. Space Res.* **11**, 221

Moos, H. W. 1990, in *Observatories in Earth Orbit and Beyond*, ed. Y. Kondo (Netherlands: Kluwer Academic Publishers) 171

Rogerson, J. B., et al. 1973, *Ap. J.*, **181**, L97

Schneider, N. M., Smyth, W. H., & McGrath, M. A. 1989, in Time-Variable Phenomena in the Jovian System, ed. M. Belton, R. West & J. Rahe, NASA SP-494.

Trafton, L. 1975, *Nature*, **258**, 690

Walker, T. P., Steigman, G., Schramm, D. N., Olive, K. A., & Kang, H.-S. 1991, *Ap. J.*, **376**, 51

The IMAPS Instrument: A New Horizon for Recording the Real Shapes of Interstellar Absorption Lines in the Far UV

Edward B. Jenkins

Princeton University Observatory

ABSTRACT

The Interstellar Medium Absorption Profile Spectrograph (IMAPS) has a very simple optical design and an efficient imaging detector that allow it to record the spectra of bright, hot stars at a wavelength resolving power $\lambda/\Delta\lambda$ of 240,000 over the important wavelength interval from 950 to 1150Å. On sounding rocket flights it provided excellent recordings of the detailed velocity structure for lines from molecules, atoms and ions in space. Far more data of better quality should come from IMAPS when it will fly on an orbital mission in 1993.

1 INTRODUCTION

Recent results from the *Hubble Space Telescope* (Savage *et al.* 1991; Cardelli *et al.* 1991; Cardelli, Savage and Ebbets 1991; Smith *et al.* 1991) together with two decades of observing with the IUE and *Copernicus* satellites have demonstrated the immense capability of ultraviolet spectroscopists to detect a wide variety of atomic and molecular species in the interstellar medium (ISM), and, moreover, to investigate their relative abundances, temperatures, ionization and dynamics in nearby parts of the galactic plane and halo (Spitzer and Jenkins 1975; Cowie and Songaila 1986; de Boer, Jura and Shull 1989; Jenkins 1989). There is no better way to grasp the usefulness of these observations than to look back on some unique insights that came out from the investigations. To name a few, there were (1) surveys of the abundance of H_2 (Savage *et al.* 1977) and its rotational excitation (Spitzer and Cochran 1973; Spitzer, Cochran and Hirshfeld 1974), (2) determinations of the fractional abundance of atomic deuterium relative to hydrogen (Rogerson and York 1973; Laurent, Vidal-Madjar and York 1979), (3) the detection of O VI (Jenkins and Meloy 1974; York 1974) which represented the co-discovery (shared with a new interpretation of the soft x-ray background: Williamson *et al.* 1974) of a pervasive hot phase of the medium, followed by further studies of this hot gas away from the galactic plane (Savage and Massa 1987), (4) the measurements of interstellar thermal pressures using the fine-structure excitation of neutral carbon (Jenkins and Shaya 1979), (5) the detection of small amounts of high-velocity gas in disturbed regions (Jenkins, Silk and Wallerstein 1976; Cowie, Songaila and York 1979), or the lack of its presence elsewhere (Cowie and York 1978), and (6) systematic surveys of the depletions of different elements from the gas phase onto dust grains (Jenkins, Savage and Spitzer 1986).

To build further on the earlier achievements, there is the promise of more advanced, major facilities, such as the *Space Telescope Imaging Spectrograph* (STIS) (Woodgate *et al.* 1986) or the *Far Ultraviolet Spectroscopic Explorer* (FUSE: see the article by Moos in this volume), that will have an even greater capacity for gathering spectral data from faint sources rapidly. The instrument discussed in this article, the *Interstellar Medium Absorption Profile Spectrograph* (IMAPS), has a modest light collecting aperture, but its thrust is toward increasing the velocity resolution to the point that the detailed structures of the lines' optical depths are resolved, or very nearly so. As outlined below, this instrument was specially designed to cover the important spectral region from 950 to 1150Å for interstellar matter research. IMAPS has already carried out successful observations on sounding rockets, and shortly it will operate on an orbital mission.

2 WAVELENGTH COVERAGE

In ultraviolet astronomy, there are two important sources of continuous opacity, each with well defined absorption edges, that are very influential in constraining the wavelength coverages of observations:

(1) The Lyman limit absorption due to atomic hydrogen in space starting at 912Å and going shortward into the very soft x-rays, and

(2) The opacity of uv transmitting optical materials in an instrument; even the best material, LiF, cuts off at about 1100Å.

In certain applications, either astronomical or instrumental, these limitations can be overcome. For the opacity in the interstellar medium, one can look at very close objects or objects situated behind voids in the ISM (Heiles 1991) – this is the objective of experiments that operate in the EUV spectral region somewhat below the Lyman limit, such as the Wide Field Camera on ROSAT (Pye *et al.* 1991), the EUV spectrograph on ORFEUS (Hurwitz and Bowyer 1991), ALEXIS (Priedhorsky *et al.* 1991), and EUVE (Bowyer and Malina 1991). Obviously, if there's too much neutral hydrogen in the direction you want to look, there's nothing you can do to overcome the absorption just shortward of the Lyman limit.

We have better control over the second source of absorption — we just need to build an instrument that has all-reflective optics and a detector that does not have a photocathode behind a supporting faceplate. This was done in the design of the spectrometer and telescope in the *Copernicus* satellite, and it will be done again when the FUSE satellite will be built. However, neither IUE nor HST can work below about 1100Å; thus ever since *Copernicus* was turned off, we have been unable to exploit this region in a routine, continuous fashion. Aside from the very low resolution data furnished by *Voyager* (Holberg 1991) and EXOSAT (White 1991), only the *Hopkins Ultraviolet Telescope* aboard the Astro-1 Shuttle mission (Davidsen *et al.* 1992) and IMAPS have been able to provide some brief glimpses of moderately distant astronomical sources below the absorption edge of transmission optical elements.

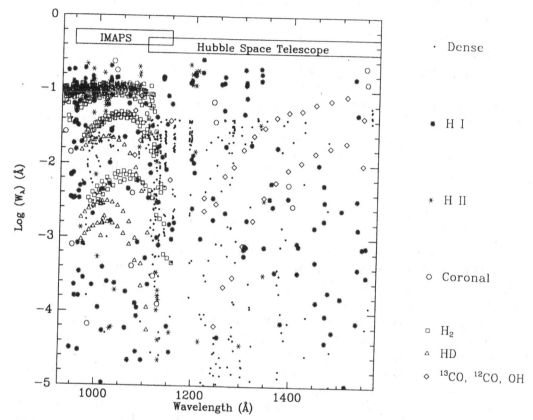

FIG. 1—Wavelengths (abscissae) and expected equivalent widths (ordinates, displayed logarithmically) of transitions of known strength arising from the ground electronic states of various molecules, atoms and ions that are expected in the interstellar medium. (From Jenkins, *et al.* 1988 © SPIE 1988)

The usefulness of the IMAPS wavelength band for studying the ISM is highlighted in Fig. 1. This diagram displays atomic and molecular transitions with known f values according to their location in wavelength and expected strength (displayed logarithmically, normalized to a neutral hydrogen column density of 10^{21}cm^{-2}). Different symbols signify types of species that predominate in the markedly dissimilar conditions in space that range from (1) dense regions that favor the existence of molecules and the neutral forms of elements that have their first ionization potential (IP) below that of hydrogen (13.6 eV), to (2) H I regions of only moderate density that have atoms primarily in their lowest ionization stage with an IP > 13.6 eV, to (3) H II regions that are populated with atoms that are photoionized to one or two stages higher than those favored in H I regions, and finally to

(4) very hot ($10^5 < T < 10^7$K), coronal-type regions that have collisionally ionized gas (or, alternatively, places where x-rays can cause appreciable ionization). It is clear from the figure that the number of lines and the variety of species that can be sampled increases dramatically below about 1150Å — a region that is missed by IUE and HST. Even if we disregard the H_2 and HD lines, more than half of the remaining (atomic) lines are in the range that is often called the "windowless ultraviolet."

3 WAVELENGTH RESOLVING POWER

In the era of the *Copernicus* accomplishments, some frustrations remained when problems could not be completely settled because the velocity details of the lines were, in most cases, not resolved. Often, there was heavy reliance on just measuring the equivalent widths several lines and then deriving column densities using the standard curve-of-growth technique. This method has its pitfalls: in spite of its surprisingly forgiving nature in some contexts for mixtures of lines that are not badly saturated (Jenkins 1986), large errors in column density can result when a small amount of material with a large velocity dispersion is mixed with much more gas that has a small dispersion and a strongly saturated absorption profile (Nachman and Hobbs 1973), causing an inflection in the flat part of the curve of growth that might not be detected (for a good example, see Fig. 4 of Snow 1977). Also, as shown by Spitzer and Jenkins (1975), different species can have different velocity dispersions along a given line of sight.

In a circumstance where there is a good signal-to-noise ratio, one can improve on a simple measurement of just an equivalent width by comparing the observed profile *shape* with trial models of a complex (i.e., not a simple gaussian) velocity profile convolved with the instrumental smoothing function (Vidal-Madjar *et al.* 1977). Even though one can attempt to minimize the χ^2 to obtain a good solution, the derived parameter sets for blends of lines are, in many instances, poorly constrained.

4 INSTRUMENT DESIGN

A detailed discussion of the IMAPS optical system and its ultraviolet image sensor has been presented by Jenkins, *et al.* (1988). The material presented in the following subsections covers only the most fundamental properties of the instrument.

4.1 Optics

The resolving power of a grating spectrograph, where the angles of incidence and diffraction both are equal to θ, is given by the formula

$$R = \lambda/\Delta\lambda = 2(f/\Delta x)\tan\theta \tag{1}$$

where f is the focal length of the camera mirror and Δx is the distance on the focal plane that corresponds to a resolution element. To achieve a large R value of

240,000, IMAPS employs an echelle grating with a large blaze angle ($\tan\theta = 2.0$), a focusing cross-disperser element with a long focal length ($f = 1800$ mm), and a detector that can resolve down to 30 μm (or smaller, if photoevents are centroided). The configuration of the two gratings is shown schematically in Fig. 2. Before reaching these elements, light entering the instrument must first pass through a mechanical collimator consisting of a stack of grids. This collimator restricts the instrument's field of view to a 1° circle in the sky (FWHM), so that stars other than the target do not interfere.

The entrance aperture of IMAPS has an area of 250 cm^{-2}. After multiplying this figure by the product of efficiencies of all the elements, including the quantum efficiency of the detector, we arrive at an effective area of about 7 cm^{-2}.

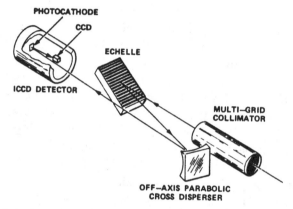

FIG. 2—Optical design of the *Interstellar Medium Absorption Profile Spectrograph* (IMAPS). A parallel beam from the target star passes through a mechanical collimator and is diffracted off an echelle grating (Ruled area: 200×400 mm). The cross-disperser grating is concave and focuses the light onto the photocathode of an electron bombarded intensified CCD detector that is very sensitive to far ultraviolet radiation (see Fig. 3 below). (From Jenkins, *et al.* 1988 © SPIE 1988)

4.2 Detector

Unlike most image sensors that operate in the windowless ultraviolet (Lampton 1991), the electron-bombarded intensified CCD detector inside IMAPS does not use a microchannel plate for the initial amplification of photoevent. As illustrated in Fig. 3, light enters the IMAPS detector and forms an image on a solid photocathode with a thin layer of KBr on its surface at the rear of the image section. The quantum efficiency of an opaque photocathode of this type, typically about 77% in our spectral range, is markedly better than the usual semi-transparent variety on detector faceplates or photocathodes that are right on the surfaces of the microchannel plates.

Electrons emitted by the photocathode are accelerated by an 18 kV electric field and focused by a magnetic field on their way to the CCD (the two fields are not exactly aligned because the electrons must be deflected away from the optical beam). When an energetic electron strikes the back-illuminated, thinned CCD, it generates several thousand secondary charges in the silicon. When the CCD is read out, each event appears as a bright spot that is easily discernible above the CCD's readout and dark current noise. The entire 320×256 pixel field of the IMAPS CCD is read out 15 times per second.

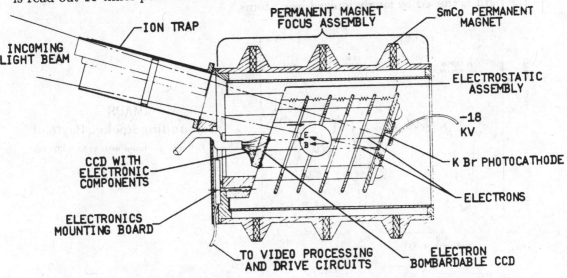

FIG. 3—The far-ultraviolet imaging detector in the IMAPS instrument. An optical image is formed on the KBr coated photocathode at the rear of an electrostatic image section inside a cylindrical permanent magnet assembly. Electrons from this photocathode are accelerated by the $\vec{E}$ field and focused by the $\vec{B}$ field so that they form an electron image on the back side of a specially thinned RCA SID-502 CCD. (From Jenkins, *et al.* 1988 © SPIE 1988)

Contrary to previous experiences with detectors having large accelerating potentials, the IMAPS image section seems remarkably immune to interference from ions (from the Earth's ionosphere) or poor vacuum conditions, even though its interior is open to the ambient environment. On one sounding rocket flight, the detector performed flawlessly even though the pressure inside the payload was about 4×10^{-4} torr for several seconds.

5 OBSERVATIONS OF π SCO ON A SOUNDING ROCKET FLIGHT

5.1 Overview

On a sounding rocket flight in April 1985, IMAPS recorded a spectrum of π Scorpii, a spectroscopic binary with a spectral classification of B1V + B2V and a V magnitude of 2.33. A picture of the spectrum is shown at the bottom of Fig. 4, shown as a montage of four images recorded in sequence as the echelle grating was tipped at different angles to show the complete free spectral range of the orders. This whole spectrum was recorded in 5 minutes, as the payload coasted on its parabolic trajectory that had an apogee of about 260 km. Just above the picture, a plot of a small portion of the spectrum in one of the echelle orders is shown. The two left-hand features are from interstellar H_2 in the $J = 2$ and 3 levels; the one on the right is caused by singly ionized iron atoms.

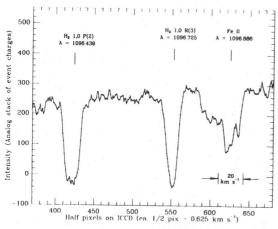

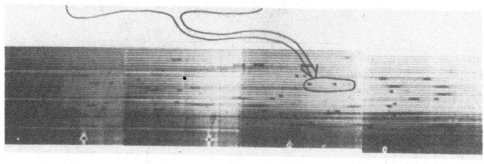

FIG. 4—The image of the echelle spectrum of π Sco (bottom) recorded by IMAPS on a sounding rocket in 1985. Just above, a plot shows a small piece of the spectrum extracted from part of one order that contains two H_2 lines and a line from Fe II. Two of these lines were recorded at much lower resolution by the *Copernicus* satellite many years ago, and they are shown at the top. (From Jenkins, *et al.* 1988 © SPIE 1988)

The following two subsections discuss some scientific highlights that have come out of an analysis of the IMAPS spectrum of π Sco.

5.2 Profiles of H_2 from $J = 2$ to 5

The importance of molecular hydrogen reaches beyond its widespread presence in interstellar regions of moderate to high density; this molecule is the starting point for a wide variety of important molecular reaction networks. UV absorption features in the Lyman and Werner bands give us an excellent opportunity to study not only the overall abundance of H_2, but also the relative populations in various stages of rotational excitation, which in turn give useful insights on interstellar conditions.

Before discussing the IMAPS results, it is useful to review the theory of what happens to H_2 in space. First, we consider its formation. It's very hard for H_2 to radiate energy, so its formation from binary collisions of 2 H atoms is fairly strongly forbidden because the formation energy can't be radiated away in time for binding to occur. Thus, H_2 must form on dust grains. In some local region, this happens at a rate equal to some rate coefficient R times the product of the neutral hydrogen atom density and the density of grains.

H_2 is destroyed by ultraviolet starlight through a compound process. The molecules can reach excited electronic levels by absorbing photons in the Lyman and Werner bands (the same features that IMAPS can record). When the molecules decay back to the ground state, they enter into various places in the vibration-rotation ladder. A small percentage of the time (about 13%, see Abgrall *et al.* 1992) they enter vibration states that are unbound ($v'' > 14$) and molecule is thus destroyed. The remaining 87% of the time, there follows a cascade down the ladder — the vibrational decays are very rapid, but measurable amounts of H_2 in excited J levels are present because the rotational decays are slow.

We see that optical pumping of H_2 into excited rotational levels and its destruction are tied together by a single process, namely, the absorption of ultraviolet photons in space. In the context that the molecules have reached an equilibrium concentration, we further note that the formation of H_2 is also tied to the other processes, although less directly. The relative concentration of H_2 in the upper J levels thus signals the rate of activity for both formation and destruction.

Unfortunately, we do not have a clear notion about how the formation energy of H_2 is dissipated when the molecules are created and leave the grains. Is it given back to the lattice of the grain? Do the newly formed molecules leave the grain with a large kinetic energy? Or is a significant amount of energy imparted into the molecule's vibrational and rotational modes? This important question might be answered by high resolution observations of H_2 in high J levels. We note from the earlier comments that the introduction of new H_2 molecules is not an important channel for populating high J levels (13% at most), but it would be interesting if one could see some high-J molecules that are moving more rapidly than those that are in thermal kinetic equilibrium and optically pumped. This becomes a possibility for high J, since the Einstein A coefficients for decay to lower levels go in proportion to J^5, and the more excited molecules will radiate to lower levels before they have a chance to slow down by colliding with other gas atoms or molecules.

In some *Copernicus* data there was evidence, in certain directions, that some high-J absorption lines were broader than their lower-J counterparts. However, a careful study of the profile shapes and positions suggested that the observed H_2 lines were blends formed by separate clouds with slightly different velocities that had different rotation temperatures, with perhaps some of the high-J levels being populated by excitation from shocks (Spitzer and Morton 1976). By virtue of its high resolution, the IMAPS spectrum of π Sco offered a new, and perhaps better opportunity to search for small amounts of H_2 at high velocities in the upper rotational levels. Jenkins, et al. (1989) analyzed a total of 70 lines with strongly differing transition probabilities in each of the levels $J = 2$ to 5 (the lines from $J = 0$ and 1 were too strong and broad to analyze).

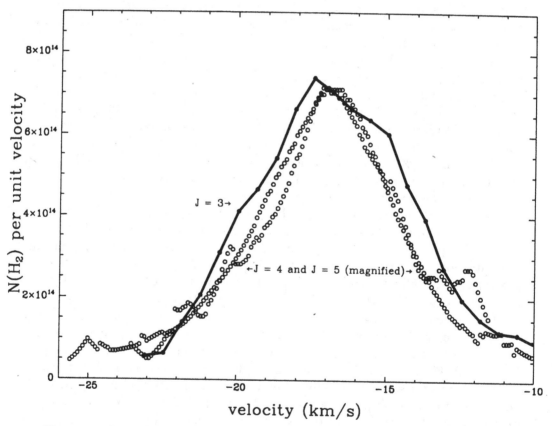

FIG. 5—Plots of the relative column densities per unit velocity for H_2 toward π Sco in three different rotational levels. The profile labeled $J = 3$ is taken from the 0-0 P(3) line, while those for $J = 4$ and 5 are from a composite of many lines. The y-axis scale applies directly to the $J = 3$ case, the $J = 4$ and 5 profiles are magnified vertically by factors of 18 and 63, respectively, so that their shapes could be compared directly to that of $J = 3$.

Surprisingly, Jenkins, *et al.* (1989) found that the higher-J lines were slightly narrower than those from lower levels. This behavior was just opposite to the effects that they were prepared to find and to those seen in the earlier *Copernicus* data! Fig. 5 shows the relative optical depths of the $J = 3$ and 4 profiles, each taken from a composite of many absorption lines. In addition to the slight difference in profile widths, Jenkins, *et al.* noted also that the velocity centroids of the profiles all agree with each other, as close as they could tell (to within about 1 km s^{-1}).

How can this be? To be sure, one can always say that π Sco is showing a circumstantial superposition of one gas complex with a large dispersion of velocities and low rotational excitation, and some other, unrelated cloud which has a narrower profile but is more excited. This view is not very appealing, since it requires that two clouds with very different uv radiation fields and densities exactly coincide in velocity to give composite profile whose centroid doesn't change with J.

It is interesting to note from a survey of 21-cm emission that the neutral hydrogen cloud in front of π Sco is isolated and small (Cappa de Nicolau and Pöppel 1986) — it is not part of the vast, general complex of gas in the Sco–Oph region. To explain the lower excitation of the fast molecules in this location, Jenkins, *et al.* proposed that their pumping radiation is partially shielded by the line absorption caused by the molecular gas with a small velocity dispersion (the low J lines have broad damping wings). What then evolves from this interpretation is the notion that some gas that is stirred up is surrounded by a large overburden of low velocity gas.

How might this unusual configuration arise? It is entirely opposite to the conventional picture that the centers of clouds are protected from the usual turmoil outside and things have settled down. The outer, less dense parts of clouds should be more exposed to the effects of passing shocks and MHD disturbances. To explain this apparent inversion, Jenkins et al. (1989) made a very speculative proposal based on some independent evidence that a shock may have passed through the region a few million years ago. (An expanding shell can be seen in 21-cm emission). When a cloud is overtaken by a shock travelling through the intercloud medium, the pressure enhancement creates another shock at the surface of the cloud that propagates inward. By now, a few million years later, this shock has reached the center of the cloud and stirred things up. Meanwhile, material on the outside has had a chance to quiet down: little high-velocity clumps near the surface have dispersed by now, but those well inside are trapped.

The unanticipated result from the study of H_2 toward π Sco may result from a very specialized condition, if the interpretation offered above is correct. If studies of other lines of sight duplicate these findings, we will need to find an explanation that is better suited to a more generalized phenomenon. Of course, as we examine the H_2 lines in the spectra of other stars, we still want to keep in mind our original objective of finding newly formed molecules at high velocity.

5.3 Using N II to Probe the Structure of an H II Region

Since the ionization potential of nitrogen is greater than that of hydrogen, we know that absorption by N II must come only from a region where the hydrogen is fully ionized. There is a multiplet of N II at 1085Å that is a particularly good probe of an H II region because the ground state from which the transitions originate is split into 3 fine-structure levels, representing the different J quantum numbers. The two excited levels are populated by electron collisions, and these excitations compete with collisional de-excitations and radiative decay. It follows that the local abundance of either of the two, excited levels (N II* or N II**) is proportional to n_e^2, while N II's zero-volt level is proportional to just n_e. By comparing the two measures, we can determine the clumpiness of ionized gas.

Observationally, the task of obtaining accurate population ratios for the levels of N II is not as easy as it first may seem. The literature of *Copernicus* observations exhibits many instances where lines from N II were observed. However, it is usually the case that if there are enough excited ions to be seen, the profile from the unexcited level is so strongly saturated that N(N II) is uncertain by 2 to 3 orders of magnitude. A major contribution to the uncertainty is the fact that one can not use a curve of growth to deal with the saturated line because there is only one line to measure!

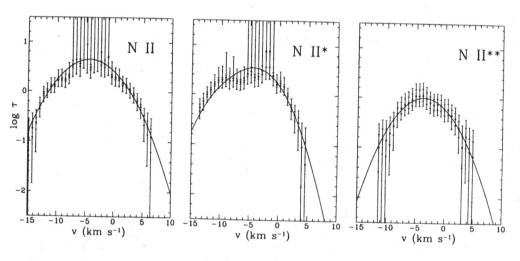

FIG. 6—Measured Optical depths of the absorption lines from the three fine-structure levels of N II, with their corresponding best fit gaussian profiles (solid parabolic curves). Inner ticks on the error bars show deviations that could result from uncertainties in the adopted background or continuum levels in the spectrum; the outer limits of the bars show the added effect of noise in the measurements.

Fig. 6 shows the optical depths of the fully resolved N II lines in the IMAPS spectrum of π Sco. (The presentation of optical depths is logarithmic, hence gaussian profiles appear as parabolas[1]). With the help of the profile shapes of the weak N II** line and the mildly saturated N II* one, we can infer a reasonable extrapolation of what the N II profile should look like in the strongly saturated region where the error bars go out of sight in the plot.

From the relative level populations, Bertoldi and Jenkins (1992) determined that most of the ionized gas must be contained within a length scale of only 0.025 pc, or, put differently, the filling factor of the gas that produces most of the N II absorption fills only about 1% of the H II region in front of π Sco. The internal density of each clump n_e averages about 40 cm^{-3}. For $T = 10^4$K, this means the pressure p/k inside each clump is substantially more than the ambient pressure of 10^3 or 10^4cm^{-3} (the H II region is old enough that it has probably reached pressure equilibrium with the surroundings).

How do we explain the existence of such over-pressured clumps with diameters of only about 0.025 pc or less? Their characteristic expansion times should be only about 1000 years. Bertoldi and Jenkins suggested that there are clumps of neutral gas that act as long-lived reservoirs that can support dense, ionized gas in photoevaporation flows. To have a reasonable probability that one of these flows could be observed, the clumps must be very close to π Sco, filling a volume much smaller than the whole H II region.

With globules so close to the star, there's another problem that arises. Much of the flow off of each globule is in one direction, i.e., toward the star. As a consequence, all of the globules should be propelled by the rocket effect (Oort and Spitzer 1955) and thus dispersed over a period of several million years if they were to consist of just simple balls of neutral gas. To overcome this problem, Bertoldi and Jenkins proposed that gravitational binding must play a role in stabilizing the positions of the clouds: Perhaps these little gas sources are really circumstellar disks around very low mass stars near π Sco, and these stars, in turn, are part of a gravitationally bound cluster. The stars are not easy to see, because they are so close to π Sco. Their evaporation flows should be conspicuous in H-α line emission or by radio continuum emission which could be detected using the VLA. Indeed, such small globules have been observed in the Orion Nebula by Churchwell, et al. (1987).

6 THE FUTURE

The two topics discussed in the above sections give some concrete insights on the types of problems that can be addressed by IMAPS. It is important to realize that only 5 minutes of observing[2] produced these results and also some conclusions,

---

[1] Note that the N II* profile is a bit lopsided: this is because there are two transitions of unequal strength, right next to each other, that are blended by the large velocity dispersion of the gas

[2] For all echelle positions taken together, to be distinguished from 5 minutes per position discussed in a later paragraph.

not covered here, about gas phase depletions for H I components at slightly different velocities along the line of sight (Joseph and Jenkins 1991). With larger observing times, we should be able to observe enough new targets and obtain better signal-to-noise ratios to pursue more ambitious goals, such as examining the stratification of cooling and recombining gases in low and high speed shocks. We could attempt to differentiate the ions and neutrals to learn how they're coupled in MHD shocks, as discussed by McKee (another chapter in this volume). It will also be interesting to study in some detail the character of boundaries between different phases of the ISM. For instance, we might learn more about conduction and evaporation between the hot and cold phases.

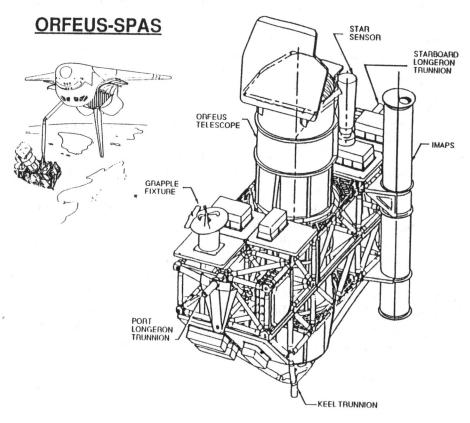

FIG. 7—The configuration of the ORFEUS-SPAS carrier to be flown in 1993, with IMAPS on the right-hand side. Inset (top left) shows orbital deployment from the Shuttle.

According to current schedule, IMAPS will fly on an orbital mission in February 1993. The IMAPS instrument will be attached to an Astrospas spacecraft (built in Germany) that will be released from the Shuttle Orbiter to operate on its own at a short distance away. The mission will last about 5 days, 1 of which will be reserved for IMAPS. The remaining 4 days will be devoted to a much larger telescope, ORFEUS, that contains two spectrographs that, like IMAPS, operate at very short wavelengths. This particular configuration of instruments on Astrospas, called ORFEUS-SPAS, is depicted in Fig. 7.

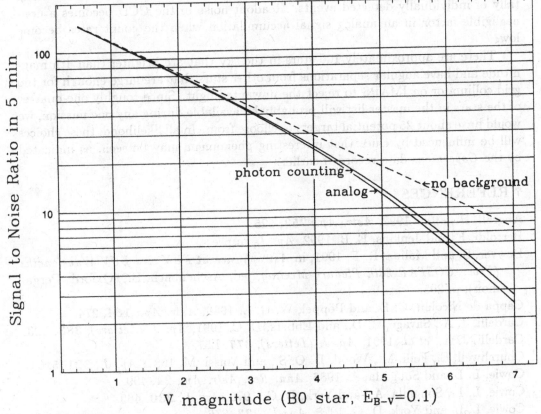

FIG. 8—Signal to Noise Ratio as a function of star's brightness for (1) photon counting with if no Lyman-α geocoronal background were present (upper dashed curve), (2) photon counting with this background (middle curve) and (3) Analog signal accumulation with the background (actual expectation for the planned mission – bottom curve).

In a single orbit, IMAPS should be able to observe a star for about 5 minutes in each of the four echelle angles, thus giving complete spectral coverage. This observing sequence would occur close to the midnight portion of the orbit, when the background from geocoronal Lyman-α radiation intensity is low. Under these circumstances, calculations of the signal-to-noise ratio (SNR) as a function of stellar magnitude for an early-type star yields the curves shown in Fig. 8. The upper dashed curve shows the SNR per resolution element that would be obtained with a photon-counting detector that had no sources of additional noise. In reality, each frame of IMAPS data will contain randomly placed counts caused by the diffuse Lyman-α glow, and statistical fluctuations in this signal will be an added source of noise that will appreciably lower the SNR for faint stars. A further, very small diminution of the SNR will result from the fact that we will allow the counts from one frame to the next to accumulate in an analog fashion, rather than a numerical tally of individually detected events. Readout noise in the CCD becomes a non-negligible factor in an analog signal accumulation when the count rates become low.

There are approximately 100 stars in the sky that are brighter than 4th magnitude and have angular separations from other stars that are large enough for the grid collimator on IMAPS to reject the unwanted light. Since roughly one-quarter of the stars in the master list will probably be available during any one mission, we would have about 25 potential targets to choose from. In all likelihood, these choices will be influenced by clues that interesting phenomena may be seen, as indicated by the *Copernicus* data recorded earlier.

7 REFERENCES

Abgrall, H., *et al.* 1992, *Astr. Ap.*, **253**, 525.

Bertoldi, F. and Jenkins, E. B. 1992, *Ap. J.*, in press.

Bowyer, S. and Malina, R. F. 1991, in *Astrophysics at FUV and EUV Wavelengths (Proc. COSPAR 28th Plenary Meeting)*, ed. Aschenbach, B., (Oxford: Pergamon), p. 205.

Cappa de Nicolau, C. E. and Pöppel, W. G. L. 1986, *Astr. Ap.*, **164**, 274.

Cardelli, J. A., Savage, B. D., and Ebbets, D. C. 1991, *Ap. J. (Letters)*, **383**, L23.

Cardelli, J. A., *et al.* 1991, *Ap. J. (Letters)*, **377**, L57.

Churchwell, E., Felli, M., Wood, D. O. S., and Massi, M. 1987, *Ap. J.*, **321**, 516.

Cowie, L. L. and Songaila, A. 1986, *Ann. Rev. Astr. Ap.*, **24**, 499.

Cowie, L. L., Songaila, A., and York, D. G. 1979, *Ap. J.*, **230**, 469.

Cowie, L. L. and York, D. G. 1978, *Ap. J.*, **223**, 876.

Davidsen, A. F., *et al.* 1992, *Ap. J.*, , in press.

de Boer, K. S., Jura, M. A., and Shull, J. M. 1989, in *Exploring the Universe with the IUE Satellite*, ed. Kondo, Y., (Dordrecht: Kluwer), pp. 485- 515.

Heiles, C. 1991, in *Extreme Ultraviolet Astronomy*, eds. Malina, R. F. and Bowyer, S., (New York: Pergamon), pp. 313-321.

Holberg, J. B. 1991, in *Extreme Ultraviolet Astronomy*, eds. Malina, R. F. and Bowyer, S., (New York: Pergamon), pp. 8-14.

Hurwitz, M. and Bowyer, S. 1991, in *Astrophysics at FUV and EUV Wavelengths (Proc. COSPAR 28th Plenary Meeting)*, ed. Aschenbach, B., (Oxford: Pergammon), p. 217.

Jenkins, E. B. 1986, *Ap. J.*, **304**, 739.

—. 1989, in *Exploring the Universe with the IUE Satellite*, ed. Kondo, Y., (Dordrecht: Kluwer), pp. 531-548.

Jenkins, E. B. and Meloy, D. A. 1974, *Ap. J. (Letters)*, **193**, L121.

Jenkins, E. B., Savage, B. D., and Spitzer, L., Jr. 1986, *Ap. J.*, **301**, 355.

Jenkins, E. B. and Shaya, E. J. 1979, *Ap. J.*, **231**, 55.

Jenkins, E. B., Silk, J., and Wallerstein, G. 1976, *Ap. J. (Suppl.)*, **32**, 681.

Jenkins, E. B., *et al.* 1988, in *Ultraviolet Technology II*, ed. Huffman, R. E., (Bellingham: The International Society for Optical Engineering), pp. 213-229.

Jenkins, E. B., Lees, J. F., van Dishoeck, E. F., and Wilcots, E. M. 1989, *Ap. J.*, **343**, 785.

Joseph, C. L. and Jenkins, E. B. 1991, *Ap. J.*, **368**, 201.

Lampton, M. L. 1991, in *Extreme Ultraviolet Astronomy*, eds. Malina, R. F. and Bowyer, S., (New York: Pergamon Press), pp. 353-363.

Laurent, C., Vidal-Madjar, A., and York, D. G. 1979, *Ap. J.*, **229**, 923.

Nachman, P. and Hobbs, L. M. 1973, *Ap. J.*, **182**, 481.

Oort, J. H. and Spitzer, L., Jr. 1955, *Ap. J.*, **121**, 6.

Priedhorsky, W. C., *et al.* 1991, in *Extreme Ultraviolet Astronomy*, eds. Malina, R. F. and Bowyer, S., (New York: Pergamon), pp. 464-477.

Pye, J. P., Watson, M. G., Pounds, K. A., and Wells, A. 1991, in *Extreme Ultraviolet Astronomy*, eds. Malina, R. F. and Bowyer, S., (New York: Pergamon), pp. 409-426.

Rogerson, J. B. and York, D. G. 1973, *Ap. J. (Letters)*, **186**, L95.

Savage, B. D. and Massa, D. 1987, *Ap. J.*, **314**, 380.

Savage, B. D., Bohlin, R. C., Drake, J. F., and Budich, W. 1977, *Ap. J.*, **216**, 291.

Savage, B. D., *et al.* 1991, *Ap. J. (Letters)*, **377**, L53.

Smith, A. M., *et al.* 1991, *Ap. J. (Letters)*, **377**, L61.

Snow, T. P. 1977, *Ap. J.*, **216**, 724.

Spitzer, L. and Cochran, W. D. 1973, *Ap. J. (Letters)*, **186**, L23.

Spitzer, L., Cochran, W. D., and Hirshfeld, A. 1974, *Ap. J. (Suppl.)*, **28**, 373.

Spitzer, L. and Jenkins, E. B. 1975, *Ann. Rev. Astr. Ap.*, **13**, 133.

Spitzer, L. and Morton, W. A. 1976, *Ap. J.*, **204**, 731.

Vidal-Madjar, A., Laurent, C., Bonnet, R. M., and York, D. G. 1977, *Ap. J.*, **211**, 91.

White, N. E. 1991, in *Extreme Ultraviolet Astronomy*, eds. Malina, R. F. and Bowyer, S., (New York: Pergamon), pp. 15-29.

Williamson, F. O., *et al.* 1974, *Ap. J. (Letters)*, **193**, L133.

Woodgate, B. E., *et al.* 1986, in *Instrumentation in Astronomy VI*, (Bellingham: The International Society for Optical Engineering), pp. 350-362.

York, D. G. 1974, *Ap. J.*, **193**, L127.

Ultraviolet Spectroscopic Instrumentation

Webster Cash

Center for Astrophysics and Space Astronomy, University of Colorado

ABSTRACT

This is a brief discussion of recent progress in the development of instrumentation for ultraviolet astronomy. Progress has been rapid in many areas, leading to new generations of very powerful instruments. The first of the new generations are only just now being launched, and are demonstrating the power of the new detectors and optics.

1 INTRODUCTION

The past decade has shown rapid advancement in spectrographs suitable for use in the Far Ultraviolet (FUV) and Extreme Ultraviolet (EUV). For the most part, however, these advances have been limited to the laboratory. While the natural timescale for new concepts to develop to flight readiness is several years, the slowdown in launch rate has stretched the timescale for effective application of new technology to a full decade or more. The result is that an entire new generation of instrumentation is still in the laboratory instead of in orbit where it would be able to perform at levels substantially beyond what we currently enjoy.

By studying spectrographs at their component level greatly improved performance in the overall systems has been achieved. Using the new graphics-oriented power of modern computers, a new generation of designs have been developed, and the power of modern optical fabrication techniques has made many of these designs, unthinkable ten years ago, fairly straightforward to build. Advances in mirror, grating and detector fabrication also have made realization of the designs possible.

The EUV and FUV mark the transition region between conventional optics which use normal incidence reflections, and the x-ray band which uses grazing incidence optics. Most spectrograph designs use one or the other of these approaches. There are cases of hybrid designs that have been discussed, but none have ever been built (McClintock and Cash, 1982). From 1100Å down to about 400Å materials with poor, but useable, reflectivity exist. Below 400Å one uses glancing incidence of necessity.

2 NORMAL INCIDENCE

Design of spectrographs at normal incidence is considered a mature field. Most of the basic concepts were demonstrated between 50 and 100 years ago. However, progress has been substantial in the last ten years. Driven partly by the unique requirements of the ultraviolet, partly by the extraordinary capabilities of modern detectors, and partly by the desire to perform new classes of observations, the design of normal incidence spectrographs has progressed.

Inevitably, progress on optical designs is linked to progress in optical fabrication techniques, and a great deal has happened to make special order optics more readily available. Certainly the two areas of greatest advancement have been in the fabrication of aspheric optics, and in the holographic ruling and blazing of gratings.

2.1 Aspherics

The Rowland Circle has been used for over a century because it provides sensitivity, broad bandpass and versatility. It remains a favorite in the ultraviolet for the same reasons. One of the main drawbacks, however, when observations of faint targets, confused fields or extended sources are needed, is the lack of long slit imaging. The Rowland circle, particularly at high dispersion, has strong astigmatism. This problem has, traditionally, been solved by moving to a toroidal grating blank. The rulings have remained uniform and parallel as viewed from above. By choosing the second dimension of curvature independent from the first, the astigmatism may be removed at one value of cos(β): two points in the spectrum. When this is accomplished, the resulting instrument provides good resolution along the slit as well as perpendicular to it. This allows for higher signal to noise, the removal of source confusion, and one dimensional imaging of extended sources.

Toroids have been in limited use for many years (Haber, 1950). In actual design and fabrication, the ellipsoid is preferred over the toroid. A toroid creates spherical aberration that grows with the speed of the telescope, while the ellipsoid is free of spherical aberration. The ellipsoid also has the advantage of having a perfect focus upon which to perform null tests during fabrication. Thus the ellipsoidal grating is truly the design of choice.

During the past few years the concept of using the Rowland for very high resolution spectroscopy has gained some acceptance. The Copernicus satellite (Rogerson et al., 1973) attained resolution of 20,000 using a spherical grating, but it had no long

Figure 1: Fringes from a null test on an asphere. The quality is not diffraction limited, but adequate to support its purpose, high resolution spectroscopy.

slit imaging or simultaneous wavelength coverage, severely limiting its sensitivity. Now, the Far Ultraviolet Spectroscopic Explorer (Cash, 1984) plans to use a modified Rowland to achieve resolution of 30,000 coupled to long slit imaging.

The problem with the stigmatic Rowland, and one approach to solving the problem were presented by Cash (Cash, 1984). When a simple ellipsoid is used at resolution of 30,000 a beam of f/20 or slower is needed to reduce the effects of type II coma (slit curvature) to an acceptable value. Two solutions were considered. Either the substrate must be modified, or the rulings must be modified. At that point in time there were no ruling machines, either holographic or mechanical, that could produce non-uniform rulings at the needed high densities, so modification of the substrate was clearly preferred. If the ellipsoid has a term proportional to xy^2 added, the type II coma is removed at one point in the spectrum, and the effective resolution of the stigmatic grating improves. Such a grating has now been manufactured by a three-axis grind and polish at Hyperfine Inc. Interferograms of the surface against an aspheric null are shown in Figure 1. While still showing strong fringes, this quality of surface is, in fact, adequate to support resolution of 25,00. It was ruled at 300 g/mm and shows resolution in excess of 25,000 in the visible portion of the spectrum. The final step was to rule the grating at 4800 g/mm and test its ultraviolet properties. The measured groove efficiency was low, prompting tests with a Scanning Tunneling Microscope (STM). This showed that while the groove quality was good, the groove shape was inefficient. The advent of the STM has become a central tool in grating manufacture, because it allows the grating ruler to directly see the groove and properly adjust the diamond tip. Ultraviolet measurements of the diffracted image quality are now being performed at the University of Colorado.

An alternative approach to the aspherics that shows particular promise of allowing inexpensive fabrication of new gratings is to use a plane grating on a thin substrate. It is then bent to the desired form and a replica copied from it (Huber, et al., 1981).

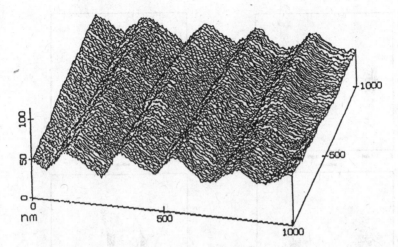

Figure 2: Scanning tunneling microscope picture of the 4800 g/mm rulings on the asphere. There was significant variation in groove quality from point to point. This was the best point studied; it shows the groove shape to have an unoptimized angle, and a shape closer to isosceles than right triangle.

2.2 Holographics

Holographic gratings are becoming more common as their capabilities improve. They have the basic advantages of low scatter and zero Rowland ghosts, which are crucial for many experiments. However, their smooth groove profiles have mostly yielded poor diffraction efficiency. With the advent of ion etching techniques, the blaze efficiency is now rising, and the increased use of holographically ruled gratings in FUV should soon follow.

Holographics can produce behavior similar to an asphere, but on a spherical substrate. Until recently, the constraints of fabrication limit the tailoring of the grooves. The result is that there is a fairly confining limit on how much improvement could be achieved. For example, a holographic grating on a sphere can create stigmatic performance at one point, but cannot maintain it over a broad band. Extensive development of such an objective holographic grating for high resolution FUV spectroscopy was demonstrated by Prange et al (Prange, et al., 1989).

Two new directions of study are currently active. First is the use of aspheric substrates. Second is the application of non-parallel wavefronts in the recording geometry. These effects are complementary and are providing some very attractive capabilities (Davila, Content and Trout, 1992). For example, Grange and Laget (Grange, 1991) have shown that the effects of type II coma can be removed from the spectrum of an ellipsoidal grating effectively duplicating the capability of a modified asphere. The difference is that the holographic can be ruled on a regular ellipsoid, one that can be fabricated using standard null tests.

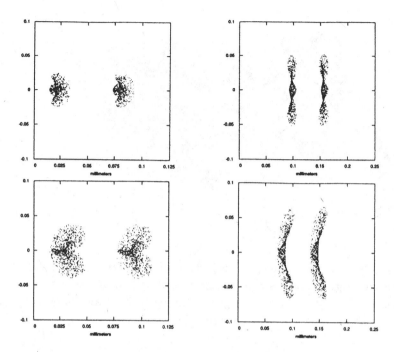

Figure 3: Comparison of high resolution Rowland designs. Above are raytraces of a type II coma - corrected asphere. 940 and 940.06Å to the left, 970 and 970.06Å to the right. Below is a comparable design using a a holographic ruling on a sphere, as prescribed by Grange (Grange, 1992).

Recently, a very exciting new class of solutions has been proposed (Duban, 1992; Grange, 1992). This solution is constrained to a spherical surface, but still allows the simultaneous high resolution and stigmatic performance produced by the asphere being developed for FUSE. Figure 3 shows a raytrace of the new Grange design compared to that of an asphere of similar capability.

2.3   Non-uniform mechanical ruling

Just as the use of holographics provides at least partial freedom from the constraints of straight rulings, the same freedoms can be achieved by use of machines which specifically tailor the grooves. In the past such machines were not successful, and little came of such efforts. Now it appears that Hitachi Inc has successfully built a ruling engine that can generate non-uniform rulings on flats and spheres (Kita and Harada 1992). The first application of the Hitachi engine is the fabrication of concave gratings for the ORFEUS mission (Hurwitz and Bowyer, 1991). This spectrograph features normal incidence concave gratings and non-uniform rulings. Unfortunately, due to the expense of building and maintaining such a special purpose machine, compared to the ease of holographic fabrication, it appears likely that this approach will soon be superseded by holographic fabrication.

3   GLANCING INCIDENCE

Prior to the 1980's the spectrograph designs in common use comprised two classes: 1) Transmission gratings in the converging beam behind grazing incidence telescopes, and 2) Rowland circle spectrographs at grazing incidence. However, both of these approaches suffer from major disadvantages.

Transmission gratings are quite difficult to work with, being fabricated from large arrays of small, thin, delicate gratings. They are inefficient, producing typically 10% of the light in each of two dispersed spectra. They are used as objective gratings, having little capability on extended sources. Nonetheless, the transmission gratings are, so far, responsible for nearly all our existing EUV spectra (Heise, et al., 1987).

The grazing incidence Rowland Circle designs use light diverging from an entrance slit, and thus can be used on extended sources. They have been used heavily in studies of the Sun. But, the astigmatism can only be cancelled at one wavelength, and grows very rapidly off that point. This limits the effective spectral coverage for astronomy. Also, in the grazing incidence Rowland geometry the diffracted light impacts the focal plane (detector surface) at grazing incidence as well, making the detector difficult to build and of low efficiency.

A notable exception is the "Type III Holographic Grating" flown by Malina et al in 1978 (Malina, et al., 1979). It was an intermediate step between the old Rowland designs and the new non-uniform ruling desings. It used an ellipsoidal substrate and contained a holographically ruled grating to be used in a grazing incidence mount. It, unfortunately had fairly poor efficiency and required a grazing incidence detector mount, but it was adequate to provide the first spectrum of HZ43.

The extreme coma that grazing incidence optics suffer is a result of the distance variations from the regions of the optic to the focal plane. This can be offset to first order by using the grating density to compensate. Hettrick and Bowyer (Hettrick and Bowyer, 1983) proposed using a variation in density across the surface of a flat to compensate for the coma. A grating can be tuned for one wavelength and angle in this way. Since it uses flat substrates and straight rulings, fabrication of this special grating is a relatively straightforward process. It has been adopted for use on the Extreme Ultraviolet Explorer.

The drawbacks for the in-plane varied-line-space gratings are substantial. First, a grating can be used in only one geometry, unlike the versatility of the Rowland which can support a range of angles. Second, the resolution of the grating suffers from major aberrations away from the central wavelength (Hettrick, 1984). This limits its ability to perform useful high resolution spectroscopy. Third, the grating scatter is preferentially in the plane of diffraction. Fourth, the geometry contains groove shadowing and is thus of lower efficiency.

In 1977 Werner (Werner, 1977) published some results on a cylindrical grating used in the extreme off-plane mount. The grating showed extremely high x-ray diffraction efficiency. This comes from a number of effects including the effects of groove shadowing and the larger defects needed to cause scattering and put power in the off-blaze orders (Neviere and Maystre, 1978).

An alternative approach was first demonstrated by Cash (Cash, 1983) and Windt and Cash (Windt and Cash, 1984). The idea was to use converging rulings to compensate for the extreme coma of a grating in a beam converging to a focus. Unlike the varied line space gratings, where the geometry is completely specified by the rulings, a radial grating can be used at any graze angle, in the same manner as a spherical concave grating. The blaze of the grating can be adjusted by changing the graze angle. Hence, one grating can be used in a variety of applications in the EUV and/or x-ray. Additionally, it was shown by Hettrick (Hettrick, 1984) that the aberrations of the radial groove are inherently superior to the those of the varied-line-space gratings.

The first practical application of this technology to astronomy was in January, 1992. An all grazing incidence rocket was flown by the University of Colorado to observe the hot white dwarf G191B2B between 200 and 400Å (Wilkinson, green and Cash, 1992). The payload used a metal optic Wolter Schwarzschild Type II telescope feeding a slitted spectrograph. The spectrograph included a grazing incidence ellipsoid which refocussed the light, and a radial groove grating to disperse the converging beam. The resulting spectrum, together with a calibration spectrum, is presented in Figure 4. It clearly shows the presence of prominent absorption edges of heavy elements. Since this spectrum was obtained in just 240 seconds, it clearly demonstrates the power of the new techniques.

4 OPTICAL COATINGS

The search for ideal ultraviolet coatings continues. It is still recognized that bare aluminum is the best material available down to about 700Å, but the problem of oxidation remains. If the aluminum is subjected to even tiny amounts of oxygen, its UV reflectivity rapidly drops. In principle this can be solved by use of in-flight coating and active gettering of the environment (Scott and Cameron, 1989), but it has yet to be demonstrated in practice.

Extensive testing of potential materials took place in the 1980's with the hope of finding better materials for reflectivity in the FUV and EUV. Windt et al (Windt, et al., 1988; Windt, et al., 1988) published a full compendium of indices of refraction based on relectance measurements of samples. They convered most available metals and a few exotic materials as well. They verified that there was no miracle element just waiting to be tried.

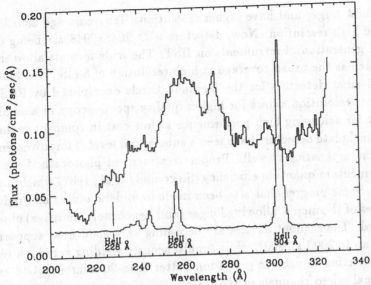

Figure 4: The spectrum of G191B2B obtained with a radial groove grating on a sounding rocket is shown aginst a laboratory calibration spectrum of the emission lines of HeII. G191B2B shows strong absorption features from heavy elements, most likely O III. See Wilkinson, Green and Cash 1992.

The best normal incidence reflector outside of unoxidized aluminum is Silicon Carbide. It has been known as an excellent FUV and EUV reflector for many years (Kelly, West and Lloyd, 1981). In the bulk material it can generate reflectivities as high as 40% at 1000Å. This capability has been utilized for a variety of experiments, but was limited in application because the material was hard and maintained its reflectivity only in bulk samples. Attempts to sputter thin films yielded poor reflectivites, most likely due to destruction of the bulk structure during the sputtering.

Then, in 1988, Kortwright and Windt (Kortright and Windt, 1988) found that SiC could be deposited in a thin film without completely destroying the optical properties. This allowed them to put SiC on gratings as well as mirrors, and the first such grating has been actively flown on a sounding rocket (Cash, et al., 1989). Studies of the degradation of the sputtered SiC (Keski-Kuha, et al., 1988) show the thin films to be fairly stable with reflectivities near 30%, a big improvement from the best metals. Sputtered SiC is already finding wide application in FUV and EUV spectrographs.

5   DETECTORS

There has been substantial improvement in detectors over the last decade. It has taken the form not of new classes of detector, but of improvements in the old detectors. The detectors are now larger and have higher resolution. Ten years ago 250x250 format was considered high resolution. Now, detectors with 2048x2048 are being developed for the second generation of intruments on HST. The wide formats allow interesting applications such as the broad coverage at high resolution of STIS (Woodgate, 1989), and the long format detectors for the Rowland Circle exemplified by FUSE (Cash, 1984). The finer resolution allows for higher quality spectroscopy in a smaller scale, also important for achieving high performance at low cost in compact geometries.

Micro-channel plate detectors have seen a substantial level of improvement in their performance characteristics as well. Proper treatment of photocathodes has led to major improvements in quantum efficiency (Siegmund, *et al.*, 1987) which is useful for spectrographs. Some progress has also been made in understanding and reducing the dark count rates of the mcp's, allowing longer (and hence more sensitive) observations to be performed. The readout devices are becoming very good now, supporting very large formats (up to 2000 x 2000), while simultaneously yielding very high resolution. They are now starting to achieve resolution better than 20 microns, good enough to resolve individual micro channels in the plate.

Finally, it should be noted that CCD's are getting better. They are now the detector of choice in both the visible and the x-ray bands. They have found limited application so far in the FUV and EUV. First, they require back illumination to record the event in a CCD, making even simple applications more difficult. Second, the light levels in the FUV are low, typically requiring photon counting as in the x-ray. However, the energy of an FUV photon is still insufficient to be separated from readout noise. Thus, CCD's will require an intensification stage (Lowrance and Joseph, 1988) until the CCD noise level drops to below 1 electron rms. Then CCD's will become more attractive.

6   SUMMARY

The results from ASTRO and from sounding rockets clearly demonstrates the power of modern Ultraviolet Spectroscopic Instrumentation. The limit on the capabilities of modern instrumentation to perform important new observations is not in the instrumentation itself, but in the support systems. We are not flying often enough, and what is flying, for the most part is severely outdated technology. This should not be allowed to continue.

The list of powerful techniques discussed in this paper is long, yet the number of missions flown is short. Since it is the spacecrafts that are limiting the flight rate, then perhaps the solution lies in building simpler, cheaper, spacecraft, and let the powerful instrumentation techniques compensate for more limited support.

REFERENCES

1. Cash, W., 1983, *Appl. Opt.*, **22**, 3971.
2. Cash, W., 1984, *Appl. Opt.*, **23**, 4518.
3. Cash, W., 1984, *Sci. and Tech. Series of the Amer. Astronautical Soc.*, **56** 107.
4. Cash, W., *et al.*, 1989, *Experimental Astronomy*, **1**, 123-143.
5. Davila, P., Content, D., and Trout, C., 1992, *Appl. Opt.*, **31**, 949.
6. Duban, M., 1992, *Appl. Opt.*, **31**, 443.
7. Grange, R., 1992, *Appl. Opt.*, in press.
8. Grange, R., Laget, M., 1992, *Appl. Opt.*, **30**, 3598.
9. Haber, H., 1950, **J. O. S. A.**, **40**, 153.
10. Heise, J., Paerels, F. B. S., Bleeker, J. A. M., Brinkman, A. C., 1987, *Ap. J.* **334**, 958.
11. Hettrick, M. C., 1984, *Appl. Opt.*, **23**, 3221.
12. Hettrick, M. C., Bowyer, C. S., 1983, *Appl. Opt.*, **22**, 3921.
13. Huber, M. C. E., Jannitti, E., LeMaitre, G., and Tondello, G., 1981, *Appl. Opt.*, **20**, 2139.
14. Hurwitz, M., Bowyer, S., 1991, "Extreme Ultraviolet Astronomy", R.F. Malina And S. Bowyer Eds., Pergamon Press, New York, 442.
15. Kelly, M. M., West, J. B., Lloyd, D. E., 1981, **J. Phys D.**, **14**, 401.
16. Keski-Kuha, R. A. M., *et al.*, 1988, *Appl. Opt.*, **27**, 1499.
17. Kita, T., and Harada, T., 1992, *Appl. Opt.*, **31**, 1399.
18. Kortright, J. B., Windt, D. L., 1988, *Appl. Opt.*, **27**, 2841.
19. Lowrance, J. L., Joseph, C. L., 1988, "Advances in Electronics and Electron Physics".
20. Malina, R.F., Bowyer, S., Finley, D., Cash, W., 1979, *Proc. Soc. Photo-Opt. Instr. Eng.*, **184**, 20, *Space Optics.*
21. McClintock, W.E., Cash, W., 1982, *Proc. Soc. Photo-Opt. Instr. Eng.*, 321.
22. Neviere, M. P. , Maystre, D., 1978, *Appl. Opt.*, **17**, 843.
23. Prange, R., *et al.*, 1989, *Appl. Opt.*, **28**, 496.
24. Rogerson, J. B., *et al.*, 1973, **181**, L97.
25. Scott, M. L., Cameron, B., 1989, *Proc. Soc. Photo-Opt. Instr. Eng.*, **1160**, 408.
26. Siegmund, O. H. W., *et al.*, 1987, *Appl. Opt.*, **26**, 3607.
27. Werner, W., 1977, *Appl. Opt.*, **16**, 2078.
28. Wilkinson, E., Green, J., Cash, W., 1992, *Ap. J. (Letters)*, submitted.
29. Windt, D., Cash, W., 1984, *Proc. Soc. Photo-Opt. Instr. Eng.*, **503**, 98-105
30. Windt, D., *et al.*, 1988, *Appl. Opt.*, **27**, 246.
31. Windt, D., *et al.*, 1988, *Appl. Opt.*, **27**, 279.
32. Woodgate, B. E., 1989, *Bull. Amer. Astron. Soc.*, **21**, 766.

NONEQUILIBRIUM ABSORPTION LINE AND EMISSION SPECTRUM DIAGNOSTICS FOR THE GALACTIC FOUNTAIN

ROBERT A. BENJAMIN and PAUL R. SHAPIRO
The University of Texas at Austin

ABSTRACT

We have calculated the time-dependent, nonequilibrium thermal and ionization history of gas cooling radiatively from 10^6 K in a one-dimensional, planar, steady-state flow model of the galactic fountain, including the effects of radiative transfer. Our previous optically thin calculations explored the effects of photoionization on such a flow and demonstrated that self-ionization was sufficient to cause the flow to match the observed galactic halo column densities of C IV, Si IV, and N V and UV emission from C IV and O III] in the constant density (isochoric) limit, which corresponded to cooling regions homogeneous on scales $D \gtrsim$ 1 kpc. Our new calculations which take full account of radiative transfer confirm the importance of self-ionization in enabling such a flow to match the data but allow a much larger range for cooling region sizes, i.e. $D_0 \gtrsim 15$ pc. For an initial flow velocity $v_0 \sim 100$ km/s, comparable to the sound speed of a 10^6 K gas, the initial density is found to be $n_{H,o} \sim 2 \times 10^{-2}$ cm^{-3}, in reasonable agreement with other observational estimates, and $D_0 \sim 40$ pc. We also compare predicted Hα fluxes, UV line emission, and broadband X-ray fluxes with observed values.

1 INTRODUCTION

Observations of UV absorption and emission lines of highly ionized atoms in interstellar gas at large distance from the galactic plane and of 21 cm emission from high velocity neutral clouds can be interpreted in terms of the galactic fountain model of Shapiro and Field (1976). In this model, galactic halo gas which originates in the disk is shock-heated to T ~ 10^6 K and flows out of the disk into the halo where it radiatively cools and recombines, loses buoyancy and falls back toward the galactic plane. This and subsequent theoretical work in the field have been summarized recently by Spitzer (1990). In order to study the physical behavior and spectral signature of such a flow, we have recalculated the nonequilibrium radiative cooling, ionization, and recombination of a gas composed of H, He, C, N, O, Ne, Mg, Si, S, Ar, Ca, Fe, and Ni. We use the abundances of Allen (1973) and the atomic data of Raymond and Smith (1977) and Raymond (1987). We have solved the ionization balance rate equations together with the equation of energy conservation for a gas cooling from 10^6 K to 10^4 K at either constant density (isochoric), constant pressure (isobaric), or the intermediate case where the cooling is initially isobaric with a transition to isochoric. The type of cooling depends upon the assumed size, D_0, of the cooling region. Large (small) regions for which the sound crossing time is longer (shorter) than the cooling time for all temperatures will be isochoric (isobaric) throughout their cooling history. The intermediate case corresponds to length scales for which the sound crossing time exceeds the cooling time at some intermediate temperature between 10^4 K and 10^6 K. We have improved upon our previous calculations described in Shapiro (1990) and Shapiro and Benjamin (1991) by taking account of optical depth effects through the flow.

For each case, we calculate the UV absorption and emission line spectrum as well as the X-ray flux and total ionizing flux for comparison with observations. This allows us to estimate the local mass flux, initial density, and initial outflow velocity required in order for such a galactic fountain flow to match the data.

2 CALCULATIONS AND RESULTS

We have considered a 1-D, plane-parallel, steady-state flow of halo gas cooling radiatively from an initial condition of coronal ionization equilibrium at $T_0=10^6$K with an initial velocity, v_0, and initial hydrogen density, $n_{H,0}$. We take account of optical depth and of diffuse emission by the cooling gas by solving the equation of radiative transfer by the method of discrete ordinates using a four-point Gaussian quadrature. The equations of energy and radiative transfer are explicitly differenced simultaneously with implicit differencing of the rate equations by a stiff rate equation solver. The galactic mass flux in this flow to each side of the disk is $\dot{M}= 24\, f\, n_{-2}\, v_{100}$ $M_\odot$/yr, where $n_{-2}=(n_{H,0}/10^{-2}$ cm$^{-3})$, $v_{100}=(v_0/100$ km/s) and f is the fraction of the galactic disk covered by such a flow such that the total area of the flow is $f\,\pi\,R_{disk}^2$, where $R_{disk}=15$ kpc.

The evolution of each fluid element in the flow is calculated by assuming the density and pressure evolve as they would if the fluid element were located within a cooling region of initial size, D_0, ranging from $D_0 \leq 0.4\,(n_{-2})^{-1}$ pc, in which case the cooling is always isobaric, to $D_0 \geq 2\,(n_{-2})^{-1}$ kpc, in which case the cooling is always isochoric. For intermediate cases, in which the cooling is initially isobaric but changes to isochoric, the initial size D_0 can be related to the size at the time of the transition, D_{tr}, by $n_{H,tr}D_{tr}^3=n_{H,0}D_0^3$. Once the gas has cooled to $\sim 10^4$ K, it gradually recombines and self-shields, thereby terminating the increase of the column densities of highly ionized species through the flow. During this phase at $\sim 10^4$ K, heating due to the ionizing flux produced in the hotter part of the flow is approximately balanced by radiative losses, and the cooling time becomes much larger than a sound crossing time. At this point the gas is subject to further gas dynamical evolution. It may, for example, compress to balance the ambient background pressure, decreasing the effectiveness of photoionization, accelerating the recombination and self-shielding and, thereby, sharply cutting off the growth of the column densities of the ions of interest. We therefore do not carry our calculation beyond this point.

We take advantage of several scaling properties. For a given value of the product $n_{H,0}D_0$, the quantity $\phi_\nu= F_\nu/n_{H,0}v_0$ is constant with respect to changes in $n_{H,0}$, D_0 and v_0, where F_ν is the emergent flux at frequency ν. The ionization parameter, $U=n_\gamma/n_H$, where n_γ is the number density of H ionizing photons in the emergent flux, can be rewritten as $U=(v_0/c)\,\phi$, where $\phi = \int \phi_\nu d\nu$. The value of ϕ, the number of H ionizing photons emitted per H atom in the flow, is only weakly dependent upon the scale of the cooling region, D_0. For given values of D_0 and v_0, the time evolution of any fluid element in the flow is the same for different values of $n_{H,0}$ if plotted against $(n_{H,0}t)$.

The column density of any species i along a line of sight perpendicular to the flow direction is given by $N_i = v_0 \int y_i \, d(n_{H,0}t)$, where $y_i = n_i/n_H$ is the number density of the ion relative to hydrogen, where the integral is over the time history of a single fluid element. As the gas cools and recombines, the column density for all non-neutral species approaches an asymptotic value. Given a value of $(n_{H,0}D_0)$, we solve for the value of v_0 (and hence U) required to match N(C IV). We then compare the predicted column densities of Si IV and N V to the observed values of Sembach and Savage (1992), N(Si IV)/N(C IV) = 0.33±0.19 and N(N V)/N(C IV) = 0.28±0.16 . We find that we can match the data for a range of $(n_{H,0}D_0)$-values, 2 kpc $\gtrsim n_{-2}D_0 \gtrsim$ 8 pc. An example of such a match is shown in Fig 1.

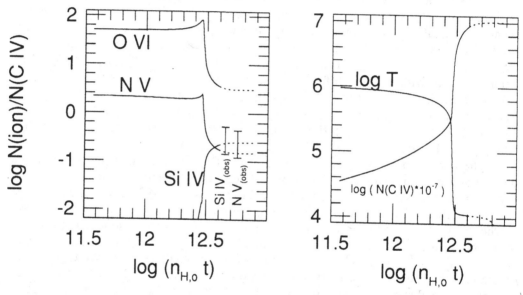

Fig. 1 — (Left panel) Column density ratios N(N V)/N(C IV), N(Si IV)/N(C IV), and N(O VI)/N(C IV) versus $(n_{H,0}t)$ (cm^{-3} sec) for a case with D_0=40 pc, v_0=100 km/s, and $n_{H,0}$=1.6x10^{-2} cm^{-3}. Evolution is stopped at the 10^4 K temperature "plateau" stage, when the cooling time becomes much longer than a sound crossing time. Dotted line indicates further evolution if the pressure in this stage remains fixed at the value of the internal pressure of the cooling region at the end of the isochoric, rapid cooling phase. The observed range for the column density ratios is indicated by vertical bars. (Right panel) Plot of temperature and log $(10^{-7}N(C IV))$ for the same case.

We found that obtaining such a match required a set value of the ionization parameter $U_{eff}\sim 10^{-3.3}$, where U_{eff} is given by

$$U_{eff} = \left(\frac{v_0}{c}\right) \phi_{eff} \, (1+\delta_\phi) \left(\frac{n_{H,0}}{n_{H,tr}}\right) \qquad (1)$$

where $\phi_{eff} = R\phi_\rho$, ϕ_ρ is the normalized flux for an isochoric flow, R=4πJ / F (where 4πJ is the mean intensity) and R~3 for cases considered here, and $\delta_\phi = (\phi/\phi_\rho)$ -1. The quantity δ_ϕ ranges from 0 (isochoric) to ~0.5 (isobaric), while $n_{H,o}/n_{H,tr}$, the inverse of the compression the gas undergoes before its cooling either becomes isochoric or else reaches the 10^4 K plateau first, ranges from 1 (isochoric) to ~100 (isobaric). Choosing an $(n_{H,o}D_o)$-value determines the value of all parameters in equation (1) except v_o. The requirement that we achieve the particular value $U_{eff} \approx 10^{-3.3}$ in order to match the observed column densities thus determines a value for v_o. The smaller the cooling region, the larger the flow velocity required. We find the required flow velocity is approximately given by $\log_{10} v_o$ (km/s) $\cong -0.33 \log_{10} n_{-2}D_o$ (pc) + 2.55 for 2 kpc $\gtrsim$ n_$_{-2}D_o \gtrsim 8$ pc, and $v_o \cong 30$ km/s for $n_{-2}D_o \gtrsim 2$ kpc. For v_o~100 km/s, an outflow velocity comparable to the sound speed of a 10^6 K gas, this gives D_o~$80(n_{-2})^{-1}$ pc. For n_$_{-2}D_o < 8$ pc, the column density ratios do not match the observed values for any v_o, so achieving $U_{eff} \approx 10^{-3.3}$ is not an adequate prescription in this limit. The difference between these results and our previous optically thin results (Shapiro 1990; Shapiro and Benjamin 1991) can be attributed to improved treatment of radiative transfer. In the previous calculations, the assumption of mono-directional flux corresponded to R=1, i.e. the diffuse emission was less efficient at ionizing the flow. Our success here in reproducing the observed column densities with a self-ionized flow explains naturally the constancy of the Si IV/ C IV and N V/ C IV ratios along different lines of sight reported by Sembach and Savage (1992). Such a flow also predicts N(O VI)/ N(C IV) ranging from 1.8 ($n_{-2}D_o$=8 pc) to 3.0 ($n_{-2}D_o \gtrsim 2$ kpc).

For each $(n_{H,o}D_o)$ value which reproduces the column densities, we solve for $n_{H,o}$ itself by requiring that the calculated emergent, upstream radiation flux for that case match the observed intensity of C IV and O III] UV emission lines (Martin and Bowyer 1990).This yields $\log_{10} n_{-2} \cong 0.52 \log_{10} D_o$ (pc) - 0.91, for 500 pc $\gtrsim D_o \gtrsim 15$ pc. For scales larger than $n_{-2}D_o \gtrsim 2$ kpc, the evolution is always isochoric and $n_{H,0}$~4×10^{-2} cm^{-3}. Such densities imply an emergent ionizing flux that exceeds the flux estimated from Hα observations of neutral clouds in the halo by Kutyrev and Reynolds (1989) by a factor of ~6. Our calculations are consistent with this, however, if the neutral halo gas observed in Hα emission was actually exposed to an attenuated or geometrically diluted version of our calculated flux. If we identify only the recombined, self-shielded part of our flow with the observed neutral clouds, for example, the calculated emergent Hα flux from *that* gas does satisfy this observational limit. Another constraint is that of the integrated Hα flux along a line perpendicular to the galactic plane derived from a survey of diffuse Hα emission (Reynolds 1990). Our predicted Hα flux from the entire flow provides ~50% of this observed flux. We also predict a value for the total emission in the O VI $\lambda\lambda$ 1033.8 doublet of $(4.6 - 6.1) \times 10^{-7}$ ergs s^{-1} cm^{-2} sr^{-1}. This exceeds by ~3 a new upper limit on the O VI flux by Edelstein and Bowyer (this volume) obtained along one line of sight. It remains to be seen whether this line of sight is typical. Future measurements of this doublet in conjunction with absorption line data will be extremely valuable.

We have calculated emergent broadband X-ray fluxes and converted them to detector countrates for comparison with the Wisconsin survey data (McCammon and Sanders 1990) and more recent ROSAT results (Burrows and Mendenhall 1991). The emergent upstream flux, unattenuated by foreground absorption, accounts for only

~25% of the observed counts in the higher energy bands (M band 0.4-1.1 keV, ROSAT 0.28-3.0 keV) while exceeding those in the lower energy bands. We find that a small but reasonable amount of attenuation (*c.f.* Martin and Bowyer 1990) due to <u>intervening</u> absorption, equivalent to N(H I) ~ 10^{20} cm^{-2}, is sufficient to bring the calculated lower energy fluxes into agreement with observations. Assuming that the entire disk of the galaxy is covered by such a flow, we derive a total X-ray luminosity (0.2-3.0 keV) of ~8 x 10^{39} ergs s^{-1}, exceeding by ~10 upper limits on X-ray emission from edge-on spiral galaxies (Bregman and Glassgold 1984). We thus require that the covering fraction, f, of such a flow be $f \leq 0.1$. This yields a mass outflow rate per side $\dot{M} \leq 2.4$ M$_0$/yr, a value in line with estimates of the mass flux rate associated with high-velocity neutral halo clouds.

We would like to thank John Raymond for making his hot gas atomic data set available to us. This work was supported by NASA Training Grant No. NGT-50849, Robert A. Welch Foundation grant F-1115, and NASA grant NAGW-2399. All of our numerical calculations were performed on the University of Texas Center for High Performance Computing Cray Y/MP.

REFERENCES

Allen, C. W. 1973, *Astrophysical Quantities* (London: The Athlone Press).
Bregman, J. N. and Glassgold, A. E. 1982, *Ap. J.*, **263**, 564.
Burrows, D. N. and Mendenhall, J.A. 1991, *Nature*, **351**, 629.
Kutyrev, A. S. and Reynolds, R. J. 1989, *Ap. J.(Letters)*, **344**, L9.
Martin, C. and Bowyer, S. 1990, *Ap. J.*, **350**, 242.
McCammon, D. and Sanders, W. T. 1990, *Ann. Rev. Astron. Astrophys.*, **28**, 657.
Raymond, J. C. 1987, private communication.
Raymond, J. C. and Smith, B. W. 1977, *Ap. J. (Supp)*, **35**, 419.
Reynolds, R.J. 1990, in *I. A. U. Symposium No. 144*, ed. H. Bloemen (Dordrecht: Kluwer Academic Publishers), p. 67.
Sembach, K. R. and Savage, B. D. 1992, Wisconsin preprint 409.
Shapiro, P.R. 1990, in *I. A. U. Symposium No. 144,* ed. H. Bloemen (Dordrecht: Kluwer Academic Publisher), p. 417.
Shapiro, P.R. and Benjamin, R.A. 1991, *P. A. S. P.*, **103**, 923.
Shapiro, P. R. and Field, G. B. 1976, *Ap.J.*, **205**, 762.
Spitzer, L. 1990, *Ann. Rev. Astron. Astrophys.*, **28**, 71.

An Imaging Extreme Ultraviolet Spectrometer

Paolo Bergamini[1], Thomas E. Berger[1], Giorgio Giaretta[1], Martin C.E. Huber[2], Giampiero Naletto[3], J. Gethyn Timothy[1] and Giuseppe Tondello[3]

[1]Center for Space Science and Astrophysics, Stanford University, ERL 315A Stanford, California 94305-4055 USA
[2]Space Science Department, ESTEC, Postbus 299, 2200AG Noordwijk, The Nethelands
[3]Istituto di Elettronica, Universita' di Padova Via Gradenigo n.6/A, I-35100 Padova, Italy

ABSTRACT

A laboratory extreme ultraviolet (EUV) imaging spectrometer has been fabricated and tested. This instrument is used to test and to characterize toroidal gratings like those which will be employed in the high-resolution spectroheliometer (HiRES) configured for flight on a sounding rocket. The imaging spectrometer will be used also for characterization and calibration of Multi Anode Microchannel Array (MAMA) detectors foreseen on the ESA/NASA Solar Heliospheric Observatory (SOHO) satellite. The spectrometer employs a concave toroidal grating illuminated at normal incidence in a 1 meter Rowland circle mounting: high efficiency is achieved because the grating is the only reflecting surface. The grating is able to produce stigmatic images over a wavelength range of about 100 Å or 200 Å centered respectively around 600 Å or 1200 Å. The source, used to illuminate sets of micron sized pinholes arranged along the spectrometer entrance slit, is a dc hollow cathode lamp: a low-pressure spark-discharge tube operating with noble gases and able to produce sharp spectral lines. The spectra are detected by mean of a MAMA detector which yields high-resolution imaging with great efficiency at EUV wavelengths. The results of the initial imaging tests and the measurements carried out are presented and discussed

1 THE SPECTROMETER

A concave toroidal diffraction grating illuminated in near normal incidence can produce exact stigmatic images in two points of the Rowland circle if the torus vertical and horizontal curvature raddii satisfy the following condition:

$$R_v = R_h \sin \alpha \sin \beta_0 \qquad (1)$$

where α and β_0 are respectively the incidence and diffraction angles (Haber 1950).The two stigmatic points are located symmetrically to the grating normal at $\pm\beta_0$ on the plain that contains the Rowland circle whose diameter is R_h. In systems where β_0 is small there is some depth of focus that enable stigmatic focusing between, and somewhat beyond the two stigmatic points. With a single toroid (moderate speed system with focal ratio ~ f/15) a wavelength range of over 100 Å in the EUV can be obtained with high image quality (Huber and Tondello 1979). The gratings we use have been fabricated using a replica technique with deformable submaster obtained from ruled spherical master (Huber *et al.*. 1981). The concave toroidal grating is replicated on a fixed zerodur substrate from the submaster deformed to the appropriate aspect ratio. An osmium coating enhances the EUV reflectivity. The characteristics of the gratings we tested are listed in table 1.

DESIRED CHARACTERISTICS OF THE F/15 TOROIDAL GRATINGS		
Horizontal radius of curvature R_h	1011.1 mm	
Vertical radius of curvature R_v	989.1 mm	
Aspect ratio R_h / R_v	0.9782	
Angle of incidence α	11.947°	
Ruled area	70 x 70 mm^2	
Ruling frequency	3600 grooves mm^{-1}	1800 grooves mm^{-1}
Wavelength at stigmatic points ($\pm\beta_0 = 0.825°$)	535 Å, 615 Å	1070 Å, 1230 Å
Wavelength on grating normal ($\beta = 0$)	575 Å	1150 Å

TABLE 1.

The spectrometer has the entrance slit, the detector and the grating located on a 1-m Rowland circle. The grating is mounted on a plateform that can be moved radially and tangentially for fine focal and spectral adjustments. The source is a DC hollow cathode lamp, a low-pressure spark-discharge tube operated with noble gases. The gas pressure in the lamp (ranging between a few tenths of a torr to a few torr depending on the gas) is maintained against the main chamber vacuum pressure (2-3×10^{-7} torr) by a differential pumping system. A set of 10 pinholes (25 μm diameter and 500 μm despaced) in an aluminum foil are located along the entrance slit. We have used the spectrometer with three different toroidal gratings that have the same aspect ratio but different ruling frequency: 3600 mm^{-1} on the first and 1800 mm^{-1} on the second and third. An open-structure MAMA detector is mounted on the output flange of the spectrometer. It is a pulse-counting imaging detector that employs an high gain curved-channel microchannel plate and a readout anode array to produce a format of 1024 x 360 pixels with pixel dimension of 25 x 25 μm^2.

2 TESTS AND RESULTS

The first grating we have mounted and aligned in the spectrometer is the 3600 mm^{-1}. The image of the spectrum produced by the hollow cathode lamp running with a mixture of Helium and Neon (20-80 %) is displayed in the figure (Fig. 1). The source discharge current was 150 mA, the gas pressure 1 *t*orr and the integration time 1000 seconds. The grating's excellent imaging properties are visible in the sharp images of the ten pinholes in every spectral component. Over the whole field of view the pinhole size is no more than 3 pixels FWHM, in both the spectral and spatial direction.

Fig. 1 - Image of the Helium-Neon spectrum obtained with the 3600 mm⁻¹ grating. The upper part of the image is fainter because an attenuating mesh is placed in front of the detector microchannel plate. The attenuator transmission coefficient is 10%.

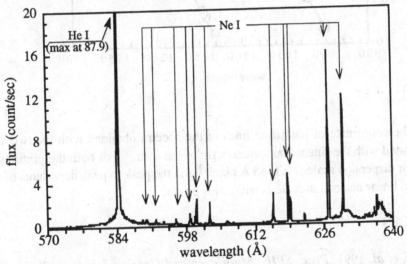

Fig. 2 - Helium-Neon spectrum obtained with the 3600 mm⁻¹ grating.

The spectrum in Fig. 2 represent the spectral profile of a single pinhole across the whole detector. The major feature is the first resonance line of He I. Through the identification of the spectral position of 13 lines of Ne I, it was possible to determine the system's linear dispersion scale (0.0693 Å/pixel). The standard deviation of residual of the measured coordinates to the linear least-squares fit result to be 0.27 pixel.

We tested in the spectrometer two 1800 mm⁻¹ gratings fabricated from the same ruled master. The first one was tested using pure Argon in the discharge lamp. In the recorded spectrum, several lines of Ar II, both in 1st and 2nd order, appear beside the principal resonance lines of Ar I (1048.22 Å, 1066.66 Å). The fact that the lamp cathode is copper made explains the presence of the Cu II lines.

The test on the second 1800 mm^{-1} was performed employing the Helium-Neon mixture. Resonance lines, both from the gas and from the cathode material of the spark-discharge lamp, appear also in the spectrum obtained with this grating. The former spectrum covers the second half of the 1800 mm^{-1} grating stigmatic range (1120-1270 Å) while the Argon spectrum (990-1130 Å) covers the first half.

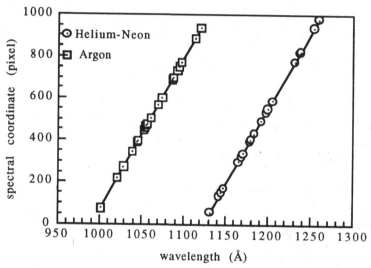

Fig. 3

In Fig. 3 the coordinate of identified lines of the spectra obtained with the two 1800 mm^{-1} are plotted with the linear least-squares fit to the data. For both the gratings the resulting linear dispersion scale is 0.1385 Å pixel^{-1} and the peak to peak departures of data from the fitted linear dispersion scale is within 1.8 pixels.

REFERENCES

Berger, T. E., *et al*, 1991, *Proc. SPIE Multilayer and Grazing Incidence X-Ray/EUV Optics*, **1546**, 446.

Haber, H. 1950, *Appl. Opt.*, **40**, 153.

Huber, M. C. E., and Tondello, G. 1979, *Appl. Opt.*, **18**, 3948.

Huber, M. C. E., Janniti, E., Lemaitre, G., and Tondello, G. 1981, *Appl. Opt.*, **20**, 2139.

HiRES: High Resolution Extreme Ultraviolet Spectroheliometer

Thomas Berger[1], Paolo Bergamini[1], Helen Kirby[1], J. Gethyn Timothy[1], Arthur B.C. Walker[1], Jagadish C. Bhattacharyya[2], Surendra K. Jain[2], Ajay K. Saxena[2], Martin C.E. Huber[3], Giampiero Naletto[4], Guiseppe Tondello[4]

[1]Center for Space Science and Astrophysics, Stanford University
[2]Indian Institute of Astrophysics
[3]Space Science Department, European Space Agency
[4]Istituto di Elettronica, University of Padua

ABSTRACT

The HiRES sounding rocket payload is designed to obtain very high spatial, spectral, and temporal resolution images of the solar chromospheric and coronal plasmas in the exterme ultraviolet (EUV) wavelength range from 500 to 650 Å. The instrument consists of a 450 mm f/15 Gregorian telescope feeding a 1-m normal incidence stigmatic spectrometer. The stigmatic spectrometer utilizes a toroidal diffraction grating formed by a unique elastic substrate deformation technique in order to achieve simultaneous spatial and spectral focusing at two points on the detector plane. Spatial resolution on the order of 0.4 arcsecond across a 3 x 3 arcmin^2 field of view is obtained. Temporal resolution of the order of milliseconds is achieved by the use of an advanced imaging Multi-Anode Microchannel Array (MAMA) detector. A hydrogen-alpha 6562.8 Å camera and a 0.25-m EUV solar irradiance spectrometer are also included in the payload.

1 INTRODUCTION

Recent high resolution sounding rocket instruments such as the High Resolution Telescope and Spectrograph (HRTS) in near-ultraviolet wavelengths [Brueckner, 1983] and the Multi-Spectral Solar Telescope Array (MSSTA) [Walker, 1988] and the Normal Incidence X-ray Telescope (NIXT) [Golub, 1990] in the soft x-ray regime have pushed the spatial resolution of solar observations towards the sub-arcsecond level and revealed coherent small scale structures in all wavelengths imaged. The improvements offered by these instruments in both spatial and spectral resolution have provided a better understanding of the structure of the chromosphere and corona and the role of the time-dependent magnetic field in the dynamics of these regions. The HiRES instrument will fill a gap in the wavelength coverage of high resolution observations by adding sub-arcsecond imaging and radiometric data in the extreme ultraviolet wavelength range between 500 and 650 Å.

2 INSTRUMENT DESCRIPTION

The HiRES instrument design is based on two key technologies: the fabrication of replicated toroidal diffraction gratings using deformable master substrates [Huber 1979, 1991] and the Multi-Anode Microchannel Array (MAMA) detector [Timothy 1989, 1991a]. A single toroidal grating produces stigmatic images which allows high-resolution imaging in the EUV by eliminating the reflectivity losses associated with multi-optic spectrometer designs. The deformable substrate fabrication method results in a significant saving of manufacturing and testing time as compared to the traditional grinding and polishing technique. The MAMA detector system to be used on the HiRES instrument is of the same design that is currently being developed for use on the ESA/NASA Solar and Heliospheric Observatory (SOHO) satellite. MAMA detectors are also being developed for the Space Telescope Imaging Spectrograph (STIS) and the Far Ultraviolet Spectroscopic Explorer (FUSE/*Lyman*) mission. The MAMA detector gives the HiRES instrument the ability to image faint, multi-spectral EUV solar features with radiometric accuracy; to image with extremely high temporal resolution; and to transmit and store the data electronically.

The instrument platform is a 22" diameter Black Brant sounding rocket. The primary elements of the instrument are a 450 mm f/15 Gregorian telescope, a 1 meter toroidal grating spectrograph, and the MAMA detector. Figure 1 shows a layout of the HiRES payload.

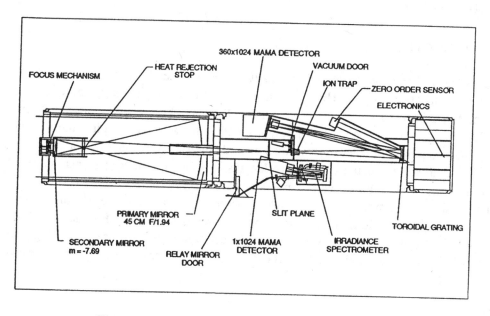

Figure 1 - Conceptual Layout of the HiRES payload

Resolution values for the HiRES instrument are as follows:

- Spatial resolution: 0.4 arcseconds
- Spectral resolution: $\lambda/\Delta\lambda = 10,000$ (0.078 Å)
- Temporal resolution: 10^{-3} sec (depending on signal levels)

Stigmatic points are located at 535 Å and 615 Å with near stigmatic imaging in between and slightly beyond these wavelengths. The MAMA detector has a nominal 360 by 1024 25 micron pixel format with integrated decode circuitry capable of centroiding the photon events to within 0.1 pixel resulting in a spectral shift sensitivity of approximately 2 mÅ (~1 km per second Doppler shifts at 600 Å). The MAMA detector uses a single curved-channel microchannel plate (MCP) for high geometric fidelity and minimal ion feedback. The spectrometer is calibrated in flight by a sealed platinum hollow cathode lamp.

The HiRES payload includes two ancillary instruments. The first is a hydrogen Balmer-alpha (6562.8 Å) CCD camera that performs real-time monitoring of instrument pointing and target selection. The second is a 0.25 meter spherical grating objectiveless spectrometer for full Sun irradiance measurements in the wavelength range of 280 Å to 1270 Å. Provision for later inclusion of a Wadsworth infrared spectrometer for simultaneous helium 10830 Å observations is built into the HiRES instrument.

Table 1 lists the dominant solar spectral emission lines in the HiRES EUV wavelength range.

Table 1 - Major Solar Emission Lines Visible to HiRES

ION	λ Å	TEMP (K)	ACTIVE INTENSITY (erg cm^{-2}s^{-1}sr^{-1})	COUNT RATE (Pixel^{-1} sec^{-1})	QUIET INTENSITY (erg cm^{-2}s^{-1}sr^{-1})	COUNT RATE (Pixel^{-1} sec^{-1})
Si XII	499.0	2.0×10^6	1284	363	66	19
He I	515.6	3.2×10^4	199	56	26	7
Si XII	521.1	2.0×10^6	720	213	33	10
He I	522.2	3.2×10^4	329	98	44	13
He I	537.3	3.2×10^4	834	253	109	33
Al XI	550.0	NA	169	48	9	3
O IV	553.7	1.6×10^5	487	138	246	70
Fe XV	568.3*	4.0×10^6	4733	76	NA	NA
He I	584.3	3.2×10^4	7200	2388	828	275
O III	599.6	1.0×10^5	74	25	43	15
Si XI	606.6*	1.6×10^6	16955	292	3049	52
He II	607.6*	8.0×10^4	91752	1582	10998	190
Mg X	609.8	1.0×10^6	943	327	136	47
Mg X	625.2	1.0×10^6	470	167	49	17
O V	629.7	2.5×10^5	1334	465	510	178

* Observed wavelength in second order. Intensity values from [Vernazza, 1978]

The HiRES instrument has two observing modes: a stationary slit mode in which the field of view is limited to the 0.4 arcsecond by 3 arcminute slit extent, and a spatial scan mode in which the sounding rocket control system is used to scan the slit across the field of view. In the stationary slit mode, the instrument collects high time resolution slit images analogous to the HRTS images. These observations are ideal for high time resolution studies of very small scale transient events such as microflares. The scan mode builds two-dimensional images perpendicular to the slit axis by scanning the slit across areas of interest on the Sun. Figure 2 is a scaled diagram of the HiRES scan mode field of view on the Sun. Each emission line in the HiRES spectral range is built into a composite image with temporal resolution limited by the scan time of 45 seconds. These images are useful for studying the morphology of the solar coronal plasmas in the distinct temperatures ranges of the ions listed in Table 1. More detailed descriptions of the HiRES instrument exist elsewhere in the literature [Timothy, 1991b; Berger, 1991].

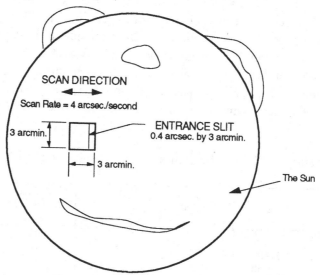

Figure 2 - Typical HiRES Image Scan

3 REFERENCES

Berger, T.E., *et al.,* 1991, *Proc. SPIE,* **1546,** 446.
Brueckner, G. and J.-D.F. Bartoe, 1983, *Ap. J.,* **272,** no. 1, 329.
Golub, L. , *et al.,* 1990, *Nature,* **344,** 842.
Huber, M.C.E. and G. Tondello, 1979, *Applied Optics,* **18,** no. 23, 3948.
Huber , M.C.E., *et al.,* 1991, *Proc. SPIE,* **1494,** 472.
Timothy, J.G., *et al.,* 1989, *Proc. SPIE,* **1158,** 104.
Timothy, J.G., *et al.,* 1991a, *Proc. SPIE,* **1494,** 394.
Timothy, J.G., *et al.,* 1991b, *Optical Eng.,* **30,** no. 8, 1142.
Vernazza, J.E. and E.M. Reeves, 1978, *Ap. J. Supp.,* **37,** 485.
Walker, A.B.C., *et al.,* 1988, *Science,* **241,** 1781.

A Search for Argon in the Atmosphere of Titan

Randy Gladstone[1], Stuart Bowyer[2], Tobias Owen[3],
and Mark Hurwitz[2]

[1]Space Sciences Laboratory, University of California at Berkeley
[2]Center for EUV Astrophysics, University of California at Berkeley
[3]Institute for Astronomy, University of Hawaii

ABSTRACT

The origin of Titan's thick nitrogen and methane atmosphere is an enigma. Detection of argon would indicate that the N_2 and CO now found in the atmosphere probably came in with ice during Titan's accretion. If there is very little argon, then we have to turn to models starting with frozen ammonia, methane and water ice, indicating a more important role for the Saturn sub-nebula, and requiring subsequent modification by photochemistry. Current estimates on the fraction of argon in Titan's atmosphere are based on indirect evidence and range from 0 to 25%. We have modeled Titan's atmosphere with argon as an added constituent. Using a photoelectron transport code, we determined the photoelectron flux in a standard model Titan atmosphere with energy degradation and transport of photoelectrons due solely to N_2. The calculated photoelectron flux was then used to excite the N_2, Ar, and CH_4 in the upper atmosphere of Titan using appropriate electron impact EUV emission cross sections. Photodissociative excitation of N_2, resulting in emissions by N and N^+ fragments, were also included. The resulting emission profiles were used to simulate the expected brightnesses of several features in the EUV dayglow spectrum of Titan. In particular, the bright resonance line of argon at 1048 Å is predicted to be present at a level of 5–10 R, depending on the assumed abundance of argon in the lower atmosphere of Titan. We then consider observability of the argon dayglow emission by the Berkeley spectrometer to be flown as part of the German/US Shuttle-borne *ORFEUS* mission. This spectrometer employs four novel spherically figured varied line-space gratings which will provide high resolution ($\lambda/\Delta\lambda > 6000$) spectra from 390 to 1200 Å. The calculated integration time required for detection of argon in the atmosphere of Titan is about 20 hours, which is, unfortunately, too long to be attempted during the 5-day *ORFEUS* mission. An accurate determination of the abundance of argon on Titan will likely have to wait for the *Cassini* orbiter and *Huygens* probe to arrive at the Saturn system.

1 INTRODUCTION

Saturn's moon Titan is the second largest satellite in the solar system (it is larger than the planet Mercury), and it has an N_2 atmosphere with 1.5 times the pressure of Earth's atmosphere. Minor amounts of hydrocarbons are also present and have led to the shrouding of Titan in a thick, smog-like haze, which is nearly featureless over the surface. Interest in argon stems from its relative cosmic abundance and its inertness, which make it a tracer of trapped interstellar gas in icy bodies now found in our solar system. Laboratory experiments have demonstrated that argon is readily trapped in amorphous ice forming at low temperatures, as are N_2,

CO, CH$_4$, and many other gases. A measurement of the Ar/N$_2$ ratio will constrain whether the nitrogen in Titan's atmosphere is primordial or derived from photolysis of ammonia. If the argon abundance is $\ll 1\%$, then it is likely that the N$_2$ is a secondary product of NH$_3$ photochemistry, whereas if the argon abundance is $> 1\%$ then the N$_2$ is more likely to be primordial (Lunine *et al.* 1989). The presence of argon in Titan's atmosphere has been inferred from *Voyager* radio occultation experiments and infrared observations, and could have abundances anywhere in the range 0–25% in the troposphere (Hunten *et al.* 1984). The *Voyager* UVS provided only a weak upper limit to the argon abundance, since the ~ 30 Å resolution of the UVS smears the expected Ar 1048 Å emission with nearby N and N$_2$ emissions.

An accurate measurement of the abundance of argon in the atmosphere would thus greatly constrain models of the origin of Titan, since it would determine the relative fractions of Titan's nitrogen inventory incorporated as N$_2$ (from the solar nebula) and NH$_3$ (from the Saturnian nebula) (Owen 1982; Prinn and Fegley 1989). With < 0.2 Å resolution at 1048 Å, the EUV spectrometer on the *ORFEUS* mission (Bowyer and Hurwitz 1990) would, in principle, be able to measure argon on Titan from Earth orbit.

2 MODEL SIMULATION

Following the general procedure of Strobel *et al.* (1991), we have simulated a number of expected EUV emission features in the spectrum of Titan in the vicinity of the argon resonance line at 1048 Å. The model Titan atmospheres used (shown in Fig. 1) are based on the studies by Lellouch *et al.* (1990) and Yung *et al.* (1984), to which we have self-consistently added two different abundances of argon (5% and 25%). Using the photoelectron transport code of Link *et al.* (1988), we determined the photoelectron flux in two model Titan atmospheres with energy degradation and transport of photoelectrons due solely to N$_2$. The calculated photoelectron flux was then used to excite the N$_2$, Ar, and CH$_4$ in the upper atmosphere of Titan using the electron impact EUV emission cross sections determined by Ajello *et al.* (1989), Ajello *et al.* (1990), and Pang *et al.* (1987), respectively. Photodissociative excitation of N$_2$, resulting in emissions by N and N$^+$ fragments, was also included, using the cross sections of Gladstone *et al.* (1990) and Meier *et al.* (1991). The study by Strobel *et al.* (1991) obtained reasonable agreement with *Voyager* UVS measurements of Titan using this method; however, they only considered two molecular emissions of N$_2$ and two N$^+$ emissions. We have included several additional emissions; in particular we calculate the emission rate of the Ar resonance line at 1048 Å resulting from photoelectron excitation. Fig. 2 presents the resulting emission rate profiles of the Ar(1048 Å), N$_2$(c_4' (0,0) 958 Å), N$_2$(b' (16,0) 871 Å), N(1200 Å,1134 Å), N$_2$(LBH), and N$^+$(1085 Å) features. The brightnesses of the features (listed in Fig. 2) were determined by integrating the calculated emission profiles with altitude, with inclusion of absorption by CH$_4$.

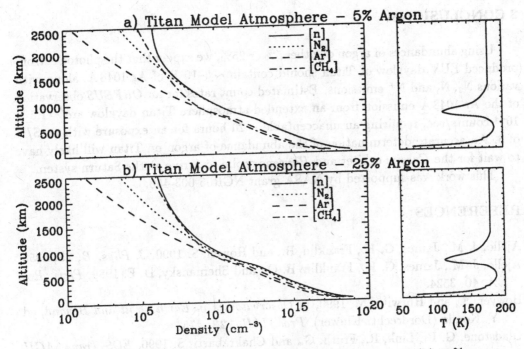

Fig. 1—Model Titan atmospheres containing (a) 5% argon and (b) 25% argon.

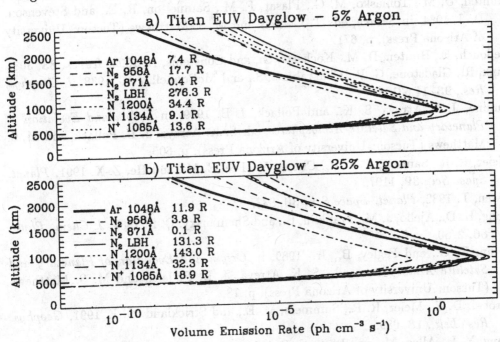

Fig. 2—Model emission rate profiles for several important EUV emission features for model Titan atmospheres containing (a) 5% argon and (b) 25% argon. The column brightness, including the effects of absorption by CH_4, is listed for each emission.

3 CONCLUSIONS

Using abundances of argon of either 5% or 25%, we expect that the photoelectron-produced EUV dayglow of Titan should contain ~ 5–10 R of Ar 1048 Å, along with various N_2, N, and N^+ emissions. Estimated count rates for an *ORFEUS* observation of the Ar 1048 Å emission from an extended-atmosphere Titan dayglow are $\sim 1.5 \times 10^{-4}$ counts/sec, requiring an unacceptable ~ 20 hours for an exposure with an S/N of 3. An accurate determination of the abundance of argon on Titan will likely have to wait for the *Cassini* orbiter and *Huygens* probe to arrive at the Saturn system.

This work was supported by NASA grant NGR05-003-450.

REFERENCES

Ajello, J. M., James, G. K., Franklin, B., and Howell, S. 1990, *J. Phys. B*, **23**, 4355.

Ajello, J. M., James, G. K., Franklin, B. O., and Shemansky, D. E. 1989, *Phys. Rev. A*, **40**, 3524.

Bowyer, S., and Hurwitz, M. 1990, in *Observatories in Earth Orbit and Beyond*, ed. Y. Kondo (Dordrecht: Kluwer), *Proc. IAU*, **123**, 475.

Gladstone, G. R., Link, R., Fruth, G., and Chakrabarti, S. 1990, *EOS Trans. AGU*, **71**, 1488.

Hunten, D. M., Tomasko, M. G., Flasar, F. M., Samuelson, R. E., and Stevenson, D. J. 1984, in *Saturn*, ed. T. Gehrels and M. S. Matthews (Tucson: University of Arizona Press), p. 671.

Lellouch, E., Hunten, D. M., Kockarts, G., and Coustenis, A. 1990, *Icarus*, **83**, 308.

Link, R., Gladstone, G. R., Chakrabarti, S., and McConnell, J. C. 1988, *J. Geophys. Res.*, **93**, 14,631.

Lunine, J. I., Atreya, S. K., and Pollack, J. B. 1989, in *Origin and Evolution of Planetary and Satellite Atmospheres*, ed. S. K. Atreya, J. B. Pollack, and M. S. Matthews (Tucson: University of Arizona Press), p. 605.

Meier, R. R., Samson, J. A. R., Chung, Y., Lee, E.-M., and He, Z.-X. 1991, *Planet. Space Sci.*, **39**, 1197.

Owen, T. 1982, *Planet. Space Sci.*, **30**, 833.

Pang, K. D., Ajello, J. M., Franklin, B., and Shemansky, D. E. 1987, *J. Chem. Phys.*, **86**, 2750.

Prinn, R. G., and Fegley, B., Jr. 1989, in *Origin and Evolution of Planetary and Satellite Atmospheres*, ed. S. K. Atreya, J. B. Pollack, and M. S. Matthews (Tucson: University of Arizona Press), p. 78.

Strobel, D. F., Meier, R. R., Summers, M. E., and Strickland, D. E. 1991, *Geophys. Res. Lett.*, **18**, 689.

Yung, Y. L., Allen, M., and Pinto, J. P. 1984, *Ap. J. Suppl.*, **55**, 465.

A Very High Resolution Spectrometer for EUV Astrophysics

Daniel M. Cotton, Brett C. Bush, James Vickers, and
Supriya Chakrabarti

Earth and Planetary Atmospheres Group
Space Sciences Laboratory, University of California at Berkeley

ABSTRACT

There are two main barriers to the development of space flight conventional interferometers that operate below 2000 Å: the lack of high quality transmission beam splitters and the need for high precision positioning and scanning mechanisms that can withstand the vibration and shock of rocket launch. We describe an interferometric spectrometer that offers several advantages over conventional high resolution grating spectrometers. One can obtain high spectral resolution easily in a compact size. The throughput (area - solid angle product) is typically 200 times larger than that of conventional spectrometers of comparable resolution, and it can be field-widened without moving parts. This new instrument can be used in a number of astrophysical applications. These applications will be discussed along with the operating principle of the optical design and recent laboratory test results.

1 INTRODUCTION

In astrophysics, the conventional technique for determining the distribution and composition of interstellar gas is the spectroscopic observation of stars at ultraviolet wavelengths. A principal result of the OAO-3 Copernicus observations is the wealth of information to be gleaned by high resolution observations of the strong resonance lines of such species as CII, SiII, CIV, NV, and OVI—lines that diagnose the ionization and thermal balance of the interstellar medium. Because the Hubble Space Telescope does not operate shortward of Lyman α, and because FUSE/LYMAN is a decade away, it is advantageous to examine alternative technologies that may open the way for very high vacuum ultraviolet wavelength resolution without the demanding arc-second optical performance characteristic of HST and FUSE.

An interferometer can, in principle, deliver excellent signal-to-noise ratio interferograms provided that it is fed by a stellar image from which most of the broadband light—the cause of the shot noise in the interferogram— has been removed. However, conventional interferometers have not been developed in the EUV and FUV primarily due to the following two technical obstacles. First, beam splitters, a key component of an interferometer, are not available in the ultraviolet. Second, precision positioning and scanning mechanisms required for operations in the ultraviolet wavelength range by conventional interferometers are extremely difficult to make. The only such device that we are aware of is a Fourier Transform Spectrometer (FTS) suitable for laboratory use (it requires a granite table for stability), using a matched pair of spectrosil plates to operate at wavelengths down to 1700 Å (Thorne et al. 1987). A LiF beam splitter can, in principle, work to approximately 1050 Å. However, interferometric quality beam splitters in these wavelengths are difficult to manufacture and have only recently been demonstrated with MgF_2 for operation at 1777 Å (Parkinson et al. 1992). Furthermore, any transmitting optical arrangement stands in the way of applicability in the EUV and soft X-Ray spectral regions.

Recently, we have developed an all-reflection interferometric technique that is applicable down to EUV wavelengths and is well suited for space flight observations. Furthermore, its optical performance has been verified using readily available components at several wavelengths including Lyman α at 1216 Å. This design along with various other configurations has been described elsewhere (Harlander et al. 1990) and been tested at 5461 and 2537 Å (Bush et al. 1991). We have also developed and begun testing a Lyman α prototype instrument. On June 19, 1991, we observed interference fringes at 1216 Å (see Chakrabarti et al. 1992). To the best of our knowledge, this is the first time that interference fringes have been observed at Lyman α.

2 THE ARIES

It is well known that interference spectrometers offer significant advantages compared to conventional grating spectrometers in the study of faint, extended sources (see, for example, Roesler 1974). The primary advantages are (1) an etendue (or throughput) typically 200 times larger than grating spectrometers operating at similar resolution, (2) compact size, especially at high resolution, (3) avoidance of high precision optics in the input and output optical systems (light buckets are sufficient) and (4) relative ease of obtaining high spectral resolution ($> 10^5$) using off the shelf components. In combination these advantages offer important economies in observing time, cost, weight, and volume in many important programs. Recently we have developed and tested an interferometric spectroscopic technique which, in addition to the advantages cited above, has no moving parts (scanning is not required), can be field widened, and has been demonstrated in an all-reflection configuration in UV wavelengths.

The instrument, the All-Reflection Instrument for EUV Spectroscopy (ARIES), is a close relative of conventional FTS, but most of the complex problems of scan and control associated with FTS are avoided. Its compact and relatively robust structure make it an attractive alternative to conventional grating spectrometers proposed for the next generation of very large telescopes and an ideal instrument for development for high resolution UV spectroscopy from space. In operation it requires stability but neither critical alignment nor scanning. It has some limitations similar to its more complex relatives, but these limitations may be avoided for most problems by suitable design and experimental technique. In this section we describe the ARIES and present results of tests we have performed in the the FUV.

The all-reflection ARIES instrument has been fully tested in the visible and ul-

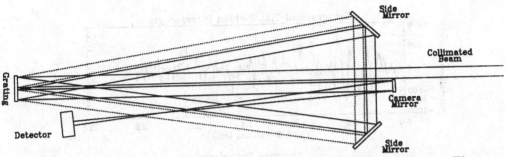

Fig. 1—The optical layout for the all-reflection, ARIES interferometer. The grating splits and recombines the two arms of the interferometer (dotted and solid lines). The camera mirror images the interferogram onto a detector.

traviolet (Harlander et al. 1990; Bush et al. 1991) and shown to have the qualities described above. However, it was not until recently that an operational model was made to work at 1216 Å (Chakrabarti et al. 1992). In this section we describe the initial tests of the FUV instrument.

The instrument layout is essentially that shown in Figure 1. The grating has a symmetric V-groove profile with a line density of 600/mm and is blazed for 1216 Å in 2nd order (Cotton et al. 1991). Second order was chosen so that initial alignment could be made with the HgI 2537 Å line in 1st order at air. To simplify alignment, a brass-board version of the space-bound instrument was developed (Tom et al. 1991a,b). The detector used was a two-dimensional imaging sealed microchannel plate detector with a wedge and strip readout. The instrument was illuminated by a H_2 gas discharge source through a grazing incidence monochromator. A histogram of the data and subsequent Fourier transform of the histogram are shown in Figure 2. The transform reveals a self-reversed line about 30 mÅ wide. This implies that the source is optically thick with a temperature around 350 K.

3 SUMMARY

We have described the All-Reflection Interferometer for EUV Studies. The ARIES design overcomes many of the obstacles that have made space-based interferometry in the ultraviolet unattainable to date. We have demonstrated its capabilities at Lyman α, and we intend to explore modified designs, including an imaging and field-widened versions. This instrument holds much promise for ultra high resolution astrophysical studies of both diffuse and point sources.

This work was supported by NASA grants NAG5–675.

REFERENCES

Bush, B., Cotton, D. M., and Chakrabarti, S. 1991, *Proc. SPIE*, **1549**, 290.
Chakrabarti, S., Cotton, D. M., Vickers, J. S., and Bush, B. 1992, *Appl. Opt.*,

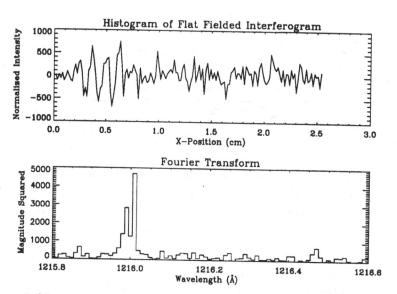

Fig. 2— A histogram of the image shown in Figure 1 along with the squared magnitude of the Fourier transform.

submitted.

Cotton, D. M., Bach, B., Bush, B., and Chakrabarti, S. 1991, *Proc. SPIE*, **1549**, 313.

Harlander, J., Roesler, F. L., and Chakrabarti, S. 1990, *Proc. SPIE*, **1334**, 120.

Parkinson, W. H., Thorne, A. P., Smith, P. L., and Yoshino, K. 1992, these proceedings.

Roesler, F. L. 1974, *Methods of Experimental Physics, vol. 12, Part A: Optical and Infrared*, Academic Press, New York.

Thorne, A., Harris, C. J., Wynne-Jones, I., Learner, R. C. M., and Cox, G. 1987 *J. Phys. E: Sci. Instrum.*, **20**, 54.

Tom, J., Cotton, D. M., Bush, B., Chung, R., and Chakrabarti, S. 1991*a*, *Proc. SPIE*, **1549**, 308.

Tom, J., Cotton, D. M., Bush, B., Chung, R., and Chakrabarti, S. 1991*b*, *Proc. SPIE*, **1549**. 302.

First Flight of an Extreme-Ultraviolet Spectrometer with a Multilayer Grating

Joseph M. Davila[1], Roger J. Thomas[1], W. T. Thompson[2], R. A. M. Keski-Kuha[1], and W. M. Neupert[1]

[1]NASA-Goddard Space Flight Center, Greenbelt, MD
[2]Applied Research Corporation, Lanham, MD

ABSTRACT

In this paper we report the first space flight of an extreme-ultraviolet (EUV) spectrograph incorporating a multilayer coated normal incidence grating in the Solar Extreme-ultraviolet Rocket Telescope and Spectrograph (SERTS). Pre-flight performance evaluation showed that the application of a 10-layer Ir/Si multilayer coating to the $3600\ l/mm$, blazed, toroidal replica grating produced a factor of nine enhancement in peak efficiency near the design wavelength around $30\ nm$ in first order over the standard gold coating used in earlier flights. In addition, a spectral resolution of better than 5000 was maintained. This technology, now proven in space flight, is applicable to most normal incidence spectographs used for astronomical observation in the ultraviolet, far and extreme ultraviolet and soft x-ray regions of the spectrum.

1 INTRODUCTION

An extensive amount of work has been done on applying multilayer coatings to telescope mirrors both for astronomical imaging applications in sounding rocket experiments (Haisch et al., 1988; Walker et al., 1989; Golub et al., 1990). In this paper we discuss the first space flight of a spectrograph utilizing this technology to enhance the efficiency of diffraction gratings in a normal incidence spectrograph. This technology allows the use of normal incidence optics in wavelength regions where only glancing incidence designs have provided acceptable efficiency in the past.

The optical concept of the Solar EUV Rocket Telescope and Spectrograph instrument (SERTS) (Neupert et al., 1981; Neupert et al., 1992) utilizes a grazing-incidence Wolter Type-2 telescope to form a real image of the sun on the entrance aperture of a stigmatic spectrograph. The entrance aperture selects a portion of the solar image and passes it onto the spectrograph grating, which then re-images it at the final focal plane simultaneously in each dispersed wavelength of the spectrum. The imaged spectra are produced by a torodial grating.

2 COATING AND LABORATORY RESULTS

The SERTS master toroid was polished directly into a Zerodur substrate. The master was ruled at Hyperfine Inc., which also made several replicas from it. These replicas were then each gold coated and tested for EUV efficiency at the Synchrotron

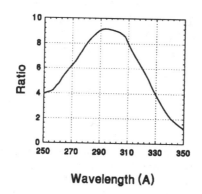

Figure 1: Enhancement in first order efficiency for the SERTS multilayer coated grating.

Ultraviolet Radiation Facility (SURF-II) of the National Institute of Standards and Technology (NIST) in Gaithersburg, Maryland (Keski-Kuha *et al.*, 1990; Thomas *et al.*, 1991). The grating was then overcoated at Goddard Space Flight Center (GSFC) with a 10-layer Ir/Si multilayer. After coating, the grating efficiency was once again measured at the NIST facility. Analysis shows that the multilayer coating produced a peak absolute efficiency of 3.3 % in first order at 31.2 *nm*. The greatest improvement occurred at 29.7 *nm*, where the grating's first order efficiency increased by a factor of 9 above that of the standard gold coating (Figure 1). The multilayer coating provided some enhancement over the entire 10 *nm* band from 25 to 35 *nm*, and at least a factor of 5 enhancement over the 6.3 *nm* band from 26.2 to 35.5 *nm*. Spectral resolution measurements were carried out in the SERTS Spectrograph Test Chamber at GSFC both before and after multilayer coating. The results showed that the spectral resolution of the grating was not changed by the multilayer coating process (Thomas *et al.*, 1991).

3 FLIGHT RESULTS

SERTS-4 was launched from White Sands Missile Range, White Sands, New Mexico, on May 7, 1991, at 12:05 MDT (1805 UT) aboard a Terrier-Black Brant booster. The flight was a complete success. The rocket pointing system quickly acquired the Sun and kept the instrument stable throughout its observing sequence. The instrument reached a maximum altitude of 203 miles (327 *km*), and obtained over six minutes worth of spectrographic data from 1806-1813 UT.

Spectra of the solar corona were obtained at two pointing positions, an active region and quiet corona. A preliminary list of the lines observed at these two positions is shown in Table 1 and Table 2. These spectra demonstrate that the multilayer coated grating maintained its spectral resolution of > 5000 even under the severe conditions encountered during the rocket flight. A detailed scientific analysis of the data is currently underway. There were an extensive set of ground-based observations taken in conjunction with SERTS-4. One of the primary objectives of this flight was to obtain coordinated VLA (Very Large Array, Soccoro, NM) observations. These will be used, along with our EUV measurements, to derive models of the coronal magnetic field in the observed regions.

4 CONCLUSIONS

The application of a multilayer coating to a large, high-density, aspheric diffraction grating replica produced a significant enhancement in first-order efficiency in the 30 nm wavelength region with no loss in its excellent spectral resolution. The flight data demonstrate that a significant improvement in throughput was achieved at wavelengths enhanced by the multilayer coating while the spectral resolution of the grating was maintained. The data from this flight demonstrate the utility of multilayer coating technology for solar observation, however similar gains can be expected in wide range of astronomical applications.

References

Golub, L., Herant, M., Kalata, K., Lovas, I., Nystrom, G., Pardo, F., Spiller, E., and Wilczynski (1990). *Nature*, **344**, 842.

Haisch, B. M., Whittmore, T. E., Joki, W. J., Brookover, W. J., and Rottman, G. J. (1988). *Proc. SPIE*, **982**, 38.

Keski-Kuha, R. A. M., Thomas, R. J., Gum, J. S., and Condor, C. E. (1990). *Appl. Opt.*, **29**, 4529.

Neupert, W. M., Epstein, G. L., Thomas, R. J., , and Feldman, U. (1981). *Sp. Sci. Rev.*, **29**, 425.

Neupert, W. M., Epstein, G. L., Thomas, R. J., , and Thompson, W. T. (1992). *Solar Phys.*, **137**, 87.

Thomas, R. J., Keski-Kuha, R. A. M., Neupert, W. M., Condor, C. E., and Gum, J. S. (1991). *Appl. Opt.*, **30**, 2245.

Walker, A. B. C., Barbee, T. W., Hoover, R. B., and Linblom, J. F. (1989). *Science*, **241**. 1781.

Table 1: Preliminary List of Lines Observed by SERTS in Active Region

No.	Ion	Wavelength	No.	Ion	Wavelength
1	Fe XIV	274.20	26	Mg VIII	339.00
2	Fe XV	284.16	27	–	341.15
3	Ni XVIII	291.98	28	Si IX	345.12
4	Si XI	303.33	29	Fe X	345.73
5	He II	303.79	30	Fe XII	346.85
6	Mn XIV	304.90	31	Si X	347.41
7	Fe XIII	312.16	32	Fe XIII	348.18
8	Co XV	312.54	33	Mg VI	349.13
–	Fe XV	312.57	34	Si IX	349.87
9	Mg VIII	313.73	35	Fe X	350.44 (2X)
10	Mg VIII	315.03	36	Fe XII	352.11
11	Si VIII	316.22	37	Fe XI	352.67
12	Mg VIII	317.01	38	Fe XIV	353.83
13	Fe XIII	318.12	39	Si X	356.03
14	Mg VII	319.02	40	Fe XI	358.62
15	Si VIII	319.85	41	Ne V	359.36
16	Ni XVIII	320.55	42	Fe XIII	359.64
17	Fe XIII	320.81	43	Fe XIII	359.84
18	Fe XIII	321.47	44	Fe XVI	360.76
19	Fe XV	321.78	45	– —	360.97 (2X)
20	Fe XV	327.03	46	Mg IX	368.07
21	Cr XIII	328.26	47	Cr XIV	389.85
22	Al X	332.78	48	Ne VI	401.95
23	Fe XIV	334.18	49	Fe XV	417.26
24	Fe XVI	335.40	50	S XIV	417.67
25	Fe XII	338.27			

Table 2: Preliminary List of Lines Observed by SERTS in Quiet Region

No.	Ion	Wavelength	No.	Ion	Wavelength
1	Si IX	296.14	21	Mg VIII	339.00
2	Si XI	303.33	22	Si IX	341.96
3	He II	303.79	23	–	344.94
4	–	305.14	24	Si IX	345.12
5	Mg VIII	313.73	25	Fe X	345.73
6	Si VIII	314.34	26	Fe XII	346.85
7	Mg VIII	315.03	27	Si X	347.41
8	Si VIII	316.22	28	Fe XIII	348.18
9	Mg VIII	317.01	29	Mg VI	349.13
10	Fe XIII	318.12	30	Si IX	349.87
11	Si VIII	319.85	31	Fe XII	352.11
12	Fe XIII	320.81	32	Fe XI	352.67
13	Fe XIII	321.47	33	Fe XIV	353.83
14	Cr XIII	328.26	34	Si X	356.03
15	Al X	332.50	35	Fe XIII	359.64
16	Al X	332.78	36	Fe XIII	359.84
17	Fe XIV	334.18	37	Fe XVI	360.76
18	Fe XVI	335.40	38	Fe XII	364.47
19	Fe XII	338.27	39	Mg IX	368.07
20	–	338.98			

A NEW METHOD FOR ANALYSING SPECTRA OBSERVED FROM LOW DENSITY AND OPTICALLY THIN PLASMAS

P. FAUCHER[1], J. DUBAU[2], M. CORNILLE[2], F.BELY-DUBAU[1]

[1]URA 1362 du CNRS, Observatoire de la Côte d'Azur, BP 229, 06304 Nice Cedex, France.
[2]UPR 176 du CNRS, DARC, Observatoire de Paris, 92195 Meudon Cedex, France.

ABSTRACT

Analysis of observed spectra can be carried out by fitting a synthetic spectrum with the observed one. The synthetic spectrum requires the calculation of many excitation rates, obtained by averaging collision strengths over a maxwellian velocity distribution. A new analysis is proposed where diagnostics can be deduced from a synthetic spectrum built directly from the collision strengths. This method is restricted to low density and optically thin plasmas.

1. INTRODUCTION

The Flat Crystal Spectrometer of the Solar Maximum Mission has recorded many soft X-ray spectra in the 1.4 - 23 Å range corresponding to active regions as well as solar flares. Some of them, obtained in the 13 - 19 Å range are of a particular interest because they correspond to emission of complex ions (Fe XVI, Fe XVII, Fe XVIII, Fe XIX, Ne IX). The theoretical analysis of these spectra requires the calculation of the atomic parameters which caracterize the physical processes responsible for the ion level population, in particular, collisional excitation rates corresponding to the numerous transitions. In general, an isothermal plasma is assumed and the atomic parameters are then introduced in the equilibrium statistical equations, which are solved in order to obtain level population. The line emissivities deduced, are dependent on electron temperature (Bely-Dubau et al. 1982). For non-isothermal plasmas, it is necessary to take account for the variation with electron temperature and electron density of the emissive volume and such a plasma is best described by a differential emission measure.

In order to overcome the difficulties due to plasma inhomogeneities, a new method is proposed in which a synthetic spectrum is built for a given electron energy. In this way it does not pre-suppose the shape of the electron energy distribution and can be applied to plasmas which may contain some non-thermal electrons. However, the method assumes that the ion population is almost entirely in the ground level and is restricted to low density plasmas.

A first application is applied to the Fe XVIII theoretical spectrum which is compared with a solar active region spectrum observed in the 13 - 19 Å by the FCS.

2. THEORY

For an isothermal plasma, the construction of a theoretical spectrum is derived from line emissivities. For an electron temperature T_e, the emissivity $\epsilon_{T_e}(\lambda_{ij})$ per ion in the line λ_{ij} corresponding to the transition $j \rightarrow i$ is :

$$\epsilon_{T_e}(\lambda_{ij}) = A^r_{ij}.N_j(T_e) \quad \text{(photon. s}^{-1}.\text{cm}^{-3}) \tag{1}$$

where $N_j(T_e)$ is the number density of the emitting level j and A_{ij} the spontaneous radiative transition probability. The determination of $N_j(T_e)$ requires the solution of the statistical equilibrium equations: $dN_j(T_e)/dt = 0$.

In our method, we assume a low density, stationary and optically thin plasma ($N_e \leq 10^{12}$cm^{-3}). In this case, the ion population is almost equal to the ionic ground level population : $N_1(T_e) = N_{\text{ION}}$, and the excited levels are populated either by electron excitation from this ground level or by cascades from upper levels and depopulated by radiative decay. For a temperature T_e this low density model gives :

$$C_{1j}(T_e) + \sum_{k>j} A^r_{jk}\, n_k(T_e) + \sum_{k<j} A^r_{kj}\, n_j(T_e) = 0 \quad j \neq 1 \tag{2}$$

where

$$n_j(T_e) = \frac{N_j(T_e)}{N_1(T_e)\, N_e} \tag{3}$$

$C_{ij}(T_e)$ is the excitation rate coefficient which describes the electronic collisional excitation process as a function of temperature T_e ; it is related to the collision strength $\Omega_{ij}(E)$ via

$$C_{ij}(T_e) = \frac{8.63\ 10^{-6}}{g_i \sqrt{T_e}} \int_{E_{ji}/kT_e}^{+\infty} \Omega_{ij}(E)\, e^{-E/kT_e}\, d(E/kT_e) \quad \text{cm}^3\,\text{s}^{-1} \tag{4}$$

where the subscripts i and j refer to initial and final target levels. $E_{ji} = E_j - E_i$ and E are respectively the transition and the impact energies in Rydberg(Ry), T_e the electron temperature in Kelvin (K), g_i the statistical weight of the initial level i and k the Boltzman constant in Ry/K.

Our method consists of replacing Eq.(2) by another set of equations depending on the energy E.

$$\Omega_{1j}(E) + \sum_{k>j} A^r_{jk}\, \tilde{n}_k(E) + \sum_{k<j} A^r_{kj}\, \tilde{n}_j(E) = 0, \quad j \neq 1 \tag{5}$$

with $\tilde{n}_j(E) = 0$ if $E < E_j - E_1$ Equation (2) is deduced from Eq.(5) after integration over energy E as in Eq.(4).

Now, we define, for each line :

$$\tilde{\epsilon}_E(\lambda_{ij}) = A^r_{ij}\, \tilde{n}_j(E), \tag{6}$$

The emissivity $\epsilon_{T_e}(\lambda_{ij})$ is related to this new parameter by a relation similar to Eq.(4) and $\tilde{\epsilon}_E$ may be considered as an energy differential emissivity. An interest of $\tilde{\epsilon}_E(\lambda_{ij})$ is that it does not depend on the shape of the electron energy distribution which, for instance, may contain some nonthermal electrons. The theoretical spectrum is then built from the $\bar{\epsilon}_E(\lambda_{ij})$ defined by (7) and corresponding to a given energy E. For each line two different broadenings are introduced : a gaussian one corresponding to a Doppler broadening and a lorentzian one representing the instrumental effect on the line profiles.

3. RESULTS AND DISCUSSION

The present method is applied to a Fe XVIII theoretical spectrum. The emissivities $\bar{\epsilon}_E(\lambda_{ij})$ are obtained by solving the set of equations (5). The atomic parameters are calculated using the University College London computational package : SUPERSTRUCTURE (Eissner et al. 1974), DISTWAV (Eissner & Seaton 1972), JAJOM (Saraph 1972)and JJOMCBE (Dubau 1988). Details of calculations are given in Cornille et al.(1992).

Table 1. Fe XVIII line energy differential emissivity $\bar{\epsilon}_E(\lambda_{ij})$ as a function of the electron impact energy E in Rydberg. Numbers in parenthesis represent powers of 10 by which entries must be multiplied.

Arrays	$\lambda(\text{Å})$	73	105	183	365
$2p^4 3d\ {}^2D_{5/2} - 2p^5\ {}^2P_{3/2}$	14.1641	1.46(-1)	1.80(-1)	2.83(-1)	3.85(-1)
$2p^4 3d\ {}^2P_{3/2} - 2p^5\ {}^2P_{3/2}$	14.1745	8.12(-2)	1.10(-1)	1.47(-1)	2.15(-1)
$2p^4 3d\ {}^2S_{1/2} - 2p^5\ {}^2P_{3/2}$	14.2302	3.22(-2)	4.05(-2)	5.09(-2)	7.68(-2)
$2p^4 3d\ {}^2D_{5/2} - 2p^5\ {}^2P_{3/2}$	14.3504	4.91(-2)	5.92(-2)	8.95(-2)	1.17(-1)
$2p^4 3d\ {}^2F_{5/2} - 2p^5\ {}^2P_{3/2}$	14.5116	3.78(-2)	4.44(-2)	6.37(-2)	8.51(-2)
$2p^4 3s\ {}^2D_{5/2} - 2p^5\ {}^2P_{3/2}$	15.6044	4.21(-2)	4.02(-2)	4.29(-2)	5.04(-2)
$2p^4 3s\ {}^4P_{3/2} - 2p^5\ {}^2P_{3/2}$	15.8157	2.61(-2)	2.46(-2)	2.44(-2)	2.80(-2)
$2p^4 3s\ {}^2P_{3/2} - 2p^5\ {}^2P_{3/2}$	15.9855	4.29(-2)	4.18(-2)	4.16(-2)	4.74(-2)
$2p^4 3s\ {}^4P_{5/2} - 2p^5\ {}^2P_{3/2}$	16.0577	8.89(-2)	6.09(-2)	3.89(-2)	3.27(-2)
$2s2p^5 3s\ {}^2P_{3/2} - 2s2p^6\ {}^2S_{1/2}$	16.1675	4.20(-2)	4.72(-2)	4.22(-2)	4.37(-2)
$2p^4 3p\ {}^2P_{3/2} - 2s2p^6\ {}^2S_{1/2}$	17.6058	9.70(-2)	1.05(-1)	8.99(-2)	8.92(-2)

Table 1 gives the variation with E of $\bar{\epsilon}_E(\lambda_{ij})$ for some intense lines of Fe XVIII, and the figure 1 shows a comparison between a Fe XVIII theoretical spectrum at 105 Rydberg and a spectrum observed from a solar active region by the FCS on May 11th 1986. This comparison allows the identification in the observed spectrum of the lines which are due to Fe XVIII. These lines are generally well separated from the other ones which are due mainly to the other Fe ions (Fe XVI, Fe XVII, Fe XIX). Some relative intensities of Fe XVIII lines are strongly dependent on electron energy, for instance, the lines in the vicinity of 16 Å corresponding to 3s → 2p transition. These lines which are well separated can be used to give a first estimation of the temperature of the emitting plasma.

On the other hand, an important theoretical line at 17.6058 Å is not reproduced with an equivalent intensity in the observed spectrum. This line which may correspond to one observed in some other solar and laboratory spectra (Feldman et al. 1973, Cohen & Feldman 1970) has an intensity which is very sensitive to configuration interaction.

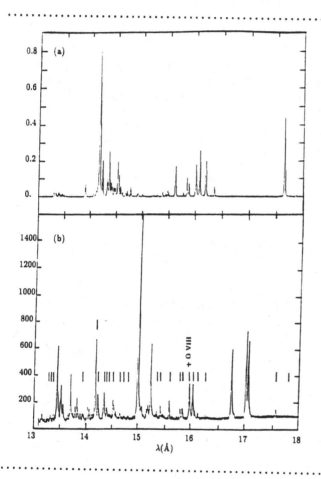

Figure 1. (a) Fe XVIII theoretical spectrum obtained at an electron energy E = 105 Ry. (b) Solar active region spectrum observed by the FCS of SMM on May 11th 1986. Vertical lines correspond to the Fe XVIII lines identified from theoretical spectrum of (a). The observed Fe XVIII line at 16 Å is blended with a O VIII line.

REFERENCES

Bely-Dubau F., Dubau J., Faucher P., et al. 1982, MNRAS, 201, 1155.
Cohen, L., Feldman, U. 1970, ApJ Letters, 160, L105.
Cornille M., Dubau J., Loulergue M., Bely-Dubau F., Faucher P., 1992, A&A, "in press".
Dubau, J. 1988,(unpublished).
Eissner, W., Seaton, M.J. 1972, J. Phys. B, 5, 218.
Eissner, W., Jones, M., Nussbaumer, H. 1974, Computer Phys. Commun., 8, 270.
Feldman, U., Doschek, G.A., Cowan, R.D., Cohen, L. 1973, J. Opt. Soc. Am., 63, 1445.
Saraph, H.E. 1972, Computer Phys. Commun., 3, 256.

Astronomy with the Deep UV Explorer Observatory

Donald H. Ferguson[1], Mark S. Giampapa[2]

[1]Donald H. Ferguson, W. J. Schafer Associates, Inc., 303 Lindbergh Avenue, Livermore, CA 94550

[2]Mark S. Giampapa, National Optical Astronomy Observatories, P.O. Box 26732, Tuscon, AZ 85726

ABSTRACT

Recent advances in control systems and sensors allow construction of an inexpensive yet high-performance orbiting observatory to collect data at ultraviolet wavelengths between 1150Å and 3000Å. The Deep Ultraviolet Explorer satellite (DUVE) will obtain all-sky imagery at various broad band wavelengths, high-resolution images, and spectra across the UV region. The DUVE program offers substantial performance advantages over current space-based observatories.

1 INTRODUCTION

The UV region is not observable from Earth and so is under-explored. Active galactic nuclei (AGNs), cataclysmic variable stars (CVs), hot white dwarfs, and of course young, hot, main sequence stellar associations all emit a substantial component of their radiation in the UV. An all-sky survey will efficiently survey the sky down to a very faint limiting magnitude, $m_{bol} = 22$, leading to a number of statistically based studies of great importance. For example, DUVE will sample the early history of galaxy formation and evolution through identification of quasars and AGNs. Mass loss in the late stages of stellar evolution will be better constrained with the large sample of degenerate stars that will be observed. Uniformity of the diffuse UV background will be measured. A large number of unique individual objects will be obtained in the high-resolution mode. Hot stellar associations will be mapped in nearby galaxy clusters, yielding clues to star formation. A comprehensive search for cataclysmic variables in nearby globular clusters will be performed. Planets will be synoptically monitored at UV wavelengths and cometary jets will be observed with heretofore unavailable detail. The high resolution and excellent sensitivity of the spectrograph aboard DUVE will expand on the highly successful work of IUE. High-excitation phenomena, especially those identified with metal resonance lines, will be studied in a wide variety of objects: AGNs, CVs, active chromosphere stars, and others. Additional UV features, such as low-z Lyman alpha absorption, forbidden and inter-combination lines, and mass loss phenomena, can be studied in detail. These are only samples of the many discoveries DUVE is expected to provide. Given the quantum jump in performance and coverage over previous instruments, shown in Table 1, experience suggests additional, unanticipated, radically new discoveries will be made by DUVE.[1]

2 DUVE CONCEPT

The DUVE program is designed to fit within the constraints of the NASA Medium Explorer program, as described in Fig. 1. Use of existing space-qualified hardware designs is made to pack a relatively large aperture and state-of-the-art instrument package within the cost, envelop, and weight restrictions. The DUVE payload is mounted atop a NASA standard spacecraft bus. DUVE features a particularly simple mechanism system – only two moving parts are required. The telescope and payload elements are modularized, further simplifying end item interfaces. Such simplicity allows the DUVE design team to reach design maturity typical of Critical Design Review before actually subcontracting out the end items, minimizing potential for expensive changes afterward. Such a procurement strategy, together with the use of pre-existing hardware wherever possible, lends itself nicely to fixed price subcontracting for major hardware elements, further reducing costs. Additionally, the DUVE management effort is controlled by a close-knit group of astronomers and very experienced spacecraft designers – performance trades will be made early and efficiently. In this manner, the DUVE team expects to easily meet the $30 million cost target. Indeed, DUVE is cost-competitive with larger ground-based observatories.

Table 1 COMPARISON OF DUVE WITH EXISTING PROGRAMS

Program	Sensitivity FLAMBDA -- ergs/s/cm2/Å (20 min Integration)		Resolution	Field of View	Duty Cycle	Cost
Spacecraft	Imagery (SNR = 2.5)	Spectra (SNR = 10)	Imagery (arcsec)	Imagery Only (arcmin)	(%)	($M)
IUE	—	3E - 14	—	—	70	300
HST	1E - 17	1.5E - 14	0.2	2.6 (WFC)	30	1,500
DUVE	1E - 17	2E - 15	0.2	170 (Wide)	85	60

3. DUVE DESIGN SYNOPSIS

HST: (imagery) -- PC at 0.15 μm
Best Sectral Resolution w/non-echelle mode
(spectra) -- FOS/"Blue"/G19OH

DUVE is different than previous orbiting observatories in three ways. First, the aperture size is very large for the cost and volume envelope constraints. Second, the control system is based on active servo compensation, as opposed to assuming a stiff, monolithic structure. Third, new sensor technology allows use of large-format Charge Coupled Devices (CCDs) that are quite sensitive in the UV.

3.1 Optics

DUVE's optical train is shown in Fig. 2. The large 0.8m Ritchey-Chretien telescope is the key to DUVE's excellent sensitivity. This telescope utilizes 80% light-weight primary and bipodal kinematical mounts for maximum stiffness relative to weight. Nevertheless, flexure is expected to be about one-half wave at 2000Å; tests on an identical design used on an unrelated program will confirm this in early 1992. This limits best resolution to about 0.2 arcseconds, as will be discussed, but collects maximum light. Higher system resolution is not needed for the wide field survey or spectrograph instruments and would seriously complicate the control system. The telescope body framework will consist of invar, and the secondary will be hard-mounted to the spider. Focus over about three waves will be provided by inserting red-leak filters of various thicknesses near the sensor. This provides a factor of five or more motion than is expected from preliminary thermal calculations. A fixed secondary results in no moving parts in the entire telescope subassembly; indeed, the only electrical interfaces will consist of temperature monitors and calibration lamp power supplies. MgF_2 overcoated aluminum will be used on the mirror surfaces with a scratch-dig specification suitable to reduce BRDF near the 2000Å scattering maximum. The telescope will be well insulated to reduce the amplitudes of thermal gradients during long exposures.

NASA Specifications	DUVE
Mass (150-300 kgs)	300 kgs
Width/Diameter (less than 200 cm, launch configuration)	120 cm
Height (under 300 cm, launch configuration)	240 cm
Power (up to 250 Watts)	50 W
Average Data Rate (less than 100 kbits/sec)	100 kbits/sec
Pointing accuracy (up to 1 arcsecond with input from science instrument, 30 arcseconds without)	$\leq 2.0"$
Lifetime needed to complete science investigation (1-3 years)	1 year, 3 years optional
Optimum orbit: - Apogee - Perigee - Inclination - Period	23,000 mi., circular Equatorial (0°) inclination 24 hr period
Time required for mission development (3-4 years)	3.5 years
Instrument payload cost (FY91 dollars)	$30M

Fig. 1– DUVE Payload Meets Medium Explorer Program Constraints

The light cone extant from the telescope enters the carrousel assembly after reflecting off the fast steering mirror. While not at a pupil, the telescope focal length is sufficient that resolution degradation due to beam walk will not impact maximum system resolution of 0.2 arcseconds. A fixed field flattener at the entrance to the carrousel assembly improves image quality over the entire 2.8 degree field of view. The carrousel consists of two twelve position wheels arranged in series. The wheels are mounted on dual concentric shafts to avoid single-point bearing failure impacts. The entrance wheel is used to select various filtering and instrument options. The sensor wheel is used for controling red-leaks and for focusing the telescope. Each "hole" in the wheel is designed to cover the wide field CCD but not the visible track CCDs located alongside, as illustrated in the focal plane layout, Fig. 2. Of course, non-red-leak portions of the focal correction filters do overhang the visible CCDs. Figure 2 also describes the filter options available with DUVE. Two options are particularly noteworthy; first, a 10 x relay is included for high resolution work across the UV band. This allows high resolution work over the central 2/3 of the large-format CCD; the remainder is masked by mounting hardware and optics, as the visible tracking CCDs must remain unobstructed. Second, a collimator-prism combination will allow for low resolution (~20Å) classification-type spectra of every object within the 2.0 degree instantaneous field of view of the wide field imager.

3.2 Sensors

The sensor package is arranged around the wide field of view UV imager. A CCD chip of 4096 x 4096 format optimized for UV performance by thinning and coating will be used. Quantum efficiency exceeds 50% and readout noise is less than 10 electrons/pixel/frame total. Optics transmission to the CCD is expected to average 12% in-band, for a total efficientcy of 8%, including losses for particulate contaminations. This is considerably higher than current spaceborne UV systems and is largely the result of higher quantum efficiency. The chip and three 1024 x 512 chips optimized for visible tracking will be mounted on a monolithic, heat-conducting substrate cooled by redundant thermoelectric coolers. A slit alongside one edge of the large format CCD will serve as the entrance to the spectrograph, whose optics consist of the usual collimator, grating, and camera mirror. The spectrograph detector will be a 2048 x 512 array of similar photodetection properties. The sensor package will be monolithically constructed from invar and thoroughly environmentally tested. There are no moving parts in the sensor subsystem; spectrograph focus is not critical in this application.

Many past UV-sensitive sensors have used red-insensitive photocathodes to reduce red-leak. With silicon-based CCDs optimized for UV performance, unfortunately relative quantum efficiencies at UV and red wavelengths are similar. Additional filtering must be performed to reduce red-leak effects. This is accomplished in three steps. First, each bandpass filter located in the entrance carrousel wheel has about 10^{-2} rejection in the red relative to UV, as these are typically dichronics. Second, each filter in the sensor carrousel wheel contains a red cutoff filter of 10^{-4} relative transmission. Finally, a second red cut-off filter located on the CCD provides 10^{-4} additional protection, for a total of 10^{-10}

relative transmission of red compared with UV light. This is adequate to give at least an order of magnitude greater signal in the UV for all but the coolest stars at the shortest wavelengths, near 1200Å, which are then of comparable signal.

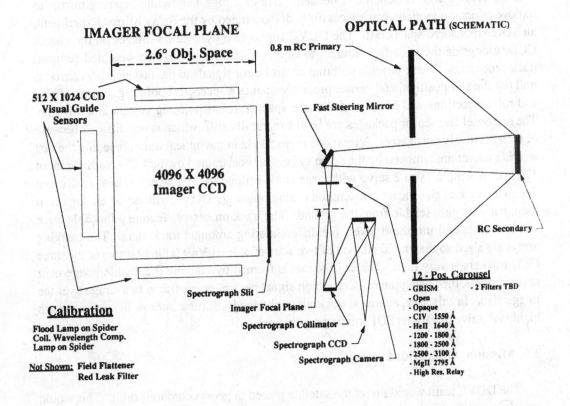

IMAGER FOCAL PLANE

2.6° Obj. Space

512 X 1024 CCD
Visual Guide
Sensors

4096 X 4096
Imager CCD

Calibration

Flood Lamp on Spider
Coll. Wavelength Comp.
Lamp on Spider

Not Shown: Field Flattener
Red Leak Filter

Spectrograph Slit
Imager Focal Plane
Spectrograph Collimator
Spectrograph CCD
Spectrograph Camera

OPTICAL PATH (SCHEMATIC)

0.8 m RC Primary

Fast Steering Mirror

RC Secondary

12 - Pos. Carousel

- GRISM   - 2 Filters TBD
- Open
- Opaque
- CIV 1550 Å
- HeII 1640 Å
- 1200 - 1800 Å
- 1800 - 2500 Å
- 2500 - 3100 Å
- MgII 2795 Å
- High Res. Relay

Fig. 2 – DUVE Schematic Optical Layout

3.3 Calibrations

Three lamp assemblies will be mounted on the secondary spider within the primarily mirror pupil. Each assembly will consist of two lamps wired in parallel so that failure of a single lamp will not jeopardize the calibrations. Two of the lamp assemblies will provide uniform diffuse illumination and a collimated wavelength comparison lamp for standard interpixel and spectroscopic comparison purposes. The fast steering mirror can be used to locate the wavelength comparison lamp either on the slit or in the wide fields imager; the latter is useful for calibration of the low resolution classification spectra obtained on the large format CCD. The third lamp assembly is a collimated visible source designed to fall on one of the visible track CCDs and adjust the servo loop gain and track processor parameters.

3.4 Control Systems

The NASA spacecraft bus will point to within 2 arcseconds. This precision is adequate to preserve the wide field imagery resolution of 2.5 arcseconds, but will degrade spectroscopic throughput somewhat and completely blur the high resolution mode imagery. A more precise control scheme is needed. Use of a high bandwidth steering mirror to remove structural motions was successfully demonstrated by the Relay Mirror Experiment, an SDIO-sponsored spacecraft. The DUVE application uses stellar images on the visible CCDs alongside the wide field imager to supply track imagery to two weighted centroid track processors. These provide pointing control error signals to the fast steering mirror to null out the star position, and hence, precisely maintain spacecraft pointing. Large offsets and roll corrections will be offloaded to the spacecraft bus pointing system at a low rate. The carrousel and sensor packages are built structurally stiff, which is not difficult because of their relatively small sizes. Vibrations up to 50 Hz in the optical train preceding the fast steering mirror are removed by the servo system, as constrained by the CCD readout rate of 120 Hz. A simple Type 2 servo with large angle offload to the spacecraft bus is adequate due to the lack of significant acceleration during pointing. DUVE will use an analog servo loop, but with gain setable from the ground. The track processors feature adjustable track window sizes and threshold levels for differentiating amongst track stars. The tracking arrays are sized so that the likelihood of two stars of at least V=9 falling on two of the three CCD trackers is very high. The error signal is formed by blurring the visible image over several pixels, then using the relatively high signal per image per frame to average over the image size. In this way, subpixel tracking is obtained. Similar approaches have proven highly effective in various SDI pointing experiments.

3.5 Mission Operations

The DUVE team would prefer the satellite placed in geosynchronous orbit. This would greatly simplify mission operations. For example, target identification would proceed in real-time, like with IUE, by downlinking the visible tracker data in the RS-170 form, with reduced resolution. Then too, only one ground site would be required. This simply must, however, be balanced by the additional cost of the larger boost system needed to reach such an orbit. Nevertheless, with on-board storage and commands provided by an enchanced data processing system, operation in low earth orbit is achievable. Such a selection is also suitable for certain earth-oriented, non-astronomical missions whose sponsors may help to defray costs. In any event, coarse pointing will be provided by the NASA-supplied spacecraft bus, greatly simplifying command and telemetry requirements for the payload. The DUVE team anticipates that data will be available at the ground site within a few tens of minutes after an observation. All raw data will be cataloged and stored, and each image and spectrum will be reduced with the most current calibrations, documented, and stored in the pipeline fashion after, of course, a quick overview by the observer. This is necessary due to the large amount of data expected from this satellite. Reduced data can then be called up separately as desired for further analysis.

4 SUMMARY AND CONCLUSIONS

The ultraviolet region between 1150Å and 3200Å has not been systematically surveyed and explored. IUE spectra and recent glimpses from the ASTRO mission indicate a great many valuable scientific discoveries await. DUVE is a space-based observatory providing all-sky imagery and high-resolution spectroscopy at wavelengths between 1150Å and 3200Å. Sky coverage, resolution, and sensitivity are each equal to or better than available on existing spacecraft. Recent advances in controls, sensors, and telescope lightweighting coupled with space-qualified designs combine to make the DUVE project affordable within the guidelines of the Medium Explorer-class mission scenario. The DUVE team is experienced and has adopted a management approach that ensures success within resource constraints.

5 REFERENCES

(1) Gillespie, B., 1991, *Space Telescope Science Institute Memorandum*, Hubble Space Telescope Cycle 2 Proposal Submissions-Updated Technical Information.

(2) Harris, A. W., Sonnenborn, G., ed. Y. Kondo, 1987, *Scientific Accomplishments of the IUE*, (Riedel: Dordrecht).

(3) Bahcall, J. N., chair, 1991, Astronomy and Astrophysics Survey Committee, National Research Council, *The Decade of Discovery in Astronomy and Astrophysics*, (National Academy Press: Washington, D.C.).

Two-Dimensional Helical Delay-Line Readout of Large-Format Microchannel-Plate Detectors for Astronomy

Peter G. Friedman[1], Judith R. Fleischman[2], and Christopher Martin[1]

[1] Department of Physics, Columbia University
[2] Department of Astronomy, Columbia University

ABSTRACT

We report progress on a new readout for microchannel-plate detectors for astronomical imaging and spectroscopy. The two-dimensional helical delay-line readout should provide a 5000×5000-pixel, 10-cm×10-cm detector. Spatial resolution (FWHM) achieved to date is ~30–40 μm. The spatial linearity residuals are 3–9 μm over 2–4 mm. We also discuss expansion of the charge cloud, ion feedback, and local aging. We are proposing this detector for a small-explorer-based far-UV all-sky survey and plan to use it on a rocket-borne far-UV survey of the interstellar medium.

1 INTRODUCTION

For over a decade microchannel-plate (MCP) detectors have been the standard for astrophysical observations in the ultraviolet band. However, the potential return from many future experiments will be fully realized only with large-format detectors that obtain the highest possible spatial resolution, linearity, and dynamic range.

We think that the two-dimensional helical delay-line MCP readout (Sobottka and Williams 1988, Friedman, Martin, and Rasmussen 1990, Williams, *et al.* 1991) is the best candidate for development of a large-format detector. It is relatively inexpensive, and should achieve resolution sufficient to make full use of MCP-pore size (~10 μm). In contrast to charge-division readouts, increasing the size of this detector, within limits, does not degrade, and may actually improve the absolute spatial resolution. The format and throughput rate can both be large. We expect that the detector will be robust in launch and space environments, and will make only modest demands on spacecraft power, volume, and weight.

1.1 Detector Goals and Applications

We are currently testing a 15-cm anode with 40-mm MCP's. Our ultimate goal is to construct a 10-cm diameter detector with 5000×5000 pixels of 20-μm FWHM resolution and ~$10^6 s^{-1}$ throughput rate. The prototype detector now has sufficient resolution to provide ~2000×2000 pixels across a 65-mm diameter. This meets the requirement for the small-explorer-based far-UV all-sky survey we are proposing.

The all-sky survey mission would perform three complementary surveys. First, a one-year all-sky survey to a limiting magnitude of ~18 in three wavelength bands: 140, 160, and 180 nm. Second, deep surveys of 10–20 selected fields to 21–22 mag in the three bands. Third, an objective spectroscopic survey to 18th mag over $\sim$2% of the sky, with $\Delta\lambda$=0.3–2.0 nm. The spectroscopic survey will permit a detailed look at a smaller, but substantial and varied sample of objects, as well as providing important basic data for the interpretation of the photometric surveys. It will furnish detailed information about the temperature, gravity, continuum slope, redshift, ionization, metallicity, stellar populations, and extinction for hundreds to thousands of sources in each class.

Within the next two years we plan to prove the space worthiness of the delay-line readout by flying three detectors on a rocket-borne far-UV survey of the interstellar medium (ISM). The large format and high throughput of delay-line detectors make it possible in a single 5-min rocket flight to map one quarter of the sky in three bands. The bands, defined by broad- or narrow-band filters, will be 140–190-nm continuum emission, CIV emission, and H_2 fluorescence. These will be the first maps made in the latter two bands. The maps will yield information on the morphology and degree of isotropy of the diffuse far-UV background and ISM, and allow us to study the interrelationships of its molecular, cold atomic, and hot phases.

2 THEORY OF OPERATION

The two-dimensional helical delay-line readout technique employs a crossed pair of flattened-helical bifilar transmission lines. Each delay line independently gives the position in one spatial dimension. Each delay line consists of a pair of bare Cu wires, one of which is biased to collect electrons. The electron cloud from a MCP event falls over several turns of each transmission line. From that region pulses propagate to timing electronics at both ends of each transmission line. Differential-mode signal amplification minimizes noise pickup and x–y crosstalk. The pulse arrival times at the ends of each delay line give the event position.

The delay-line readout achieves its results by requiring pulses to travel $\sim$300 times as far as they would in a straight path across the anode. This reduces the effective propagation velocity of the pulse, thereby reducing by the same factor the spatial uncertainty derived from a timing measurement.

3 RESULTS

3.1 Spatial Resolution and Linearity

The pinhole-array image in Fig. 1a gives an indication of the detector's performance. We have also measured the spatial resolution and linearity by precisely positioning a single nominal 10-μm UV spot on the front MCP. The gain of the Z-stack MCP is $\sim$1$\times10^8$. Spatial linearity residuals are 3–9 μm rms over 2–4 mm. The FWHM resolution is 30 μm in one dimension (x, upper wire plane) and 40 μm in the other (y, lower plane), including the contribution of the UV spot. This is an improvement over the 60-μm FWHM resolution quoted in Friedman, Martin, and Rasmussen 1990. We believe that the earlier, poorer, resolution resulted from extracting large amounts of charge from the MCP at the positions of the UV spots during setup (sec. 3.4).

3.2 Charge-Cloud Variance

We have found that MCP pulse-to-pulse charge-cloud shape variances (as well as ion feedback—sec. 3.3), are largely responsible for the difference between actual and theoretical spatial resolution. The effect of such variances on the spatial resolution should increase with the time available for the charge cloud to spread out as it travels from the MCP to the anode. To test the hypothesis that charge-cloud-shape variances are a dominant contribution to degradation of spatial resolution in this, and possibly in charge-division devices, we reduced the MCP-to-anode travel time by 42%. (This also reduced by ~20% the rise time of the wave form from which the timing information is extracted. However, this could not explain improved resolution because the system is not noise limited.) The resolution improved by 25−30%, a quadrature improvement of 30−40 μm. This supports the above hypothesis.

3.3 Ion Feedback

The funnel shape in Fig. 1b shows that the resolution degrades with increasing pulse height. This is in contrast to the noise-limited case in which resolution improves with signal amplitude. The MCP's used thus far were not electron scrubbed (or burned in), a preconditioning process which removes adsorbed gas and stabilizes aging. We hypothesize that the cause of the funnel shape is ion feedback. This causes after-pulses, seen in anode wave forms, that overlap with the initial pulse before it reaches the end of the delay line. If the hypothesis is correct, scrubbing the plates should improve the resolution by up to a factor of two.

3.4 Local Fatigue of the MCP

We are studying local aging effects that are intrinsic to MCP's. We have discovered that in unscrubbed plates local aging can significantly degrade local spatial resolution and linearity. Fig. 2 shows a 40% decrease in local MCP gain, accompanied by a factor of 2.7 degradation of the local spatial resolution, during the extraction of 0.5 C/cm^2 from a 2000-μm^2 area of the rear MCP. Fig. 1c shows the resulting spatial nonlinearity in the region around two fatigued spots, probed by a pinhole array. Large nonlinearity is apparent. We expect these effects to be independent of readout method and to be substantially reduced in scrubbed MCP's. However, the effects require further study to permit acquisition of fields containing bright stars, with pore-limited resolution, even using a "long-life" scrubbed detector.

5 DISCUSSION

The detector resolution is still dominated by ion feedback and MCP charge-cloud variance, not by electronics. These contribute ~20 μm to the spatial resolution, but can be reduced by electron scrubbing and further closing the MCP-to-anode gap. A smaller gap is permissible because there is no evidence of undersampling nonlinearity, which would be caused if the charge cloud covered too few wires. Our experience with ion feedback and local MCP aging underscores the importance of electron scrubbing even for laboratory prototype detectors. Finally, the resolution is already sufficient for the small-explorer far-UV survey mission.

6 REFERENCES

Friedman, P.G., Martin, C., and Rasmussen, A. 1990, *SPIE Proc.*, **1344**, 183.
Sobottka, S. E. and Williams, M. B. 1988, *IEEE Trans. Nucl. Sci.*, **35**, 348.
Williams, M.B. *et al.* 1991, *Nucl. Instrum. Methods*, **A302**, 105.
This work supported by NASA grants NAGW-1887, NAGW-2593 and NAG-5-642.

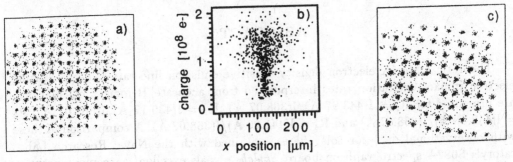

Fig. 1—a) Image of pinhole array with 10-μm pinholes on a 250 μm grid. b) Scatter plot of charge in event vs. measured position. c) Local linearity around fatigued area of MCP probed by pinhole array.

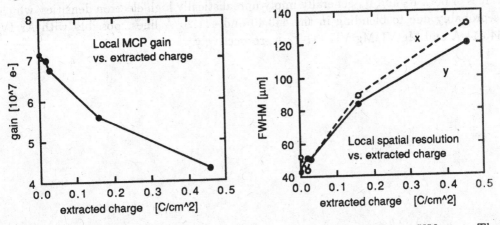

Fig. 2—Evolution of a small MCP region stimulated by a 10-μm UV spot. The plots show degradation in MCP gain and spatial resolution.

EUV Lines Of *Mg IX* As n_e-Diagnostics For High Density Flares

L. K. Harra[1], F. P. Keenan[1], K. G. Widing[2] and E. S. Conlon[1]

[1]Department of Pure and Applied Physics, The Queen's University of Belfast, Belfast BT7 1NN, Northern Ireland, UK

[2]Code 4174W, E. O. Hulburt Center for Space Research, U. S. Naval Research Laboratory, Washington DC 20375–5000

ABSTRACT

Theoretical Mg IX electron density sensitive emission line ratios, derived using electron impact excitation rates interpolated from accurate **R**-matrix calculations, are presented for R_1 = I(443.97 Å)/I(368.07 Å), R_2 = I(439.17 Å)/I(368.07 Å), R_3 = I(443.40 Å)/I(368.07 Å) and R_4 = I(441.20 Å)/I(368.07 Å). A comparison of these with observational data for solar flares, obtained with the Naval Research Laboratory's S082A spectrograph on board *Skylab*, reveals excellent agreement between theory and observation for R_1 and R_2, which confirms the usefulness of these ratios as N_e-diagnostics for solar flares, as well as providing experimental support for the accuracy of the atomic data adopted in the line ratio calculations. However the observed values of both R_3 and R_4 generally imply unrealistically high electron densities, which is probably due to blending in the 443.40 and 441.20 Å lines, possibly with Ar IV 443.44 Å and Mg VI/Mg VII 441.22 Å, respectively.

1. INTRODUCTION

Emission lines arising from transitions among the $2s^2$, $2s2p$ and $2p^2$ levels of ions in the beryllium isoelectronic sequence are often detected in solar ultraviolet spectra. They may be used to infer the electron density (N_e) and temperature (T_e) of the emitting plasma through diagnostic line ratios, although to determine these reliably accurate atomic data must be employed in the calculations, especially for electron and proton impact excitation rates and oscillator strengths.

In this paper we employ the most recent atomic physics data for Mg IX to derive theoretical line ratios for this ion, and show by comparison with XUV solar observational data from *Skylab* that they are useful N_e-diagnostics for high electron density solar flares.

2. THEORETICAL RATIOS

The model ion for Mg IX consisted of the six LS states with the two outer electrons in the $n = 2$ shell, i.e. $2s^2$ $^1S^e$, $2s2p$ $^3P^o$, $^1P^o$; $2p^2$ $^3P^e$, $^1D^e$ and $^1S^e$, making a total of ten levels when the fine structure splitting in the ^3P terms is included. The theoretical Mg IX emission line ratios

$$R_1 = I(2s2p\ ^3P^o_2 - 2p^2\ ^3P^e_2)/I(2s^2\ ^1S - 2s2p\ ^1P^o),$$
$$R_2 = I(2s2p\ ^3P^o_1 - 2p^2\ ^3P^e_2)/I(1s^2\ ^1S - 2s2p\ ^1P^o),$$
$$R_3 = I(2s2p\ ^3P^o_1 - 2p^2\ ^3P^e_1)/I(2s^2\ ^1S - 2s2p\ ^1P^o),$$

and

$$R_4 = I(2s2p\ ^3P^o_0 - 2s^2\ ^3P^e_1)/I(2s^2\ ^1S - 2s2p\ ^1P^o),$$

were calculated as a function of electron density at the temperature of maximum Mg IX fractional abundance in ionisation equilibrium, log $T_{max} = 6.0$, plus ± 0.2 dex about this value, where the fractional abundance has fallen to $\simeq 10^{-2}$. The ratios are found to be relatively insensitive to variations in the electron temperature with, for example, R_1–R_4 varying by only $\sim 30\%$ between log $T_e = 5.8$ and 6.2. However they are very sensitive to changes in the electron density when $N_e \geq 10^{12}$ cm^{-3}, and hence should be useful as N_e-diagnostics for high density solar flares.

3. RESULTS AND DISCUSSION

Mg IX emission lines were observed in the solar spectrum by the Naval Research Laboratory's XUV slitless spectrograph (S082A) on board *Skylab* (Dere 1978). This instrument covered the wavelength region 171–630 Å, with a maximum spectral resolution of ~ 0.1 Å and a spatial resolution of 2″. The following Mg IX transitions were identified: $2s^2\ ^1S - 2s2p\ ^1P^o$ (368.07 Å), $2s2p\ ^3P_1^o - 2p^2\ ^3P_2^e$ (439.17 Å), $2s2p\ ^3P_0^o - 2p^2\ ^3P_1^e$ (441.20 Å), $2s2p\ ^3P_1^o - 2p^2\ ^3P_1^e$ (443.40 Å) and $2s2p\ ^3P_2^o - 2p^2\ ^3P_2^e$ (443.97 Å).

We have measured Mg IX line ratios from S082A spectra traced from photographic plates. Table 1 lists the observed line ratios for selected solar flares. An inspection of Table 1 shows that the electron densities deduced for the 1973 December 2 flare from R_1 and R_2 are compatible, and furthermore are in good agreement with that derived from Ne VII, which confirms the usefulness of R_1 and R_2 as N_e-diagnostics. In addition, we note that one would expect (from the Fe XII and Ne VII results), R_1 and R_2 in the 1973 June 15 and 1973 August 9 flares to be in the low density limit, and indeed the observed values of R_1 and R_2 in these events are within a few percent of those calculated for log $T_e = 6.0$ ($R_1 = 0.026$; $R_2 = 0.009$). This provides experimental support for the accuracy of the atomic data adopted in the present calculations, as R_1 and R_2 in the low density limit are in the coronal approximation and hence depend principally on the ratio of the relevant electron impact excitation rates.

In the case of R_3 and R_4, the observed line ratios in Table 1 imply unrealistically high electron densities, apart from R_4 in the 1973 June 15 event. This is found to be due to both the 443.40 and 441.20 Å lines appearing to be blended, probably with Ar IV 443.44 Å and Mg VI/Mg VII 441.22 Å, respectively. Higher spectral resolution observations of Mg IX (perhaps with the *Coronal Diagnostic Spectrometer* on board SOHO) would therefore be of great interest to investigate if the discrepancies between theory and observation could be resolved.

References

1. P. Lee, A.J. Lieber, A.K. Pradhan, and Y. Xu, Phys. Rev. A **34**, 3210 (1986).

2. D.L. McKenzie, P.B. Landecker, U. Feldman, and G.A. Doschek, Astrophys. J. **289**, 849 (1985).

3. A.H. Gabriel and C. Jordan, Case Stud. At. Phys. **2**, 209 (1972).

4. F.P. Keenan, S.S. Tayal, and A.E. Kingston, Mon. Not. R. Astron. Soc. **207**, 51P (1984).

5. F.P. Keenan, S.S. Tayal, and A.E. Kingston, Solar Phys. **92**, 75 (1984).

Table 1– The relevant plate numbers are listed and the observed values of the line ratios for the following solar flares: 1973 December 2 at 1517 UT , 1973 June 15 at 1428 UT and 1973 August 9 at 1555 UT. These data should be accurate to approximately ±30% . Also listed in the table are the values of log N_e derived for these flares using the log T_{max} = 6.0 calculations in Figures 1–4, along with the electron densities found from line ratios in Ne VII and Fe XII, which are formed at similar electron temperatures to Mg IX (log T_{max}(Ne VII) = 5.7; log T_{max}(Fe XII) = 6.2.

TABLE 1

OBSERVED Mg IX LINE RATIOS AND THE DERIVED LOGARITHMIC ELECTRON DENSITIES

Flare	Plate	λ_1(Å)	λ_2(Å)	I(λ_1)/I(λ_2)	log N_e	log N_e(other)
1973 December 2	3A–045	443.97	368.07	3.6–2 (R$_1$)^a	12.5	12.3 (Ne VII)
1973 December 2	3A–045	439.17	368.07	1.5–2 (R$_2$)	12.6	12.3 (Ne VII)
1973 December 2	3A–045	443.40	368.07	1.9–2 (R$_3$)	14.5	12.3 (Ne VII)
1973 December 2	3A–045	441.20	368.07	1.6–2 (R$_4$)	13.1	12.3 (Ne VII)
1973 June 15	1A–333,334	443.97	368.07	2.7–2 (R$_1$)≤11.5		10.3 (Fe XII)
1973 June 15	1A–333,334	439.17	368.07	1.0–2 (R$_2$)≤12.0		10.3 (Fe XII)
1973 June 15	1A–333,334	443.40	368.07	8.0–3 (R$_3$)	12.6	10.3 (Fe XII)
1973 June 15	1A–333,334	441.20	368.07	7.3–3 (R$_4$)≤10.5		10.3 (Fe XII)
1973 August 9	2A–029	443.97	368.07	2.7–2 (R$_1$)≤11.5		10.3 (Fe XII)

^{a}A–B implies A × 10^{-B}

Methods for Absolute Intensity Calibration of Survey Spectrometers in the 5nm - 150nm Spectral Region

N C Hawkes[*], K D Lawson[*] and N J Peacock[*]

JET Joint Undertaking , Abingdon, Oxfordshire, OX14 3EA, England.
[*]Culham Laboratory (UKAEA/EURATOM Fusion Association) Abingdon Oxon. OX14 3DB England.

ABSTRACT

A toroidal grating, flat field spectrometer employing a microchannel plate (MCP) detector and image intensifier is widely used for surveying the XUV emission from fusion plasmas (Fonck, Ramsey and Yelle 1982; Behringer et al. 1986). We report on diverse techniques to form a composite relative intensity calibration for such an instrument in the wavelength region 5nm - 150nm. The calibration is placed on an absolute basis using transfer radiation standards in the UV and VUV.

Calibration techniques using the emission from the fusion plasma itself are less prone to errors arising from source polarisation, F-number matching and time response. In this respect we have made use of Y-rast transitions arising from charge-exchange collisions between plasma ions in tokamaks and injected atomic beams . Model calculations of transitions in O V and O VI and in He II (3p -4d /1s -4p) and C VI (2p -3d /1s -3p) have also proved useful in constructing a composite relative inverse sensitivity curve $S^{-1}(\lambda)$ over most of the spectrometer bandwidth. In the case of the hydrogenic transitions, complications arise due to the precise l-state populations following charge exchange and electron impact collisions. Knowing the emissivity of the plasma in the visible, $S^{-1}(\lambda)$ can be placed on an absolute basis using the branching ratio $1s2s\,^3S_1$ -$1s2p\,^3P_1$ / $1s^2\,^1S_0$-$1s2p\,^3P_1$ in Be III and B IV. The $\Delta n = 1$, long wavelength transitions in these particular two He- like ions offer the enormous advantage of being spectrally isolated from their neighbouring triplet components. The absolute sensitivity is in satisfactory agreement with that derived from the separate theoretical efficiencies of the elemental components of the spectrometer.

1. Introduction

The present paper describes several calibration techniques performed periodically on two similar VUV survey spectrometers which have been used on various tokamaks such as JET, ASDEX and DITE. The spectrometers, designated "A" and "B", are compact, toroidal-grating, grazing-incidence instruments, of the type first described by Fonck *et al.* (1982) and are designed to produce a flat-field focal plane with a microchannel plate (MCP) and image intensified detector. Unless stated otherwise, the calibration curves refer to "A". The operating parameters of the detector are the MCP standing volts, V_{MCP}, typically 650V <=> 1000V and the phosphor voltage relative to the MCP, V_{ph}, typically 5000V.

A schematic layout of a survey spectrometer coupled to the JET torus is illustrated in Fig.1 The latest version of these instruments (Model 251 manufactured by McPherson) is equipped with three remotely interchangeable gratings with ruling frequencies of 290, 450 and 2100 grooves/mm. Each grating represents a compromise between spectral coverage and resolution; see table1,

TABLE 1 WAVELENGTH COVERAGE AND RESOLUTION (FWHM) FOR THE SURVEY SPECTROMETER WHEN USING EACH OF THE THREE AVAILABLE GRATINGS.		
Grating	Coverage	Resolution
290 g.mm^{-1}	150-1700 Å	4 Å
450 g.mm^{-1}	100-1100 Å	3 Å
2100 g.mm^{-1}	90-330 Å	0.5 Å

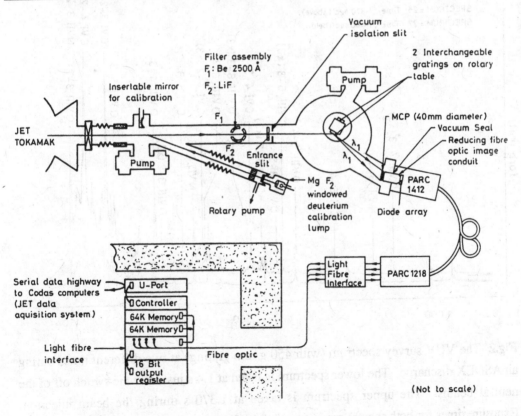

Fig.1 Schematic layout of survey spectrometer on JET.

2. Charge Exchange (CX) Spectra

State-selective CX collisions between injected atomic beams (H^0, D^0, He^0, etc with energies $E_0 \geq 1$ au) and background plasma ions, especially the nuclei of light elements eg O^{8+}, C^{6+}, Be^{4+}, He^{2+}, give rise to Y-RAST trasitions which are prominent features of the VUV spectrum. Following Isler and Langley (1985), the CX line intensities, $\varepsilon_{CX}(\lambda)$, when compared with their *effective* cross sections (including *l*-state mixing and cascades), may be used to determine the spectral sensitivity curve for the survey instrument,

$$\varepsilon_{CX}(\lambda) = N(H^0)\, N(z) < \sigma_\lambda(nl,n'l')_{CX}\, V_b >$$

where $\sigma_\lambda(nl,n'l')_{CX}$ is the *effective* cross section for CX excitation of the nl-n'l' transition in the recombined ion $N(z-1)$, $N(z)$ is the density of the recombining ions and $eN(H^0)V_b$ is the beam current.

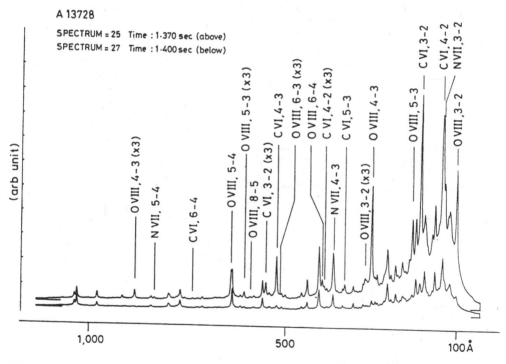

Fig.2 The VUV survey spectrum (with 450 g/mm grating) at two different times during an ASDEX discharge. The lower spectrum is taken at 1.4 s just after the switch off of the neutral beams. The upper spectrum is taken at 1.370 s during the beam injection. Exposure times for both spectra are 16 ms. All the features annotated are charge-exchange-recombination lines which appear only during the beam-injection period.

The inverse sensitivity $S^{-1}(\lambda)$ for the 290 g/mm grating, Fig.3, is placed on an absolute basis at wavelengths between 1100Å and 1700Å by reference to a transfer standard D_2 lamp calibrated at NPL, Carolan et al.(1987). The 450 g/mm grating, Fig.4, can then be determined by reference to the 290 g/mm CX data, assuming that the impurity concentrations and beam attenuation do not change throughout a sequence of discharges.

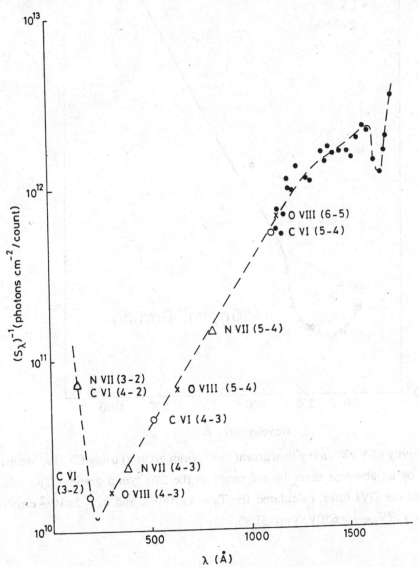

Fig.3 Sensitivity of VUV survey instrument (290 g/mm grating) based on the relative intensities of the CX lines at short wavelengths and normalized to the absolute sensitivity using a D_2 standard lamp at $\lambda > 1150$ Å ; o......C VI ,Δ.......N VII, x.......OVIII, •.........D_2 lamp.

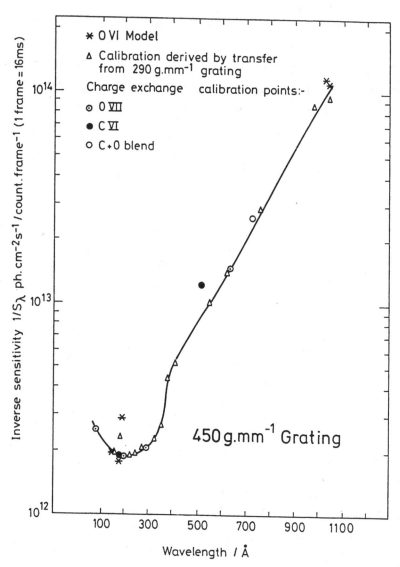

Fig.4 Sensitivity of VUV survey instrument (450 g/mm grating) using CX line intensities and placed on an absolute bases by reference to the 290 g/mm grating. The relative intensities of the OVI lines, calculated for $T_e = 1 \times 10^6$ °K and $n_e = 5 \times 10^{12}$ cm^{-3}, are indicated by $*$. ($V_{MCP} = 680V$, $V_{PH} = 5kV$).

3. Model Calculations Of Line Intensities eg OV, OVI

Line emission from oxygen ions in the VUV region have been studied experimentally and modelled theoretically (Huang *et al.* 1988; Kato, Masai and Sato 1985; Kato, Lang and Berrington 1990). In ungettered tokamaks, OV and OVI lines are prominent features throughout the VUV and XUV. Typical values of the electron density (n_e) and temperature (T_e) at the location of, for example, the OV emission shell near the plasma edge in a tokamak are $n_e \approx 10^{12}$ cm^{-3} and $T_e \approx 50 \Rightarrow 100$ eV. An analysis of spectra from JET and ASDEX using the derived $S^{-1}(\lambda)$ shape for the 450 g/mm grating can be compared to the theoretical predictions (Kato, Masai and Sato 1985; Kato, Lang and Berrington 1990) with the following results:

DENSITY-DEPENDENT LINES IN OV:

The intensity ratios of the of lines excited from the metastable 2s2p $^3P_{0,1,2}$ levels to those excited from the 2s^2 ^{1}S ground state are density dependent in the range $10^{10} < n_e < 10^{13}$cm^{-3}. The 2s2p ^{3}P - 2p^2 ^{3}P multiplet at 760Å / 2s^2 ^{1}S - 2s2p ^{1}P$_1$ at 630Å is just such a ratio.

TABLE 2 DERIVATION OF ELECTRON DENSITY FROM OV

SHOT #	n_e(vol. av.)	$\mathcal{E}$(760Å)/ $\mathcal{E}$(630Å)	n_e derived for location of OV emission layer near plasma edge†
JET 10817	3x10^{13}cm^{-3}	0.72	2x 10^{12}cm^{-3}
- 10826	1x10^{13}cm^{-3}	0.53	0.7x 10^{12}cm^{-3}
ASDX 13728		0.7	1.7x 10^{12}cm^{-3}

† $T_e \approx 100$eV assumed with inner-shell ionisation of OIV contributing to the metastable state populations.

TABLE 3 TEMPERATURE DEPENDENT LINE RATIOS IN OV

SHOT #	T_e(vol. av.)	$\mathcal{E}$(193Å)/$\mathcal{E}$(760Å)	T_e(eV) derived for location of OV emission line ratio
JET 10817	1.4 keV	0.49 x 10^{-1}	20
- 10826	1 keV	0.113	70
ASDX 13728		0.82 x 10^{-1}	47

The relative line intensities of the VUV lines from OVI have been modeled (Summers and Wood 1988) for a range of steady-state electron parameters . The data shows a best-fit to the calibration curve in Fig.4 for $T_e = 1 \times 10^6 \,^\circ K$ and $n_e = 5 \times 10^{12} \, cm^{-3}$.

4 Branching Ratios

While there are several possibilities for branching ratios in the VUV, the intercombination lines of He-like ions of the light elements , especially Be and B, are particularly attractive in having unblended multiplet components at widely separated wavelengths, see for eg., Figs. 5 & 6. For these elements the ratio $A(2s2p \,^3S_1 - \,^3P_1) /$ $A(1s^2 \,^1S_0 - 2s2p \,^3P_1)$ is $A(101.69\text{Å}) / A(3720\text{Å}) = 1/84$ in Be and $A(61.09\text{Å}) /$ $A(2826.66\text{Å}) = 1/11$ in B.

Ordinarily, the branching ratio would require only two instruments, one operating in the visible with the other in the VUV, both viewing the same plasma chord. At typical concentrations levels of $\approx 1\% n_e$ in tokamaks, the intercombination line in BeIII at 101.69Å is a weak feature relative to the allowed resonance line at 100.26Å and is, furthermore, blended in the spectrum of the VUV survey instrument. It was thus convenient to use the allowed line $\mathcal{E}_A$ (photons $cm^{-2} \, s^{-1}$) as a reference emmisivity over a wide range of tokamak paramaters. The allowed / intercombination line intensity ratio $\mathcal{E}_A/\mathcal{E}_i$ was therefore studied both theoretically (Summers *et al.* 1991) and experimentally, (Hawkes, Peacock and Lawson 1991) , using a higher resolution XUV spectrometer, similar to that described by Schwob *et al.* (1987), again viewing the plasma, Fig.7, along the common line-of-sight.

Having established the relative sensitivities of the VUV and XUV instruments at 100Å, it is convenient to measure $\mathcal{E}_A$ with the VUV instrument and $\mathcal{E}_i$ from the triplet component $\mathcal{E}_{j=1}$ at 3720Å using the branching ratio of the A_{ij} values. It is clear from a comparison of the theoretical predictions (Summers *et al.* 1991) $\mathcal{E}_A/\mathcal{E}_i \approx 10$, with the experimental values $\mathcal{E}_A/\mathcal{E}_i \approx 60$, that an equilibrium situation for the BeIII ion is invalid and that a transiently ionising model is more appropriate.

Following the above proceedure, $S^{-1}(\lambda)$ for the VUV survey instrument can be written

$$S^{-1}(l = 100\text{Å}) = G \, \mathcal{E}_A \, / \, C_r$$

, where C_r is the count rate of the allowed line and G is the μ-channel plate gain.

$$\mathcal{E}_A = \mathcal{E}_{j=1} . \frac{1}{84} . \frac{\mathcal{E}_A}{\mathcal{E}_i}$$

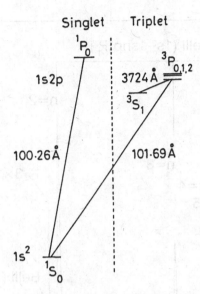

Fig.5 Partial energy level diagram for BeIII showing the transitions used in this calibration.

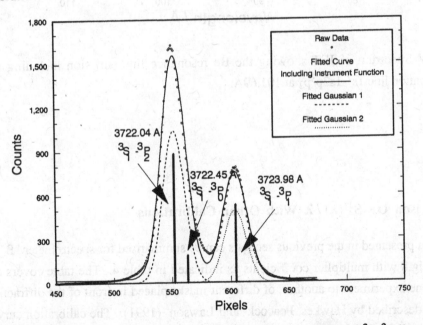

Fig.6 Spectrum, recorded with the 1 m spectrometer, of the BeIII ^3S-^3P lines at 3724 Å. The figure includes an idealised spectrum and two Gaussian components resulting from a least squares fit to the measured data. The sum of the Gaussians is shown solid. the poor fit, at the wings in particular, is due to an instrument function which is slightly too narrow. The intensity ratios of the fitted Gaussians 2.08 : 1 correspond to the statistical weights of the transitions, 5:1:3 (where the components of intensity 5 and 1 are blended).

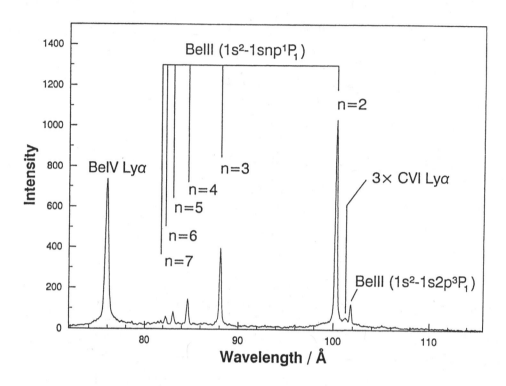

Fig.7 XUV Spectrum of JET showing the Be resonance line emission including the intercombination line $1s^2-1s2p^3p_1$ at 101.69Å.

5 Comparison OF $S^{-1}(\lambda)/\lambda$ With Other Calibrations

The data presented in the previous sections can be summarised for spectrometer "B" as shown in Fig.8 with multiplier coefficients as indicated in table 4. The table covers the use, from one experiment to another, of different masks placed in front of the diffraction gratings as described by Hawkes, Peacock and Lawson (1991). The calibration curves for the two spectrometers, "A" using charge exchange lines and "B" using a VUV calibration lamp and branching ratios, Fig.9, show striking agreement in the overall shape of their response curves. That the absolute sensitivities of the spectrometers differ is to be expected since the Gain / Voltage characteristics of the two MCP's are different, Fig.10.

TABLE 4   COEFFICIENTS APPLICABLE TO CALIBRATION CURVES (FIG.8)

JET SHOTS	290g/mm and 450g/mm	2100g/mm	
21575-21030	1.52 X10^9	-	Instrument "B"; 9.25x9.25mm^2 masks installed
21031-21959	3.28 X10^9	3.86 X10^{10}	Spectrometer "B"; 3.72x2.4 mm^2 mask installed to allow use of the 2100g/mm grating. Sensitivity of all gratings consequently reduced.
21962-23500	3.29 X10^9	2.21 X10^{10}	Larger grating mask reinstalled on 290g/mm and 450g/mm gratings with the smaller mask on the 2100 g/mm grating

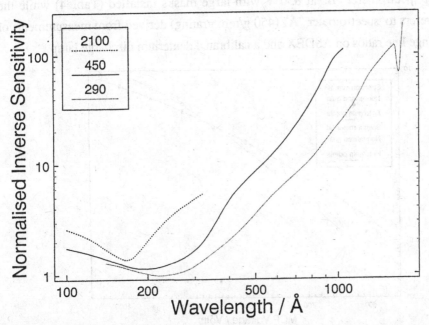

Fig.8   Normalised spectrometer "B" inverse sensitivity curves. To use these curves select the coefficient applicable from table 4 and multiply the relevant curve by this factor. The shape of the curves have been derived from the hollow cathode lamp data below 584 Å and charge exchange data. The region 1100-1700 Å has been measured with a deuterium lamp as part of the charge exchange work.

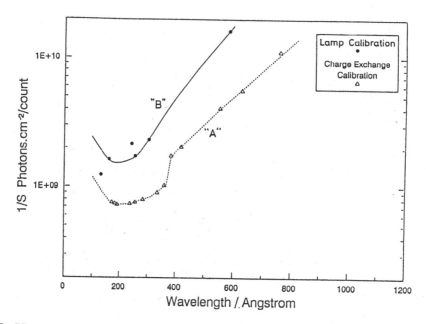

Fig.9 Upper curve represents the inverse sensitivity, from branching ratio and lamp data combined, of spectrometer "B" at 850 V with large masks installed (Table4) while the lower curve refers to spectrometer "A" (450 g/mm grating) derived from measurements of charge exchange line ratios on ASDEX and a calibrated deuterium discharge lamp.

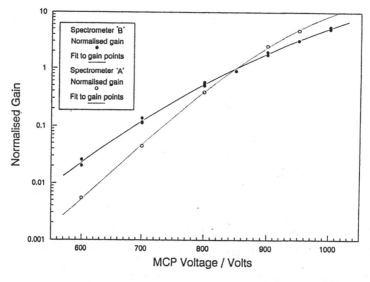

Fig.10 Gain variation of detector as a function of MCP voltage for spectrometer "B", solid curve, showing a second order polynomial fit to the data, and dashed line, spectrometer "A", (Hawkes, Peacock and Lawson 1991). Both curves have been normalised to unit gain at 850 volts.

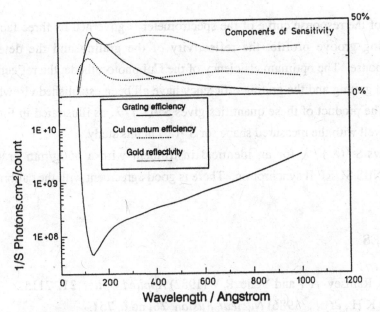

Fig.11 Lower curve is theoretically calculated inverse sensitivity curve for the family of gratings, 290, 450 and 2100 gmm[-1], gold coated and with a CuI coated photocathode. Upper curves are the contributions from grating efficiency and gold reflectivity[12]; CuI photocathode quantum efficiency[13,14].

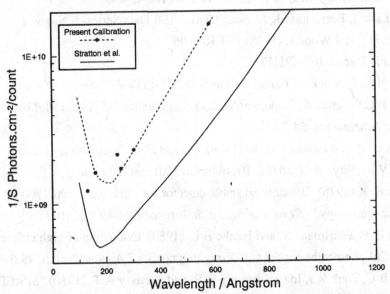

Fig.12 Calibration curve (solid line) for a 450 g/mm grating in an identical instrument (Stratton *et al.* 1986), with V_{MPC} adjusted to 850V, obtained on the NBS SURF II synchrotron, compared with the absolute calibration reported here (dashed line) using branching ratios on spectrometer "B" with the large aperture mask.

The shape of the response curve of the spectrometer is governed by three factors, namely the grating groove profile, the reflectivity of the grating and the detector photocathode response. The optimum efficiency of the CuI photocathode, the reflectivity of the gold coated grating and the grating efficiency have all been established elsewhere, see references. The product of these quantities gives $S^{-1}(\lambda)/\lambda$, as illustrated in Fig.11, which compares well with the measured shape for $S^{-1}(\lambda)$ in this study.

Fig. 12 shows $S^{-1}(\lambda)/\lambda$ for an identical instrument with a 450g/mm grating, measured on the NBS SURF II synchrotron. There is good agreement with the calibration reported here.

6. REFERENCES

1   Fonck R J, Ramsey A T and Yelle R V (1982) *Applied Optics* **21**, 2115.

2   Behringer K H , *et al*.. (1986) *Nuclear Fusion* **26**, no.6, 751.

3   Isler R C and Langley R A (1985) *Appl. Optics* **24**, 254.

4   Carolan P G, Duval B P *et al* .(1987) *Phys. Rev.* A **35,** 3454.

5   Huang L K, Lipmann S *et al*.. (1988) *Phys Rev* A **37**, No 10, 3927.

6   Kato T, Masai K and Sato K (1985) *Phys Lett* **108A**, 259.

7   Kato T, Lang J, Berrington K E (Mar 1990) NIFS Data Series-2, Nagoya,

8   Summers H P and Wood L (1988) JET R88-06

9   Summers H P *et al*. JET-(91)12

10   Hawkes N C, Peacock N J and Lawson K D JET-(91)07

11   Schwob J L,Wouters A, Suckewer W S and Finkenthal M (1987) M *Review of Scientific Instruments* **58** 9 1601

12   Williams G P and Howells M R (1979) Photon Flux Calculations at the NSLS UV Facility. *BNL 26121* Brookhaven National Laboratory,

13   Richards R K (1978) "Broadband stable detector for ultraviolet (300-1700 Å) plasma spectroscopy." *Review of Scientific Instruments* **49** 8 1210.

14   Saloman E B, Pearlman .S, and Henke B L (1980)"Evaluation of high efficiency CsI and CuI photocathodes for soft X-ray diagnostics." *Applied Optics* **19** 5 749.

15   Stratton B C, Fonk R J, Ida K, Jaehnig K P, and Ramsey A T (1986) "SPRED spectrograph upgrade: high resolution grating and improved absolute calibrations.' *Review of ScientifiC Instruments* **57** 8 2043.

C IV Emission Lines in an Active Region Spectrum obtained with SERTS

F. P. Keenan[1], R. J. Thomas[2], W. M. Neupert[2], E. S. Conlon[1], and V. M. Burke[3]

[1]Department of Pure and Applied Physics, The Queen's University of Belfast, Belfast BT7 1NN, Northern Ireland
[2]Laboratory for Astronomy and Solar Physics, Code 680, NASA–Goddard Space Flight Center, Greenbelt MD 20771, U. S. A.
[3]Department of Applied Mathematics and Theoretical Physics, The Queen's University of Belfast, Belfast BT7 1NN, Northern Ireland

ABSTRACT

Theoretical line ratios involving $2s\ ^2S - 3p\ ^2P$, $2p\ ^2P - 3s\ ^2S$ and $2p\ ^2S - 3d\ ^2D$ transitions in C IV between 312 and 420 Å are presented. A comparison of these with observations of a solar active region obtained during a rocket flight by the *Solar EUV Rocket Telescope and Spectrograph* (SERTS) reveals good agreement between theory and experiment, with discrepancies that average only 19%. This provides experimental support for the accuracy of the atomic data adopted in the line ratio calculations, and also resolves discrepancies found previously when the theoretical results were compared with solar data from the S082A instrument on board *Skylab*. The potential usefulness of the C IV line ratios as electron temperature diagnostics for the solar transition region is briefly discussed.

1 INTRODUCTION

Emission lines arising from transitions in ions of the lithium isoelectronic sequence are frequently observed in the spectra of astrophysical objects, such as the solar transition region and corona (Phillips *et al.* 1982). They may be used to derive the electron temperature (T_e) and density (N_e) of the emitting region through diagnostic line ratios, as discussed by, for example, Kunc (1988). However to calculate reliable theoretical ratios, accurate atomic data must be employed, especially for f-values and electron impact excitation rates (Dufton and Kingston 1981).

Recently, Keenan *et al.* (1992) used electron impact excitation rates calculated with the **R**-matrix code for transitions in Li-like C IV to derive diagnostic line ratios for this ion, which were compared with intermediate spectral resolution solar observations from the S082A instrument on board *Skylab*. Unfortunately, the observed and theoretical line ratios often showed large discrepancies (of up to a factor of 3.1), which was suggested might be due to blending in the S082A data.

In this paper we compare the theoretical C IV line ratios with higher quality solar observations obtained with the *Solar EUV Rocket Telescope and Spectrograph* (SERTS), to investigate if the discrepancies noted above can be removed.

2 THEORETICAL RATIOS

The model ion adopted for C IV has been discussed by Keenan *et al.* (1992), where details of the line ratio calculations may be found. In Figure 1 we plot the theoretical emission line ratios $R_1 = I(2s\ ^2S - 3p\ ^2P_{1/2,3/2})/I(2p\ ^2P_{3/2} - 3s\ ^2S)$, $R_2 = I(2p\ ^2P_{1/2} - 3d\ ^2D_{3/2})/I(2p\ ^2P_{3/2} - 3s\ ^2S)$ and $R_3 = I(2p\ ^2P_{3/2} - 3d\ ^2D_{5/2})/I(2p\ ^2P_{3/2} - 3s\ ^2S)$, for a range of electron temperatures about that of maximum C IV fractional abundance in ionization equilibrium, $\log T_e = \log T_{max} = 5.0$ (Arnaud and Rothenflug, 1985). The calculations in the figures were performed for an electron density of 10^{11} cm^{-3}, although we note that the line ratios are density insensitive for $N_e \leq 10^{13}$ cm^{-3}.

3 OBSERVATIONAL DATA

The solar spectrum analysed in the present paper was that of an active region obtained by the *Solar EUV Rocket Telescope and Spectrograph* (SERTS) during a rocket flight on 1989 May 5. This instrument covered the wavelength region 235–450 Å, with a spectral resolution better than 80 mÅ (FWHM) and a spatial resolution of 6 arc sec. The instrument and rocket observations (Neupert *et al.* 1992) are discussed in more detail in the papers in this proceedings by Davila *et al.*, Neupert *et al.*, and Thomas *et al.*

We have identified the following C IV emission lines in the SERTS active region spectrum: $2s\ ^2S - 3p\ ^2P_{1/2,3/2}$ at 312.43 Å, $2p\ ^2P_{1/2} - 3d\ ^2D_{3/2}$ (384.03 Å), $2p\ ^2P_{3/2} - 3d\ ^2D_{5/2}$ (384.17 Å) and $2p\ ^2P_{3/2} - 3s\ ^2S$ (419.72 Å). Relative intensities for these lines (which have typical errors of ±20%) were determined by fitting gaussian profiles to the spectrum using preliminary photometric calibration results. The quality of the observational data are illustrated in Figure 2, where we plot the active region spectrum between 312 and 313 Å. In particular, we can see from the figure that the C IV 312.43 Å line is resolved from Fe XV 312.55 Å, which is not the case in solar spectra obtained with the S082A instrument on board *Skylab* (see Keenan *et al.* 1992).

4 RESULTS AND DISCUSSION

In Table 1 we list the observed C IV emission line ratios $R_1 = I(312.43\ \text{Å})/I(419.72\ \text{Å})$, $R_2 = I(384.03\ \text{Å})/I(419.72\ \text{Å})$ and $R_3 = I(384.17\ \text{Å})/I(419.72\ \text{Å})$, along with the theoretical results at $\log T_e = 5.0$ from Figure 1. An inspection of the table reveals that agreement between theory and observation is good, with discrepancies that average only 19%. This provides experimental support for the accuracy of the atomic data adopted in the line ratio calculations, as the transitions involved are in the coronal approximation, and hence the values of R_1–R_3 depend principally on the ratio of the relevant electron impact excitation rates. However, more importantly, it also resolves the serious discrepancy between theory and observation found by Keenan *et al.* (1992) for line ratios involving the $2s$ $^2S - 3p$ $^2P_{1/2,3/2}$ lines at 312.43 Å, which was probably due to blending of this feature with Fe XV 312.55 Å in the lower spectral resolution S082A observations.

Finally, we note that the ratios in Figure 1 are quite sensitive to variations in the electron temperature, and hence in principle may be useful as T_e–diagnostics. For example, R_2 varies by a factor of 2.8 between log $T_e = 4.6$ and 5.4. However these ratios would need to be determined to a much higher degree of accuracy than is possible with the SERTS instrument for reliable temperatures to be derived. (The typical ±35% error in the C IV line ratios deduced from SERTS data would lead to *at least* a factor of 3 uncertainty in the derived T_e). In the future accurate measurements for C IV should be possible using the *Coronal Diagnostic Spectrometer* on board SOHO.

Acknowledgements: ESC and VMB are grateful to the SERC for financial support. This work was supported by NATO travel grant 0469/87 and the Nuffield Foundation.

REFERENCES

Arnaud, M. and Rothenflug, R. 1985, *Astr. Ap. Suppl.* **60**, 425.

Dufton, P. L. and Kingston, A. E. 1981, *Adv. Atom. Molec. Phys.* **17**, 355.

Keenan, F. P., Conlon, E. S., Harra, L. K., Burke, V. M. and Widing, K. G. 1992, *Ap. J.* **385**, 381.

Kunc, J. A. 1988, *J. Appl. Phys.* **63**, 656.

Neupert, W. M., Epstein, G. L., Thomas, R. J., and Thompson, W. T. 1992, *Solar Phys.* **137**, 87.

Phillips, K. J. H., *et al.* 1982, *Ap. J.* **256**, 774.

Table 1. Observed and theoretical C IV emission line ratios

λ_1(Å)	λ_2(Å)	$R_{obs} = I(\lambda_1)/I(\lambda_2)$	R_{theory}^a
312.43	419.72	0.90 (R_1)	0.80
384.03	419.72	0.57 (R_2)	0.48
384.17	419.72	0.60 (R_3)	0.85

[a]Calculated at the electron temperature of maximum C IV fractional abundance in ionization equilibrium, log T_e = log T_{max} = 5.0.

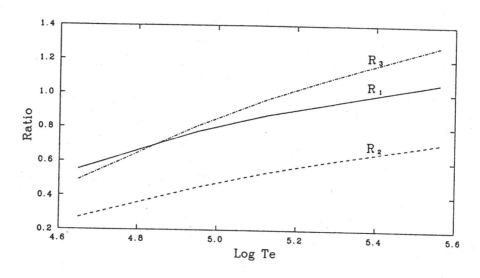

Figure 1. The theoretical C IV emission line ratios $R_1 = I(2s\ ^2S - 3p\ ^2P_{1/2,3/2})/I(2p\ ^2P_{3/2} - 3s\ ^2S) = I(312.43\ \text{Å})/I(419.72\ \text{Å})$, $R_2 = I(2p\ ^2P_{1/2} - 3d\ ^2D_{3/2})/I(2p\ ^2P_{3/2} - 3s\ ^2S) = I(384.03\ \text{Å})/I(419.72\ \text{Å})$ and $R_3 = I(2p\ ^2P_{3/2} - 3d\ ^2D_{5/2})/I(2p\ ^2P_{3/2} - 3s\ ^2S) = I(384.17\ \text{Å})/I(419.72\ \text{Å})$, where I is in units of photon numbers, plotted as a function of T_e at $N_e = 10^{11}$ cm^{-3}.

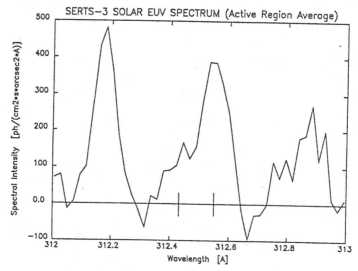

Figure 2. Plot of the active region spectrum obtained with SERTS in the wavelength interval 312–313 Å. The C IV 312.43 Å and Fe XV 312.55 Å features are marked on the spectrum by lines.

Electron Density Diagnostics Applicable to IUE Spectra of Gaseous Nebulae

F. P. Keenan[1], W. A. Feibelman[2], L. K. Harra[1], E. S. Conlon[1] and K. M. Aggarwal[3]

[1]Department of Pure and Applied Physics, The Queen's University of Belfast, Belfast BT7 1NN, Northern Ireland
[2]Laboratory for Astronomy and Solar Physics, Code 684.1, NASA–Goddard Space Flight Center, Greenbelt MD 20771, U. S. A.
[3]Department of Physics and Astrophysics, University of Delhi, Delhi – 110 007, India

ABSTRACT

Observed values of the emission line ratios $R = I(3s^2\ ^1S - 3s3p\ ^3P_2)/I(3s^2\ ^1S - 3s3p\ ^3P_1) = I(2660\ \text{Å})/I(2669\ \text{Å})$ in Al II, $R_1 = I(2s^2\ ^1S - 2s2p\ ^3P_2)/I(2s^2\ ^1S - 2s2p\ ^3P_1) = I(1907\ \text{Å})/I(1909\ \text{Å})$ in C III, and $R_2 = I(3s^2\ ^1S - 3s3p\ ^3P_2)/I(3s^2\ ^1S - 3s3p\ ^3P_1) = I(1883\ \text{Å})/I(1892\ \text{Å})$ in Si III, measured from high resolution spectra obtained with the *International Ultraviolet Explorer* (IUE) satellite, are presented for several planetary nebulae and symbiotic stars. Electron densities deduced from these ratios in conjunction with new theoretical R, R_1 and R_2 diagnostics (which are significantly different from those calculated by previous authors), are found to be compatible, and are also in good agreement with those derived from line ratios in other species. This provides observational support for the accuracy of the atomic data adopted in the line ratio calculations.

1 INTRODUCTION

The forbidden (J = 2) and intercombination (J = 1) components of the $2s^2\ ^1S - 2s2p\ ^3P_J$ and $3s^2\ ^1S - 3s3p\ ^3P_J$ multiplets in C III and Si III, respectively, have been extensively observed in the *International Ultraviolet Explorer* (IUE) spectra of low density astrophysical plasmas, such as planetary nebulae and symbiotic stars (see, for example, Feibelman 1982; Hayes and Nussbaumer 1986). Nussbaumer and Schild (1979) and Dufton, Keenan, and Kingston (1984) noted that the flux ratios of these components, namely $R_1 = F(1907\ \text{Å})/F(1909\ \text{Å})$ in C III and $R_2 = F(1883\ \text{Å})/F(1892\ \text{Å})$ in Si III, may be used to derive the electron density of the emitting plasma. These authors presented calculations of R_1 and R_2, and illustrated their usefulness over the electron density range $N_e \simeq 10^3 - 10^6\ \text{cm}^{-3}$ by a comparison with IUE observations of planetary nebulae and symbiotic stars.

In this paper we extend the above work by deriving theoretical emission line ratios (R) involving the analogous transitions in Al II, namely $3s^2\ ^1S - 3s3p\ ^3P_2$ and $3s^2\ ^1S - 3s3p\ ^3P_1$ at 2660 Å and 2669 Å, respectively, and also use improved atomic data for C III and Si III to rederive the R_1 and R_2 ratios. Our theoretical values of R, R_1 and R_2 are subsequently compared with IUE observations of planetary nebulae and symbiotic stars to investigate their accuracy.

2 THEORETICAL EMISSION LINE RATIOS

The model ion for Al II has been discussed in detail by Keenan *et al.* (1992), and those for C III and Si III by Keenan, Feibelman, and Berrington (1992). These papers also summarize the theoretical R, R_1 and R_2 emission line ratios. We note that the current Al II results are approximately a factor of 3 smaller than those of Johnson, Smith, and Parkinson (1986), while the C III and Si III diagnostics are up to $\sim$20% different from those of Nussbaumer and Schild (1979) and Dufton, Keenan, and Kingston (1984), respectively. These discrepancies are due to the adoption of improved electron impact excitation rates (for Al II and C III) and A-values (for Si III) in the present analysis.

3 OBSERVED EMISSION LINE RATIOS

These have been measured from high resolution ultraviolet spectra obtained with the IUE satellite. Unfortunately, only two objects could be identified in which the R ratio in Al II could be reliably measured, while in the case of C III and Si III the dataset was restricted to those objects for which both R_1 and R_2 could be determined, and also only included spectra that did not show the Fe II doublet at 1880 and 1884 Å, as these masked the Si III 1883 Å feature. These objects (the planetary nebulae NGC 6572 and NGC 7027, and the symbiotic stars RR Tel and AX Per) are listed in Table 1, along with the IUE images used in the analysis. Also given are the derived emission line fluxes, which were determined using the Goddard Regional Data Reduction Facility. The quality of the observational data are illustrated in Figures 1 and 2.

4 RESULTS AND DISCUSSION

In Table 2 we summarise the electron densities derived from the observed values of R, R_1 and R_2 in conjunction with the line ratio calculations of Keenan *et al.* (1992) and Keenan, Feibelman, and Berrington (1992), along with adopted electron temperatures from references given in these papers. Also listed are the values of log N_e previously determined for these objects from line ratios in other species. An inspection of the table reveals that the electron densities deduced from the R_1 and R_2 ratios are consistent, with differences of only 0.3 dex in both cases. In addition, the densities determined from all three diagnostics are similar to those derived from line ratios in other species. This provides observational support for the accuracy of the Al II, C III and Si III atomic data adopted in the present analysis.

The examples discussed above indicate that the R, R_1 and R_2 ratios are reliable density diagnostics. For IUE observations, they are useful for plasmas with $10^2 \leq N_e \leq 10^6$ cm^{-3}, but for instruments with a larger dynamic range (such as the *Goddard*

High Resolution Spectrograph on board HST) they could be applied to observations of higher density plasmas.

Acknowledgements: LKH and ESC are grateful to the Department of Education for Northern Ireland and SERC, respectively, for financial support. This work we supported by the Nuffield Foundation and by NATO travel grant 0469/87.

REFERENCES

Dufton, P. L., Keenan, F. P., and Kingston, A.E. 1984, *M. N. R. A. S.* **209**, 1P.

Feibelman, W. A. 1982, *Ap. J.* **258**, 548.

Hayes, M. A. and Nussbaumer, H. 1986, *Astr. Ap.* **161**, 287.

Johnson, B. C., Smith, P. L., and Parkinson, W. H. 1986, *Ap. J.* **308**, 1013.

Keenan, F. P., Feibelman, W. A., and Berrington, K. A. 1992, *Ap. J.* (in press).

Keenan, F. P., Harra, L. K., Aggarwal, K. M., and Feibelman, W. A. 1992, *Ap. J.* **385**, 375.

Nussbaumer, H. and Schild, H. 1979, *Astr. Ap.* **75**, L17.

Table 1. Observed emission line ratios R = I(2660 Å)/I(2669 Å) in Al II, R_1 = I(1907 Å)/I(1909 Å) in C III and R_2 = I(1883 Å)/I(1892 Å) in Si III.

Object	IUE Image No.	R	R_1	R_2
NGC 6572	LWR 4213	–	0.750	0.970
	SWP 14919			
NGC 7027	LWR 2571	0.720	–	–
RR Tel	LWR 16187	0.034	–	–
AX Per	LWR 5793	–	0.060	0.190
	SWP 21443			

Table 2. Derived Al II, C III and Si III logarithmic electron densities.

Object	T_e(K)	log N_e(R)	log N_e(R_1)	log N_e(R_2)	log N_e(other)[a]
NGC 6572	10000	–	4.5	4.2	4.4
NGC 7027	14000	4.2	–	–	4.5
RR Tel	13000	5.8	–	–	6.2
AX Per	12000	–	5.7	5.4	–

[a]Determined from line ratios in N II. O II. O III and S II.

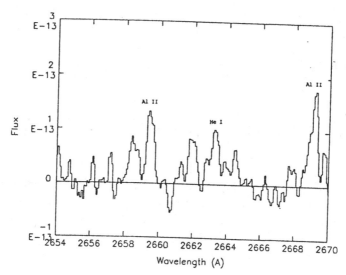

Figure 1. High resolution IUE spectrum of the planetary nebula NGC 7027 in the wavelength region 2654–2670 Å, where the flux is in units of ergcm^{-2}s^{-1}. The Al II $3s^2$ ^{1}S – $3s3p$ ^{3}P$_2$ and $3s^2$ ^{1}S – $3s3p$ ^{3}P$_1$ transitions at 2660 Å and 2669 Å, respectively, are clearly labelled in the figure.

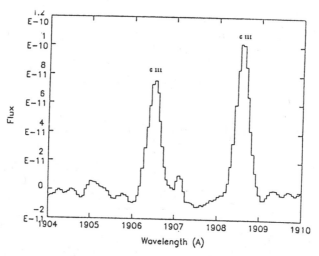

Figure 2. High resolution IUE spectrum of the planetary nebula NGC 6572 in the wavelength region 1904–1910 Å, where the flux is in units of ergcm^{-2}s^{-1}. The C III $2s^2$ ^{1}S – $2s2p$ ^{3}P$_2$ and $2s^2$ ^{1}S – $2s2p$ ^{3}P$_1$ transitions at 1907 Å and 1909 Å, respectively, are clearly labelled in the figure.

Solar O IV and S IV lines from the High Resolution Telescope and Spectrograph (HRTS) and the S082B spectrograph on board Skylab

F. P. Keenan[1], J. W. Cook[2], J. G. Doyle[3], P. L. Dufton[1], M. A. Hayes[4] and A. E. Kingston[5]

[1]Department of Pure and Applied Physics, The Queen's University of Belfast, Belfast BT7 1NN, Northern Ireland
[2]Code 4163, E.O. Hulburt Center for Space Research, Naval Research Laboratory, Washington DC 20375-5000, U. S. A.
[3]Armagh Observatory, Armagh BT61 9DG, Northern Ireland
[4]Daresbury Laboratory, Warrington WA4 4AD, England
[5]Department of Applied Mathematics and Theoretical Physics, The Queen's University of Belfast, Belfast BT7 1NN, Northern Ireland

ABSTRACT

A comparison of theoretical O IV electron density diagnostics with observational data from a quiet solar region, a sunspot and an active region obtained with the *High Resolution Telescope and Spectrograph* (HRTS), and a flare observed with the S082B instrument on board *Skylab*, reveals that the O IV 1404.8 Å line is not only blended with the S IV 1404.8 Å transition, but also with an unknown feature that contributes a significant amount to the blend. It is therefore suggested that line ratios which include the O IV 1404.8 Å transition should not be employed as density diagnostics.

1 INTRODUCTION

Flower and Nussbaumer (1975) first noted that the $2s^2 2p\ ^2P - 2s 2p^2\ ^4P$ intercombination multiplet of O IV at $\sim$1400 Å should be useful as electron density diagnostics, through emission line intensity ratios such as $R_1 = I(^2P_{3/2} - {}^4P_{1/2})/I(^2P_{3/2} - {}^4P_{5/2})$ $= I(1407.4\ \text{Å})/I(1401.2\ \text{Å})$ and $R_2 = I(^2P_{3/2} - {}^4P_{1/2})/I(^2P_{3/2} - {}^4P_{3/2}) = I(1407.4$ Å$)/I(1404.8\ \text{Å})$. Subsequently, Nussbaumer and Storey (1982) calculated improved theoretical line intensity ratios for this ion, based on more accurate atomic physics data than were available to Flower and Nussbaumer.

Although the $^2P_{3/2} - {}^4P_{3/2}$ component of the O IV intercombination multiplet at 1404.8 Å is known to be blended with the S IV $3s^2 3p\ ^2P_{1/2} - 3s 3p^2\ ^4P_{1/2}$ transition, several authors have assumed that the latter does not contribute substantially to the total line intensity, so that O IV line ratios involving 1404.8 Å may safely be used as density diagnostics (see, for example, Nussbaumer and Storey 1982). In this paper we analyse the O IV and S IV emission lines in solar spectra obtained with both HRTS and the S082B spectrograph on board *Skylab*, to investigate if this assumption is valid.

2 THEORETICAL EMISSION LINE RATIOS

The model ion for O IV has been discussed in detail by Cook *et al.* (1992). This paper also summarize the theoretical R_1 and R_2 emission line ratios as a function of electron density at the temperature of maximum O IV fractional abundance in ionization equilibrium, $T_e = T_{max} = 1.4 \times 10^5$ K, although we note that these data are relatively temperature insensitive. For example, changing T_e by a factor of two leads to a $\leq 10\%$ variation in R_1 or R_2.

3 OBSERVED EMISSION LINE RATIOS

Observational data analysed in the present paper consist of a HRTS sounding rocket spectrum of a quiet region, two positions on a sunspot umbra and an active region (discussed in detail by Cook *et al.* 1992), and the August 9 1973 flare spectrum aquired by the S082B spectrograph on board *Skylab* (Doyle and Cook 1992). The quality of the observational data are illustrated in Figures 1 and 2, where we plot the HRTS spectrum in the 1394–1410 Å and 1410–1426 Å wavelength regions, respectively.

Emission line intensities were determined from the above spectra by profile fitting. The subsequent O IV R_1 and R_2 ratios are listed in Table 1, where we have used the measured intensity of the S IV 1406.1 Å transition in conjunction with the theoretical line ratio calculations of Dufton *et al.* (1982) to remove the S IV 1404.8 Å flux from the O IV/S IV 1404.8 Å blend. We also list values of the S IV $R_1 = I(1416.9$ Å$)/I(1406.1$ Å$)$ and $R_2 = I(1423.9$ Å$)/I(1416.9$ Å$)$ line ratios where measurable.

4 RESULTS AND DISCUSSION

In Table 2 we summarise the electron densities derived from the observed values of R_1 and R_2 in both O IV and S IV using the theoretical diagnostics listed by Cook *et al.* (1992) and Dufton *et al.* (1982), respectively. An inspection of the table reveals that R_1 and R_2 in O IV do not give compatible results, with discrepancies of typically 1.4 dex. Although the S IV electron densities are very limited, they tend to be in better agreement with the R_1 values in O IV, suggesting that the problem lies with R_2, and in particular the 1404.8 Å transition. The most likely explanation is that there is another line present in this blend besides O IV and S IV, although an inspection of lines lists reveal no suitable candidates. Until a possible blending species can be identified (possibly from laboratory observations), we strongly suggest that line ratios which include the O IV 1404.8 Å transition should not be employed as density diagnostics.

Acknowledgements: The work of JWC was supported by NASA under DPR W–

14,541 and by the Office of Naval Research. This work was supported by the Nuffield Foundation and by NATO travel grant 0469/87.

REFERENCES

Cook, J. W., *et al.* 1992, *Ap. J.*, submitted.

Doyle, J. G. and Cook, J. W. 1992, *Astr. Ap.*, in press.

Dufton, P. L., Hibbert, A., Kingston, A. E. and Doschek, G. A. 1982, *Ap. J.*, **258**, 548.

Flower, D. R. and Nussbaumer, H. 1975, *Astr. Ap.*, **45**, 145.

Nussbaumer, H. and Storey, P. J. 1982, *Astr. Ap.*, **115**, 205.

Table 1. Observed O IV and S IV emission line ratios.

Feature	R_1 (O IV)	R_2 (O IV)	R_1 (S IV)	R_2 (S IV)
Quiet sun	0.41	0.52	–	–
Active region	0.34	0.91	0.74	0.48
Sunspot umbra A	0.33	0.92	0.45	–
Sunspot umbra B	0.27	0.42	1.15	–
Flare, 15:54:17 UT	0.45	1.25	–	–
Flare, 15:54:47 UT	0.46	1.47	–	–
Flare, 15:55:34 UT	0.40	1.72	–	–
Flare, 15:56:00 UT	0.37	1.19	–	–
Flare, 15:56:46 UT	0.34	1.82	–	–

Table 2. Derived O IV and S IV logarithmic electron densities.

Feature	R_1 (O IV)	R_2 (O IV)	R_1 (S IV)	R_2 (S IV)
Quiet sun	12.3	9.7	–	–
Active region	11.3	10.3	L[a]	12.0
Sunspot umbra A	11.3	10.3	11.5	–
Sunspot umbra B	10.8	9.4	L	–
Flare, 15:54:17 UT	H[a]	10.6	–	–
Flare, 15:54:47 UT	H	10.8	–	–
Flare, 15:55:34 UT	12.0	10.9	–	–
Flare, 15:56:00 UT	11.5	10.5	–	–
Flare, 15:56:46 UT	11.3	11.0	–	–

[a]indicates that the observed line ratio is in the low (L) or high (H) density limit.

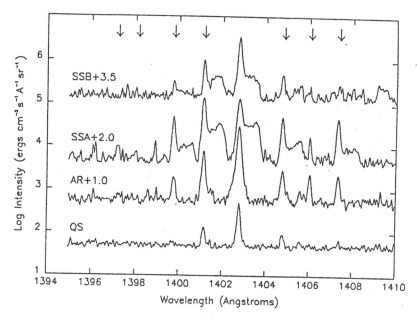

Figure 1. HRTS observations of the O IV and S IV lines in the 1394–1410 Å wavelength region, from slit positions covering a quiet solar area (QS), an active region (AR), and two positions within a sunspot umbra (SSA and SSB).

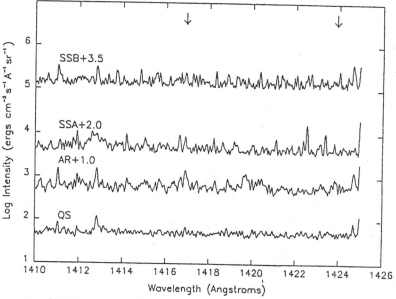

Figure 2. HRTS observations of the O IV and S IV lines in the 1410–1426 Å wavelength region, from slit positions covering a quiet solar area (QS), an active region (AR), and two positions within a sunspot umbra (SSA and SSB).

Far and Extreme Ultraviolet Spectrophotometry of the Hot DA White Dwarfs G191-B2B and HZ43 with the Hopkins Ultraviolet Telescope

Randy A. Kimble[1], Arthur F. Davidsen[2], Knox S. Long[3], Charles W. Bowers[2], Gerard A. Kriss[2], David S. Finley[4], and Detlev Koester[5]

[1]Laboratory for Astronomy and Solar Physics, Goddard Space Flight Center
[2]Center for Astrophysical Sciences, The Johns Hopkins University
[3]Space Telescope Science Institute
[4]Center for EUV Astrophysics, University of California, Berkeley
[5]Department of Physics and Astronomy, Louisiana State University

ABSTRACT

In-flight calibration of the sensitivity of the Hopkins Ultraviolet Telescope has been based on model atmosphere calculations for the hot DA white dwarf star G191-B2B. We describe here the close agreement of that calibration with the sensitivity derived from laboratory measurements and observations of a second hot DA white dwarf, HZ43.

1 INTRODUCTION

The Hopkins Ultraviolet Telescope (HUT) flew aboard the shuttle Columbia as part of the *Astro-1* mission for nine days during December 1990. During that flight, HUT carried out 3 Å resolution spectrophotometry in the 912-1860 Å range of the far ultraviolet for a wide variety of astronomical objects; a few nearby objects were observed in the 420-912 Å range of the extreme ultraviolet as well.

As described by Davidsen *et al.* (1992), calibration of the in-flight sensitivity of HUT was performed by comparing the spectrum observed for the hot DA white dwarf G191-B2B with a model atmosphere calculation. In light of the historically contentious calibration debate for wavelengths between Lyman α and the Lyman limit, we present here the confirmation of the adopted in-flight calibration by laboratory measurements and by observations of a second (cooler) white dwarf, HZ43.

2 COMPARISONS WITH THE IN-FLIGHT CALIBRATION

In Figure 1, we show the results of pre- and post-flight laboratory calibrations of the HUT instrument. Figure 1a shows the pre-flight effective area measured at a number of wavelengths through the HUT first order far UV wavelength range, along with a smooth curve (4th order cubic spline) which has been fit to the measured points.

Figure 1b presents the results of a post-flight calibration of the HUT spectrograph which was carried out at 7 wavelengths, 6 of which span the 912-1860 Å far UV

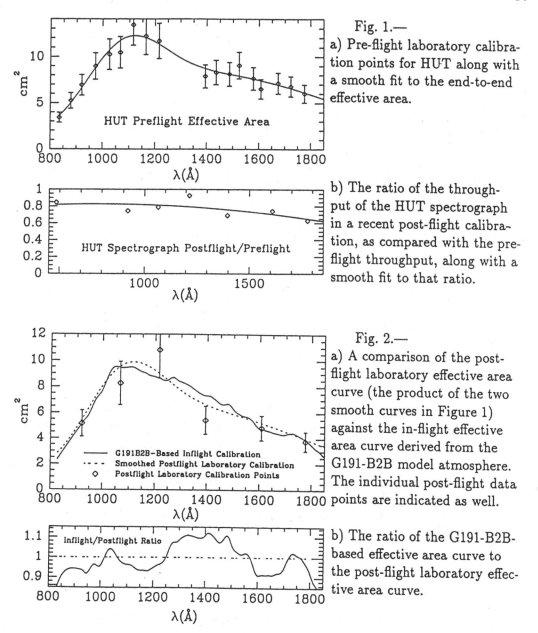

Fig. 1.—

a) Pre-flight laboratory calibration points for HUT along with a smooth fit to the end-to-end effective area.

b) The ratio of the throughput of the HUT spectrograph in a recent post-flight calibration, as compared with the pre-flight throughput, along with a smooth fit to that ratio.

Fig. 2.—

a) A comparison of the post-flight laboratory effective area curve (the product of the two smooth curves in Figure 1) against the in-flight effective area curve derived from the G191-B2B model atmosphere. The individual post-flight data points are indicated as well.

b) The ratio of the G191-B2B-based effective area curve to the post-flight laboratory effective area curve.

range. The resulting ratio of the post-flight spectrograph throughput to the pre-flight throughput has also been fit by a low order curve in order to smooth over the observational scatter. The adopted curve is both physically reasonable (declining more in the long wavelength tail of the CsI photocathode than at short wavelengths) and in good agreement with our previous experience with similar detectors within the HUT program.

The product of the two laboratory calibration curves of Figure 1 yields the post-flight laboratory effective area curve which is shown in Figure 2a in comparison with the in-flight effective area curve derived from the G191-B2B model atmosphere. As the ratio of the two curves shown in Figure 2b indicates, the agreement between them is remarkable. *With no arbitrary scaling,* the mean level of the curves agrees to within 1% (fortuitously better than the quoted error of the NIST standard photodiodes to which the laboratory calibration is referenced), with an rms dispersion of only 6.7% and maximum excursions of no more than 12% over the 920 to 1777 Å range of the laboratory data points.

HUT observations of HZ43 provide a further assessment of the accuracy of the adopted in-flight calibration. The observed far UV spectrum, fluxed using the G191-B2B-based calibration, can be compared with the model atmosphere prediction for HZ43 based on stellar parameters derived from its Balmer line spectrum.

To the extent that the model atmospheres for the two stars are similar, this comparison only addresses the stability of the HUT calibration and not its absolute accuracy. However, the optical data indicate that HZ43 is considerably cooler than G191-B2B, with T_{eff} = 50600 (Finley, Koester, and Basri 1992), as compared with the temperature of 59250 K determined for G191-B2B (Holberg *et al.* 1991). The model atmosphere for the HZ43 parameters predicts a sharper turnover from Lyman α to the Lyman limit as compared with G191-B2B, with a relative drop of 20% by 1040 Å and a total decline of 25-30% by the Lyman limit. The comparison thus does offer some leverage for confirming the absolute accuracy of the calibration as well.

The comparison between the appropriate HZ43 model atmosphere and the HUT spectrum as fluxed using the G191-B2B-based calibration is shown in Figure 3a, and the ratio of the observed flux to the model prediction is presented in Figure 3b. Because of erratic pointing during the HZ43 observation, the stellar flux was not always directed into the spectrograph entrance slit. We have forced the long wavelength tail of the observed spectrum to agree with the model, in order to assess the level of agreement in spectral shape shortward of Lyman α; the normalization correction required was only 8% and is consistent with the amount of pointing jitter observed.

Excellent agreement between the observed and model fluxes for HZ43 is found from the long wavelength end of the HUT spectrum down to 1000 Å. Below that, a slight excess is found near 990 Å (most likely attributable to residual 989 Å airglow in the HZ43 spectrum); the observed spectrum then dips below the model prediction by approximately 15% from 975 Å down to the Lyman limit. This discrepancy could indicate that G191-B2B is slightly hotter than the adopted temperature or that HZ43 is slightly cooler; alternatively, the modest disagreement may simply reflect greater uncertainties in modelling the region of the converging Lyman series, where overlap of the strong pressure-broadened lines is severe.

In any case, the excellent agreement over most of the wavelength range further reinforces the conclusion that no systematic errors beyond the 10% level exist in the HUT in-flight calibration, down at least to 975 Å, and uncertainties down to the Lyman limit are only slightly greater.

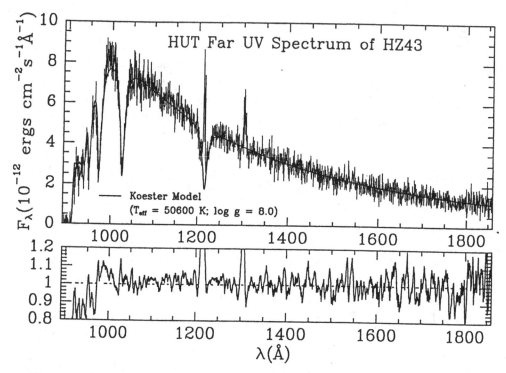

Fig. 3.—a) The HUT spectrum of HZ43, as fluxed using the G191-B2B-based sensitivity curve, plotted against the model atmosphere spectrum for HZ43. b) The ratio of the observed flux to the model-predicted flux for HZ43.

3 CONCLUSIONS

The HUT in-flight sensitivity calibration, based on a G191-B2B model atmosphere, is well-confirmed by post-flight laboratory measurements and by observations of the cooler DA HZ43. These results provide powerful confirmation that a) there are no significant systematic errors in the HUT flux calibration, even in the heretofore problematic region shortward of Lyman α, and b) that G191-B2B model atmosphere calculations provide an excellent flux standard in the far ultraviolet.

The Hopkins Ultraviolet Telescope project is supported by NASA contract NAS 5-27000 to the Johns Hopkins University.

REFERENCES

Davidsen, A. F., Long, K. S., Durrance, S. T., Blair, W. P., Bowers, C. W., Conard, S. J., Feldman, P. D., Ferguson, H. C., Fountain, G. H., Kimble, R. A., Kriss, G. A., Moos, H. W., and Potocki, K. A. 1992, *Ap. J.*, in press.
Finley, D. S., Koester, D., and Basri, G. 1992, in preparation.
Holberg, J. B., Ali, B., Carone, T. E., and Polidan, R. S. 1991, *Ap. J.*, **375**, 716.

Temperature diagnostic in the EUV with broad band photometry.

M.Landini[1] and B.C.Monsignori Fossi[2]

[1]Astronomy and Space Science Department, Florence University, Italy
[2]Arcetri Astrophysical Observatory, Florence, Italy

ABSTRACT

The extreme ultraviolet region of the spectrum (100 - 300 Å) is particularly useful to study thin plasmas in the temperature regime between 10^5 and 10^6 K and it is available for the investigation of astrophysical plasmas due to space experiments devoted to EUV research. Simulations of power emitted by low density and optically thin plasmas, using the spectral emissivity code of Landini and Monsignori Fossi (1990) and the effective areas of broad bands detectors on board of ROSAT satellite are performed. A numerical method able to deduce the temperature and differential emission measure distribution is analyzed. Limits and uncertainties of the method are outlined.

1. Introduction

Modelling of astrophysical plasmas in the high temperature regimes is possible by evaluating the differential emission measure distribution (d.e.m.) that fits in the best way the observed singnals in the EUV an X-ray spectral region. In the following sections the capability of one inversion tecnique for the d.e.m. determination is analyzed and its limits discussed. The mathematical problem is described in section 2. and some numerical simulations are shown in section 3. Conclusions are summarized in section 4.

2. The mathematical problem.

The signal measured from an optically thin plasma is given by:

$$I = k \int_{\lambda} \int_{V} G(\lambda, T, n_e)\epsilon(\lambda)n^2 d\lambda dV$$

where: k is a constant due to the properties of the instrument (collecting area, type of detection, etc) and to the source distance; $n^2 dV (cm^{-3})$ is the volume emission measure; the function G is the power emitted per unit emission measure and is a function of wavelength, temperature and of density: $\epsilon(\lambda)$ is the efficiency of the system.

Introducing the "differential emission measure" (d.e.m.):

$$d.e.m. = f(T) = \frac{n^2 dV}{dT}$$

that specifies the emission measure in the temperature interval between T and T+dT, the signal becomes:

$$I_{es} = k \int_{T} G_s(T)f(T)dT$$

with

$$G_s(T) = \int_\lambda G(\lambda, T, ne)\epsilon(\lambda)dl$$

where the f(T) function is assumed to be described by a cubic spline function with n (few) selected reference points T_i for which the proper $f_i = f(T_i)$ is specified.

If I_{os} is the observed signal from detector s and σ_s is the error, the quantity:

$$\chi^2 = \sum_s \frac{(I_{es} - I_{os})^2}{\sigma_s^2} \qquad for \ \ s = 1, ..., m$$

is minimized with respect to f_i and in order to stabilize the f(T) solutions different a priori conditions may be used.

We assume as external condition, the costancy of the quantity:

S = $\sum_i f_i \ln \frac{f_i}{w_i}$ = "entropy of information $< f >$"

The function to be optimized when f_i change is:

$$F = S + \alpha\chi^2$$

where α is a Lagrange multiplier.

The optimizing operation $[\frac{dF}{df_i} = 0]$ leads to the iterative solution of the equations:

$$\ln f_j = \ln w_j - 1 - \alpha \sum_s \frac{(I_{es} - I_{os})}{\sigma_s^2} \frac{d\int_T G_s(T)f(t)dT}{df_j} \qquad j = 1...n$$

where the weights are assumed: $w_i = ef_i$ The smoothness, or the details of the d.e.m. solution depends from the number of selected mesh points (n).

This procedure has been shosen since it gives only positive d.e.m. values (a rather important physical constraint) andit has proved to be quickly converging.

3. The simulation.

The simulation procedure is performed through the following steps: analytical functions have been tested to simulate "reasonable" $G_s(T)$ functions shapes; the "simulated signals" are genereted by numerical integration and the procedure to evaluate the retrieved d.e.m. has been extensivily investigated.

The method has been tested both without and with random noise added to the signals and comparison is performed between the use of a large number and a small number of signals.

A rather satisfing retrieval is obtained if a large number of signals is available also if noise is present. When a small number of signals is available the d.e.m. may be only partially recovered and problems increase in presence of noise.

In order to give some examples we have selected a set of very broad $G_s(T)$ shapes generated using the functions:

$$\frac{G_s(T) = s^4 T^2}{\exp(Ts) \qquad s = 1,...25}$$

which peaks at about $T_p = 3/s$ and attains 10% of the peak value in about two orders of magnitude in T. The integration routine has been checked using the d.e.m. distribution:

$$f(T) = 15T$$

in order that the signal

$$I_s = \int_T G_s(T) f(T) dT = \pi^4$$

A large number of $G_s(T)$ has been generated in the range $0 \leq \log T \leq 4$, (fig. 1a) and the inversion technique has been applied; one example of the results is shown in fig. 1b.

A more reliable test is given using the broad band EUV filters on board of ROSAT together with one soft X-Ray detector. The $G_s(T)$ functions are shown in fig 2a; they have been obtained by "convolution" of the plasma emissivity evaluated by the authors (Landini and Monsignori Fossi, 1990) with the effective area of the detectors. A typical quiet sun d.e.m. has been used for the simulation and the result of the inversion procedure is shown in fig 2b. The retrieval technique easily recovers the high temperature peak and cut-off of the d.e.m. distribution, but many very different solutions in the low temperature regime fits equally well the simulated signals.

4. Conclusions

An inversion technique of broad band observations is analyzed performing exstensive simulations by means of analytical kernels.

The d.e.m. used for the simulation is resolved in detail only when a large number of "signals ", properly distributed in temperature, is available; random noise makes the inversion problem more difficult, but a large number of well distributed signals allows a good recovering also in presence of noise and non sistematic errors.

When that same technique is applied to broad band photometry in the EUV, one may conclude: broad band photometry in the EUV may be used for plasma diagnostic and may give general information on the differential emission measure (d.e.m.) distribution also in the high temperature regimes but care must be used in the definition of the ranges of temperature where the adopted filters are sensitive to the expected d.e.m. distributions; extensive simulations must be performed using inversion techniques to investigate accuracy and limits of the "recovered" d.e.m. distribution; the effect of noise must be carefully verified; it may give completely useless results, where the temperature information is less accurate.

REFERENCES

Landini M. and Monsignori Fossi B.C., (1990) *Astron. Astrophys. Suppl.Ser.*, **82**,229

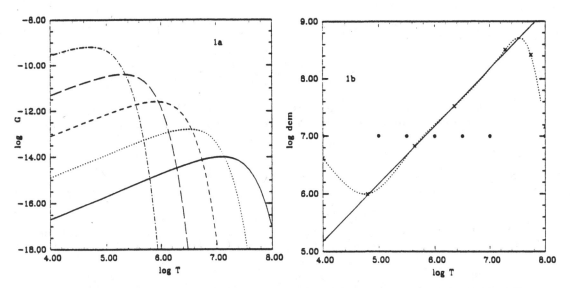

fig.1a: A selection of $G_s(T)$ functions used for the "analytical" simulations: the peaks scale as s^2 and the width at 1/2 of the peak value is about d logT = 2. **fig. 1b**: The original d.e.m. (f(T) = 15 T) (solid line) and the retrieved (dash line) solution using all the 25 signals. The trial start mesh points are shown by dots. The procedure may change both the value and the positions of the mesh points (cross).

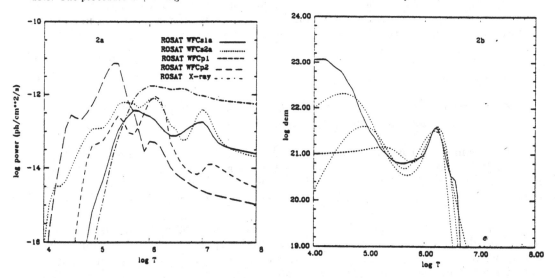

fig 2a: The $G_s(T)$ functions used for the ROSAT WFC and soft X-Ray detector. The power emitted by the plasma at different temperatures has been evaluated according to Landini and Monsignori Fossi (1990). **fig 2b**: Comparison of the "source function" (solid line)) and the retrieved solution (dash line) using the ROSAT WFC and X-ray detector. A quiet solar d.e.m. distribution is used (peak temperature at $\log T = 6.2$) and the signals are affected by noise.

Design of Compact High-Resolution Far-Ultraviolet Spectrographs equipped with a Spherical Grating Having Variable Spacing and Curved Grooves

Takeshi Namioka[1] and Masato Koike[2]

[1]Universities Space Research Association, NASA/Goddard Space Flight Center
[2]Center for X-Ray Optics, Lawrence Berkeley Laboratory, University of California

ABSTRACT

Spherical ruled gratings with variable spacing and curved grooves were designed for a FUSE type and an Eagle type spectrograph using a merit function which closely represents the rms spread of the ray-traced spots. The performance of the designed gratings in the spectrographs was evaluated by means of spot diagrams and line profiles. The results show resolving power of ~ 50000 for the FUSE type and ~ 80000 for the Eagle type, over a wavelength range of 91-103 nm. However, astigmatism is not sufficiently corrected for the purpose.

1 INTRODUCTION

The performance of Rowland circle mountings has been reinvestigated lately in connection with the development of a compact high-resolution far-ultraviolet spectrograph for a next-generation space observatory, such as the Lyman/FUSE (Moos 1989). A normal-incidence type Rowland circle mounting equipped with a conventional concave grating can provide a very high resolving power, $e.g.$, $\sim 3 \times 10^5$ in a wavelength range of 50 - 200 nm with a 6.65-m, 1200 grooves/mm grating (Ito $et\ al.$ 1986). However, small aberrations remained in the mounting become no longer tolerable for space oriented high-resolution spectrographs of the next generation because of their stringent requirements for the compactness.

To overcome this difficulty, many investigators have proposed various schemes for the reduction of the residual aberrations by incorporating gratings with variable spacing and curved grooves and/or aspheric blanks (Cash 1984; Duban 1989 and 1991; Content $et\ al.$ 1990 and 1991; Grange and Laget 1991). These proposals are made in expectation of rapid technological advances in the fabrication of gratings and aspheric mirrors, such as ion-etched holographic gratings with 6000 grooves/mm, modified ellipsoidal grating blanks, and highly coherent short-wavelength lasers for holographic recording. Furthermore, the design concept of these new types of gratings is based on individual minimization of certain aberration terms of the light path function over a given wavelength range or on cancellation of certain terms for fixed wavelengths. These procedures often are incapable of meeting the requirements for highly sophisticated spectrographs for space research, because spectral aberration cannot be represented by a linear combination of the aberration terms in the light path function.

Recently, we have developed a new design method that takes into account all possible aberrations (up to the 3rd order) in spectral images, the ruled area, and the slit height (Namioka and Koike 1992). The method utilizes analytical formulas for the spot diagrams, as well as a merit function which closely represents the rms spread of the spots formed when an infinite number of rays are traced. We have applied this new method to the design of compact high-resolution far-ultraviolet spectrographs of near normal incidence Rowland circle mounting. In these designs, we assumed mechanically ruled spherical gratings with variable spacing and curved grooves of high density, because technology for ruling such gratings is well established (Hurwitz *et al.* 1990). It was the intent of this study to see whether this realistic assumption can provide a spectrograph design that meets the requirements set for the Lyman/FUSE spectrograph.

2 DESIGN OF GRATINGS FOR THE SPECTROGRAPHS

We designed two types of Rowland circle spectrographs: (1) a FUSE type spectrograph which has a similarity to the current Lyman/FUSE spectrograph design (Moos 1989) in the use of normal spectrum formed on both sides of the grating normal and (2) an in-plane Eagle type spectrograph for possible improvement of the FUSE type spectrograph.

In designing these spectrographs, we assumed the following scheme for ruling gratings with variable spacing and curved grooves (see Fig. 1): (1) curved grooves are ruled by constraining the movement of the diamond tool in the

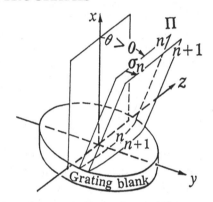

Fig. 1—Scheme for ruling a grating with variable spacing and curved grooves. See text for details.

reference plane Π which makes an angle θ with the xz plane, the x axis being the grating blank normal and (2) variable spacing is achieved by advancing the reference plane from the ruling position for the nth groove to the next by a predetermined amount σ_n in the y direction, the zeroth groove being the one passing through the origin, *i.e.*, the vertex of the grating blank (Harada and Kita 1980). The amount of advance σ_n is determined by the effective grating constant σ and the ruling parameters a, b, and c through the equation (Namioka and Koike 1992)

$$\sigma_n = \sigma + a - b + 2an + 6bn^2 + 4cn^3 . \tag{1}$$

Also assumed are: spherical gratings with a radius of curvature of 1631.3 mm, $\sigma = 1/6000$ mm, and a ruled area of 200(W) $\times$ 163(H) mm^2; a self-luminous entrance slit of 0 mm (W) $\times$ 1 mm(H); a wavelength range of 91−103 nm in +1st order; and angles of incidence of 35.5913° for the FUSE type and 20° for the Eagle type.

For minimization of the aberrations, we applied a damped least squares method to the merit function derived by the authors (Namioka and Koike 1992), which represents the rms spread of the spots averaged over a given wavelength range. The results obtained are: $a = -6.0919 \times 10^{-12}$ mm, $b = -2.4132 \times 10^{-25}$ mm, and $c = 3.1994 \times 10^{-28}$ mm, and $\theta = 35.5849°$ for the FUSE type and $a = -2.6889 \times 10^{-12}$ mm, $b = 3.0213 \times 10^{-25}$ mm, $c = 1.7158 \times 10^{-29}$ mm, and $\theta = 17.5273°$ for the Eagle type.

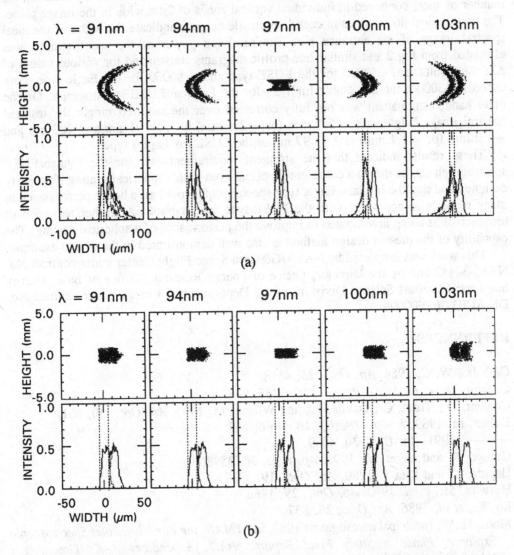

Fig. 2—Spot diagrams and line profiles constructed for spectrographs of the FUSE type (a) and the Eagle type (b). $\Delta\lambda$'s are $\lambda/50000$ in (a) and $\lambda/80000$ in (b).

3 EVALUATION BY RAY TRACING

Figure 2 shows spot diagrams and line profiles for (a) the FUSE type and (b) the Eagle type spectrograph equipped with the respective designed-gratings and an entrance slit of 10 μm x 1 mm. These spot diagrams were obtained by tracing 1000 rays through the individual spectrographs for each one of two adjacent wavelengths λ and $\lambda + \Delta\lambda$. The values of λ are indicated in the figure and $\Delta\lambda$'s used are $\lambda/50000$ for the FUSE type (a) and $\lambda/80000$ for the Eagle type (b). The line profiles were constructed by counting the number of spots contained in individual vertical zones of 2μm wide in the image plane. The two vertical dotted lines in each line-profile diagram indicate the width of the ideal spectral image of the entrance slit. The resolving power of the spectrographs was estimated from Fig.2 and similar line profile diagrams constructed for various values of $\Delta\lambda$: the results are ~ 50000 for the FUSE type and ~ 80000 for the Eagle type, both exceeding 30000, the acceptable limit set for the Layman/FUSE spectrograph. On the other hand, astigmatism was not fully corrected over the required range: the resulted spectral image heights are ~ 5.1 mm (or ~ 2.1 mm) at $\lambda = 91$ nm and 103 nm and ~ 1.2 mm (or ~ 1.1 mm) at $\lambda = 97$ nm in the FUSE (or Eagle) type.

These results indicate that the stringent requirements set for the Lyman/FUSE spectrograph can be met to a considerable extent even with a spherical grating if properly designed and that the in-plane Eagle type spectrograph would give better performance in many respects as compared with the FUSE type spectrograph. A further reduction of astigmatism is being investigated to improve the performance of a spherical grating. The capability of the present design method is also well demonstrated in the design example.

This work was supported by NASA/Goddard Space Flight Center under contract No. NAS5-30442 and by the Director, Office of Energy Research, Office of Basic Energy Sciences, Materials Science Division, of the Department of Energy under contract No. DE-AC03-76SF00098.

REFERENCES

Cash Jr., W. C. 1984, *Ap. Opt.*, **23**, 4518.
Content, D., *et al.* 1990, *Proc. SPIE*, **1235**, 943.
Content, D., Trout, C., Davila, P., and Wilson, M. 1991, *Ap. Opt.*, **30**, 801.
Duban, M. 1989, *J. Opt. (Paris)*, **20**, No.6, 269.
———— . 1991, *Ap. Opt.*, **30**, 4019.
Grange, R. and Laget, M. 1991, *Ap. Opt.*, **30**, 3598.
Harada, T. and Kita T. 1980, *Ap. Opt.*, **19**, 3987.
Hurwitz, M., *et al.* 1990, *Ap. Opt.*, **29**, 1866.
Ito, K., *et al.* 1986, *Ap. Opt.*, **25**, 837.
Moos, H. W. (principal investigator) 1989, in *LYMAN, the Far Ultraviolet Spectroscopic Explorer, Phase A Study Final Report*. Vol.2, *Appendices A–K* (Greenbelt : NASA Goddard Space Flight Center), p. E-1.
Namioka, T. and Koike, M. 1992, *Nucl. Instr. Meth. A.*, in press.

Extreme Ultraviolet Observation of Mass Flow in the Low Corona over a Large Sunspot

Werner M. Neupert[1], Jeffrey W. Brosius[2], Roger J. Thomas[1], and William T. Thompson[3]

[1]NASA/Goddard Space Flight Center
[2]Hughes STX Corporation
[3]Applied Research Corporation

ABSTRACT

We have used an extreme ultraviolet (EUV) imaging spectrograph (SERTS) covering the spectral range from 235 to 450 A to study velocity fields in the low solar corona. During a flight in May, 1989, we obtained emission line profile measurements along a chord through an active region on the Sun, including the corona over a sunspot and the initial stage of a small flare. Relative Doppler velocities were measured in the lines of Mg IX, Fe XV, and Fe XVI with a sensitivity of 2-3 km s^{-1} at 350 A. The only significant Doppler shift observed was in the emission line of Mg IX at 368.1 A over the umbra of the large sunspot. The maximum detected shift corresponded to a peak velocity toward the observer of 14 +/- 3 km s^{-1} relative to the mean of measurements in this emission line made elsewhere over the active region. The magnetic field in the low corona was aligned to within 10° of the line of sight at the location of maximum Doppler shift. Depending on the closure of the field, such a mass flow could either contribute to the solar wind or re-appear as a downflow of material in distant regions on the solar surface.

1 INTRODUCTION

One objective of the SERTS (Solar EUV Rocket Telescope and Spectrograph) program is the study of the dynamics of the solar corona and the application of such knowledge to understand the physical processes that energize the coronal plasma. The SERTS uses a Wolter Type II glancing incidence telescope to form a real image of the Sun on the entrance aperture of a near-normal incidence toroidal grating spectrograph (Neupert et al. 1992a) designed to provide near stigmatic imaging of the spectrum from 235 A to 450 A. This range that encompasses emission lines formed at temperatures from 4×10^4 K to 20×10^6 K. Spatial resolution was about 7 arc s, adequate to distinguish major features in the solar corona. The plate scale on the EUV-sensitive photographic emulsion (Eastman-Kodak 101-07) was 2.2 mA/micron. The angular field of view normal to the plane of dispersion was 21 arc m. An auxiliary optical system with a broad-band filter (3805-3865A) and another film camera recorded features in the solar image surrounding the spectrograph entrance aperture, so observations could later be correlated with ground-based data.

2 OBSERVATIONS AND DATA ANALYSIS

Observations were made on May 5, 1989, when the instrument was flown to a height of 320 km over White Sands, New Mexico, by a Terrier-boosted Black Brant rocket. The target was a region of solar activity, NOAA Region 5464, located at S18 W49. Spectra were recorded across the central portion of the region and could be associated with various emitting features in the corona using the spectroheliograms also obtained (Neupert et al. 1992a). Four spectral lines that had photographic densities in the linear portion of the D-log E curve were selected for this analysis. These were two lines of Fe XVI at 335.40 and 360.76 A, Fe XV at 284.16 A, and Mg IX at 368.07 A.

Two methods were used to measure the positions of spectral lines: The first consisted of measuring the locations of 200 micron long (18 arc s on the Sun) line segments directly, with a Grant Series 800 Comparator-Microdensitometer. In the other, the flight images were digitized with a Perkin-Elmer PDS Scanning Microdensitometer and the central wavelengths of spectral line segments 100 microns long determined with a Gaussian fitting technique. Laboratory calibration spectra with comparable photographic densities were used to establish the noise level due to grain in the photographic emulsion, which limited the minimum detectable Doppler velocity shift to about 3 km s^{-1} at 350 A.

3 RESULTS

The wavelengths of coronal spectral lines have estimated accuracies of 4-20 mA (Behring et al. 1976) corresponding to line of sight velocity uncertainties of 4-20 km s^{-1} at 300 A. Therefore, small, localized deviations from mean line positions (wavelength shifts) can only be converted to line of sight velocities relative to a zero mean line of sight velocity for the active region as a whole. The emission lines of Fe XV and Fe XVI (originating at 2-3 x 10^6 K) did not show significant wavelength shifts at any location over the active region (Fig.1). A significant (5 standard deviation) blue shift, corresponding to a velocity of 14 km s^{-1} was, however, observed in the Mg IX line (originating at 1.1 x 10^6 K) over the umbra of a large sunspot. At that location, the coronal magnetic field, estimated using a potential field code (Sakurai 1982), was aligned to within 10° of the SERTS line of sight, so that plasma flow along field lines could have been observable as a Doppler shift.

Depending on the closure of the magnetic field, the mass outflow over the umbra could either contribute to the solar wind or re-appear as a down-flow of material in distant regions of the solar surface. Since the slit passed over many coronal loop systems, none of which exhibited velocity shifts, we suggest that the upward-moving plasma was moving along open field lines into interplanetary space. A more complete report of this work has been submitted to the Astrophysical Journal (Neupert at al. 1992b). This research has been supported by NASA's Space Physics Division under RTOP 170-38-51.

REFERENCES

Behring, W. E., Cohen, L., Feldman, U., and Doschek, G. A. 1976, Ap. J. 203. 521.

Neupert, W. M., Epstein, G. L., Thomas, R. J., and Thompson, W. T. 1992a, Solar Phys., 137, 87.

Neupert, W. M., Brosius, J. W., Thomas, R. J., and Thompson, W. T. 1992b Ap. J., submitted.

Sakurai, T. 1982, Solar Phys., 76, 301.

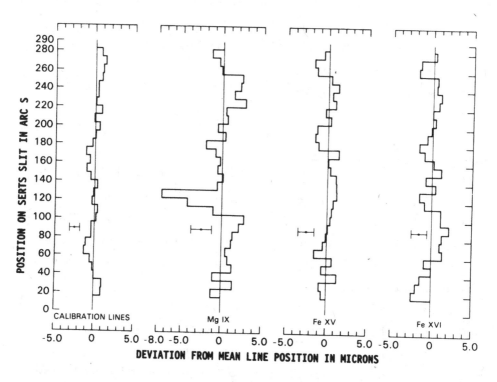

Fig. 1. Deviations from a nominal straight line slit image for each position along the SERTS slit for the average of five calibration lines of Ne II and for coronal lines of Mg IX, Fe XV, and Fe XVI. The error bar for the calibration lines is one standard deviation (+/- 1 sigma) of the five-line average. For the coronal lines, the error bars represent standard deviations of measurements on individual laboratory lines comparable in photographic density to the weakest portion of the coronal line. Blue-shifted emission corresponding to a Doppler shift of 14 km s^{-1} in the Mg IX line at position 129 coincided with the umbra of a large sunspot.

A Vacuum Ultraviolet Fourier Transform Spectrometer

W. H. Parkinson[1], A. P. Thorne[2], Peter L. Smith[1], and K. Yoshino[1]

[1]Harvard-Smithsonian Center for Astrophysics, Cambridge, MA USA
[2]Blackett Laboratory, Imperial College, London, UK

1 INTRODUCTION

Fourier transform spectroscopy is a well-established technique for high-resolution spectroscopic measurements at infrared, visible, and ultraviolet (UV) wavelengths. The instrumentation is usually based on the Michelson interferometer, and the spectrum is obtained by performing a discrete Fourier transform on a interferogram produced by scanning the moving mirror of the interferometer at a constant rate. However, no Fourier transform spectroscopy had been done at vacuum ultraviolet (VUV) wavelengths (120-200 nm), until 1985 when the first laboratory VUV Fourier transform spectrometer (FTS) was developed (Thorne *et al.* 1987) at the Physics Department, Imperial College (IC), London, England. This first VUV-FTS led to a second generation FTS that was developed and marketed by Chelsea Instruments, Ltd., (Surrey, KT8 0QX, England). We have used both instruments to obtain emission and absorption spectra with a resolving power of 2×10^6 in the UV and VUV. Recent work is aimed at quantitative spectroscopic measurements which improve the databases for astrophysically and aeronomically important species.

In Fourier transform emission spectroscopy, the noise from all lines in the optical bandpass of a spectrum is distributed throughout the spectrum as white noise. However, in absorption measurements with an FTS, the signal-to-noise ratio is inversely proportional to the square root of the band width of the background continuum used for the absorption. To narrow the wavelength bandpass of the continuum background we have used a 0.3 m Czerny-Turner spectrometer, McPherson, Model 213 (Acton, MA, 01720, USA).

2 EMISSION

Cobalt, an astrophysically important transition element, is currently being studied in emission (Pickering and Thorne 1991). Spectra of Co I and II have been obtained using a water-cooled, cobalt, hollow-cathode lamp with the FTS. The resolution of blends and hyperfine structure is limited by the Doppler linewidths alone (0.2 cm^{-1} in the UV). Figure 1 illustrates the measured and fitted curves of the hyperfine structure in the Co II transition $z^5P_3 - a^5P_2$ at 220.69 nm.

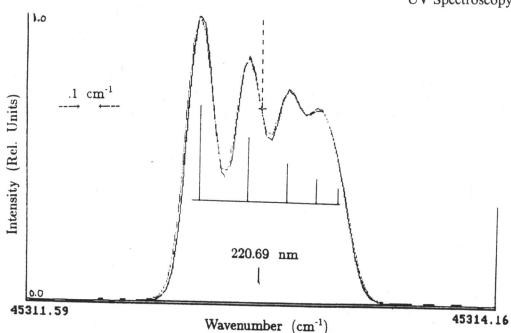

Fig. 1.—Co II transition with fitted curve. The hyperfine structure splitting of the Co II transition $3p^7(^4P)$ 4p z^5P_3 - $3d^7(^4P)$ 4s a^5P_2.

3 ABSORPTION

The photodestruction of NO is recognized as an important stratospheric process. All previous quantitative photoabsorption measurements of the VUV NO bands have been performed at insufficient spectroscopic resolution of the line profiles. Even the 6.6-m grating spectrometer (Yoshino, Freeman, and Parkinson, 1980) that achieved the uniquely high resolution (2×10^5) required for absolute cross section measurements is not high enough in its resolution to measure the absolute cross section of sharp bands like the δ-bands of NO. We have been therefore engaged in a program to use the VUV-FTS to measure absolute cross sections of the δ-bands of NO.

Our studies have been performed at an instrumental width of 0.04 cm^{-1}, which is small compared with the Doppler widths of the rotational lines at 295 K. Absorption cross sections of the $\delta(0,0)$ and $\delta(1,0)$ bands of NO have been measured at 295 K with the 6.6-m grating spectrometer and with the VUV-FTS (Fig. 2a & 2b) and at 78 K with the VUV-FTS (Fig. 2c).

4 MgF$_2$ BEAMSPLITTER

The VUV-FTS at IC and the commercially available one from Chelsea Instruments employ a unique beamsplitter configuration. The splitting of the input beam and the recombining of the interfering beams occur on different portions of a single, plane-parallel, disk of fused silica. The wedge angle, which is usually applied to

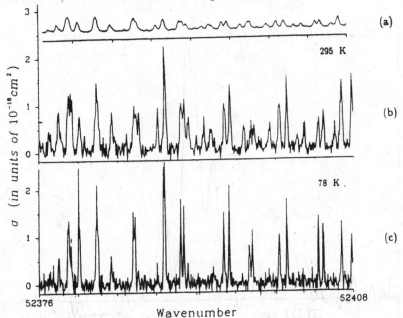

Fig. 2.—The absorption cross sections of the $\delta(0, 0)$ band of NO at 295 K (a) measured with the 6.6-m grating spectrometer, and at 295 K (b) and at 78 K (c) measured with the VUV-FTS.

beamsplitters in order to divert troublesome back-surface reflections, is obtained in this new configuration by optically contacting a matched pair of rectangular wedges made from a single piece of fused quartz to opposite sides of the plane-parallel beam-splitter. This beamsplitter has been proven to work well for $\lambda > 179$ nm. However, for studies at shorter wavelengths, other optical materials are required. We decided to use MgF_2 because it is commercially available, known to transmit light below 120 nm, can be optically worked to the tolerances required, and is not as hygroscopic as other possible materials (*eg.* LiF).

We obtained a beamsplitter made by Bernhard Halle Nachfl. Co. (Berlin, Germany) from a single, 60-mm diameter magnesium fluoride (MgF_2) crystal. The over-all dimensions were set to match those of the fused quartz beamsplitter of the other VUV-FTS's. The quartz and the MgF_2 beamsplitters were coated by Acton Research Corp. (Acton, MA, 01720, USA) for optimum transmission and reflectivity.

Both beamsplitters were tested in the IC UV-FTS. The quartz beamsplitter was used first as a benchmark for performance of the UV-FTS under 'standard' condition of adjustment. Then, test measurements of modulation and resolving power were made with the MgF_2 beamsplitter. Initial measurements of modulation yield at 257.3 nm and 193.0 nm values of 50% and 30%, respectively for MgF_2, compared with the modulation for the quartz beamsplitter at 253.7 nm of 70%. We then used a platinum hollow-cathode to examine Pt II lines near 178 nm. A resolution of 10^6 was measured by examining the rich hyperfine structure of the 177.7 nm line (see Figure 3). The wavelength limit of the UV-FTS was found to be near 175.0 nm and was set

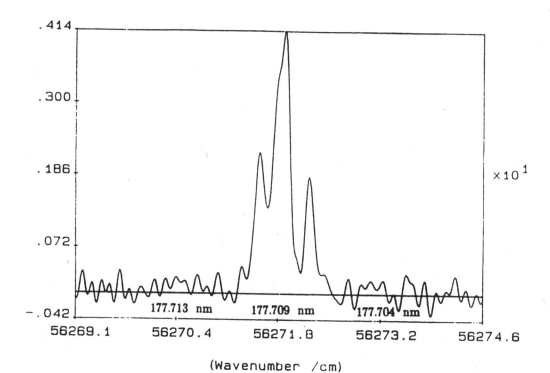

Fig. 3.—The hyperfine structure splitting of the transition 6s $^4F_{9/2}$ - 6p $^4G^o_{11/2}$ of Pt II at 177.71 nm.

by the quartz windows on the photomultiplier.

In order to examine the light transmission and optical characteristics of the whole interferometer, we have equipped it with VUV reflecting and transmitting optics throughout. Hamamatsu type R1259 and R1220, side-on, MgF_2, windowed photomultipliers, with CsI photocathodes (Bridgewater, NJ, 08807, USA) have also been installed for the next series of tests and measurements.

Supported in part by NSF Grant ATM-87-17875, by a NATO Grant for Collaboration in Research, by the Paul Instrument Fund of the Royal Society, and by Newton Optical Technologies, Ltd.

REFERENCES

Thorne, A. P., Harris, C. J., Wynne-Jones, I., Learner, R. C. M., and Cox, G. 1987, *J. Phys. E: Sci. Instrum.*, **20**, 54.

Pickering, J. and Thorne, A. P. 1991, private communication.

Yoshino, K., Freeman, D. E., and Parkinson, W. H. 1980, *Appl. Opt.*, **19**, 66-71.

High Spectral Resolution O VI Emission Line Mapping of the Cygnus Loop Supernova Remnant

Andrew P. Rasmussen and Christopher Martin

Columbia Astrophysics Laboratory and Physics Dept., Columbia University

ABSTRACT

We present a far–ultraviolet spectrophotometric emission line mapping of the Cygnus Loop supernova remnant. These are results from the first flight of the rocket borne, High Resolution Emission Line Spectrometer (*HIRES*). The lines were measured with high significance, greater than 9 and 4σ for the doublet lines $\lambda\lambda$ 1031.9, 1037.6 respectively. Redistributing the collected O VI photons onto the sky reveals significant limb brightening, and suggests a shell structure to the emitting region. The emission line profiles, which are broader than the instrument resolution, were fit to simple, expanding shell models. Best fit values give an expansion velocity to the emissive region of 185 ($\pm$19) km s^{-1} and a corrected average surface brightness of 8.8 ($\pm$3.6) $\times$ 10^{-6} erg s^{-1} cm^{-2} sr^{-1} in the doublet. Comparison of the observed brightness with predictions of both radiative and non-radiative shock models provide constraints for the global blast wave ram pressure of the O VI shock. Cooling time constraints based on the age and evolution of the Cygnus Loop do not rule out either emission mechanism at this time, suggesting that the remnant is approaching its radiative phase. However, the similarity in morphology between the O VI map and the soft X–ray picture would argue for a nonradiative origin for most of the observed emission, and the presence of a widespread, intermediate velocity ($\sim$ 250 km s^{-1}) blast wave is inferred.

1 INTRODUCTION

According to radiative shock models (Hartigan, Raymond and Hartmann 1987), O VI ($2S_{1/2}$-$2P^{\circ}_{1/2,3/2}$) emission is copiously produced and becomes a dominant cooling mechanism for shocks with velocities $\gtrsim$150 km s^{-1}. Consequently, the presence of O VI line radiation in shocks of the Cygnus Loop has long been anticipated. We present a spectral line image from the first data to unambiguously identify the O VI emission doublet from the Cygnus Loop* (Rasmussen and Martin, 1990; 1992).

HIRES is a long slit spectrograph with double dispersion design (see Figure 1), featuring high spectral resolution and excellent airglow rejection and optimized for discovering and mapping in single rocket flights, faint emission lines from the diffuse ISM (Rasmussen, 1992). Our wide field of view and moderate angular resolution permitted mapping in low resolution (3′×18′), the O VI doublet ($\lambda\lambda$ 1032, 1038) to reveal the image of a limb brightened object (see Figure 3). The velocity of the emissive region was determined directly, by fitting expanding shell models to the line profiles. Figure 2 shows the spectra obtained from the Cygnus Loop.

---

* The recent observation made by the Hopkins Ultraviolet Telescope (*HUT*) aboard the Astro-1 platform also detected these emission lines for two pointings in the Cygnus Loop (Blair *et al.*, 1991).

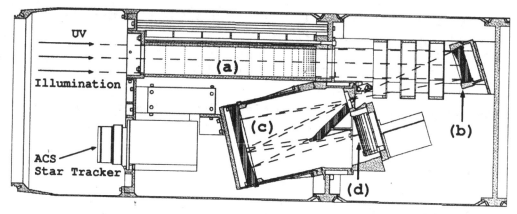

Fig. 1—Schematic diagram of the *HIRES* instrument. The principal components are labeled. (a)-Mechanical collimator, F.O.V. 8° (FWZI) × 3′ (FWHM). (b)-Parabolic objective grating, mechanically ruled with $1/d = 3000$ l/mm. (c)-Ebert-Fastie spectrometer configuration. The entrance slit width limits the bandpass into the spectrometer. The spherical mirror standing opposite the echelle collimates the incoming beam onto the grating, and again acts as a camera to focus the dispersed spectrum onto the detector. The echelle ($1/d = 300$ l/mm) is used in high order to provide 0.5Å resolution over the 12.7Å bandpass. (d)-MCP imaging detector with KBr photocathode and wedge-and-strip anode.

2 THE O VI PRODUCING SHOCK

The complimentary intensity and velocity measurements provide a unique opportunity to compare the predictions of radiative and nonradiative shock models, to interpret this observation. This approach is unique because O VI emission in a nonradiative shock is neither sensitive to poorly understood preshock conditions (since charge exchange is negligible), nor strongly dependent on the electron–ion equilibration rate in the postshock region (because O VI is produced after a large number of electronic collisions). In this approximation, the emissivity is dependent only on the postshock electron temperature and on the oxygen abundance in the ISM. In comparison, the radiative models (HRH) predict a larger emissivity, and with simple calculations, we predict the compressions and 'cooling times' for the postshock gas to become emissive in O VI. We find that the observed O VI emission could be produced with either a radiative shock ($v_s = 210 \pm 22\,\mathrm{km\,s^{-1}}$), or with a nonradiative mechanism with $v_s = 250 \pm 26\,\mathrm{km\,s^{-1}}$. The cooling time associated with the nonradiative shock is long ($5 - 7 \times 10^3$yr), and may not yet be cooling rapidly. Conversely, the cooling time for the radiative model ($\gtrsim 1.5 \times 10^3$yr) is short enough to develop a partially complete radiative shock. Neither mechanism is therefore ruled out. However, the radiative model predicts emission in localized, radiating filaments, with a coverage of at most 3%. In contrast, the nonradiative model requires a widespread ($\lesssim 30\%$) coverage. The similarity in appearance to the X–ray picture (Ku *et al.* 1984; Seward, 1990), together with the velocity information may provide the strongest clue as to the origin of this emission.

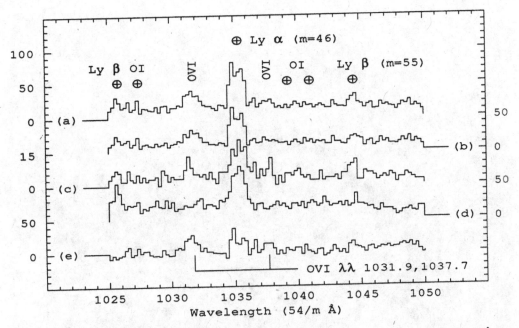

Fig. 2—Spectra from the Cygnus Loop. The spectra shown are oversampled (0.25Å binning *vs.* 0.5Å spectral resolution). (a)–'On-source' spectrum extracted from MCP events collected from the 3.0° diameter disk centered on $(\alpha, \delta)_{1950} = (20h50m, 31.0°)$. Positions for anticipated geocoronal emission lines are labelled. The dominant Ly α peak falls in a different order (m=46) between the O VI $\lambda\lambda$ 1031.9, 1037.7 peaks. (b)–Spectrum of the inner (2.6°) portion of the Cygnus Loop. (c)–Spectrum of the outer (2.6°-3.0°) annulus of the Cygnus Loop. Here, the O VI lines are significantly narrower than in (a) or (b). (This is consistent with the notion of an expanding shell.) (d)–'Off-source' spectrum taken from a region that excludes the 3.0° 'source' section. (e)–'Difference' spectrum between (a) and (d). Note the clear enhancement of the O VI peaks.

This suggests a large nonradiative contribution to the measured O VI luminosity ($L_{O\ VI} \sim 1.2 \times 10^{36}$ergs s^{-1}), and the presence of a widespread, intermediate velocity ($\sim 250\,\mathrm{km\,s^{-1}}$) adiabatic blast wave in the Cygnus Loop.

This work was supported by NASA grants NAG-5-642 and NGT-50198. A. R. was funded by a NASA Graduate Student Researcher Fellowship. This is contribution number 485 of the Columbia Astrophysics Laboratory.

REFERENCES

Blair, W. P. *et al.* 1991, *Ap. J.* (*Letters*), **379**, 33.
Hartigan, P., Raymond, J. and Hartmann, L. 1987, *Ap. J.*, **316**, 323.
Ku, W. H. M., Kahn, S. M., Pisarski, R. and Long, K. S. 1984 *Ap. J.*, **278**, 615.
Rasmussen, A. 1992, Ph.D. thesis, Columbia University.
Rasmussen, A. and Martin, C. 1990, *Bull. AAS*, **22**, 1272.
Rasmussen, A. and Martin, C. 1992 *Ap. J.* (*Letters*), submitted.
Seward, F. D. 1990, *Ap. J. Suppl.*, **73**, 781.

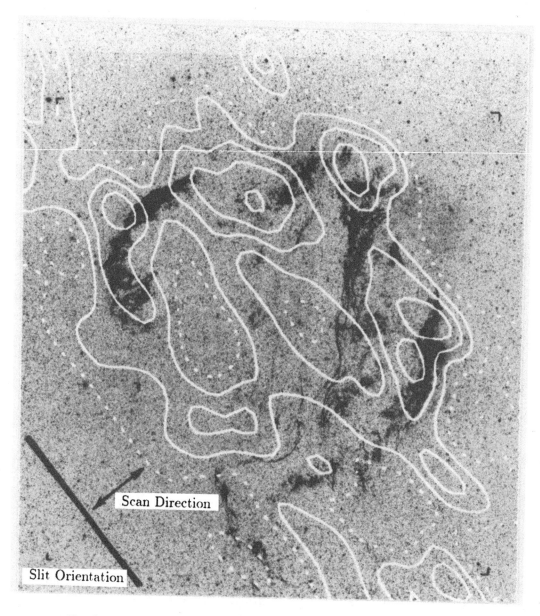

Fig. 3—The O VI line detection projected onto the sky plane. The spectral line contour map was prepared with a series of simple operations using *IRAF*. A source-photon map, consisting of ∼210 'signal' photons plus background events, was normalized with an exposure map. This was smoothed to the instrumental point spread function, and a background field was subtracted from it. Finally, a correction for instrumental vignetting was performed using an illumination function taken from the measured Lyα field. Additional smoothing was performed in the scan direction to produce comparable resolution in that direction. The O VI contour map overlaid over the *POSS* E plate. The contours represent steps in surface brightness: The dashed line is zero, and solid lines are increments in units of 2×10^{-6} ergs s^{-1} cm^{-2} sr^{-1}. The effective exposure of this map is slightly over 4 seconds.

The Cleanliness Control Program for SUMER/SOHO

Udo Schühle

Max-Planck-Institut für Aeronomie, W-3411 Katlenburg-Lindau, FRG

ABSTRACT

An overview of the cleanliness control activities is given for the solar EUV instrument SUMER/SOHO. Results of contamination effects studies, both theoretical and experimental, are summerized, and from these, cleanliness requirements for the instrument and the EUV mirrors in particular are derived. Specific design characteristics implemented to comply with these requirements, aiming at prolonged performance under irradiation by solar photon and particle flux, are highlighted. Monitoring the contamination of flight hardware and facilities make use of advanced techniques. For verification of mirror efficiencies and reflection characteristics of materials an EUV reflectometer has been built. Unique feature if this apparatus is a high photon flux light source in a UHV environment.

1 INTRODUCTION

SUMER is a high-resolution far-ultraviolet spectrometer which is designed to operate as part of the Solar and Heliospheric Observatory (SOHO) in the wavelength range of 50 to 160 nm. The joint ESA/NASA mission is to be launched in 1995 with a baselined observation period of 2 years. The instrument combines good image quality and high spectral resolution using the optical scheme depicted in Fig. 1 (see also Wilhelm et al. 1989). A single mirror telescope images the Sun onto the entrance slit of the spectrometer. Light from the slit is collimated and directed to a scan mirror and a spherical concave grating which produces a stigmatic image of the slit in the detector plane. Mirrors are made from SiC substrates with SiC/CVD surface cladding for best reflectivity at normal incidence in the far ultraviolet range. The two-dimensional imaging detector (Multi Anode Microchannel Array Detector) has a pixel size corresponding to 1 arcsec in spatial direction and 4 pm spectral resolution in first order (2 pm in second order) of the grating.

Many solar ultraviolet instruments developed during the last 15 years have suffered significant performance degradation after deployment in space due to contamination. Once an instrument is in orbit, it is not easily possible with current technology to reclean the optics or to take any measures to recover the instruments performance.

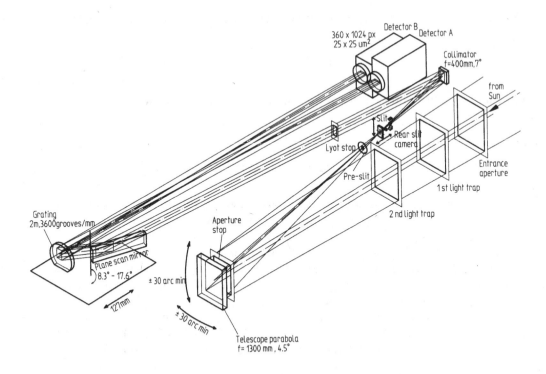

Fig. 1 – Schematic optical layout of the SUMER instrument.

The optical design of SUMER, using two normal-incidence mirrors, one grazing-incidence mirror and a concave grating, renders this instrument very sensitive to contamination of particulate or molecular nature. In addition, the combined effects of organic species and high flux of solar radiation is known to produce irreversible deposition of contaminants on sunlit surfaces by photochemical activation and subsequent polymerization (Steward 1989). The telescope mirror, therefore, is in general the most sensitive part of a solar EUV instrument, and the strong degradation of previous solar ultraviolet instruments can be attributed to this effect (Lemaire 1991).

Prevention and control of contamination is thus a major effort, which has to be carried through the entire program, yet different program phases require specific strategies: Most importantly, at the design phase preventive design considerations can still be implemented that avoid or at least reduce contamination that mainly occures at the last phase, when the instrument is delivered to the space environment. Here, cleanliness oriented operation can significantly reduce contamination by, e.g., closing an aperture door during inactive periods, thereby reducing the total time of UV irradiation while venting can still proceed.

The contamination control program carried out for SUMER included several studies to assess the effects on the optical performance imparted by contamination expected during exposure to solar EUV flux and solar wind particles, self-contamination by dust particles and outgassing organic condensables, as well as combined effects of the above. The results led to the implementation of preventive measures in the design of the instrument and stringent contamination control activities. Here we present the results of contamination studies, and we outline specific measures to prevent and control the contamination of the EUV instrument.

2 CONTAMINATION EFFECTS STUDIES

In a first, theoretical, study all possible sources of contaminants have been considered and their expected degrading effect on the instrument performance been estimated. According to these results cleanliness requirements were established. The components most sensitive to degradation by contaminants are the mirrors, and the expected degrading effects set the limit to the acceptable amount of contaminants. The most severe degrading effect on the instruments performance is change of reflectance of the mirrors by absorption, obscuration, and increase of scatter. Absorption and obscuration simply reduce the instruments throughput, while scattering degrades the point spread function and reduces the contrast. Both, sensitivity and image quality of the instrument can be acutely affected by various kinds of contamination. The following possible contaminants have been identified (Krueger 1989):

- dust particles
- organic condensables
- organic condensables and UV radiation
- solar wind
- organic condensables and solar wind

Spectral and angular resolution may be degraded by roughening of the optical surfaces by impact of particles from space environment, e.g., solar wind, micrometeoroids, but also by dust particles, increasing scattering from surfaces, thus degrading image quality and contrast. Sensitivity and contrast are also suffering from molecular contamination, especially condensable organic species, building a strongly absorbing overlayer on optical surfaces. This is more severe for normal incidence than for grazing incidence optics, as a contamination layer may still have good reflectivity at grazing angles. Gratings used near normal incidence, however, may be doubly effected as the contaminant layer may change the diffraction efficiency as well (Koide 1987).

The deposition of organic contaminants may be dramatically enhanced when the contaminated surface is exposed to ultraviolet radiation of solar intensity by photochemical reactions leading to polymerization of deposited material. This is expected to be the prime degradation process of space optical UV instruments. In addition, the irradiation by solar wind particles, i.e. protons and alpha particles, may, although at a much smaller rate, contribute to this polymerization process (Gillette 1971; Shimizu 1979). However, the radiation damage, physical alteration of the surface profile, roughening of the surface profile, is probably more deteriorating.

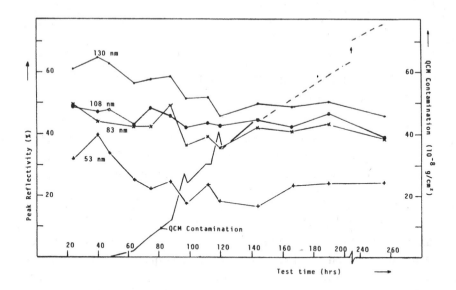

Fig. 2 – Peak reflectivity versus exposure time at the four wavelengths 53 nm, 83 nm, 108 nm, 130 nm. This is related to the mass accumulation on the QCM.

Experimental studies have been performed simultaniously in an attempt to quantitatively confirm theoretical predictions. Samples of SiC/CVD mirrors were exposed under vacuum to organic contaminants while they were irradiated by far ultraviolet light, the intensity being close to solar intensity at 160 nm. Intermittent measurements of reflectivity at different wavelengths in the EUV have shown only weak, but steady reduction of mirror efficiency correlating with the amount of collected contaminants (see Fig. 2), yet the amount of reduction of reflectivity was less than predicted. A microscopic investigation of the contaminated mirror surface was undertaken which showed that contaminants do polymerize under the UV exposure while agglomerating to island like spots (see Fig. 3). Thus, reflectance loss was not as high as expected, because no homogenious,absorbing film has been built, but increase of scattering must be considered.

Other samples of SiC/CVD mirrors have been exposed to an ion beam source run with H_2^+ and He^+ in order to simulate the effect of solar wind proton and alpha particles. The irradiated dose was varied to represent a total exposure of a 1 to 4 years mission time. Specifically, the H_2^+ dose was chosen to be between 2 and 8×10^{16} cm^{-2} @ 1 keV, the He^+ dose was between 0.5 and 2×10^{15} cm^{-2} @ 4 keV. Reflectivity was measured after each exposure. Inspection of the samples showed that the exposure has, even at the lowest dose, visibly changed the reflection properties. The EUV-reflectivity was found to be reduced by 20 % to 50 % proportional to the dose of primary particles at every wavelength measured (Dornier 1989).

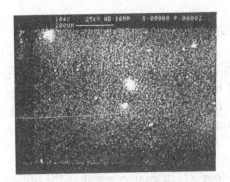 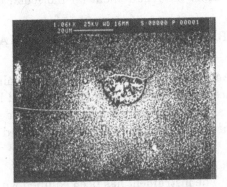

Fig. 3 – Microscopic photograph of the contaminated surface: The contaminants polymerize with UV light in islands (rather than a homogenious film).

3 CLEANLINESS REQUIREMENTS

The development of cleanliness requirements must be accomplished, using different methods for particulate and molecular contamination, by assessment of the degrading effects. The least tolerable degradation is dictated by the scientific objectives, and from these the requirements must be inferred.

Observations very close above the solar limb require a very low scatter telescope mirror to suppress the much brighter solar disk, while high image quality requires a very narrow piont spread function. For the SUMER telescope mirror we require scatter properties such that the diameter of 80 % encircled energy is less than 1 arcsec, and no more than 2×10^{-5} of the solar brightness is scattered outside a 2 arcmin radius. Using the particle size distribution of MIL-STD-1246B the angular distribution of scattered light from dust particles can be calculated (Ray 1990). For the SUMER telescope and spectrometer compartments we derive required particle cleanliness of Level 200 and Level 300, respectively, at the end of the integration phase, assuming that thereafter no more particles can be introduced to the optical housing (Schühle 1990).

Molecular contamination on mirrors have been assumed to be highly absorbing in the EUV, whether they were UV polymerized or not. In fact, it has been shown for some substances commonly used, that the produced contamination film becomes opaque already at a thickness of 15 nm (Muscari 1981). For SUMER, a 15 % loss of reflectivity at each of the four mirrors at end-of-life is the maximum degradation that could be accepted. Accordingly, this only allowes a contaminant layer of 1.0 nm thickness, or as much as 0.1 μg/cm^2 (of material with density 1 g/cm^3). This is the end-of-life requirement for the optical housing of the instrument. While this requirement is very hard to accomplish, it is clear that before launch the contamination level must be considerably lower since outgassing is an ongoing, though declining, process. Evidently, without special measures that prevent molecular outgassing and deposition the requirement for molecular cleanliness cannot be fulfilled.

4 PREVENTION OF CONTAMINATION

According to the results of the contamination studies some significant measures to reduce contamination below the required level have been implemented, most important among others are design changes and selection of proper materials. Some design features have especially been implemented aiming at the reduction of contamination:

- Aperture Door
  The instrument has been equiped with a door which guarantees that the optical compartment is always closed except when EUV measurements are undertaken. The spring loaded door serves also as an outlet for purge gas which is continuously supplied during ground operations and storage. After launch the door is partially opened for venting and to facilitate the exhaust of outgassing material without delivering any UV light onto the mirrors. This period of time shall last a few months. During this phase, energy is transmitted through a UV blind window which is implemented inside the aperture door. This way a temperature of the primary mirror of 45° C is achieved.

- Passive Heating of Primary Mirror
  The primary mirror shall always be at the highest temperature level in order to prevent condensation of molecular contaminants and to make residence times of contaminants small and desorption rates high. A passive heating solution was chosen to accomplish this, making use of the solar radiation input in the visible and IR. The mirror will be thermally isolated from the structure and will reach 75° C. When the aperture door is closed, a UV-black window in it will result in a mirror temperature of 45° C.

- Solar Wind Deflector
  Deflection plates inside the optical entrance baffle, held at a potential of -2 kV, will prevent protons and alpha particles with energies up to 5 keV and 10 keV, respectively, from impacting the telescope mirror.

- Detector Cover
  To protect the detector photocathode coating from moisture and the microchannel plates from condensible hydrocarbons one detector tube is hermetically sealed and held under vacuum by an ion pump. The tube can be opened by an actuator mechanism that opens the ultra high vacuum valve.

- External Electrical Components
  Only the front face of the detector is connected to the optical housing. The

body of the detector head assembly, containing all electronic parts, is isolated from the optics via a seal around the rim of the detector front plate.

- **Clean Optical Compartment**
  Organic material has been minimized restrictively inside the optical cavity and been placed outside if possible. All electrical boards and components are kept outside, and an exeption has only been made for the motors and encoders of the various mechanisms. Electrical supply runs via feedthroughs to motors and encoders.

- **Dry Lubrication of Mechanisms**
  Mechanisms use lubrication-free devices like, e.g., flexural pivots or ceramic ball bearings, or make use of dry lubrication on MoS2 basis.

- **Ultra High Vacuum Motors**
  Stepper motors have been specially designed for ultra high vacuum use and extensively conditioned and verified for low outgassing before integration.

- **Material Selection**
  Organic material has been avoided whenever possible, otherwise high temperature materials or components have been chosen. The selected components, and all organic materials, inside the optical housing have been tested for outgassing using a GC/MS technique. Conditioning of the item under test consisted of oven baking under constant clean gas purging and was continued until the outgassing level was eventually found to be acceptable or the item was rejected. The conditioning procedure performed with the test specimen is then applicable to all components of this kind before integration.

- **Cleaning Procedures**
  Precision cleaning is mandatory for every piece of flight hardware. Ultra clean water is used for cleaning with different kind of detergent solutions, depending on the material, and ultra sonic bath is applied when possible.

- **Assembly and Test Provisions**
  After the cleaning procedure, piece parts are kept in a clean, laminar air flow area and assembly is taking place in a Class 100 cleanroom. The cleanroom is equiped with active charcoal filters inside the upper plenum which keep the level of organic contaminants in the air at a minimum (verified regularly by gas chromatography). Vacuum systems for conditioning, tests and EUV calibration are oil-free ultra high vacuum systems. They are integrated into the cleanroom facility such that ports can be opened for loading and unloading from the cleanroom area. Fig. 4 shows the floor plan of the SUMER Cleanroom and Test- and Calibration Facility.

- Venting and Purging

  Venting of vacuum systems and purging of the instrument is performed with clean, filtered nitrogen supplied by a central distribution system with triple filters at the point of use. After assembly, alignment and calibration the instrument will be continously purged until launch.

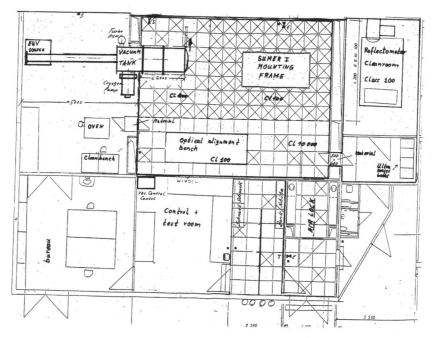

Fig. 4 – Floorplan of the SUMER cleanroom, test and calibration facility.

5 VERIFICATION OF CLEANLINESS

The cleanroom facility is being monitored continuously by particle counters and witnes plate fallout monitors. Vacuum systems, although with oil-free pumps, are known to be a source of contamination because of easier distribution of species outgassing from the internal equipment, use mass analysers and quartz cristal microbalances to monitor molecular constituents of residual gas and their condensation.

For verification of reflectivity of optical mirrors at EUV wavelengths a reflectometer has been constructed with an oil-free pumping system and commercial light source, monochromator, and sample manipulator. It will allow measurements of the bidirectional reflection distribution function at any angle of incidence as well as diffraction grating efficiencies throughout the SUMER wavelength range. It consists of a 700 mm diameter UHV vessel that can be loaded through a front door which opens into a Class 100 area. A photograph of the reflectometer is shown in Fig. 5.

The light source is a capillary discharge lamp (UPS lamp) and the monochromator is a 0.2-m concave grating with adjustable slits, both systems are UHV design. Operation of the lamp with different noble gases provides a number of roughly 30 lines distributed inside the wavelength range of SUMER with a resolution of 0.1 nm. The intensity of the system is surprisingly good: At a slit width of 0.07 mm a count rate in the Megahertz range is achieved at the principal lines. Samples of mirrors or other material to be investigated can be rotated and translated by a manipulator with 5° of freedom and 200 mm vertical motion capability.

Fig. 5 – The EUV reflectometer system.

6 Acknowledgement

This work has been financially supported by the German BMFT/DARA and the MPG.

7 REFERENCES

Dornier GmbH 1989, Contamination Study for SUMER, Final Report SUM-MPAE-SS-020000-00-00.

Gillette, R. B., Kenyon, B. A. 1971, *Appl. Opt.*, **10**, 545.

Koide, T., et al. 1987, *Appl. Opt.*, **26**, 3884.

Krueger, F. 1989, Cleanliness Theoretical and semiempirical study for SUMER on SOHO, Final Report SUM-MPAE-SS-020001-00-00.

Lemaire, P. 1991, *ESA Journal*, **15**, 237.

Muscari, J. A. 1981, Absorption Spectra of Typical Space Materials used in the Vacuum Ultraviolet, Proc SPIE Technical Symposium East '81., Washington D.C.

Ray, D. C., Welch, B. Y., Malina, R. F., Battel, S. J. 1990, Contamination Management for EUV Space Optics, Microcontamination 90 Conference, Publ. No. 439.

Schühle, U., 1990, Particle Contamination Specification for SUMER, SUM-MPAE-SL-020001-00-00.

Shimizu, K., Kawakatsu, H. 1979, in *Surface Contamination*, ed. K. L. Mittal, (New York/London: Plenum Press), p. 113.

Steward, T. B., Arnolg, G. S., Hall, F., Marten, H. D. 1989, Photolysis of Spacecraft Contaminants, Report SD-TR-89-45, The Aerospace Corporation, El Segundo, CA.

Wilhelm, K., et al. 1989, in *The SOHO Mission – Scientific and Technical Aspects of the Instruments*, ESA SP-1104, p. 31.

Planar Delay Line Readouts for High Resolution Astronomical EUV/UV Spectroscopy

Oswald.H.W. Siegmund, J. Stock, R. Raffanti, D. Marsh and M. Lampton

Experimental Astrophysics Group, Space Sciences Laboratory,
University of California, Berkeley, CA 94720 USA

ABSTRACT

We have developed two dimensional planar delay line image readout schemes for microchannel plate detectors applied to high resolution astronomical EUV/UV spectroscopy. Various anode formats from 95mm x 60mm, to 65mm x 15mm have been fabricated and evaluated. Imaging tests with spot images and bar patterns show that spatial resolutions of ≈15µm FWHM in X and ≈30µm FWHM in Y have been achieved. The global image non-linearities are of the order of a resolution element, and the images are stable to a few microns under various operating conditions. Event rates of >10^5 sec^{-1} have been attained, and the flat field response is stable and correctable to the limit of detection statistics.

1. DELAY LINE DETECTOR

Recent investigations (Siegmund et al, 1991a,b. Williams et al, 1989. Keller et al, 1987) have shown delay line image readout techniques to be useful in photon counting, imaging MCP detectors. In this paper we present results obtained with double delay line (DDL) anode structures. The DDL anode type (Fig. 1) has two sets of interleaved wedge shaped electrodes that are a half period out of phase, defining the active area.

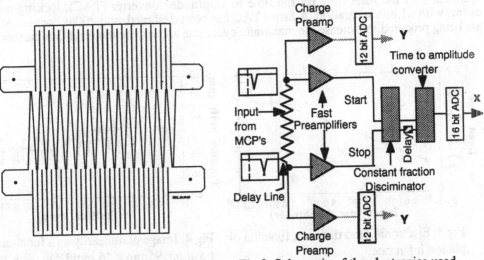

Fig 1. Schematic of a double delay line anode.

Fig 2. Schematic of the electronics used for photon position encoding

Two serpentine delay lines are connected to the bases of the wedges on each side. The charge from the MCP is divided between the wedges only, and propagates to the anode ends along the delay lines. Determination of the Y and X photon event centroid coordinates are by charge division ratios between the opposing wedge sets, and by difference of signal arrival times at the two ends of the anode respectively. A set of three MCP's (12.5μm diameter channels, 13° bias, on 15μm centers, channel length to diameter ratio of 80:1) in a back to back Z configuration were used to detect and amplify incoming radiation. The delay line anodes were mounted ~7mm behind the MCP stack, with an accelerating potential applied between the anode and the back MCP.

Table 1. Delay Line Anode Specifications

Anode	Anode type	R (Ω)	Delay (ns)	Periods	X FWHM	Y FWHM
DDL012	65x15mm DDL	15Ω	32 ns	95	15μm	30μm
DDL016	95x25mm DDL	16Ω	38 ns	115	25μm	60μm

Delay line schemes of different sizes and formats have been evaluated (Table 1, Siegmund et al, 1991b). Anodes are etched on 6010 RT/Duroid substrates (ceramic doped PTFE, ε ≈10.5, loss coefficient 0.0023 @ 10 GHz) coated with 36μm Cu to optimize the anode delay and minimize the signal attenuation and dispersion. The 95mm x 25mm DDL format is currently being used in the ORFEUS-ASTROSPAS ultraviolet spectrometer detectors, and the 65mm x 15mm format is being investigated for the Far Ultraviolet Spectroscopy Explorer (FUSE) mission.

The electronic scheme used to encode the single photon event centroid positions for both the anodes is shown in Fig. 2. Digitized position data is shipped to a Sun Sparcstation via an IEEE 488 link for data storage and image analysis. The electronic system previously used (Siegmund et al, 1991b) was limited to photon event rates of a few x 10^4 sec^{-1} due to the the speed (8μs convert) of the analog to digital converter (CS5101). A new electronic system, that employs 1μs convert ADC's has been built and tested. Results show that the new system accommodates global photon event rates in excess of 10^5 sec^{-1} (Fig.3). The non-paralysible dead time curve corresponds to an effective dead time of ≈1.8μs, which is dominated by the reset time for the time to amplitude converter (TAC), locking out all events with ≤1.8μs separation. A faster TAC has been designed, and 200ns convert ADC's are being procured to increase the maximum event rate and reduce the dead time further.

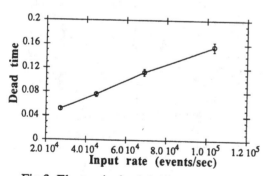

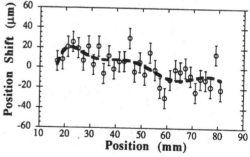

Fig 3. Electronic dead time as a function of photon input rate.

Fig 4. Image nonlinearity as a function of position. 95mm x 15 mm DDL, 2 x 10^7 gain, 2mm pinhole test array.

2. IMAGING PERFORMANCE

The performance of the delay line anodes was evaluated with a mercury (2537Å) lamp, and either a mask containing small (≈10μm) pinholes (2 x 2mm period array), or an Ealing cassegrain microscope objective (Model # 25-0506) with a set of slit, or spot, masks.

The linearity of the delay line anode readouts was measured from the centroid positions of the images of the mask pinholes and comparison to the expected mask array positions. This shows (Fig.4) that the maximum deviations are less than 20µm ($\approx$1 resolution element) and that the distortion is a slow function of position. The mask pinhole positions are also only accurate to $\approx$10µm! The stability of the delay line imaging was also investigated (vs MCP gain, event rate, MCP/anode potential). The pinhole image centroids were observed to remain stable to better than 5µm from <1 event pore^{-1} s^{-1} to 10 event pore^{-1} s^{-1}. Variations in the gain and drift potential in normal operation give $\approx$1µm shifts and are thus negligible.

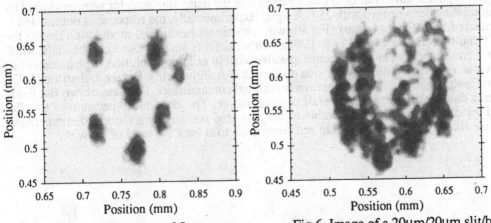

Fig 5. Image of an array of 5µm spots 65mm x 15mm DDL. 2 x 10^7. 2537Å.

Fig 6. Image of a 20µm/20µm slit/bar pattern. 65mm x 15mm DDL, 2 x 10^7.

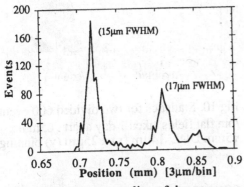

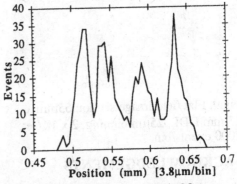

Fig 7. X histogram slice of the top row of 5µm spot images shown in Fig.5.

Fig 8. X histogram slice of the 20µm /20µm slit/bar pattern shown in Fig.6. 25µm high Y slice.

The resolution of DDL anodes is determined (Siegmund et al, 1991a,b) by the event timing error for the X coordinate, and the charge amplifier noise and partition noise for the Y coordinate. Timing error is determined by a number of factors, including signal attenuation and dispersion, preamplifier noise ($\approx$3mV FWHM), discriminator walk (<5ps) and jitter (<5ps FWHM), TAC noise ($\approx$5ps FWHM), and ADC errors (16 bits ±0.5 LSB). The predicted limiting electronic resolution is then $\approx$13µm FWHM.

Images of a pattern of 5μm spots (Fig.5) and of slit patterns (Fig.6) allow the determination of the point spread function (PSF) and MTF. Pinhole image histograms (Fig.7) demonstrate that the raw X and Y PSF's are ≈15 μm FWHM, and 30μm FWHM, respectively. Slit pattern X histograms (Fig.8) indicate 65%-75% MTF (12μm - 15μm FWHM PSF) at 25 lp/mm, correlating well with the spot data. This is impressive since the MCP pores are 12.5μm on 15μm centers! Note that image 'quantization' due to the MCP pore size causes significant image distortions on the 15μm level.

3. FLAT FIELD RESPONSE

The high resolution flat field characteristics of the delay line detector were evaluated by accumulating deep images with 2537Å light. Experimentally, the response is not flat and is dominated by MCP variations (Fig 9), with prominent hexagonal modulation due to the MCP multifibers (Siegmund et al 1991a). Other defects such as blocked channels, MCP defect zones and enhanced response are also visible at high resolution. Enhancement of response in some multifibers is due to the 2537Å light being longer λ than the MCP photoelectric cutoff, and thus sensitive to residual contaminants. Division of two flat fields taken a day apart gives a flat overall residual image. The amplitude spectrum of this flat field residual (Fig.10) is comparable to the expected statistical variations, demonstrating that the fixed pattern noise is stable and correctable to at least the level of a few %.

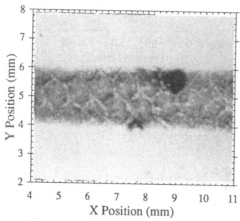

Fig.9. Flat field image section,65mm x 15 mm DDL, 25μm binning, 2 x 10^7 e⁻, 1000 events/bin.

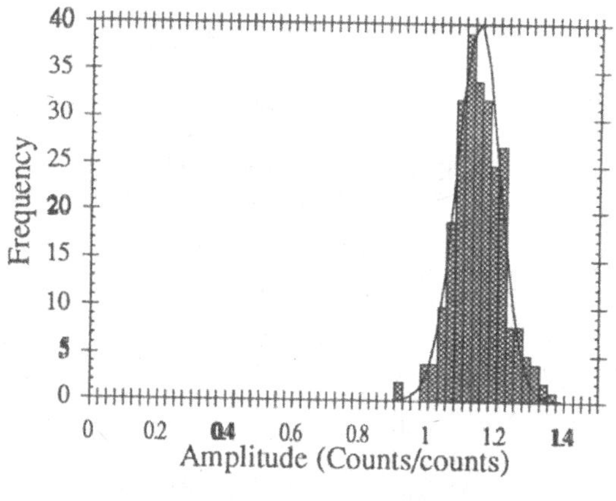

Fig 10. Statistics for two divided 600 event /bin flat fields taken 1 day apart . 65mm x 15mm DDL, 14μm (x) x 25μm (y) binning

4. ACKNOWLEDGEMENTS

This work was supported by NASA grants, NAGW-1290, NAGW-2032, NAGW-1879, and NGR-05-003-450.

5. REFERENCES

Keller, H., Klingelhofer, G., and Kankeleit, E., 1987, *Nucl. Instr. Meth.* **A258**, 221.
Siegmund, O.H.W., Lampton, M., Raffanti, R., and Herrick, W. 1991a, *Nucl. Instr. Meth.* **A310**, 311.
Siegmund, O.H.W., Raffanti, R., Stock, J., Herrick, W., and Lampton, M., 1991b, in *Photoelectronic Image Devices* , ed. B.L. Morgan (IOP), p.123.
Williams, M.B., and Sobottka, S.E., 1989, *IEEE Trans. Nucl. Sci.* **NS-36**. 227.

Performance Characteristics of High-Gain Curved-Channel Microchannel Plates

D. C. Slater[1], H. Kirby[1], M. Pertsova[1], J. G. Timothy[1], and B. N. Laprade[2]

[1]Center for Space Science and Astrophysics, Stanford University, USA
[2]Galileo Electro Optics Corporation, USA

ABSTRACT

We report on the development and performance characteristics of 25-mm-diameter format single high-gain curved-channel (C^2) microchannel plates (MCPs). C^2 MCPs are currently being developed for the Multi-Anode Microchannel Array (MAMA) detector systems for both the SUMER and UVCS payloads on board the Solar and Heliospheric Observatory (SOHO) satellite. Gain, dynamic range, dark noise and imaging performance characteristics are presented for C^2 MCPs fabricated using two different channel shearing techniques developed at Galileo Electro Optics Corporation: the conventional 'pin-shear'; and a new technique called 'spin-shear'. A brief description of each of these shearing techniques is also presented.

1 INTRODUCTION

The microchannel plate (MCP) is a thin semiconducting glass plate composed of $>10^6$ hollow channels capable of providing amplification of UV photons, ions, electrons, and soft x-rays. The many closely spaced channels within the MCP's active area allow it to be used in applications requiring high spatial resolution imaging. MCPs are commonly used as amplification stages in image intensifier devices and other related photon sensitive detectors such as the Multi-Anode Microchannel Array (MAMA) detector system (Timothy and Bybee 1977). MAMA detectors utilizing 25-mm diameter format high-gain curved-channel (C^2) MCPs are slated for use on board the ESA/NASA Solar and Heliospheric Observatory (SOHO) space program. This format MCP has a rectangular active area of 10 x 28 mm^2, and is comprised of 12.5-μm diameter channels on 15-μm centers with a length-to-diameter (L/D) ratio of 120:1. It will feed a proximity focused MAMA anode array composed of 25 x 25 μm^2 pixels in a (360 x 1024)-pixel rectangular format (Timothy 1991).

C^2 MCPs (Timothy 1981) provide moderately high gain saturation levels of 10^5 - 10^6 electrons pulse^{-1}, narrow output pulse-height distributions, excellent ion-feedback suppression and smaller sized output charge clouds compared to the output of "chevron" or "Z-plate" stacks. Because the C^2 MCP is inherently a single plate, it requires only a single high-voltage power supply to operate. These advantages make the C^2 MCP the optimal plate for use in the MAMA detector system. In addition, because each C^2 MCP channel is electrically isolated from its neighbors, allows for a much higher achievable dynamic range for a given plate resistance than that of either the "chevron" or "Z-plate" stacks.

2 C² MCP FABRICATION

In order to achieve the necessary channel curvature to properly suppress ion-feedback in C² MCPs requires a processing step in the manufacture called "shearing". Two shearing techniques have been developed at Galileo Electro Optics Corporation that have produced MCPs with good channel curvature and output uniformity. These two techniques are known as the 'pin-shear' and 'spin-shear' techniques. With the 'pin-shear' technique the plate is placed in the shearing machine and heated from below. When the glass reaches the proper viscosity at high temperature a linear force is applied to the top surface which causes the channels to bend outwards in the direction of the force. With the 'spin-shear' technique the plate is placed on a rotating disk and spun while at elevated temperature. The centrifugal force causes the channels to bend outwards at their centers, with the top and bottom edges held firm by the rigid glass.

3 PERFORMANCE CHARACTERISTICS

3.1 Output Gain and Dark Count Rate

Figure 1 shows the pulse-height distribution (PHD) at an applied MCP voltage of 2200 V of a representative SOHO configuration C² MCP. This data was acquired using 2537 Å photons from a Hg "penray" lamp. The moderately high modal gain (gain at the

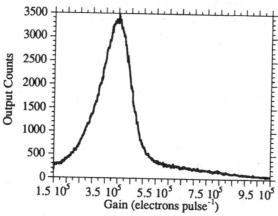

Figure 1. The pulse-height distribution of a representative SOHO format C² MCP at an applied MCP voltage of 2200 V. The modal gain is 4 x 10^5 electrons pulse^{-1}, and the PHD resolution is 35%.

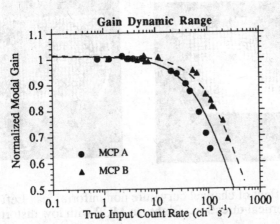

Figure 2. Modal gain as a function of the true input count rate per MCP channel measured with two representative SOHO format C^2 MCPs. The curve through the data points is a least squares linear fit of the data to the theoretical response.

peak of the distribution) of 4×10^5 electrons pulse^{-1}, narrow shape (FWHM/modal gain = 35%) and low ion-feedback tail on the high-gain side of the peak are characteristic of well-sheared C^2 plates.

The dark count rate for SOHO format MCPs that have been properly baked and conditioned is generally < 10 cm^{-2} s^{-1} at full operating voltage. The dark PHD shows the characteristic negative exponential shape as expected with darks generated primarily by radioactive beta decay of certain elements in the glass composition (Fraser, Pearson and Lees 1987).

3.2 Dynamic Range

Figure 2 shows measured modal gain values as a function of the input count rate per MCP channel measured with two representative SOHO MCPs. As the input count rate increases, the output gain drops as a result of channel saturation. The time it takes to replenish charge to depleted channels is proportional to the product of the channel resistance and capacitance (Wiza 1979). SOHO format MCPs manufactured by Galileo Electro Optics Corporation have channel resistances that range between $3.3 - 5.3 \times 10^{13}$ Ω. These plates have demonstrated gain drops of 10% at input count rates of >40 counts channel^{-1} s^{-1}. When used in MAMA imaging detectors, the detection efficiency has been measured to drop 10% at count rates exceeding 150 counts pixel^{-1} s^{-1}.

3.3 Imaging Performance/Spatial Linearity

One of the major difficulties in constructing C^2 MCPs is in achieving a highly uniform shear across the entire active area. A highly uniform shear is required in order to

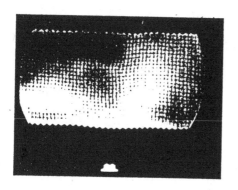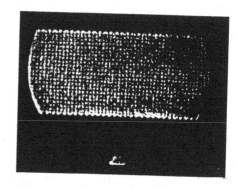

Figure 3. Imaging effect of channel curvature nonuniformities. Left–high distortion in the vertical direction. Right–an MCP with low distortion.

maintain uniform output gain for photometric stability and to minimize geometric distortion in the images produced by the detector.

Figure 3 shows the effects of channel curvature uniformity in the imaging performance of C^2 MCPs. These images were produced by stimulating two operating 'pin-sheared' MCPs with 2537 Å photons with a grid pattern laid on the front surface of both plates. The left image shows the effects of severe geometric distortion in the vertical direction (direction of shear) due to nonuniformities in channel curvature. The right image shows a plate with much less distortion.

Low geometric distortion is highly desirable to maintain high spatial output linearity. Measurements made with SOHO format C^2 MCPs have demonstrated output linearity with deviations of only ±8.8 μm (1σ) across the entire horizontal axis (25.6 mm) of the MCP.

4 CONCLUSIONS

High-gain C^2 MCPs in 25-mm diameter format configurations have demonstrated moderately high modal gains, low PHD resolutions, low dark noise, excellent spatial output uniformity and high spatial linearity.

5 REFERENCES

Fraser, G. W., Pearson, J. F., and Lees, J. E. 1987, *Nucl. Instr. & Meths. in Phy. Res.*, **A254**, 447.

Timothy, J. G. 1991, *SPIE Proceedings*, **1549**, 221.

Timothy, J. G. 1981, *Rev. Sci. Instrum.*, **52**, 1131.

Timothy, J. G., and Bybee, R. L. 1977, *Solid State Imaging Devices*, **116**, 24.

Wiza, J. L. 1979, *Nucl. Instrum. & Meth.*, **162**, 587.

Calibrated Solar EUV Spectrum from SERTS

Roger J. Thomas[1], Werner M. Neupert[1], and William T. Thompson[2]

[1]Laboratory for Astronomy and Solar Physics, Code 680,
  NASA–Goddard Space Flight Center, Greenbelt MD 20771
[2]Applied Research Corporation, Landover MD 20785

ABSTRACT

The Solar EUV Rocket Telescope and Spectrograph (SERTS) provides spatially imaged high-resolution spectra at wavelengths of 235 – 450 Å, including many emission lines formed over the temperature range of $4.7 \leq \log T \leq 7.5$ characteristic of the corona and upper transition region. The instrument utilizes a grazing-incidence Wolter-2 telescope feeding a normal-incidence toroidal grating spectrograph which is quasi-stigmatic throughout its broad bandpass. Photometric calibrations have now been performed for each of the optical components carried on the 1989 and 1991 flights of this experiment. Calculations have also been carried out to determine the residual atmospheric extinction in the EUV at rocket altitudes during each flight, resulting in relative intensity measurements good to $\pm 25\%$ over the full spectral range. An absolute photometric scale accurate to within a factor of 2 was then derived by fitting our solar observations to reported values for the average He II quiet sun flux at 304 Å. In addition, an absolute wavelength scale for the spectrograph was established by fitting postflight laboratory measurements to precisely known EUV standards. Solar wavelengths determined in this way should be accurate to better than 10 mÅ for most lines. A catalog is being developed from these calibrated observations that will list absolute wavelength, intensity, and width for all lines detected in the averaged EUV spectrum of a solar active region. Where known, atomic transitions and formation temperatures will also be given.

1 INTRODUCTION

The SERTS solar rocket experiment combines good spatial and spectral resolutions ($\approx$ 5 $arcsec$ and 50 mÅ) over a wavelength range of 235 – 450 Å and a field of view nearly 5 $arcmin$ long, while simultaneously providing EUV spectroheliograms over an even larger FOV (Neupert $et\ al.$ 1992). The spectrograph has a dispersion of 2.2 Å mm^{-1} and plate scale of 11.3 $\mu m\ arcsec^{-1}$. Its flight on 5 May 1989 produced observations of a large solar active region (NOAA AR5464 at S18W49). The 7 May 1992 flight obtained spectra of primarily quiet areas, extending across the solar limb, and featured the first use of a multilayer-coated toroidal grating (Thomas $et\ al.$ 1991, Keski-Kuha $et\ al.$ 1991). Before and after each flight, the spectrograph recorded laboratory spectra of He II, Ne II, and Ne III in order to verify focus and alignment, and to establish instrumental parameters. At present, special EUV-sensitive photographic film (Kodak 101-07) is used to record the near-stigmatic spectral images, which are then digitized with a Perkin-Elmer PDS Scanning Microdensitometer. The film's D − log E curve is derived independently for each flight or laboratory run by comparing EUV images of the same source region made with different exposure times.

For the initial spectral catalog, we selected the longest available exposure (246 s), a solar active region spectrum taken on the 1989 flight, and spatially averaged it along the entire slit length (first correcting for a slight tilt of the slit relative to the dispersion plane). The instrument's full spectral resolution is preserved by using 5-μm pixels in wavelength, corresponding to 11 $m\text{Å}$. After background subtraction, a wavelength varying noise level was estimated from fluctuations in the remaining baseline over several spectral regions apparently devoid of emission lines. Any feature that exceeded this noise level by at least a factor of three was fit by either a single or double Gaussian profile, with each Gaussian providing a measured position, amplitude, and width. A few such features were rejected as plate flaws since their fitted profiles are narrower than the minimum possible instrumental width. The rest form the basis of the EUV spectral catalog described below.

2 WAVELENGTH SCALE

The wavelength scale is derived from laboratory EUV spectra of a hollow-cathode gas-discharge lamp, recorded with the SERTS instrument shortly after its flight and reduced by procedures similar to those outlined in the previous section. Positions are determined by Gaussian fits for eleven Ne II lines and for the He II line at 304 Å, all of which have absolute wavelengths that are known to better than 0.3 $m\text{Å}$ (Kaufman and Edlen 1974). The derived scale is given by a third-order polynomial (in pixel position relative to the He II centroid) fitted to the known laboratory wavelengths. Residuals from this fit are ≤ 0.5 $m\text{Å}$ for all twelve lines, clearly demonstrating the intrinsic accuracy of the SERTS spectrograph. The laboratory wavelength scale is then used for flight data by referencing all measured solar line positions to that of the solar He II 304 Å line.

One complication arises from the fact that the 304 Å line is saturated in the 1989 long-exposure flight frame over most of the slit length, and so we could use only a small part of the slit at one end with the lowest intensity in order to set the wavelength scale reference position. Thus, the accuracy of the absolute scale depends on the assumption that the wavelength of the laboratory He II line is the same as that of the solar line from a location on the edge of the observed solar active region. Wavelength shifts due to line-of-sight velocity differences in the rocket's motion during an exposure or due to solar rotation are both less than 2 $m\text{Å}$ and are neglected here. The derived wavelength scale is obviously most accurate over the range that was fitted to the measured laboratory standards, namely 304 − 448 Å. It becomes less certain as it is extrapolated toward the spectrograph's short wavelength limit of 235 Å. Considering all sources of error, the EUV wavelengths determined from the averaged solar active region spectrum should be accurate to better than 10 $m\text{Å}$ for most lines in the catalog.

3 PHOTOMETRIC CALIBRATION

Direct photometric calibrations in the EUV have been carried out for each of the SERTS optical components, including the Wolter-2 grazing-incidence telescope, aluminum blocking filter, and normal-incidence toroidal grating. The telescope reflectance was measured at 304 Å, and its gradual variation with wavelength was modeled by theoretical calculation. The absolute efficiencies of the filter and grating were both measured over the full wavelength range at the Synchrotron Ultraviolet Radiation Facility (SURF-2) of the National Institute of Standards and Technology. Combined with the geometric collecting area of the telescope, these efficiencies give the instrument's effective area as a function of wavelength, which was found to range from 0.01 cm^2 at 250 Å to about 0.1 cm^2 beyond 380 Å.

In flight, the response of the instrument is also affected by EUV extinction along the line of sight due to the Earth's atmosphere at rocket altitudes. Calculations of this effect averaged over each film exposure were made from the detailed MSIS-86 thermospheric model of Hedin (1987), using laboratory cross sections for O (Angel and Samson 1988), N_2 (Samson et al. 1987), and O_2 (Samson et al. 1982). We find that atmospheric extinction typically varies by around $\pm$ 8% of its mean value over the SERTS spectral range for any given exposure.

Wavelength differences in the photometric response of the flight film were examined by using the above results to compare predicted intensity ratios of line pairs that are known to be density-*insensitive* with those measured by SERTS, a technique described by Neupert and Kastner (1983). This comparison shows that the photon response of our film does not vary significantly with wavelength over the observed range (*i.e.*, a given number of incident photons will cause the same film darkening at all wavelengths). The comparison also provides a check on the overall relative calibration of the instrument, indicating an accuracy within $\pm$25% from 250 to 450 Å, and better than that over shorter spectral intervals.

Finally, the relative calibration obtained in this way was put onto an absolute scale by forcing our observations of the quiet sun He II emission at 304 Å to agree with an average of reported values for that flux available in the literature, namely 6730 ± 650 erg $cm^{-2}s^{-1}sr^{-1} = 2420 \pm 235$ ph $cm^{-2}s^{-1}arcsec^{-2}$. As a check, we then used the resulting absolute calibration to determine a value for the quiet sun He II emission at 256 Å as observed by SERTS, which turned out to be within 4% of the average of other reports. An inversion of this scale can also be used to derive the absolute response of our flight film, with the result that a film density of 1.0 corresponds to an illumination rate of 1.3×10^9 ph cm^{-2} at any EUV wavelength in the SERTS range. Based on this calibration, the initial SERTS spectral catalog has a 2σ-sensitivity level of 14 ph $cm^{-2}s^{-1}arcsec^{-2}$ at 304 Å, improving to around 3 ph $cm^{-2}s^{-1}arcsec^{-2}$ above 380 Å. We estimate that absolute intensities determined for most of the lines in the catalog should be good to within a factor of 2.

4 EUV SPECTRAL CATALOG

There are 174 emission lines that meet our selection criteria in the averaged SERTS spectrum of a solar active region. Of these, some 156 have at least preliminary identifications based on comparisons with previous observations (Behring *et al.* 1976, Dere 1978, Kelly 1987). The lowest temperature lines are of He II (formed near log T = 4.7), C IV (5.0), and Ne III (5.1). Then there is complete temperature coverage between Ne V (at log T = 5.5) and Ca XVIII (6.8). Four elements have sequences of at least 4 consecutive stages of ionization: Mg V–IX, Si VII–XI, S XI–XIV, and Fe IX–XVII. There are six isoelectronic sequences represented by at least 4 members: Li-like, Be-like, B-like, C-like, Na-like, and Mg-like. And there are nine ions which each have more than 6 lines in the catalog: Mg VII (9 lines), Mg VIII (7), Si IX (7), Si X (8), Fe XI (7), Fe XII (9), Fe XIII (13), Fe XIV (12), and Fe XV (7). Thus, this calibrated EUV spectrum will be ideal for numerous detailed studies of solar plasma characteristics, such as differential emission measure, spectroscopic density or temperature diagnostics, line formation mechanisms, and abundance variations.

We are presently investigating the proposed identification of each line by comparing its relative intensity against predictions. A final version of the catalog is being prepared for publication. It will list absolute wavelength, intensity, and width, along with all fitting uncertainties. Where known, atomic transitions and formation temperatures will also be given.

This work has been supported under RTOP grant 879-11-38 from the Solar Physics Office of NASA's Space Physics Division.

REFERENCES

Angel, G. C., and Samson, J. A. R. 1988, *Phys. Rev. A*, **38**, 5578.

Behring, W. E., Cohen, L., Feldman, U., and Doschek, G. 1976, *Ap. J.*, **203**, 521.

Dere, K. P. 1978, *Ap. J.*, **221**, 1062.

Hedin, A. E. 1987, *J. Geophys. Res.*, **92**, 4649.

Kaufman, V., and Edlen, B. 1974, *J. Physical and Chemical Reference Data*, **3**, 825.

Kelly, R. L. 1987, *J. Physical and Chemical Reference Data*, **16**, Suppl. 1, 1371.

Keski-Kuha, R. A. M., Thomas, R. J., and Davila, J. M. 1991, *SPIE*, **1546**, 614.

Neupert, W. M., and Kastner, S. O. 1983, *Astron. Astrophys.*, **128**, 181.

Neupert, W. M., Epstein, G. L., Thomas, R. J., and Thompson, W. T. 1992, *Solar Phys.*, **137**, 87.

Samson, J. A. R., Rayborn, G. H., and Pareek, P. N. 1982, *J. Chem. Phys.*, **76**(1), 393.

Samson, J. A. R., Masuoka, T., Pareek, P. N., and Angel, G. C. 1987, *J. Chem. Phys.*, **86**(11), 6128.

Thomas, R. J., Keski-Kuha, R. A. M., Neupert, W. M., Condor, C. E., and Gum, J. S. 1991, *Applied Optics*, **30**, 2245.

Imaging Detector Systems for use at Ultraviolet and Soft X-ray Wavelengths

J. Gethyn Timothy
Center for Space Science and Astrophysics
Stanford University
Stanford, CA 94035-4055

ABSTRACT

The Multi-Anode Microchannel Arrays (MAMAs), a family of imaging, pulse-counting detector systems, are under active development for use at ultraviolet and soft x-ray wavelengths both on the ground and in space. MAMA detector systems with pixel dimensions of 25 x 25 microns^2 and formats as large as 2048 x 2048 pixels, and with pixel dimensions of 14 x 14 microns^2 and a format of 224 x 960 pixels, are now under evaluation. In this paper the performance characteristics of the different MAMA detector systems are briefly reviewed and the use of custom Application Specific Integrated Circuits (ASICs) to improve the dynamic range, uniformity-of-response, and the spatial resolution is described.

1 INTRODUCTION

We are currently fabricating and characterizing Multi-Anode Microchannel Array (MAMA) detector systems for use on a number of space ultraviolet astrophysics missions at far-ultraviolet (FUV) and extreme-ultraviolet (EUV) wavelengths between about 300 and 28 nm. MAMA detector systems will be used in the Solar Ultraviolet Measurements of Emitted Radiation (SUMER) (Wilhelm *et al.* 1988) and the Ultraviolet Coronagraph Spectrometer (UVCS) (Kohl *et al.* 1988) instruments on the ESA/NASA Solar and Heliospheric Observatory (SOHO) (Domingo and Poland 1988) scheduled for launch in 1995. Prototype very-large-format MAMA detector systems are currently under test and in fabrication for the NASA Goddard Space Flight Center's *Hubble* Space Telescope Imaging Spectrograph (STIS) (Woodgate *et al.* 1986) a second-generation instrument scheduled for in-orbit installation in 1997. Proof-of-concept MAMA detector systems are also under test as part of the Far Ultraviolet Spectroscopic Explorer (FUSE/*Lyman*) Phase B study (Moos *et al.* 1989).

2 MAMA DETECTOR SYSTEM

Details of the construction and mode-of-operation of the MAMA detector system have recently been presented in the literature (Timothy 1989, 1991). The components of a MAMA detector consist of, first, the tube assembly and, second, the associated analog and digital electronic circuits. The MAMA detector tube, which can be sealed with a window or used in an open-structure configuration, contains a single, high-gain, curved-channel microchannel plate (MCP) electron multiplier with the photocathode material deposited on, or mounted in proximity focus with the front surface. Electrodes are mounted in proximity focus with the output surface of the MCP to detect and measure the positions of the electron clouds generated by single photon events (see Fig. 1).

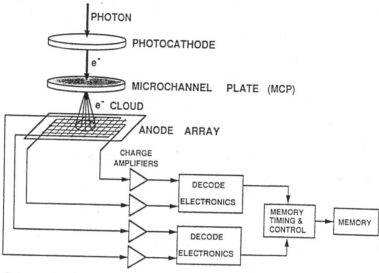

Fig. 1 - Schematic of the imaging MAMA detector system.

The charge collected on the anode electrodes is amplified and shaped by high-speed amplifier and discriminator circuits. Digital logic circuits respond to the simultaneous arrival of the shaped signals from several of these electrodes in each axis, which are arranged in groups to uniquely identify a x b pixels in one dimension with only a + b amplifier and discriminator circuits.

3 SOHO MAMA DETECTOR SYSTEMS

The MAMA detectors for the SUMER and UVCS instruments on the SOHO mission, fabricated by Ball Electro-Optics/Cryogenics Division (BECD), Boulder CO, have a high degree of commonality, but also significant differences dictated by the scientific needs of the two instruments. The key parameters are listed in Table 1. The compact size of the detector head assembly (see Figure 2) is made possible by the use of Application Specific Integrated Circuits (ASICs) for the amplifier/discriminator (Smeins, Stechman and Cole 1991) and decode (Kasle and DeMicheli 1991) circuits. These ASICs produce a system pulse-pair resolution of less than 170 ns and a total power consumption of less than 7W. In addition, the uniformity of response has been significantly improved and the spatial resolution improved by almost a factor of two using the new decode algorithm.

Table 1 - Key Parameters of the MAMA detectors for the SUMER and UVCS Instruments on the SOHO Mission.

	SUMER		UVCS	
	Detector A	Detector B	Detector 1	Detector 2
Pixel Format	360 x 1024	360 x 1024	360 x 1024	360 x 1024
Pixel Dimensions	25 x 25 μm^2	25 x 25 μm^2	25 x 25 μm^2	25 x 25 μm^2
Anode Array Active Area	9.0 x 25.6 mm^2	9.0 x 25.6 mm^2	9.0 x 25.6 mm^2	9.0 x 25.6 mm^2
MCP Active Area	10 x 27 mm^2	10 x 27 mm^2	10 x 27 mm^2	10 x 27 mm^2
MCP Pore Size	12 microns	12 microns	12 microns	12 microns
Number of Amplifiers (including analog output)	105 (104+1)	105 (104+1)	105 (104+1)	105 (104+1)
Photocathode Material	MgF_2 and KBr	MgF_2 and KBr	CsI	KBr
Hybrid Amplifier and Discriminator	Yes	Yes	Yes	Yes
Gate Array Decode Circuits	Yes	Yes	Yes	Yes
Configuration	Openable cover	Open	Sealed	Openable cover

Fig. 2 - Engineering model of the SUMER open MAMA detector system head assembly.

4 STIS MAMA DETECTOR SYSTEMS

The NASA Goddard Space Flight Center's *Hubble* Space Telescope Imaging Spectrograph (STIS) is a multi-mode instrument designed for echelle spectroscopy, long-slit spectroscopy and direct imaging. Two MAMA detectors are used to cover the wavelength range from 300 nm down to the short wavelength limit at 115 nm. The original STIS (2048 x 2048)-pixel detector is fabricated from four contiguous (1024 x 1024)-pixel arrays with a 3-pixel dead space. The other characteristics of the detector are listed in Table 2. Recently, the STIS instrument has been forced to descope, and the (1024 x 1024)-pixel MAMA detector system is now baselined for this instrument.

Table 2 - Key Characteristics of the MAMA detectors for STIS and FUSE/*Lyman*.

	STIS	FUSE
Pixel Format:	2048 x 2048 (4 x 1024 x 1024)	728 x 8096 (4 x 728 x 2024)
Pixel Dimensions:	25 x 25 microns^2	22 x 16 microns^2
Anode Array Active Area:	51.2 x 51.2 mm^2	16.0 x 32.4 mm^2 (x 4)
MCP Active Area:	52 x 52 mm^2	17 x 33 mm^2 (x 4)
MCP Pore Size:	10 microns	8 microns
Number of Amplifiers: (including analog output)	529 ([4 x 132] + 1)	577 ([4 x 144] + 1)
Photocathode Material:	CsI and Cs$_2$Te	KBr
Hybrid Amplifier and Discriminator	Yes	Yes
Gate Array Decode Circuits:	Yes	Yes
Openable Cover:	No	Yes

5 FUSE/*LYMAN* MAMA DETECTOR SYSTEM

The prime EUV spectrograph of the FUSE instrument employs an aspheric concave diffraction grating in a Rowland circle mounting. The detector for this spectrograph must have a long rectangular format to cover the spectral range from 90 to 120 nm. In addition, the pixel size must be smaller than 25 microns in order to obtain the required spectral resolution. Accordingly, we now have under test proof-of-concept sealed and open MAMA detectors with formats of 224 x 960 pixels and pixel dimensions of 14 x 14 microns^2. The key characteristics of the FUSE Phase A detector concept are listed in Table 2.

6 ACKNOWLEDGEMENTS

I am happy to acknowledge the efforts of the many members of the MAMA program at BECD. This work is supported by NASA contracts NAS5-29389 and NAS5-30387, and NASA grants NAGW-540 and NAG5-664.

7 REFERENCES

Domingo, V. and Poland, A. I., 1988, *European Space Agency*, Vol. **SP-1104,** p. 7.
Kasle, D. B. and DeMicheli, G., 1991, *IEEE Computer Society Press*, Vol. **367**, p. 86.
Kohl. J., *et al.*, 1988, *European Space Agency*, Vol. **SP-1104**, p. 49.
Moos W., *et al.*, 1989, *NASA Phase a Study Final Report*, Project NAS5-30339.
Smeins, L.G., Stechman J.M. and Cole E.H., 1991, SPIE, Vol. **1549**, p. 59.
Timothy, J. G., 1989, *SPIE*, Vol. **1158**, p. 104.
Timothy, J. G., 1991, *SPIE*, Vol. **1494**, p. 394.
Wilhelm K., *et al.*, 1988, *European Space Agency*, Vol. **SP-1104**, p. 31.
Woodgate, B. E., *et al.*, 1986, *SPIE*, Vol. **627**, p. 350.

Modeling of UV Lines from Cataclysmic Variable Winds

Peter Vitello[1] and Isaac Shlosman[2]

[1]Physics Department, Lawrence Livermore National Laboratory
[2]Department of Physics and Astronomy, University of Kentucky

ABSTRACT

Ultraviolet resonance lines appear to be powerful tools for diagnostics and study of the outflows from cataclysmic variable stars. The origin of these winds and their geometry are not known at present. In order to compare observed lines from the *IUE* database with synthetic line profiles from accretion disk winds, we have developed a numerical radiation transfer model for 3D non-radial, rotating winds from an accretion disk. Using this model we have studied line spectra from a number of cataclysmic variables. We find that good fits to observed spectra can be obtained using disk winds for mass loss rates much smaller than the accretion rates.

1 INTRODUCTION

Observations by *IUE* and *Voyager* show clear evidence for high-velocity winds from cataclysmic variables (CVs), and suggest their accretion disk origin (see, *e.g.*, Drew 1990). Clues to the wind geometry come from the sensitivity of the UV high-ionization line spectra (CIV λ1549, SiIV λ1397 and NV λ1240) to the inclination angle of the system. P Cygni profiles with blueshifted absorption and redshifted emission are prominent in the low-to-intermediate ($\lesssim 65°$) inclination systems, especially those with high accretion rates. During the low-luminosity state of dwarf novae or in high inclination systems, the absorption component vanishes and the UV lines appear broad and highly symmetric.

CV wind mass loss rates have recently been calculated from synthesized model UV resonance line profiles which include the non-spherical nature of the disk radiation field (Drew 1987; Mauche and Raymond 1987; Drew, Hoare and Woods 1991). These studies, however, share the assumptions that the wind is radially driven from the WD. The detailed study by Mauche and Raymond (1987) require mass loss rates which are very large fractions of the accretion rate, *e.g.* $\dot{M}_w/\dot{M}_a \gtrsim 1/3$, in order to explain observations. We find that if one considers mass loss from the extended accretion disk surface, then greatly reduced mass loss rates can be obtained.

This paper makes use of the *IUE* database to compare observed line shapes of two CVs, RW Sex and RW Tri, with the synthetic line profiles from bi-conical rotating accretion disk winds calculated using a 3D radiation transfer in the Sobolev approximation (Rybicki and Hummer 1978; 1983). Full details of the model will be published in a forthcoming paper. We find that the introduction of a non-radial rotating wind provides a more realistic representation, and results in good fits to the observed line profiles simultaneously in absorption and emission.

2 SYNTHETIC LINE MODELING

Wind streamlines are taken to be 3D spirals which start at the disk surface and continue at the constant angle to the rotation axis, *i.e.* their projections onto the rz-plane are straight lines. The rz projection of the velocity along each streamline is assumed to be linear with length scale R_v or a power law function ($v_{rz} = v_o + (v_\infty - v_o)/(1 + (R_v/l)^\alpha)$). Here l is the distance from the disk in the rz-plane, and v_∞ is the asymptotic velocity along a streamline which is assumed to be a factor β greater than the *local* escape velocity at the base of the streamline. The angular velocity is determined assuming conservation of angular momentum and an initial keplerian motion. The mass loss per unit surface of the disk in the direction of a streamline is taken to be constant.

In calculating the CV emission the disk is assumed to be optically thick as long as its local blackbody temperature exceeds 8,000 °K, which defines also the outer disk radius. Limb darkening effects are included. In our models, the wind originates in the part of the disk where the temperature is between 10,000–50,000 °K (Abbott 1982; Vitello and Shlosman 1988). The boundary layer (BL) was assumed optically thick and emitting also as a blackbody. The size of the boundary layer is taken from the model of Patterson and Raymond (1985), but we leave its effective temperature, T_{BL}, as a free parameter. Ionization in the wind is calculated by assuming a constant temperature $\sim 20,000$ °K and local ionization equilibrium, including photoionization by UV photons from the disk, the BL and the white dwarf.

The line spectrum is found by evaluating the scattered luminosity from the wind and the net unscattered luminosity from the disk separately. The scattered contribution to the line luminosity is calculated using the 3D formalism developed by Rybicki and Hummer (1978; 1983). The source function is obtained by assuming a single resonant velocity surface. Absorption is modeled by tracing rays from the disk in the direction to the observer and reducing the intensity whenever the ray crosses a resonant velocity surface. Because the disk is taken as being optically thick, part of the wind is obscured at small and intermediate aspect angles, enhancing line asymmetries. The doublet structure of CIV line is ignored, and to allow for the *IUE* instrumental broadening our profiles were smoothed with a Gaussian of FWHM 4.5Å.

3 RESULTS

We consider two representative nova-like CV systems: RW Sex, which is a non-eclipsing system which shows strong P Cygni line profiles, and the high inclination, eclipsing system RW Tri. The *IUE* spectra for RW Sex show significant variations

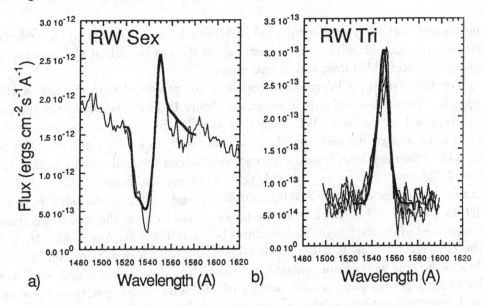

Fig. 1—Fitted *IUE* spectra. Panel a) shows the RW Sex synthetic spectrum (thick line) fit to SWP 22705 (thin line). In panel b), *IUE* spectra SWP 16041, SWP 16063, SPW 16064, and SPW 17621 for RW Tri (thin lines) are compared to our model profile (thick line).

in the continuum and line shapes so we will consider one representative spectrum only. Figure 1a shows our synthetic CIV line for RW Sex superimposed on the *IUE* spectrum SWP 22705. In modeling RW Sex we use the following fixed system parameters: $M_{WD} = 0.8 M_\odot$ and $R_{WD} = 7 \times 10^8$ cm. The best fit, if the boundary layer is ignored, is obtained for inclination $i = 60°$, $\dot{M}_a = 1.3 \times 10^{-9}$ $M_\odot \text{yr}^{-1}$, $\dot{M}_w = 3 \times 10^{-11} M_\odot \text{yr}^{-1}$, wind angular range 20–65°, wind base temperature range on the disk 30,000–50,000 °K, and a powerlaw velocity profile with lengh scale 50 R_{WD}, $\alpha = 2$, and $\beta = 1.3$. Both emission and absorption parts of the line are in good agreement. It is important to note that our model requires a mass loss rate that is only a few per cent of the accretion rate. The wind is initially very optically thick, but becomes optically thin before reaching terminal velocity. The ionization fraction of CIV peaks roughly at a distance 10 R_{WD} from the white dwarf at 76%. It becomes much less than unity both close to the disk and at large distances. The absorption component of the synthetic line is weakly dependent upon i for $i \simeq 60°$ as long as the observer views the disk through the wind (*i.e.* $i < 65°$). The ratio of the emission component to the continuum, however, is strongly dependent upon i. Boundary layer temperatures $\lesssim 50,000$ °K do not modify the modeled spectrum for these parameters. For larger BL temperatures, $T_{BL} \gtrsim 75,000$ °K, BL photoionization becomes important, and the mass loss rate must be greatly increased ($\gtrsim 10$) to fit the observed spectra. Studies using a spherical wind model show that for all values of T_{BL} a radial wind requires a greater mass loss rate than a disk wind. A second distinction between our disk wind spectra and that from a spherical wind is increased line broadening in the disk wind due to rotation. This effect is not observable in the low resolution *IUE* spectra, but should be discernible under higher resolution.

For RW Tri, the CIV emission region is not projected onto the disk surface, hence the line is observed only in emission. Figure 1b gives the our model spectrum superimposed on the four SWP images for the CIV line. The values for the white dwarf mass and radius used are $M_{WD} = 0.6 M_{\odot}$ and $R_{WD} = 8 \times 10^8$ cm. The best fit, if the boundary layer is again ignored, is obtained for inclination $i = 78°$, $\dot{M}_a = 1 \times 10^{-8} M_{\odot} yr^{-1}$, $\dot{M}_w = 1 \times 10^{-10} M_{\odot} yr^{-1}$, wind angular range 10–65°, wind base temperature range on the disk 30,000–50,000 °K, and a linear profile with lengh scale 100 R_{WD} and $\beta = 1.5$. Because of the larger mass loss rate, the model spectrum is insensitive to boundary layer temperatures below 100,000 °K. Again, $\dot{M}_w / \dot{M}_a << 1$. The *IUE* spectra show that the line is asymmetric, with a slower drop off in the red than in the blue wing. Emission asymmetrics occur naturally in the disk wind model due to multiple resonant velocity surface effects. The model spectrum, however, is much more symmetric than the observed profiles. This may be due to approximations in our current disk wind model, such a the single velocity surface treatment for the source function, or limitations in our kinematic representation of the wind.

We find that good fits to *IUE* spectra for RW Sex and RW Tri can be obtained using disk winds for mass loss rates much smaller than the accretion rates. Disk winds appear to require lower mass loss rates than would be required for a spherical outflow. Inherent in a disk wind is rotational broadening which should be observable using high resolution instruments.

ACKNOWLEDGEMENTS. This work was performed under the auspices of the US Department of Energy by LLNL under contract number W-7405-ENG-48 and supported in part by NASA grant NAG5-1387.

REFERENCES

Abbott, D.C. 1982, *Ap. J.*, **259**, 282.

Drew, J. 1987, *M. N. R. A. S.*, **224**, 595.

Drew, J. 1990, in Proc. IAU Coll. 122 on *Physics of Classical Novae*, eds., A. Cassatella and R. Viotti (Springer-Verlag, Berlin), p. 228.

Drew, J.,E., Hoare, M.G., and Woods, J.A. 1991, *M. N. R. A. S.*, **250**, 144.

Mauche, C.W. & Raymond, J.C. 1987, *Ap. J.*, **323**, 690.

Patterson, J. & Raymond, 1985, *Ap. J.*, **292**, 550.

Rybicki, G.B. & Hummer, D.G. 1978, *Ap. J.*, **219**, 654.

Rybicki, G.B. & Hummer, D.G. 1983, *Ap. J.*, **274**, 380.

Vitello, P.A.J. & Shlosman, I. 1988, *Ap. J.*, **327**, 680.

Spectral Imaging of the Chromosphere of AR Lacertae

Frederick M. Walter[1], James E. Neff[2], Isabella Pagano[3], and
Marcello Rodonò[3]

[1]Astronomy Program, Department of Earth and Space Sciences, State University
of New York at Stony Brook
[2]Department of Astronomy, Pennsylvania State University
[3]Institute of Astronomy, Catania University and Astrophysical Observatory

ABSTRACT

The technique of spectral imaging utilizes the velocity information in a broadened
stellar line to map out the surface brightness distribution in the line. We discuss the
technique, and an application to the active cool star(s) AR Lac. Our scientific goal is to
investigate the stability and migration of chromospheric plages on other stars.

1 INTRODUCTION: SPECTRAL IMAGING

Stars are not the unblemished orbs they were once thought to be. We need look
no further than our Sun to see the wealth of atmospheric structures, from spots,
plages and prominences to the coronal loops. Nonetheless, much of our modelling of
the atmospheres of cool stars implicitly assumes a single atmospheric component. If
we could see and map the atmospheric structures on other stars, we could determine
the filling factors of various atmospheric components. But stars are far away, and
their surfaces cannot be resolved by filled-aperture telescopes. Until optical and UV
interferometry capable of μarcsec resolution is a reality, we must resort to indirect
imaging techniques.

Spectral Imaging is a variant of the Doppler Imaging technique (Vogt, Penrod,
and Hatzes 1987), which utilizes the velocity information in a resolved line profile to
map the distribution of brightness over the surface of the star. One assumes that
the observed, rotationally broadened line profile is the summation of unresolved lines
with the same intrinsic profile from every point on the stellar surface. The star is
assumed to rotate as a rigid body, so that every spatial location has a known radial
velocity. A region of contrasting surface brightness will show up as excess emission
on the line profile at the rotational velocity shift of the feature; *the line profile is a
one-dimensional map of the surface brightness distribution of the star.*

Spectral Imaging is a more general technique we use to map the 2 and 3 dimensional structure in the stellar chromosphere. We assume neither an intrinsic line
profile, nor invariance of the profile across the star. We allow for possible radial
extent of the features above the photosphere. We solve iteratively for

- The longitude θ, from the time the feature crosses the line center.

- The product of the latitude ϕ and the height h above the photosphere, from the maximum observed velocity $V_{max}=V_{rot}\sin i \cos(\phi)\, h/R_*$, where i is the inclination of the stellar rotation axis, and R_* is the stellar radius.

- The longitudinal extent of the feature, from its width in velocity space.

2 IMAGING THE AR LACERTAE SYSTEM

AR Lacertae is a bright, well studied, eclipsing RS CVn system, with K0 III and G2 IV components. G.E. Kron (1947) first suggested starspots on the K star to explain brightness differences during totality. The orbital period is 1.983 days; rotation is synchronous with the orbit. The stellar radii are 2.8 and 1.5 $R_\odot$; $V_{rot}\sin i=72$ and 38 km s^{-1}, respectively. The system undergoes total eclipses; the 87° inclination leads to a north-south ambiguity in latitude determinations.

We first applied the spectral imaging technique to the AR Lac chromosphere in October 1983 (Walter et al. 1987), using 8 Mg II k spectra obtained around 0.8 of an orbit to generate a map of the brightness distribution of the lower chromosphere. Most of the flux from the system comes from rotationally broadened lines which follow the stellar radial velocity centroids. We identify this quiescent flux with a global network, but cannot distinguish this from a uniform distribution of active regions. We identified 2 localized regions of enhanced Mg II emission (plages) on the K star, and saw a large radio flare, which we associated with a brightening of the k line from the K star. We obtained upper limits to the sizes of the plages, and showed that the surface fluxes in the plages exceeded that of the quiescent network by about a factor of 5. The plages were near the equator.

The system as observed in September 1985 (Neff et al. 1989) was quantitively similar to that seen in 1983. Identifying the 2 large plages with those seen in 1983 (the relative intensities and angular separation are similar to those seen in 1983), the plages have migrated around the star by about 120° with respect to the binary phase. The more complete phase coverage let us identify a third, high latitude plage on the K star. Surprisingly, the leading hemisphere of the G star appeared nearly black in Mg II k. The radial velocity amplitudes suggest that the Mg II k emission from the plages originates about 0.1R_* above the photosphere of the K star.

We obtained 34 high dispersion spectra of Mg II in September 1987, covering much of 2 orbital periods. Preliminary results have been given by Neff (1990) and Pagano et al.(1992). The same plages appear still to be visible, with some migration in longitude. The September 1989 data are not yet fully analyzed (Pagano et al. 1992). The plage migration seems to have continued. We have good photometry at this epoch; one starspot seems to coincide in longitude with a plage. The mean migration rate of the plages is about -50° yr^{-1}.

3 DOING IT RIGHT: THE 1991 CAMPAIGN

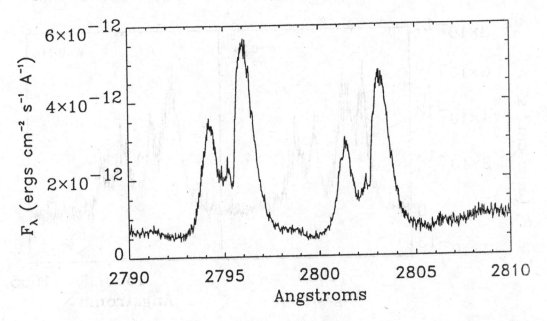

Fig. 1—Mg II k and h near quadrature (orbital phase 0.34), as observed with the GHRS on 1991 December 30. The K star is redshifted; the G star is blueshifted. The interstellar h and k lines are on the blue side of the K star line.

For the 1991 campaign we had 102 hours of continuous coverage with the IUE, in addition to 48 hours of multi-frequency monitoring with the VLA and the usual ground-based photometry and spectroscopy. What makes this campaign especially exciting was our ability to obtain simultaneous high S/N spectra with the HST/GHRS (Goddard High Resolution Spectrograph) of both Mg II and C IV. For the first time we can look for atmospheric structures in the transition region. We also had ROSAT PSPC coverage simultaneous with the GHRS observations.

The 8 GHRS observations taken at about 5.5 hour intervals give us excellent Mg II profiles (Figure 1), and for the first time, spectrally resolved C IV lines. These permit a probe of the higher temperature gas in the transition region. These spectra show that the ratio of the line intensities from the K and G star does not change significantly at Mg II, yet the ratio changes by some 30% for the C IV lines (Figure 2). There are clearly large differences between hemispheres of the G star.

4 SUMMARY

The UV spectral imaging data permit us to observe the active regions on these stars. Our long term goals are to:
- Locate the plage regions. Do they coincide with the photospheric starspots? Are the Mg II-bright plages seen at C IV, and vice-versa?
- Estimate filling factors and true surface fluxes for the plage regions.

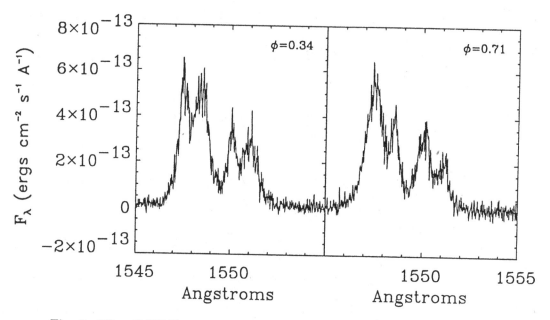

Fig. 2—**The C IV line profiles at opposite quadratures,** as observed with the GHRS on 1991 December 30-31. Each line in the $\lambda\lambda 1548,1551$Å doublet is itself doubled. At orbital phase $\phi=0.34$, the G star is to the blue. Note that the flux in the C IV lines of the G star decrease by 30% during this time. A later observation at $\phi=0.15$ showed a return to the higher flux levels.

- Measure lifetimes of the plages, and study their long-term motions across the star. With latitudinal information, and assuming exactly synchronous rotation, we can observe the surface differential rotation. This has been determined from photospheric spots to be of order 10^{-5} rad s^{-1} deg^{-1} (Pagano *et al.* 1992).
- Study long-term flux variability and watch the stellar cycles.

High dispersion spectroscopy is truly a powerful technique for imaging spatially unresolved astronomical objects. IUE and GHRS have opened up stellar chromospheres to scrutiny; someday AXAF will permit imaging of stellar coronae.

REFERENCES

Kron, G.E. (1947) PASP **59**, 261.

Pagano, I., Rodonó, M., and Neff, J.E. (1992) in *Surface Inhomogeneities in Late-Type Stars*, eds. P.B. Byrne and D.J. Mullan, (Springer-Verlag), in press.

Neff, J.E. (1990) in *Active Close Binaries*, ed. C. Ibanoğlu, (Kluwer), p. 805.

Neff, J.E., Walter, F.M., Rodonò, M., Linsky, J.L. (1989). A&A, **215**, 79.

Vogt, S.S., Penrod, G.D., and Hatzes, A.P. 1987. Ap.J. **321**, 496.

Walter, F.M., Neff, J.E., Gibson, D.M., Linsky, J.L., Rodonò, M., Gary, D.E., Butler, C.J. (1987) A&A, **186**, 241.

X-Rays of IC443 – Remnant of Tang Dynasty Supernova [1]

ZhenRu Wang
Department of Astronomy, Nanjing University, Nanjing 210008, PRC

ABSTRACT

Hard X-rays with energies up to 20 keV were observed from IC443 by the X-ray satellite Ginga. The X-ray flux below 6 keV is found consistent with that of earlier observations with Einstein and HEAO 1, and the X-ray spectrum smoothly extends to 20 keV. The feature of Fe K line is not conspicuous; an upper limit of the equivalent width for its emission is 250 eV. It is likely that the hard X-rays are emitted from a shock-heated plasma with a temperature higher than 10 keV and a number density smaller than 0.1 cm^{-3} which is probably located in the SW and W regions of IC443. This model predicts the age of IC443 to be about 1000 years. It is suggested that IC443 is the remnant of a supernova in AD 837.

IC443 has been studied extensively in radio, infrared, optical and X-ray. It shows a radio, optical and soft X-ray enhancement in its NE and a radio and optical shell with a diameter of about 40', as noted by Petre et al. (1988). The enhancement of its NE is generally believed as an encounter between the shock and a HI dense cloud (De Noyer 1978). The distance of IC443 is considered to be about 1.5 kpc (Fesen 1984). The strong infrared emission from IC443 was observed by IRAS, with the infrared luminosity of 10^{37} erg s^{-1}, much higher than the luminosities in other wavelength ranges (Mufson et al. 1986, Braun & Strom 1986).

IC443 and its background were observed with the large area proportional counters (LAC) on board Ginga 8 and 9 Oct 1990 for the accumulation time of 18688 and 14208 respectively. During the observation the effective area of LAC was 3000 cm^2. The pulse height of the spectrum for IC443 after subtracting the background is shown in Fig. 1. The Ginga observed results compared with previous X-ray observations are listed in Table 1.

The X-ray spectrum of IC443 is fitted by non-equilibrium model of Masai (1984) with two thermal components, i.e., $kT_s = 0.9$ keV for the soft and $kT_h = 14.3$ keV for the hard component as shown in Fig. 1. The problem of power-law fit and its difficulties will be described in another paper (Wang et al. 1992a). From its soft

Table 1. X-ray observed results of IC443

	Ginga	HEAO 1	Einstein
Energy range(keV)	2-20	2-10	1-4.5
Flux(10^{-11}erg cm^{-2} s^{-1})	9	6.7	30
Line feature	not apparent		Si, S
	$W_{Fe}(6-7 \text{ keV}) < 250$ eV		(He-like line)
Spectrum fit	0.9 keV	1.05 keV	0.2 keV
	14.3 keV		0.95 keV

[1] Project supported by the National Science Foundation of China.

1

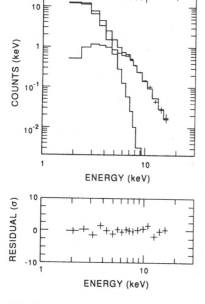

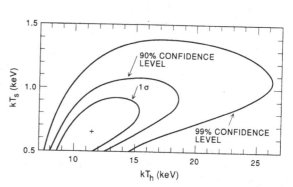

Fig. 1. (left) The observed spectrum and its fit with two thermal components $T_h = 14.3$ keV and $T_s = 0.9$ keV, where $n_e t = 2.6 \times 10^{11}$ cm^{-3}s, $\log N_H = 22.2$, and the abundances relative to the normal values, Fe:0.32, Si:0.95, S:0.73. The set of parameters gives the reduced χ^2 value of 1.215.

Fig. 2. (right) The allowed region of electron of temperatures for the two-component thermal bremsstrahlung model. Here T_s and T_h are correlated with each other.

X-ray IPC map (Petre et al. 1988, Petre 1992), we know that the soft X-rays are mainly from its NE. Subtracting the soft component, we can fit the spectrum of the hard component. Their temperatures, T_h and T_s, are correlated. The allowed region in the $T_h - T_s$ plane is shown in Fig. 2.

It is natural that the brighter the X-rays the area emits, the higher the number density and the lower the temperature it has. This is just the case of the brighter NE area of IC443 in its IPC map. Hence the hard X-ray emission areas seem to be located in the SW and the W regions of IC443, i.e. the lower surface brightness region in its X-ray IPC map. This conclusion is consistent with the locations of the error boxes of HED3 (2-60 keV) and MED (2-40 keV) indicated by the HEAO 1 A-2 observation and the IPC spectra hardness ratio map (Petre et al. 1988). It is also consistent with the morphology of its multiwavelength map. The curvature radius of the NE denser area is smaller than the curvature radius of the rarer SW area (Petre et al. 1988). It is just because the rarer area has the higher shock velocity and results in larger curvature radius. The interaction with molecular clouds results in various morphologies of SNRs and the not unique shock velocities of a SNR, CTB109 is another wonderful example (Wang et al. 1992b). The theoretical explanation of the weakness of the iron emission line and more information from the two components will be in another paper (Wang et al. 1992a)

It is beyond the scope to construct a detailed model of IC443. We have only to point out the existence of the hard component which can be emitted from a shock-heated plasma of a temperature as high as 10-15 keV. The multitemperature model for IC443 indicates an inhomogeneity in density, as demonstrated by the association of shock -heated clouds with a variety of density inserted in the shock-heated diffuse

ISM. Since the shock velocity corresponding to the hard component is as high as $2.5 - 3.5 \times 10^3$ km s^{-1}, this gives the shock crossing time to be 1000-1400 yr. Thus explosion of the supernova may have been recorded by careful observers.

It was suggested by Shklovsky (1954) and Shajn & Gaze (1954 a,b) that IC443 could be the remnant of an ancient guest star in AD 837, because of its positional coincidence. However, Xi (1955) argued against this identification and considered that this object should be the comet Halley. Moreover IC443 to be an old supernova remnant was supported by pre-Ginga observations. Now the discovery of the hard component by Ginga gives the age of IC443 to be 1000-1400 yr, which seems consistent with IC443 being the remnant of the guest star AD 837. Thus it is very necessary to investigate the Chinese records on it.

Several records of a guest star are found in Xin-Tang-Shu (Ouyang and Song 1061) and Weng-Xian-Tong-Kao (Ma 1254). Let us first pick up the following record of the Guest star: "on a Jia-Shen day (April 29) in the 3rd month of the 2nd year of the Tang-Wen-Zong Kai-Cheng reign-period AD837), a guest star appeared below the Dong-Jing (Gem., 22nd lunar mansion). On a Bing-Wu day in the fourth month (May 21), the guest star below the Dong-Jing went out of sight." (Ouyang Xiu and Song Qi 1061; Ma Duan Lin 1254). See Figs. 3 and 4. This Guest star is included in the supernova catalogues of Biot (1846), Williams (1871), Lundmark (1921), Ho (1962) and Xi & Bo (1965).

Some records are also found for the comet Halley, e.g., there is one for it in Xin-Tang-Shu as follows: "On a Gui-wei day (April 28), it was about 3 ft long. It then disappeared at the right of Xuan Yuan (Leo). It is normal for a comet to point westward in the morning and eastward in the evening" (Ouyang Xiu and Song Qi 1061). See Fig. 4. Since it was not possible for the comet Halley found on 28 April in Leo then on 29 April in Gem, i.e., across the distance of 45° for one day (Xi and Bo 1965) and stay at a fixed position Gem for 3 weeks, our opinion is that the ancient records in AD837 can be at least divided into two parts as mentioned above. Thus

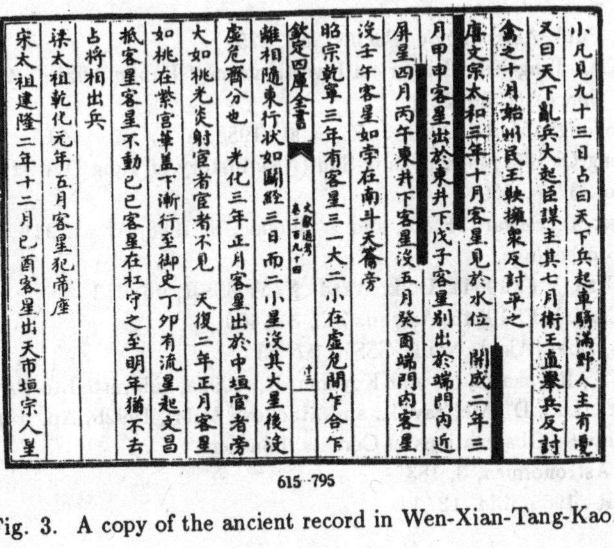
小凡見九十三日占曰天下兵起車騎滿野人主有憂
又曰天下亂兵大起臣謀主其七月衛王直舉兵反討
禽之十月始州民王鞅擁衆反討平之
唐文宗太和三年十月客星見於水位　開成二年三
月甲申客星出於東井下戊子客星別出於端門內近
屏星四月丙午東井下客星沒五月癸酉端門內客星
沒壬午客星如孛在南斗天籥旁
昭宗乾寧三年有客星三一大二小在虛危間乍合乍
欽定四庫全書　文獻通考 卷二百九十四　十三
離相隨東行狀如闘經三日而二小星沒其大星後沒
虛危齊分也　光化三年正月客星出於中垣宦者旁
大如桃光炎射宦者宦者不見　天復二年正月客星
如桃在紫宮華蓋下漸行至御史丁卯有流星起文昌
抵客星客星不動己巳客星在杠守之至明年猶不去
占將相出兵
梁太祖乾化元年五月客星犯帝座
宋太祖建隆二年十二月己酉客星出天市垣宗人星

615 - 795

Fig. 3. A copy of the ancient record in Wen-Xian-Tang-Kao.

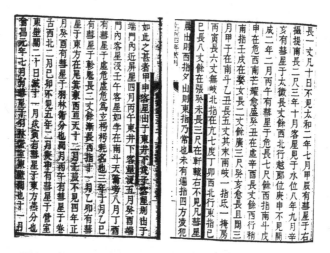

Fig. 4. A copy of the ancient records in Xin-Tang-Shu.

we conclude that the guest star AD 837 which appeared in Gem is identified with a SN which has evolved to IC443, because its age derived from our hard X-ray observation is 1000-1400 yr. Their historical evidence is shown in Figs. 3 and 4.

Thanks to Hayakawa, Asaoka and Koyama for their cooperation in the work on IC443, and to Masai for his computer code and helpful discussion. Thanks to the publications office at JILA for help in preparing this manuscript, and finally thanks for the visiting award from Nagoya Univ. for my short term visit.

Biot, M. E. 1846, Connais Temps, Additions, 60

Braun, R. and Strom, R. G. 1986, A. Ap. **164**, 193

DeNoyer, L. K. 1978 , M.N.R.A.S., **183**, 18

Fesen, R. A. 1984, Ap. J. **281**, 658

Ho, P. Y. 1962, Vistas in Astronomy, **5**, 127

Lundmark, K. 1921, Publ. Astron. Soc. Pacific., **33**, 225

Ma, D. 1254, Wen-Xian-Tang-Kao (Historical Investigation of Public Affairs), 294

Masai, K. 1984, Ap. Space Sci, **98**, 367

Mufson, S. L., McCollough, M. L., Dikel, J. R., et al., 1986, A. J., **92**, 1349

Ouyang Xiu, and Song Qi 1061, in Xin-Tang-Shu (New history of Tang Dynasty),
    Astronomy 2, vol. 32, p. 839

Petre, R., Szymkoviak, E., Seward, F. D., and Willingale, R. 1988, Ap. J. **216**, 320

Petre, R. 1982 this proceedings.

Shajn., G. A., and Caze, V.F. 1954a, Dolk. Akad. Nauk SSSR, **96**, 713

Shajn., G. A., and Caze, V.F. 1954b, Astron. Zh., **31**, 409

Shklovsky, I. S. 1954, Dokl. Akad. Nauk. SSSR, **97**, 53

Wang, Z.R., Asaoka, I., Hayakawa, S., and Koyama, K., 1992a, submitted to PASJ

Wang, Z.R., Qu, Q.Y., Luo, D., McCray, R. and MacLow, M.M., 1992b, Ap. J. (in press)

Willimas, J. 1871, Chinese Observations of Comets, London

Xi, Z. Z. 1955, Acta Astronomica, **3**, 183

Xi. Z. Z. and Bo. S. R.. 1965. ibid. **13**. 1

Judy Lepine 1992

X-Ray Spectroscopy

RECENT RESULTS FROM X-RAY SPECTROSCOPY OF TOKAMAK PLASMAS

Elisabeth Rachlew-Källne
Department of Physics I, Royal Institute of Technology, S-10044 Stockholm, Sweden

ABSTRACT

X-ray spectroscopy continues to be one of most the important diagnostic tools for the central core of the hot fusion plasmas to measure parameters such as ion temperature and plasma rotation. Simultaneously, other spectroscopic tools e.g., visible spectroscopy based on observations of spectra produced by charge exchange processes provides parallell and complementary information.

The increasing rapid variations in temperatures from e.g., 3 to 20 keV, also requires increased experimental accuracy in the line profile measurements both temporally and spatially. Furthermore, with the many different operational modes of the plasma, the impurity behaviour must be studied over a wide wavelength range in order to cover the highly ionised high Z species, such as Ni^{26+}, as well as the low Z elements such as Be, C and O in their various degrees of ionisation throughout the plasma radius.

In this paper examples of recent results of applications of x-ray spectroscopy to diagnostics of the hot plasmas produced by the large tokamaks JET and TFTR are given.

1. INTRODUCTION

High resolution x-ray spectroscopy has become an undisputable and indispensable diagnostic tool at the large fusion experiments with a breakthrough that can be dated back more than ten years.

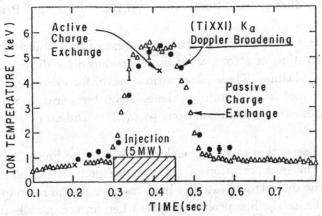

Fig.1 Ion temperature measured from x-ray line spectra and charge exchange during a plasma discharge heated by NBI at the PLT tokamak (Bitter et al,1979a)

At this time the successful heating of plasma ions with auxilliary heating was measured with high resolution x-ray spectroscopy from the Doppler broadening of highly charged impurity ions, in this case Fe^{24+}. The resonance line $1s^2\ ^1S_o$ - $1s2p$ $^1P_1$, was chosen to measure the spectacular increase in ion temperatures at the Princeton Large Torus (PLT) tokamak (Bitter et al.,1979a).

The first comparison with other diagnostic techniques was also made to verify the results from the x-ray line profile measurements. As can be seen from Fig.1 the various techniques show a good agreement in the measurement of the ion temperature variation throughout the plasma discharge.

Since then, many new observations of x-ray spectra from tokamak plasmas have been published with much progress in the analysis of the data and in the application of experimental techniques which will be discussed in this paper with examples from the two current large fusion experiments, the Euratom JET project located in England and the US TFTR project in Princeton.

2. SPECTROSCOPY

2.1 Analysis of the He-like spectra.

A result of the application of high resolution x-ray spectroscopy to most major tokamak experiments is that the helium-like spectra for ions over a large iso-electronic sequence have been measured with high precision and thereby furthered the development of atomic theory for the processes involved in the production of these spectra. It is worthwhile to observe that not only have spectral intensity variations in the line spectra revealed shortcomings in the theoretical predictions but it has also been possible to measure wavelengths with high precision to compare with current predictions on Lamb shift calculations (Källne et al,1984; Marmar et al,1986; Bitter et al,1985; TFR Group et al,1985;Drake,1988). There are also many recent calculations of these spectra (e.g.,Vainshtein and Safronova,1978 and 1985; Sampson and Zhang,1988; Bombarda et.al,1988; Belic and Pradhan,1987; Dubau,1992).

Of particular importance, however, in the context of the subject of this paper, has been the observation of strong x-ray lines produced by dielectronic recombination in the hot plasma. These observations have, for example, given a new accurate diagnostic tool for determining the electron temperature of the plasma using the results from the new calculations (e.g., Bely-Dubau et al,1983;Rice et al,1986).

However, the observation of strong dielectronic satellite lines also introduced a complication to the interpretation of the measurement of Doppler broadened line profiles since the dielectronic satellite lines merge into the resonance line and considerably complicate the line profile analysis when an accuracy better than 10 percent is required. The first analysis of the influence of these unresolved satellite lines observed from the PLT tokamak dates back ten years (Bitter et al. 1979b and 1981) and recently further comparison with observations and new calculations have been presented (Decaux et al,1991). These new results give the possibility to determine with higher accuracy, as is required in today's fusion experiments, the ion temperature as well as rotation velocities as will be discussed further in this paper.

2.2 Analysis Techniques.

In most of these observations of high resolution x-ray spectra from the hot plasma, the spectra are recorded as line-of-sight observations through some chord of the plasma. Consequently, the recorded spectra are built up of from a manifold of spectral lines produced by collisional processes occurring over the full range of temperatures and densities throughout the plasma. With the development of theoretical calculations for the relevant collisional processes the atomic data necessary to simulate the complete spectra have become available for testing against the experimental data. With independent measurements of temperatures and densities from other diagnostic techniques the radial variation of temperatures and densities can be taken into account to produce a simulated spectrum and to compare this with observations (Zastrow, Källne and Summers, 1990, Bombarda et al.,1988, Hsuan,H. et al,1987). An example of such a comparison is given in Fig.2.

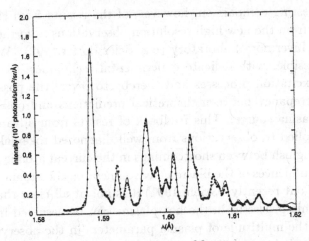

Fig.2 Measured x-ray spectra from Ni^{26+} from the JET tokamak with a synthetic spectrum fitted(Bombarda et al,1988)

The analysis of relative intensities of the main diagram lines of the helium-like spectrum of nickel observed from the JET tokamak revealed a systematic difference between observations and theoretical predictions for the intercombination transition $1s^2\ ^1S_0$-$1s2p^3P_2$ observed from Ni^{26+}, while the other lines in the spectrum showed good agreement with predictions (Zastrow,Källne and Summers,1990) as shown in Fig.3.

The observation of this discrepancy has stimulated further observations from the TFTR plasma (Bitter et al,1991) in order to clarify the possible cause for the observed differences.

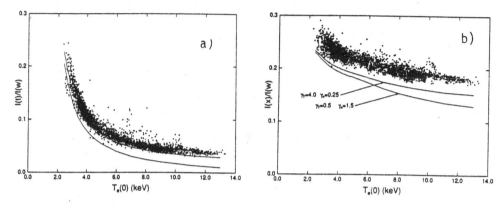

Fig.3 Line ratios from a) the dielectronic and inner shell satellite line,t, to the resonance line and b) the intercombination line,x, to the resonance line, for Ni26+ as a function of electron temperature from the JET tokamak (Zastrow et al,1990)

A most interesting comparison, however, of the results from the plasma observations comes from the new high resolution observations from the EBIT source at the Lawrence Livermore Laboratory (e.g.,Beiersdorfer,1992). With the EBIT ion trap it is possible, with delicate experimental techniques, to separately activate different excitation processes and thereby to reveal the possible cause for the observed discrepancy between theoretical predictions and observations from the hot fusion plasma source. This feedback of results from advanced ion beam experiments , applied to observations from well diagnosed tokamak plasmas will enable us to distinguish between shortcomings in the current atomic data used and possible identify influences of the plasma environment on the atomic processes. As has been pointed out recently (Griem,1991) it is not at all clear that our current modelling of atomic processes in the hot plasma can be assumed independent of the variations of the multitude of plasma parameters in the observational region of the hot plasma. It is rewarding to notice, however, that the achieved increase in accuracy of the observed high resolution x-ray spectra from the well diagnosed tokamak plasmas will undoubtedly further the diagnostics of more distant and uncontrolled plasma sources, such as for example solar flares (Phillips et al.,1992)

3. LINE PROFILES AND LINE POSITIONS

The resonance line in the spectrum from helium-like ions from inherent metal impurities in the plasma is the strongest emission line even over the wide range of electron temperatures, from ca 2 keV up to 12 keV, presently being produced in the large fusion experiments. Since the resonance line is merged with a manifold of dielectronic satellite lines, a line profile analysis must take these contributions into account for an accurate determination of position and width of the resonance line. A synthetic spectrum can be calculated using the atomic data for the dielectronic satellite lines and also using the measured values from e.g., Thomson scattering and laser interferometry for the electron temperature and density and the radial distributions of these parameters. However, it is also necessary, in the comparison with the observed spectra, to use as input parameters ion temperature, toroidal velocity and their radial distributions in order to find the best fit to the observed spectrum. In this way a line position can be obtained with an accuracy of much less than a tenth of the line width.

3.1 Accuracy in Toroidal Rotation Measurements.

From the first observations of x-ray spectra from Fe^{26+} from the PLT tokamak during neutral beam heating it was evident that there was a large induced toroidal rotation of the plasma ions (Bitter et al,1979a).

Fig.4 shows the toroidal velocity data with a velocity component of 10^7 cm/s,i.e., comparable to the observed x-ray line width from the PLT measurements.

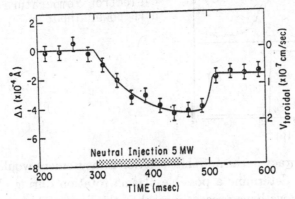

Fig.4 Toroidal rotation velocity measured from Fe^{24+} from the PLT tokamak during NBI (Bitter et al,1979a)

In present day experiments, however, some crucial scientific questions are related to the coupling of radio frequency waves to the plasma and the possible induced toroidal rotation. Fig.5 shows a recent example from the JET tokamak where, using the increased accuracy in the line profile analysis, as has been described above, line shifts up to hundred times smaller than those observed during neutral beam injection have been observed during radio frequency heating only (Eriksson et al, 1992).

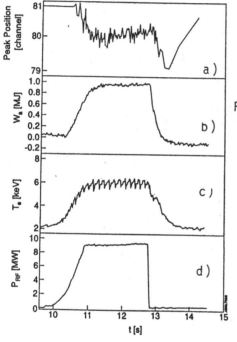

Fig.5 Toroidal rotation velocity measured during RF heated plasma discharge at the JET tokamak(Eriksson et al,1992) a)toroidal rotation(1ch $2 \cdot 10^5$cm/s) b)anisotropic energy content c)electron temperature d)RF power input

Even further accuracy in the line shift position measurement would be required, however, in order to determine a possible plasma rotation during ohmic heating only. Here, however, we have precently reached the limit of accuracy set by the required statistics in a line profile analysis simultaneously with a required purity of the plasma, i.e., the emission from the inherent metal impurities has reached such a low level that a minute seeding of the plasma with e.g., an inert gas would be necessary to give sufficient intensity in the high resolution x-ray spectra recorded with high temporal resolution.

3.2 Comparison with Spectral Information from Charge Exchange Observations.

With the existence of several complementary techniques for measuring ion temperature and toroidal velocity and with the increased accuracy in the analysis methods it has become possible to compare the results from different techniques to a level where small systematic errors can be revealed. An example of this is given from the measurements using visible charge exchange spectroscopy and high resolution x-ray spectroscopy from the JET tokamak. In the measurements of line position from the charge exchange spectra a predicted systematic shift caused by the energy dependence of the charge exchange cross section and the viewing geometry could be revealed in comparison with the accurate x-ray spectroscopy data (Danielsson et al.,1992). This is an interesting result since it reveals a systematic line shift caused by the measurement technique and not being related to any toroidal plasma rotation. It is also important to note that this shift, once revealed through these results, can easily be corrected for and thereby decreases the error in the toroidal rotation velocity measurements.

4. DEVELOPMENT OF NEW INSTRUMENTATION

The high resolution x-ray spectrometers at the large fusion experiments have to comply with the hostile environment of large neutron fluxes, remote access and limited viewing aperture. One of the urgent requirements has been to be able to directly observe the radial distributions of the impurities in order to gain information on the tranport of particles in the plasma. To meet this requirement several approaches have been tried, of them the first to scan the x-ray crystal spectrometer along a radial line-of-sight (Källne et al, 1985). At TFTR (Bitter et at,1988) multichord crystal spectrometers have been installed. A new design, a scanning double crystal spectrometer, has been installed on top of the JET tokamak (Schumacher et al,1989) to scan, with limited spectral resolution, the wavelengths from 1 to 20 Å, i.e., to be able to record radial distributions from highly ionised metal impurities as well as from low Z elements such as oxygen. The first results from the radial scan of the x-ray line spectra showed that the x-ray line emission extends even to the outer, cool regions of the plasma (Källne et al,1984; Rice et al,1987). These astonishing observations have challenged atomic plasma spectroscopists to develop more advanced models for the interpretation of the prediction of the helium-like line spectra (e.g., Kato et al, 1991). Furthermore, the radial profile observations showed that x-ray spectroscopy could infer new diagnostic data on the neutral particle distribution in the plasma (Rice et al,1986).

The formidable task to span the highly charged high Z elements as well as low Z elements such as C,O,Be has been challenged with a single line-of-sight double crystal spectrometer (Barnsley et al,1991) and, recently, also with a rotating multi-crystal spectrometer (Barnsley et al,1992). In these approaches the radial distribution information is sacrificed in order to be able to scan a wide wavelength range with high spectral and high temporal resolution. With these new instruments it has been possible to follow the impurity behaviour under very different plasma conditions in the JET experiment (Coffey et al 1992a,b).

5. FUTURE OUTLOOK

In spite of the enticing developments in x-ray spectroscopy techniques, e.g., the use of multilayer mirrors, imaging detectors, advanced collimators, the requirements of the x-ray spectroscopy at the large fusion experiments have reached such a limit that the instrumentation can not easily comply with the hostile environment the current fusion experiments present. Mainly two factors contribute to the future limitations, the large neutron fluxes and the very limited x-ray line emission caused by the cleanliness of the plasma. Both these factors are results of success of the goal of the fusion experiment. Therefore, it is realistic to state that the diagnostics of these hot ignited plasmas must be designed in accordance with the successful conditions, and with the detection of the most abundant diagnostic species, e.g., neutron spectroscopy will be more important as well as visible spectroscopy using advanced fiber guide techniques. The high resolution x-ray spectroscopy has contributed considerably towards the understanding and the control of the plasma operation in the fusion experiments and will certainly continue to be an important diagnostic technique in the "smaller" non-burning plasma experiments. Of course, such a statement is a look into the future, where one important issue is the confinement of the $alfa$-particles in the next stage of the fusion experiments with hot ignited plasmas. The high resolution x-ray spectroscopy will without doubt continue to contribute considerably to the knowledge of the more remote x-ray sources of astrophysical origin.

6. ACKNOWLEDGEMENTS

I am grateful to M.Bitter at TFTR who supplied me with recent data, to T.Kato from Nagoya who reported news from JT60 and to my colleagues involved in the work at JET, among them R.Barnsley, U.Schumacher, K.-D.Zastrow and H.P.Summers. This work has been supported by the Swedish Natural Science Research Council (NFR) and the author acknowledges support from the Conference Organisers.

7. REFERENCES

Barnsley,R.,Schumacher,U.,Källne,E.,Morsi,H.W.,and Rupprecht,G.,1991, Rev.Sci. Instrum.62,889
Barnsley,R.,Lea,S.N.,Patel,A.,and Peacock,N.J.,1992, This Proceedings
Beiersdorfer,P.,1992, This Proceedings
Belic,D and Pradhan,A.K.,1987, Comm.At.Mo.Phys.20,317
Bely-Dubau,F. et al,1983, Physics Lett.93A,189
Bitter,M., et al,1979a, Phys.Rev.Lett.42,304
Bitter, M., et al, 1979b, Phys.Rev.Lett.43,129
Bitter,M.,et al,1981, Phys.Rev.Lett.47,921
Bitter,M.,1985, Phys.Rev.A32,3011
Bitter,M.,et al, 1988, Rev.Sci.Instrum.59,2131
Bitter,M.,et al,1991, Phys.Rev.A44,1796
Bombarda,F., et al, 1988, Phys.Rev.A37,504

Coffey,I.H., Barnsley,R., Lawson,K.D., Peacock,N.J., and Keenan,F.P.,1992a, This Proceedings

Coffey,I.H. et al, 1992b, This Proceedings

Danielsson,M., et al, 1992, Rev.Sci.Instrum, in press

Decaux,V., et al, 1991, Phys.Rev.A44,R6987

Drake,G.W.F.,1988, Can.J.Phys.66,586

Dubau,J.,1992, This Proceedings

Eriksson, L.-G., Giannella,R., Hellsten,T.,Källne,E., and Sundström,G., 1992, J.Plasma Physics, in press

Griem,H., et al, 1991,Atomic Processes in Plasmas, in "Future Research Opportunities in Atomic,Molecular, and Optical Physics", Lawrence Berkeley Laboratory,1991, PUB-5305,

Hsuan,H., et al, 1987, Phys.Rev.A35,4280

Källne, E.,Källne,J., Richard,P., and Stöckli,M., 1984a, J.Phys.B 17, L115

Källne,E.,et al, 1984b, Phys.Rev.Lett.52,2245

Kato,T., et al, 1991, Phys.Rev.A44,6776

Keenan,F.P.,1992, This Proceedings

Marmar,E.S., Rice, J.E., Källne,E., Källne,J., and LaVilla,R.E., 1986, Phys.Rev.A33, 774

Phillips,K.J.H., Keenan,F.P., Harra,L.K. and McCann, S.M.,1992,This Proceedings

Rice, J.E., Marmar, E.S., Källne, E., and Källne,J., 1986a, Rev.Sci.Instrum.57,2154

Rice,J.E., Marmar,E.S.,Terry,J.L., Källne,E.and Källne,J., 1986b, Phys.Rev.Lett.56,50

Rice,J.E.,Marmar,E.S., Källne,E., and Källne,J., 1987, Phys.Rev.A35, 3033

Sampson,D.H. and Zhang,H.L.,1988, Phys.Rev.A37,3765

Schumacher,U., Källne,E., Morsi,H.W., and Rupprecht,G.,1989, Rev.Sci.Instrum.60,562

Vainshtein,L.A. and Safronova,U.I.,1978, Atomic Data and Nuclear Data Tables 21,49

Vainshtein,L.A. and Safronova,U.I.,1985,Physica Scripta 31,519

Zastrow,K.-D., Källne,E., and Summers,H.P., 1990, Phys.Rev.A41,1427

Zastrow,K.-D., et al, 1991, J.Appl.Phys.70,6732

Highlights of the BBXRT Mission

R. Petre, P .J. Serlemitsos, F. E. Marshall, K. A. Jahoda, E .A. Boldt, S. S. Holt, R. L. Kelley. J. H. Swank, A. E. Szymkowiak, K. A. Arnaud, and the BBXRT Science Team

NASA/Goddard Space Flight Center

ABSTRACT

The Broad Band X-Ray Telescope (BBXRT) was designed to carry out sensitive, moderate resolution spectroscopy of cosmic X-ray sources in the 0.3-10 keV band from the Space Shuttle. The BBXRT flew on the 9-day Astro-1 mission in December, 1990, observing approximately fifty X-ray sources of all varieties. We present some of the more interesting BBXRT results, focusing our attention on those in which plasmas in an astrophysical setting were observed.

1 THE BBXRT INSTRUMENT

The Broad-Band X-Ray Telescope (BBXRT) is the first instrument to carry out high sensitivity, moderate resolution spectroscopy of cosmic X-ray sources over the broad energy range 0.3-10 keV (Serlemitsos et a. 1991). It was designed to extend the types of investigation carried out using the very successful Solid State Spectrometer (SSS) on the *Einstein* Observatory to weaker sources, and to the Fe K complex at high energy and the O K complex at low energy. The instrument consisted of a pair of coaligned conical imaging X-ray mirrors with a segmented Si(Li) detector at the focus of each. The conical mirror, pioneered at GSFC, is the key to the success of the BBXRT. Given that spectroscopy of isolated point sources does not require the high spatial resolution of conventional imaging mirrors like those on *Einstein* and ROSAT, it was seen that a substitution of a large number of light weight, simple conical foils for the small number of thick cross section, precisely figured, heavy glass mirrors would produce an enormous increase in X-ray collecting power, especially at high energies. The resulting mirror utilizes 60 percent of its aperture (contrasted with the 10 percent typical of Wolter-type mirrors), and still has an intrinsic spatial resolution of about 10 arc seconds. While the BBXRT mirrors did not achieve the intrinsic spatial resolution (the half-power diameter is ~2.6 arc minutes), they do have the large predicted collecting area. The spatial resolution as realized is a good match to the detector. As shown in Figure 1, each detector consists of five discrete elements: a central circular detector of radius 2.25 arc minutes, surrounded by an outer annulus to a radius of 8.5 arc minutes divided into four quadrants. All elements are separated by a 1.5 arc-minute wide, X-ray opaque, mask. Two key features of the detector make possible more sensitive broad band spectroscopy than previously achieved. First, the detector has on-orbit spectral resolution of 90-100 eV at 0.5 keV and 155-170 eV at 6 keV, making possible more accurate determination of line properties for plasma diagnostics and measurements of previously undetectable or unresolvable lines. Second, the five elements behave as truly independent detectors, and via a variety of anticoincidence techniques make possible a reduction of the non-X-ray background to a level nearly two orders of magnitude below that of the SSS. In addition, as we highlight below, the independent operation of the elements also facilitates spatially resolved spectroscopy of extended objects such as supernova remnants and clusters of galaxies.

During its nine-day flight on the Space Shuttle Astro-1 mission in December, 1990, the BBXRT observed X-ray sources of all types, including stars, X-ray binaries, supernova remnants, active galaxies, clusters of galaxies, and the X-ray background. Virtually every observation provided some new spectral information. Below we highlight some of those results regarding plasmas in astrophysical settings: supernova remnants, clusters of galaxies, the Carina nebula, and the Seyfert II galaxy NGC 1068.

2 SUPERNOVA REMNANTS

Supernova remnants are obvious targets for an X-ray spectrometer. The spectrum of most remnants consists largely of line emission from ejecta and interstellar medium material, shock heated to typical temperatures of $1-5 \times 10^7$ K. At these temperatures the strongest line emission above 0.3 keV comes from He-like and H-like ions of the elements oxygen through nickel, and from the L blend of iron. As the shock-heated material is in a highly dynamic state, coronal equilibrium models are largely inapplicable, and we turn for plasma diagnostics to time-dependent ionization models such as those of Hamilton, Sarazin and Chevalier (1983).

2.1 Tycho

Tycho, the remnant of a Type I supernova whose explosion was observed in 1572, is one of the most extensively studied of all supernova remnants. The current best model of the X-ray spectrum, which used data from a number of instruments to cover the entire X-ray band, invokes a time-dependent ionization model with two components, one representing the propagating blast wave, and the other a reverse shock encountering layered ejecta (Hamilton, Sarazin and Szymkowiak 1986). With the BBXRT it was possible to observe simultaneously the high and low energy components and thus remove any systematic uncertainties associated with instrumental cross calibrations. The BBXRT pointing location is shown in Figure 1. The central element observed a bright portion of the X-ray shell (the spectrum of which is compared in Figure 2 with that from a much longer SSS exposure); the opposite side of the shell was observed by an outer element. The two BBXRT spectra appear virtually identical, despite the fact that they represent emission from regions separated by a distance of approximately 6 pc. This suggests that the supernova event was symmetric and that it is expanding into a uniform local medium. A number of strong lines are visible to the eye, including some that were previously never resolved (for example, the "beta" transition of He-like silicon at 2.04 keV). Merely measuring the energy centroids and the intensity ratios of the prominent lines provides some new and interesting information. For instance, one of the sources of controversy based on the results of lower resolution spectrometers has been the centroid energy of the Fe K line (Hamilton, Sarazin and Szymkowiak 1986, and references therein). BBXRT measures this centroid directly, without the need for accounting for the slope of the underlying continuum. The measured value of 6.41 ± 0.02 keV indicates a very low ionization state (Fe XIX or lower).

Using diagnostics for non-equilibrium plasmas (Hamilton, Sarazin, and Chevalier 1983), we can constrain the parameters characterizing the plasma: Ts, the shock temperature, and η, the product of the square of the ambient preshock density and the explosion energy. As shown in Figure 3, the parameter values allowed by diagnostics using the lighter elements (Mg, Si, and S) are disjoint from those allowed by the Fe K line energy. We conclude that the environment in which the Fe is embedded, the innermost shell of ejecta, is hotter and less dense than the outer shells. While the possibility of this was suggested by Hamilton, Sarazin, and Szymkowiak (1986), BBXRT provides a direct measurement almost trivially.

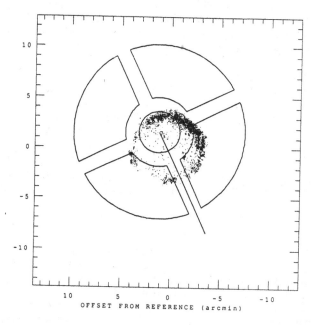

Figure 1: BBXRT observation of Tycho SNR, showing individual detector elements.

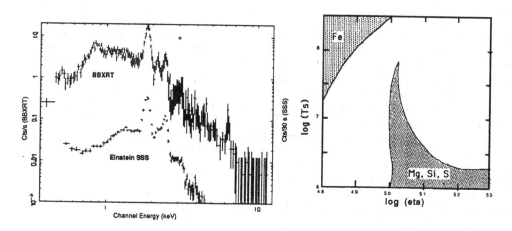

Figure 2 (left): BBXRT spectrum of Tycho, compared with *Einstein* SSS spectrum. The SSS data have bee rescaled for clarity of illustration. **Figure 3** (right): Allowed 90 percent confidence ranges of shock temperature (Ts) and η from Tycho line diagnostics.

2.2 Cas A

BBXRT also observed Cas A, the well-studied remnant of a Type II supernova which occurred in 1604. While the Cas A spectrum in Figure 4 displays the same general properties as that of Tycho, namely strong lines of Mg, Si, S, and Fe, the line shapes and relative strengths are different. In particular, the Si is not as prominent as in Tycho, the Fe is substantially stronger, and all the lines are intrinsically broadened. Markert et al. (1983) ascribed the shape of the Si line to Doppler broadening by a ring of ejecta observed at a range of velocities. In Figure 5 we focus attention upon the Fe K region. Unlike the line from Tycho, the average line energy in Cas A is around 6.6 keV, more consistent with mostly stripped Fe. Also, BBXRT reveals for the first time that its shape is inconsistent with a single, narrow line. It can be fit as a single broad line at 6.6 keV of width 84 eV, or as two narrow lines with energies 6.46 keV and 6.65 keV. If we assume a single, broad line, then the centroid is larger than any value on the grid of models for a single Sedov-Taylor blast wave constructed by Hamilton, Sarazin, and Chevalier (1983). This suggests the need to model Cas A as the sum of multiple non-equilibrium models, as was done for Tycho.

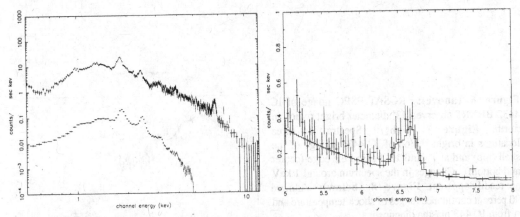

Figure 4 (left): BBXRT spectrum of Cas A (top trace) compared with *Einstein* SSS spectrum (bottom trace). **Figure 5 (right):** Fe K region of Cas A spectrum. Line is fit using two narrow components, at 6.46 and 6.65 keV.

2.3 IC 443

A ROSAT image of the evolved remnant IC 443 is shown in Figure 6. Its global morphology is strongly influenced by the presence of a molecular cloud in the line of sight (Petre et al. 1988). Bright X-ray emission, relatively unobscured by the molecular cloud, arises in a region of enhanced density in the vicinity of (but not associated with) bright optical filamentation in the northern portion of the remnant. The goal of the BBXRT observation was to perform spatially resolved spectroscopy of the bright region, whose 13 arc minute extent is well matched to the 17 arc minute BBXRT field of view. In particular, variations of column density (due to the molecular cloud), temperature, and ionization conditions were being sought. In Figure 7 we show three spectra taken simultaneously in different detector elements: one from the edge of the shell, and others centered approximately 7 and 10 arc minutes (3 and 4.5 pc) behind the shell. Spectral differences are quite noticeable. There is a clear trend with increasing distance behind the shell towards lines of higher ionization state in the band around 1 keV (a region dominated by Ne IX and Ne X), the Mg K region around 1.35 keV, and the Si K band around 1.8 keV.

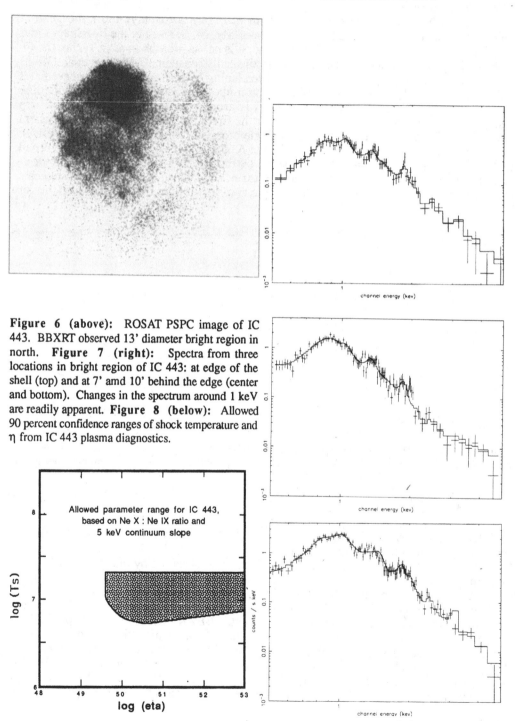

Figure 6 (above): ROSAT PSPC image of IC 443. BBXRT observed 13' diameter bright region in north. **Figure 7 (right):** Spectra from three locations in bright region of IC 443: at edge of the shell (top) and at 7' amd 10' behind the edge (center and bottom). Changes in the spectrum around 1 keV are readily apparent. **Figure 8 (below):** Allowed 90 percent confidence ranges of shock temperature and η from IC 443 plasma diagnostics.

Allowed parameter range for IC 443, based on Ne X : Ne IX ratio and 5 keV continuum slope

Compared with the spectra of the younger remnants, the lines from any of the IC 443 spectra are not nearly as strong. In fact, model fitting suggests that all the elements from Ne through Fe are underabundant with respect to solar. Even with the relatively weak line emission, the spectral resolution and broad bandpass of the BBXRT experiment make possible for the first time on this remnant the kind of plasma diagnostics performed previously on only the bright, young remnants. In Figure 8 we show the allowed region of η -Ts space from applying two non-equilibrium diagnostics: the ratio of the best fit values for the Ne Heα and Lyα line strengths, and the slope of the continuum around 5 keV. The allowed values of η and Ts are in a region where the differences between non-equilibrium and equilibrium models become small, suggesting a rather evolved remnant. While this might seem at odds with the suggestion that IC 443 is the result of a supernova recorded in Chinese recorded in the year A.D. 837 (Wang 1992), both of these facts might be accommodated by the idea suggested primarily by H I data that IC 443 is actually a complex of three distinct but connected structures produced by three supernovae, each one of which not only created its own shell, but reenergized the previously existing one(s) (Braun and Strom 1986).

3 η CARINAE AND THE CARINA NEBULA

η Carinae is a mysterious object which underwent a substantial outburst in 1843. It is currently thought to be a massive star emanating a strong wind, which in turn is colliding with a slowly moving dust shell that was ejected during the outburst. It is known to be a source of hard X-rays. The region immediately surrounding h Car contains a number of OB associations, the Wolf-Rayet star WR 25, and much diffuse emission. In Figure 9, we show the BBXRT spectra of this complex region accumulated simultaneously in the 5 detector elements. The leftmost spectrum contains η Car itself, with the noticeably harder emission and strong Fe line. The central element observed the star WR 25; its spectrum is somewhat harder than the others, which sample mostly the diffuse emission from the region. Every element requires at least two temperature components; the one containing η Car requires four: two for the diffuse emission, and two for η Car itself. The spectrum of η Car supports the hypothesis that the central star is expelling a wind that forms an optically thin 6×10^7 K shock as it encounters the dust shell.

4 CLUSTERS OF GALAXIES

The X-ray emission in clusters of galaxies arises largely from a hot ($\sim 10^8$ K) intracluster medium. The temperature and density structure of this gas largely follows the cluster potential, and is the best means by which the gravitational mass of the cluster can be inferred. While the origin of the hot gas has not yet been satisfactorily established, the presence of metals, and iron in particular, indicates that the gas has been processed through stellar nucleosynthesis. Its capability for position resolved spectroscopy allowed the BBXRT to carry out the most sensitive searches to date for spatially dependent variations of temperature and abundances in the central regions of clusters of galaxies. In particular it was possible to look for evidence of a cooling flow in the central regions.

4.1 Perseus

Even though Perseus is one of the brightest and best studied clusters, there is still some controversy regarding the spatial distribution of iron and the temperature structure of the cluster as a whole (Mushotzky 1991). A series of BBXRT observations allowed tracing of

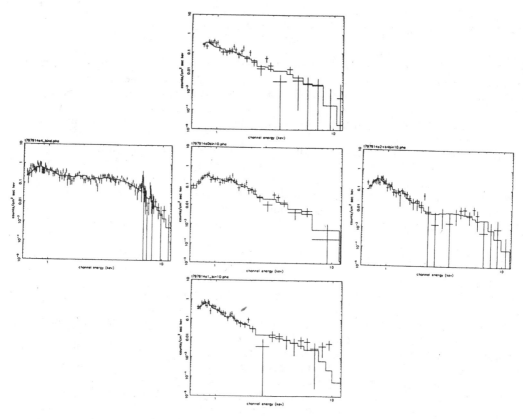

Figure 9: Spectra from five detector elements of the Carina Nebula. Leftmost spectrum contains obvious Fe K lines and excess of hard emission associated with η Carinae. Excess hard flux in central element is associated with star WR 40. Remainder of elements detect diffuse flux from nebula and OB associations.

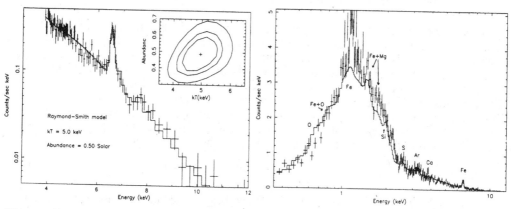

Figure 10 (left): Fe K region of Perseus Cluster spectrum, with confidence contours for temperature and Fe abundance. **Figure 11 (right):** Perseus spectrum with best fit to data above 2 keV. Excess emission at low energies due to cooling flow is apparent.

both out to a radius of 30 arc minutes (1 Mpc). Thermal Fe K emission is observed everywhere, and there is no evidence of a sharp abundance gradient. The abundance everywhere is consistent with a value of 0.5 solar. As shown in Figure 10, the energy of the strongest line observed is consistent with the Heα line, with a redshift of 0.0182 (the optical redshift to Perseus). In fact, the observation of Perseus and other clusters establishes X-ray spectroscopy as a means for determining the redshift of clusters without the need to worry about the velocity dispersion of the component galaxies. BBXRT also observed unambiguous evidence of a cooling flow. In Figure 11, we show the BBXRT spectrum of the central 6 arc minutes (200 kpc) of the cluster. Superimposed on the data is the best fit to the spectrum above 2 keV. The strong excess line emission below 2 keV indicates the presence of a substantial amount of cooler material. This excess of soft emission is absent elsewhere in the cluster. In Figure 12 we show the ratio between the best fit cooling flow model and the data below 1 keV. A dearth of counts above ~500 eV is apparent. This feature can be fit by an absorption edge due to cold oxygen, at the redshift of the cluster. The confidence limits on the edge energy are shown in Figure 13. The equivalent "excess" hydrogen column density is approximately 1×10^{21} cm^{-2}. While the source of this excess absorption has not yet been determined, its apparent physical association with the cluster suggests that we are observing matter that has cooled out of the cooling flow. The filamentary structures observed in the cooling flows of other clusters by the ROSAT HRI (Sarazin et al. 1992) lends credence to the notion that the cooling flows consist of clumpy material at a variety of temperatures.

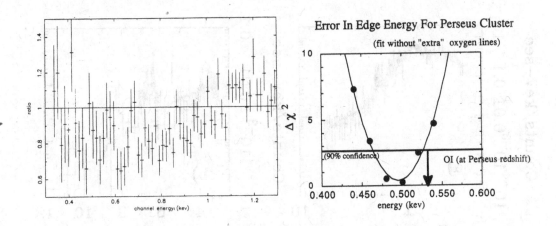

Figure 12 (left): Ratio of data to cooling flow fit for Perseus, showing decrement of counts associated with oxygen absorption. **Figure 13 (right):** Confidence bounds for energy of absorption edge in Perseus, consistent at ~90 percent confidence with the redshift of the cluster.

4.2 Abell 262

BBXRT observed a cooling flow in A262 similar to the one in Perseus. Figures 14 and 15 show the analogs of Figures 11 and 13 for A262: the clear excess of counts below 2 keV and the confidence limits on the energy of an absorption edge due to oxygen at the cluster redshift.

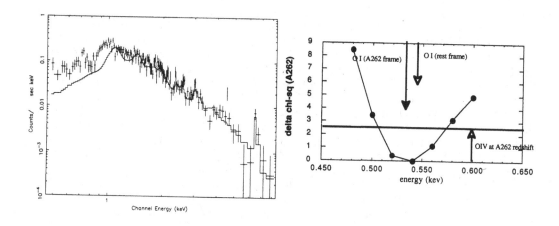

Figure 14 (left): Spectrum of A262 with best fit to data above 2 keV. Excess emission at low energies due to cooling flow is apparent. **Figure 15 (right):** Confidence bounds for energy of absorption edge in A262.

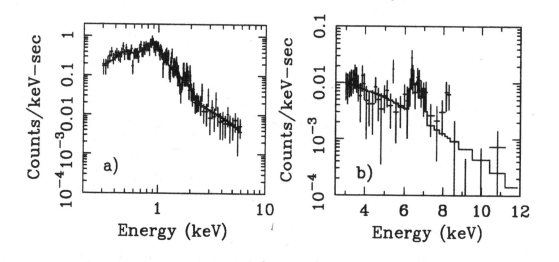

Figure 16 (left): The spectrum of NGC 1068 as seen at low energies. Peak is due to strong Fe L emission. **Figure 17 (right):** High energy spectrum of NGC 1068 highlighting complex Fe K line.

5 NGC 1068

The BBXRT observation of the Seyfert type II galaxy NGC 1068 (Marshall et al. 1991) supports in general the standard picture of an obscured nucleus whose radiation is Compton scattered to the observer by warm electrons. As shown in Figure 16, multiple Fe

K and Fe L lines were detected, indicating a broad range of ionization states. The Fe L lines and the Fe K component at 6.4 keV probably arise in a "warm" component, which is also responsible for the polarized emission detected in optical and ultraviolet light. The H-like and He-like recombination lines of Fe are thought to originate in a previously undetected "hot" component, which probably has a temperature close to the Compton temperature of the radiation from the nucleus. Based on the lack of a detectable oxygen recombination line, we infer that the oxygen is at least a factor of six underabundant with respect to Fe in the "warm" component.

6 OTHER SIGNIFICANT RESULTS

The space here is too brief to cover all of the significant results obtained by BBXRT. Some of the more interesting ones are: **NGC 1399:** The line emission observed in the spectrum of this giant elliptical galaxy offers the first unequivocal evidence that the X-rays from these objects arise in a massive halo. The temperature measured by BBXRT requires a mass-to-light ratio of larger than 40, providing new and rigid constraints on the amount of dark matter necessary to gravitationally bind this object. **PKS 2155-304:** BBXRT confirmed with high statistical significance the existence of a broad absorption feature at ~500 eV in the spectrum of the BL Lac object PKS 2155-304. The energy of the feature is consistent with O VIII resonant absorption blueshifted from the object (Madejski et al. 1991). **M81:** The spectrum of this nearby spiral galaxy is dominated by the emission from its nuclear region. It is best fit by a simple power law of slope $\Gamma = 0.8$, typical of Seyfert I galaxies, with moderate absorption intrinsic to the host galaxy.. The X-ray spectrum of this object, combined with the fact that both the X-ray luminosity and the mass limit on a central object from dynamical studies, suggests that the mechanism producing X-rays in Seyfert galaxies, typically two orders of magnitude more luminous, is also at work here.

7 REFERENCES

Arnaud, K. A., et al. 1991, *Proc. 28th Yamada Conference on Frontiers in X-ray Astronomy*, in press.

Braun, R., and Strom, R. G. 1986, *Astr. Ap.*, **164**, 193.

Hamilton, A.J.S., Sarazin, C. L., and Chevalier, R. A. 1983, *Ap. J. (Suppl.)*, **51**, 115.

Hamilton, A.J. S., Sarazin, C. L., and Szymkowiak, A. E. 1986, *Ap. J.*, **300**, 713.

Madejski, G. M., et al. 1991, *Proc. 28th Yamada Conference on Frontiers in X-ray Astronomy*, in press.

Markert, T. H., Canizares, C. R., Clark, G. W., and Winkler, P. F. 1983, *Ap. J.*, **268**, 134.

Marshall, F. E., et al. 1991, *Proc. 28th Yamada Conference on Frontiers in X-ray Astronomy*, in press.

Mushotzky, R. F. 1991, in *Clusters of Galaxies*, ed. A. Fabian, in press.

Petre, R., Szymkowiak, A.E., Seward, F. D., and Willingale, R. 1988, *Ap. J.*, **335**, 215.

Sarazin, C. L., O'Connell, R. W., and McNamara, B. R. 1992, *Ap. J.*, in press.

Serlemitsos, P. J., et al. 1991, *Proc. 28th Yamada Conference on Frontiers in X-ray Astronomy*, in press.

Wang, Z.-R. 1992, these proceedings.

Spectroscopic Results from ROSAT

Bernd Aschenbach

Max-Planck-Institut für extraterrestrische Physik
D-8046 Garching, Germany

ABSTRACT

On June 1 1990 the X-ray astronomy satellite ROSAT was launched into a near-earth orbit and has been in operation since then. ROSAT carries two grazing incidence telescopes covering the XUV ($\sim$ 50-750 Å) and the soft X-ray ($\sim$ 5-120 Å) spectral regions. The X-ray telescope is equipped with a channel plate detector of high spatial resolution delivering images of a few arcsec resolution and alternatively, with position sensitive proportional counters (PSPC) of $\sim$ 25 arcsec resolution. The PSPC has a spectral resolution of $\sim$ 41% at 1 keV, with which the energy spectra of point-like sources are measured and spectrally resolved images of extended sources are taken. The spectra taken with the PSPC - although of moderate resolution - have delivered new insight of the physics of many classes of cosmic X-ray sources. The present paper concentrates on some aspects related to the physical state of galactic supernova remnants.

1 INTRODUCTION

X-Ray astronomy is now moving towards the end of the third decade of observational research since the discovery of the first non-solar cosmic X-ray source in 1962. Collimated counters have been used exclusively in the first 16 years, and they are still the only way to perform X-ray observations at energies above about 20 keV. At lower energies imaging telescopes offer the opportunity to take true pictures of celestial sources at a significantly reduced level of background noise and angular resolution of a few arcseconds up to now. In this respect X-ray astronomy has come close to the performance delivered by ground-based optical telescopes and radio telescopes.

In 1978 the Einstein Observatory was launched and the first imaging X-ray observations from a satellite were performed. At the end of the mission it had been demonstrated that X-ray emission from cosmic sources is not the exception but is the rule. Almost every class of astronomical objects emits X-rays, a fact which promises a rich scientific return from an all-sky survey with an imaging telescope.

On June 1 1990, 17:48 local time, after about 15 years of research, development and construction the German X-ray astronomy satellite ROSAT was launched from Cape Canaveral by a Delta-II rocket into a circular near-Earth orbit of 580 km altitude and 53° inclination. The satellite carries the largest and most precise X-ray telescope put into orbit so far with 3 focal plane instruments. A second telescope, the Wide Field Camera, extends the energy range of the X-ray telescope from the soft X-ray band longward into the XUV and EUV.

The scientific objectives of the ROSAT mission are twofold. The primary objective of ROSAT was to perform the first all-sky survey with imaging telescopes in the X-ray and EUV spectral bands. This objective has been accomplished and the sky coverage is > 99%. The all-sky data have been processed once in an automated and conservative way requiring relatively high signal to noise. In total about 60,000 new X-ray sources have been detected, about 25% of which have been readily identified by positional comparison with astronomical catalogues. For 90% of the sources the position accuray is better than 30 arcsec radius, which eases the identification significantly. About 25 to 30% of the objects are of galactic origin, mostly normal stars; about half of the objects are active galactic nuclei and quasars and the rest is made up of clusters of galaxies.

Apart from the detection and location of point sources ROSAT has mapped the diffuse galactic soft X-ray emission. These all-sky maps show spatial structure on all scales ranging from more than 100° down to the survey limit of a few arcminutes. The largest coherent objects are the North Polar Spur (d $\sim$ 116°), the Monogem Ring (d $\sim$ 20°), the Cygnus Super-Bubble (d $\sim$ 13°), the Cygnus Loop (d $\sim$ 3° $\times$ 4°), the Vela SNR (d $\sim$ 7.3°), e.g. They have been imaged completely for the first time, such that their full extent and the associated structure has become evident.

The majority of the X-ray sources discovered during the all-sky survey are too faint for detailed spectral studies but spectral hardness ratios are available, which are quite useful for classification and statistical investigations. Detailed spectroscopy is possible for about 500 to 1,000 sources. Spatially resolved spectroscopy of extended sources like supernova remnants and clusters of galaxies is another key element of the ROSAT all-sky survey, which will probably remain the only study of the largest objects, at least, for many years to come.

The second phase of the ROSAT mission is being carried out exclusively in the framework of an international guest observer program. More than 1,300 pointed observations have already been made to either take images, or to take energy spectra or to study temporal variability of all kinds of cosmic X-ray sources. Up to now more than 800 Principal Investigators are involved in ROSAT studies. Of course there is no way to provide a comprehensive overview of the scientific results obtained with ROSAT so far. Instead, I will concentrate on a few examples to illustrate the capability and performance of ROSAT. After a brief introduction of the spectroscopic means of ROSAT I will report about some new discoveries made from ROSAT observations of the Cygnus Loop, the Vela and Puppis-A supernova remnants.

2 ROSAT INSTRUMENTATION

ROSAT carries two independent but co-aligned imaging telescopes. The X-ray telescope (XRT) working between 0.1 to 2.4 keV, consists of a 4-fold nested Wolter type I mirror assembly of 2.4 m focal length and a maximum aperture of 835 mm (Aschenbach 1988). Extraordinary performance values have been achieved. The angular resolution is estimated to 3.6 arcsec, and the residual microroughness of the mirror surfaces is 2.8 Å.

The focal plane instrumentation of the XRT, developed and built at the MPE, includes a carousel with three X-ray detectors, each of which can be rotated into the focus of the telescope. Two detectors are redundant Position Sensitive Proportional Counters (PSPC's), which are operated in gas flow mode (Pfeffermann et al. 1986). The PSPC provides spatial and spectral resolution over the full field of view of 2° which vary slightly with photon energy E. The angular resolution is limited by the PSPC to about 25 arcsec on-axis and degrades to a few arcminutes for field angles larger than about 20 arcmin due to the geometric aberrations of the telescope. The energy resolution is $\Delta E/E = 0.41/\sqrt{E_{keV}}$. The effective collecting area including the PSPC efficiency is about 220 cm^2 at 1 keV for on-axis rays and decreases slightly with increasing field angle.

The third detector, the High Resolution Imager (HRI) built at the SAO Cambridge, USA, is a microchannel plate detector with an intrinsic spatial resolution of about 20 μm. With the HRI in the focus of the telescope the angular resolution is expected to be dominated by the point spread function of the mirror assembly. A value close to 4 arcsec is expected but still to be confirmed.

One major factor, which determines the sensitivity of astronomical instruments to detect point like sources and to measure brightness distributions of extended objects, is the amount of unwanted background. For X-ray astronomy measurements two components of background radiation are to be considered, which are due to the photons of the diffuse celestial X-ray background emission and charged particles, the latter of which mimic X-ray photons in the detector. Whereas the study of the diffuse X-ray background is an astrophysically interesting objective in itself, the detector background induced by charged particles is usually not, and the rate of events accepted by the detector is minimized by anti-coincidence techniques and shielding. Fig. 1 shows the pulse height spectrum of the charged particle background and the diffuse X-ray background measured with the ROSAT PSPC in orbit during a pointing at β Leonis. The spectrum of the particle background is essentially flat, and the total amount of accepted background events amounts to 3×10^{-5} counts s^{-1} arcmin^{-2}. This very low background means that pointed PSPC observations with a typical resolution element of 25 arcsec diameter are not background limited for exposures less than 10^5 sec. For the survey the resolution element is about 2 arcmin in diameter, and PSPC observations become particle background limited only for times exceeding 10^4 sec, which is available only in regions close to the ecliptic

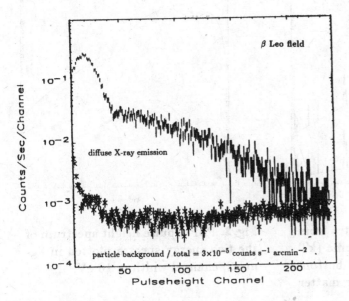

Fig. 1. PSPC pulse height spectra of the diffuse X-ray background (top data) and the charged particle background (bottom data); 1 channel $\sim$ 10 eV.

poles. Therefore the majority of the ROSAT PSPC observations both in the all-sky survey and the pointing programme are essentially free of particle background contamination, and the majority of the recorded events are true X-ray photons.

The spectrum of the diffuse X-ray background differs from the spectrum of the charged particle background and has a significantly higher brightness for energies lower than about 2 keV (see Fig. 1), so that the mapping of the X-ray background by ROSAT is essentially limited by the brightness of the X-ray backround emission itself. It is estimated that the morphology of the diffuse emission can be mapped at an angular resolution of about 6 arcmin over the whole sky. The energy spectrum (see Fig. 1) clearly shows two components, at least. The prominent bump at energies below 0.5 keV is attributed to galactic emission, whereas the flux above 1 keV is very likely to be of extragalactic origin with some mix for energies in between.

It is mentioned that under bad viewing conditions of the telescope close to the Earth horizon scattered solar X-rays provide a third source of background.

The spectral resolution of the ROSAT PSPC is illustrated by the following two sample spectra. Fig. 2 shows the energy spectrum emitted from an optically thin thermal plasma at a temperature equivalent to 0.25 kev and 1 keV, respectively. The elemental abundances have been fixed at their solar values, and the spectra have been computed following the code developed by Raymond and Smith (1977) and its recent updates. The appearance of the spectrum is dominated by emission lines from the various ionic constituents with an underlying continuum, the slope of which is

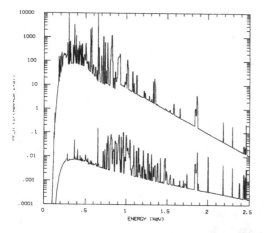

Fig. 2. Soft X-ray energy spectra from an optically thin thermal plasma (kT = 0.25 keV top data, 1 keV bottom data) modified by interstellar matter absorption.

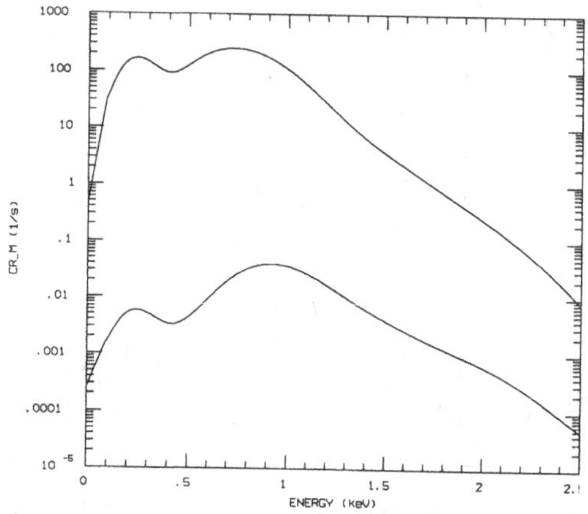

Fig. 3. PSPC pulse height spectrum of the two energy spectra plotted in figure 2; channel width $\sim$ 10 eV.

characteristic of the electron temperature. At the low energy end the spectrum is modified by photoelectrical absorption of the source photons by interstellar matter. In the two sample spectra shown the total amount of matter integrated along the line of sight is equivalent to a hydrogen column density of $N_H = 3 \cdot 10^{20}$ cm^{-2}. Of course, most of the absorption above $\sim$0.25 keV is due to elements heavier than hydrogen (Morrison and McCammon, 1983).

Figure 3 shows the two PSPC pulse height spectra corresponding to the two energy spectra displayed in figure 2. Clearly, the spectral resolution of the PSPC is insufficient to resolve any of the emission lines, but the shape of the spectra depends on the presence and strengths of the emission lines such that the PSPC can easily discriminate between the two temperatures. The discrimination is supported by the differing high energy slopes. Furthermore, the turn-over of the spectrum towards lower energies which is determined by the amount of interstellar absorption can be measured. In general, despite the lack of high spectral resolution the ROSAT PSPC can disstinguish between thermal and non-thermal source spectra, and it can measure the corresponding spectral power law indices and temperatures, respectively, which is a useful tool to characterize and classify cosmic sources.

3 THE CYGNUS LOOP

The Cygnus Loop is generally called a middle-aged SNR of $\sim$ 10,000 to 20,000 years age at a distance of $\sim$ 770 pc. X-ray emission from the loop has been discovered already in 1970 (Grader et al.); the first X-ray image of a few arcmin spatial resolution has been obtained with the Einstein IPC by Ku et al. (1984). During the ROSAT all sky survey the Cygnus constellation was scanned in November 1990 and a PSPC image of the Cygnus Loop was obtained as shown in Fig. 4. The effective spatial resolution is $\sim$ 1 arcmin half power radius, uniform across the image.

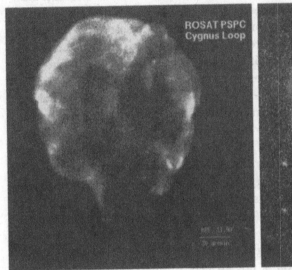 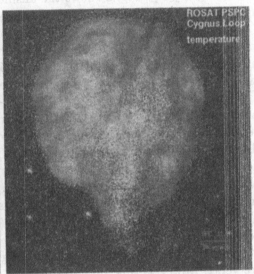

Fig. 4. X-ray image of the Cygnus Loop obtained during the ROSAT all-sky survey.

Fig. 5. Temperature distribution in the Cygnus Loop. Dark grey corresponds to $1 \cdot 10^6$ K, white to $10 \cdot 10^6$ K.

The surface brightness is highest in the north-east, which corresponds to NGC 6992. The brightness is lowest in the center part and it varies along the rim. The X-ray bright regions tend to resolve into filamentary structure, such that the X-ray image is very reminiscent of the optical H_α image. However, there are also optical filaments, which have no X-ray counterpart and vice versa. The X-ray prominent V-type structure close to the south-west rim is not visible in the optical image.

The claim that the X-ray emitting plasma in the Cygnus Loop is not isothermal has first been made by Kayat et al. (1980) using data from a sounding rocket flight. Temperature variations were confirmed by Charles et al. (1985) using Einstein IPC data. Leahy et al. (1990) analysed non-imaging data from the HEAO 1 A-2 experiment and find that a two-temperature equilibrium plasma of $\sim 2.3 \cdot 10^6$ K and $12.6 \cdot 10^6$ K with very small variations across the remnant fits the data best. However, they claim a significant variation of low energy absorption by interstellar matter across the Loop.

The ROSAT PSPC data have been used to determine the temperature of the X-ray emitting plasma in each individual image pixel. A measure for the temperature is the energy per X-ray count, the calibration of which has been computed from a one-temperature Raymond-Smith emission code assuming standard cosmic elemental abundances. Of course, the calibration changes with the amount of interstellar column density N_H. Fig. 5 shows the unsmoothed temperature distribution assuming a uniform N_H. The temperature ranges from $1 - 10 \cdot 10^6$ K. The temperature is lowest around the rim and at the very bright spots on the eastern and western boundary. The hottest regions with ~ 20 arcmin diameter are located very close to the center. Generally speaking the interior of the remnant appears to be the hottest part, whereas the regions with bright filamentary structure are significantly cooler.

Given this wide temperature spread it is obvious that spatially non-resolved energy spectra, like the HEAO 1 A-2 measurements (Leahy et al.), cannot be fitted with a one-temperature model, but by models of two temperatures at least. Although detailed spectral fitting to the ROSAT data has not yet been performed there is evidence that N_H variations across the remnant, which would change the calibration between temperature and energy per X-ray count, are not significant. The rim, which is most likely to be at the same temperature along the circumference, shows indeed a value for the energy per X-ray count which is constant all around the circumference.

Generally, there is a tendency that bright regions show low temperatures and vice versa. However, this anticorrelation is not strictly fulfilled as it would be expected from exact pressure equilibrium. A map of the thermal pressure associated with the X-ray emitting plasma has been computed from the brightness per pixel, which is proportional to the square of the matter density times the length of the line of sight through the emitting region, and the temperature determined from the energy per X-ray count. The resulting map, which is pressure times square root of the depth of the emission region, shows large variations of up to a factor of ~ 50. Some of the variations may be produced by variations in the depth of the emission region from one image pixel to the other, a major fraction is due to variations of the thermal pressure, though. The lowest pressure is found in the fainter parts of the remnant; the filaments are slightly higher in pressure and the highest pressure appears in general in the brightest parts of the filaments. Exceptions, however, exist as well. The pressure in the bright V-type structure close to the south-west rim is not constant, but it is significantly higher in the eastern wing of the V. Obviously, the Cygnus Loop as a whole is not in pressure equilibrium as far as the thermal pressure of the X-ray plasma is concerned.

Pressure variations are expected from the interaction of the supernova blast wave with interstellar clouds of densities higher than the ambient intercloud medium. McKee and Cowie (1975) calculate thermal pressure variations in a cloudy medium of up to a factor of 3.15 for freshly shocked clouds. Using this interpretation the high pressure clouds must have been shocked only very recently. A detailed analysis of the ROSAT observations is still to be done. In particular the amount of pressure variation and the dependence on density / temperature is still to be understood.

The determination of the density, pressure and temperature of the Cygnus Loop has been performed assuming a hot and optically thin equilibrium plasma of cosmic elemental abundances. Of course, these assumptions have some impact on the plasma emissivity, which is a dominant factor in the determination of the plasma state. Three possible deviations come to mind:

i.) The emitting plasma is not in collisional equilibrium ionization. A plasma instantaneously heated by a shock wave needs time to settle to ionization equilibrium, which is about $n_e \cdot t \sim 10^{12} \text{cm}^{-3} \cdot \text{s}$; n_e is the electron density. Vedder et al. (1986) have reported ionization non-equilibrium conditions in the northern area of the Cygnus Loop, which indicates either a low electron density or rather recently shocked matter. The latter hypothesis is in line with the assumption that recently shocked interstellar clouds are responsible for the X-ray bright regions in the Cygnus Loop. It has been shown that a two temperature collisional ionization equilibrium (CIE) spectrum can mimic a non-equilibrium ionization (NEI) spectrum. Usually the electron density of the NEI spectrum required to explain the observed flux is much lower than the corresponding densities of the two temperature CIE spectrum, with differing results among the investigators, though (Aschenbach 1988a). But it seems that CIE models overestimate the electron density quite a bit with a corresponding overestimate of the thermal pressure. On the other hand this effect is counteracted by the increased temperature required for the NIE models, so that the net effect on pressure is still unclear.

ii.) For the analysis it has been assumed that cosmic elemental abundances are constant throughout the remnant. If the X-ray bright regions were enhanced in heavy elements the volume emissivity would increase and the density and the pressure would be overestimated. Such enhancements are expected for the stellar ejecta but they should not dominate the X-ray emission of relatively old SNRs like the Cygnus Loop.

iii.) The temperature derived from the X-ray spectrum is the electron temperature. If the electrons are heated by Coulomb collisions it is likely that the temperature of the ions heated by the shock wave is significantly higher. In such a case the thermal pressure is clearly underestimated since the partial pressure of the electrons is much lower than that of the ions. Since the pressure is observed to be low in the hot and

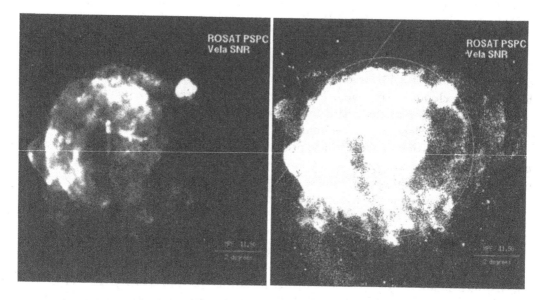

Fig. 6. X-ray image of the Vela SNR obtained during the ROSAT all-sky survey; Puppis-A shows up overexposed at the north-west boundary.

Fig. 7. X-ray image of the Vela SNR enhanced to show the faintest structures; the circle is centered on the Vela pulsar position; the straight line delineates the galactic plane.

dilute inner parts of the remnant these regions are suspected not to be in thermal equilibrium. This hypothesis is in agreement with the fact that the time scale to reach thermal equilibrium between electrons and ions by Coulomb collisions is proportional to $T^{3/2} \cdot n^{-1}$, n being the plasma density.

4 THE VELA AND PUPPIS-A REGION

4.1 The Extent of the Vela SNR

The Vela supernova remnant is one of the rare examples of SNRs, which show a thermal, large diameter partial X-ray shell containing a pulsar and a small synchroton nebula probably powered by the pulsar (Harnden et al. 1985). An X-ray image of the Vela SNR has been obtained by a series of 36 mosaic pointings with the Einstein IPC (Kahn et al. 1985). The Vela constellation was scanned by ROSAT in November 1990 and the image obtained during the all sky survey is shown in Fig. 6. The image also contains the SNR Puppis-A at the north-west edge of Vela. The grey scale for the image has been chosen to display just the brighter parts, such that the filamentary structure of the emission shows up.

Similiar to the Cygnus Loop quite a number of individual bright filaments are visible. In contrast to the Cygnus Loop the boundary of the Vela remnant is not that well defined, in particular the southern region, which apears rather irregular in shape. Of particular interest is the eastern rim, from which a wedge type structure protrudes almost by 1° showing a bright patch on its outward boundary. A similar wedge type structure, though much fainter, is visible further up north offset from the rim by ~ 1.5°.

Figure 7 is a different display of the ROSAT Vela SNR image which emphasizes the low brightness regions. From this figure the full extent of the Vela SNR becomes evident. The very faint western parts are enclosed by a long arc like filament running south from the Puppis-A position. The existence of this arc has already been mentioned by Seward (1991), who associated the arc with the Vela SNR based on its soft Vela type IPC X-ray spectrum. An eye-ball fit circle has been included in Fig. 7, which outlines the boundary of the Vela SNR. The remnant fits nicely into this circle, which has a diameter of 7.3°. Obviously the remnant is significantly larger than 5.5° previously derived from radio and optical images. Fig. 7 shows additional patchy emission outside of the circle, particularly in the west and southeast, and it is not clear whether this emission is associated with the Vela supernova event. Possibly this emission is a signal from the Gum nebula, seen for the first time in X-rays. However, further analysis on a much larger sky area around Vela is required to demonstrate the existence of X-ray emitting plasma associated with the Gum nebula.

As described in the previous chapter on the Cygnus Loop temperature and pressure maps have been built for the Vela remnant. Temperature variations across the remnant are evident as well as pressure variations. However, the range of pressure variations is less for the Vela remnant and particularly high pressure is found only for small size regions within the bright filament area in the north and southeast of the remnant. As in the case of the Cygnus Loop further analysis is required to understand the results in the framework of the interaction of the blast wave with a cloudy medium.

4.2 Partial Absorption of Puppis-A by the Vela SNR

The SNR Puppis-A is one of the brightest X-ray sources in the sky. The supernova went off probably 3,700 years ago at a distance of about 2 kpc (Winkler et al. 1988). Figure 8 shows the ROSAT X-ray image obtained during the all-sky survey. Clearly visible is the sharp delineation of the rim along the south-east north-west direction which runs approximately parallel to the galactic plane. Like other middle-aged remnants Puppis-A shows a patchy brightness distribution which is most likely due to shocked interstellar clouds.

In Figure 9 the spectral variations across the remnant are plotted, and unlike Vela or the Cygnus Loop Puppis-A does not show such strong changes in spectral

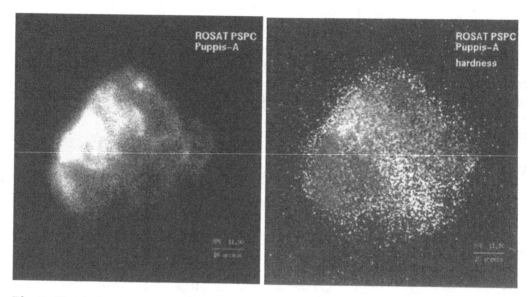

Fig. 8. X-ray image of Puppis-A obtained during the ROSAT all-sky survey.

Fig. 9. X-ray spectral hardness across Puppis-A; dark grey corresponds to a softer spectrum than white; note the white central bar.

hardness, except in a broad bar-like region, which runs right across the remnant from the south-west to the north-east. This bar-like structure seems to be aligned with the most north-west thin filament of the Vela SNR supposed to delineate its outer boundary. The impression from Fig. 7 is that this filament, which approaches Puppis-A from the southwest direction, continues across Puppis-A. The bar-like structure follows an extrapolation of that filament across Puppis-A but it is slightly off-set to the north. The emission of that portion of the filament crossing Puppis-A is extremely difficult to detect because of its very low surface brightness compared with Puppis-A. However, the energy spectra are supposed to be significantly different based on the much higher interstellar absorption towards Puppis-A. Fig. 10 shows the PSPC spectrum of a circular cut out of Puppis-A, which is centered on the extrapolation of the Vela filament (centre: $\alpha(2000) = 8^h\ 23^m\ 49^s$, $\delta(2000) = $ -43° 14' 0"; r = 10 arcmin), but outside of the spectrally hard bar-like structure. A fit to the spectrum reveals that the high energy component can be represented by a two-temperature Raymond-Smith spectrum ($kT_1 \sim 0.24$ keV, $kT_2 \sim 1.1$ keV) subjected to photoelectrical absorption equivalent to $N_H \sim 1.2 \cdot 10^{21}$ cm^{-2}. The spectral component seen below about 0.3 keV is consistent with a second thermal spectrum, which has a temperature $kT_s \sim 0.12$ keV and an absorption of $N_{H,s} \sim 2.5 \cdot 10^{20}$ cm^{-2}. Since the spectrum of the filament seen outside of Puppis-A is in agreement with such a low temperature spectrum it is concluded that the filament indeed continues

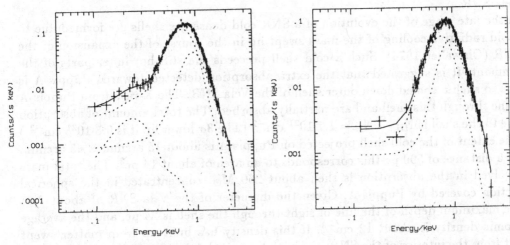

Fig. 10. PSPC spectrum of a circular region in the south of Puppis-A.

Fig. 11. PSPC spectrum of a circular region in the centre of Puppis-A.

across Puppis-A. In particular the low absorption which is about a factor of 5 less strengthens the view that the filament is much closer than Puppis-A and that it is associated with the Vela SNR.

A second PSPC spectrum has been cut out from a circular region of Puppis-A, which is centered on the spectrally hard bar-like structure (Fig.11; centre: $\alpha(2000)$ = 8^h 22^m 5^s, $\delta(2000)$ = -43° 3' 0"; r = 8 arcmin).

From this spectrum several conclusions can be drawn. Since the spectrum does not show a low energy precursor like the spectrum of the filament region displayed in Fig. 10, the soft X-ray emission from the filament is confined to the region south of the bar-like structure, i.e. the filamentary topology outside of Puppis-A is maintained across Puppis-A. Secondly, the harder spectrum of the bar-like region is not due to a different intrinsic spectrum of Puppis-A but it is due to a higher photoelectrical absorption indicated by a low energy cut-off which is shifted towards higher energies. An acceptable fit with a two temperature Raymond-Smith spectrum is achieved with $kT_{1,h} \sim 0.22$ keV, $kT_{2,h} \sim 1.1$ keV and $N_{H,h} \sim 2.9 \cdot 10^{21}$ cm^{-2}. The temperatures found are identical between the two Puppis-A areas studied here, underlining that there are little intrinsic spectral variations. A major difference, however, is observed in the values for the absorption. Of course, the changes of N_H could arise from anywhere along the line of sight towards Puppis-A, although no molecular clouds have been found in this particular region (Murphy & May, 1991), which would indicate some extra absorption. However, the alignment of the bar-like structure with the Vela filament and the existence of this filament across the Puppis-A region suggest that the absorption is physically associated with the Vela SNR.

In the late stage of the evolution of a SNR cold dense gas shells are formed due to rapid radiative cooling of the mass swept up in the course of the expansion of the SNR (Chevalier 1974). Such a cold shell preceeds the still hot inner parts of the remnant. It is suggested that the extra absorption detected towards Puppis-A is due to such a cooled down outer shell of the Vela SNR. The X-rays from Puppis-A shine through this shell and are partially absorbed. The total amount of absorption due to the shell is $N_{H,h}$ - $N_H = 1.7 \cdot 10^{21}$ cm^{-2}.(The 3σ lower limit is $0.8 \cdot 10^{21}$ cm^{-2}.) The extent of the cold shell projected on Puppis-A is about 15 arcmin $\times$ 45 arcmin. At a distance of 500 pc this corresponds to an area of about 14 pc^2. The total mass involved in the absorption is then about 260 $M_\odot$, concentrated in the spherical section covered by Puppis-A. Given the diameter of the Vela SNR of about $7.3°$ the maximum depth of the line of sight through the shell is 24 pc, and the average atomic density is about 12 cm^{-3}. If this density has built up from matter swept out from the interior of the SNR, the initial density of the interstellar medium, into which the Vela supernova was expanding, amounts to about 2.4 cm^{-3}, which is not unlikely for a distance about 30 pc above the galactic plane. The X-ray image (Fig. 6, 7) seems to support the view that the interior of the remnant close to the south-east of the filament is deficient of mass by the fact that the X-ray surface brightness is extremely low in that area. Whether this process has been at work on the remnant as a whole is difficult to state. Absorption structures similiar to that found in front of Puppis-A have not been detected at other locations along the Vela SNR perimeter so far, although this finding is not conclusive as there is no other background source as bright as Puppis-A close to the perimeter of the Vela SNR.

5 CONCLUDING REMARKS

In the previous chapters ROSAT PSPC observations of some supernova remnants have been used to illustrate the spectroscopic capability of ROSAT. Of course, PSPC spectra have been taken for other classes of objects in the context of a large variety of astrophysical themes, which include: the diffuse galactic X-ray emission and the quest for a hot galactic halo; obscuration / shadowing of the diffuse galactic X-ray emission by interstellar clouds; temperature and emission measure distribution of stars throughout the Hertzsprung-Russell diagram; temperature of the atmosphere of white dwarfs; surface temperature of neutron stars; spectral components and emission mechanisms in X-ray binaries and cataclysmic variables; supersoft X-ray sources and their relation to low mass X-ray binaries; the hot interstellar medium in the Large Magellanic Cloud and the Andromeda Nebula; the spectral components in Active Galactic Nuclei (AGN) and the systematic search for a soft excess in their spectra; spectral changes in AGN with redshift; search for absorption troughs in the spectra of BL Lacertae objects; spectrum of the extragalactic background emission and the relation to the spectra of high redshift AGN; the spatial temperature distribution in clusters of galaxies; temperature measurements of cooling flows; search for sub-structures in clusters of galaxies both spatial and thermal; and many more.

With almost two years in operation the ROSAT mission has collected a wealth of new astrophysical data already, the analysis and interpretation of which have just started. An avalanche of scientific results is expected to roll in the next few months when the existing data will be distributed to the international guest observer community.

REFERENCES

Aschenbach, B. 1988, *Appl Optics* 27, 1404.

Aschenbach, B. 1988a, in
 IAU Colloquium Vol. **101**, *Supernova Remnants and the Interstellar Medium*,
 ed. R. S. Roger and T. L. Landecker, Cambridge University Press, p. 99.

Charles, P., Kahn, S. and McKee, C. 1985, *Ap. J.* 295, 456.

Chevalier, R. A. 1974, *Ap. J.* 188, 501.

Grader, R., Hill, R. and Stoering, J. 1970, *Ap. J. (Letters)* 161, 45.

Harnden, F. R., Jr., Grant, P. D., Kahn, S. M. and Seward, F. D. 1985,
 Ap. J. 299, 828.

Kahn, S. M., Gorenstein, P., Harnden, F. R. Jr. and Seward, F. D. 1985, new-line/indent *Ap. J.* 299, 821.

Kayat, M., Rolf, D., Smith, G. and Willingale, R. 1980, *M.N.R.A.S.* 191, 729.

Ku, W., Kahn, S., Pisarski, R. and Long, K. 1984, *Ap. J.* 278, 615.

Leahy, D. A., Fink, R. and Nousek, J. 1990, *Ap. J.* 363, 547.

McKee, C. F. and Cowie, L. L. 1975, *Ap. J.* 195, 715.

Morrison, R. and McCammon, D. 1983, *Ap. J.* 270, 119.

Murphy, D. C. and May, J. 1991, *Astr. Ap.* 247, 202.

Pfeffermann, E., et al. 1986, *Proc. SPIE* 733, 519.

Raymond, J. C. and Smith, B. W. 1977, *Ap. J.Suppl.* 35, 419.

Seward, F. D. 1991, in *Imaging X-Ray Astronomy*, ed. M. Elvis, Cambridge
 University Press, p. 241.

Vedder, P. W., Canizares, C. R. and Markert, T. H. 1986, *Ap. J.* 307, 269.

Winkler, P. F., Tuttle, J. H., Kirshner, R. P. and Irwin, M. J. 1988, in
 IAU Colloquium Vol. **101**, *Supernova Remnants and the Interstellar Medium*,
 ed. R. S. Roger and T. L. Landecker, Cambridge University Press, p. 65.

Iron K Line Diagnostics in Astrophysical Sources

Luigi Piro

Istituto Di Astrofisica Spaziale, C.N.R., Frascati, Italy

ABSTRACT

A brief review on the mechanisms of iron K line formation and the use of this feature as diagnostics of the emitting medium in diverse astrophysical setting is presented, with particular regard to the case of Active Galactic Nuclei. Some future perspectives for the missions of the coming decade are illustrated.

1 INTRODUCTION

The iron K line at 6.4-6.9 keV is a common, often sole, feature in the X-ray spectra of several classes of X-ray sources, from normal and nearby objects like stars to distant and exotic ones like Active Galactic Nuclei (AGN).

The main reason of the ubiquity of this feature is the large iron abundance, in fact the greatest among the (heavy) elements which can produce a line at energies greater than 2-3 keV. For example, in the case of fluorescent emission, typical of accretion-driven X-ray sources like X-ray binaries and AGN, such an abundance, along with the values of the photoelectric cross section and the fluorescent yield, ($\omega_k \simeq 0.34$ to 0.7 depending on the ionization state), can account for equivalent width (EW) of hundreds eV. Lines with even larger EW ($\simeq 1$ keV) can originate in hot thermal plasma, such as in clusters of galaxies and stellar coronae, where the line formation mechanism is collisional excitation.

The iron line represents an invaluable tool for understanding the physical, geometrical and kinematical properties of the medium in X-ray sources. One of the most significant example was the discovery of the iron line at 6.7 keV in clusters of galaxies (Mitchell et al. 1976, Serlemitsos et al. 1977) that clearly established the thermal origin of the X-ray emission of these objects.

In this paper, after a brief review of the mechanisms of iron line production in astrophysical settings, I will describe the use of iron line as diagnostics of the medium in X-ray sources, with particular regard to AGN, and point out some future perspectives opened by recent observational and theoretical work.

2. MECHANISMS OF IRON LINE FORMATION

The most important physical processes of iron K line formation include photonization of the K shell followed by fluorescence for atoms in ionization state I-XXIV, collisional eccitation followed by radiative decay in Fe XXV-XXVI and recombination of electrons with FeXXVI-XXVII. The first mechanisms is the main process of line formation in photoionized gas (although recombination may play an important role if the medium is heavily ionized), whereas collisional eccitation is the main mechanism in coronal plasma.

2.1 Coronal plasma

In this instance the heat input is coupled directly to ions and electrons (that is $T_e = T_i$), line and continuum emission are optically thin and the density is low enough that all ions are in the ground state. The ionization stage is controlled by the balance between electron collisional ionization and radiative or dielectronic recombination. Radiative de-excitation following collisional eccitation is the main mechanism of line emission.

Spectra of coronal plasma have been calculated by several workers (e.g. Raymond and Smith 1977, Mewe et al. 1985, 1986). This emission mechanism is typical of stellar coronae, old supernova remnants, hot gas in elliptical galaxies and hot intracluster gas in clusters of galaxies. Lines of He like iron (E=6.7 keV) with EW $\simeq 1$ keV are commonly observed.

2.2 Non-equilibrium plasma in young supernova remnants

Heat input, usually produced by a shock, is coupled directly only to ions. Due to the extreme low densities, the time scale of thermalization of electrons by Coulomb collisions is of the order of 10^5 years, much greater than the age of the remnant. On the other hand, observational evidence of a continuum with a temperature of several keV (Pravdo and Smith 1979) indicates that, in fact, Te=Ti, arguing for additional and stronger mechanisms of electron - ion interaction. A non - equilibrium situation develops in the ionization structure because the timescale for collisional ionization is much greater than the age of the remnant. The gas is therefore underionized relatively to the electron temperature, that explains the presence of strong emission lines, typical of low temperature plasma, along with a hot continuum component. As the ionization stage increases towards the equilibrium , the centroid of iron K line shifts from 6.4 to 6.9 keV (Preite-Martinez and Fusco-Femiano 1983, Kaastra and Bleeker 1990).

2.3 Photoionized medium

Heat input is provided by Compton heating and photoionization by X-ray photons produced from an external source. The ionization equilibrium is obtained when photoionization balances recombination.

At variance with the previous case, photoionization of energetic photons makes the gas overionized compared to the electron temperature. For example FeXXV (He like) can be the predominant state for $T=10^6$ K in a photoionized medium (Kallman and McCray 1982), whereas a temperature about 30 times greater is necessary in a coronal plasma in order to have the same ionization stage.

Iron K line is mostly produced by fluorescence of FeI-XXIV and recombination for higher ionization states. The status of the gas is well described by the ionization parameter, defined as $\xi = L/(n_e R^2)$, where L is the luminosity of the ionizing continuum and R the distance of the medium from the X-ray source (Tarter, Tucker and Salpeter 1969, Kallman and McCray 1982).

Iron K lines observed in all accretion-powered systems like X-ray binaries, cataclysmic variables, AGN and galactic black hole candidates are produced by this mechanism. Line energy ranges from 6.4 keV of neutral iron, typical in AGN, X-ray pulsars and cataclysmic variables to 6.7-6.9 keV of He-like H-like iron, in low mass X-ray binaries.

3 IRON LINE DIAGNOSTICS

Several information on the medium can be obtained from the line energy, width and intensity, that is from parameters that can be rather well determined with proportional counters and GSPC's ($\Delta E/E \simeq 10\text{-}20\%$), the typical detectors of X-ray astronomy. With the incoming generation of solid state detectors the profile of the line will provide a further tool to probe the status of the medium.

3.1 Line energy

The line centroid is a slowly increasing function of the ionization state, going from 6.40 keV in FeI to 6.45 in FeXVII, and then increasing steeply to 6.7 keV in FeXXV (He like) and 6.9 keV in FeXXVI (H like). The measurement of the line energy thus give a direct indication of the ionization state of the medium.

The line centroid is also affected by Doppler and gravitational redshift, that can be substantial in the case of line emission by an accretion disk around a black hole (Fabian et al. 1989, Stella 1990, Matt, Perola, Piro 1991, Matt et. al 1992, Laor 1990), as might be the case of AGN and galactic black hole candidates. The magnitude of the effect depends on the disk inclination, with a redshift as large as 0.3 keV for a face on configuration (Matt, Perola, Piro 1991). Some authors (Hayakawa 1991, Matsuoka et al. 1991) argue that in the Seyfert 1 galaxy NGC 6814, given the vicinity of the line emitting medium to the central X-ray source (Kunieda et al. 1990), the matter could be highly ionized and the line emitted around 6.7 keV would be then gravitationally redshifted at the observed energy of 6.4 keV.

Another effect that can produce a considerable energy shift (and line broadening - see below) is Compton scattering. In each scattering the photon energy shift is $\Delta \epsilon = \epsilon(4kT_e - \epsilon)/(m_e c^2)$. The ultimate energy shift of a photon escaping from the medium will depend also on the optical depth of the medium, that gives the number of scattering before escaping, and on the geometrical configuration. Detailed models have been worked out by several authors (see Kallman and White 1988 for a review). An approximate estimation can be obtained in the cold case by multiplying the average shift per scattering ($\Delta\lambda = \Delta\lambda_{max}/2 = \lambda_c = h/m_e c = 0.024$ A, corresponding to $\Delta\epsilon = 80$ eV at 6.4 keV) by the number of scattering before escaping which is $n \simeq \tau^2/3$ for a source located at the center of an optically thick sphere. For example, for $\tau=4$ we expect a redshift of about 0.4 keV.

Comptonization and Doppler/gravitational shifts have been proposed in alternative to account for the very broad (FWHM=1-2 keV) and redshifted (E=5.9-6.2 keV, but see Tanaka 1990) iron lines observed in galactic black hole candidates Cyg X-1 (Fabian et al. 1989) and 4U1543-47 (van der Woerd, White and Kahn 1989).

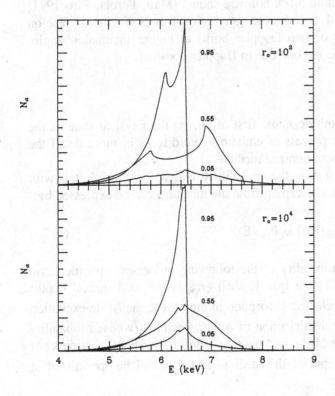

Fig.1- Line profiles from an accretion disk around a Schwarzschild black hole as a function of cosine of inclination and for two different values of the outer radius r_o (units of GM/c^2, from Matt et al. 1992)

3.2 Line width

Several processes can cause line broadening.

- Unresolved blending of different narrow line components from different ionization states, as could be the case of the Seyfert 2 galaxy NGC 1068 observed by BBXRT (Petre, this conference)

- Thermal broadening. This effect is usually negligible, being $\Delta E=E(kT/Am_pc^2)^{1/2}$ (e.g=0.05 eV for T=2 10^4 K)

- Compton broadening due to multiple scattering by free electrons, described in the previous section, that can account for widths as large as 3 keV for kT=10 keV and τ=5. In the case of fluorescence by an optically thick cold slab seen in reflection energy shift and line broadening are negligible (Matt, Perola, Piro 1991)
- Doppler and gravitational broadening due to the bulk motion of the medium around a collapsed object. In the case of Keplerian motion $\Delta E \simeq (r/50r_s)^{-1/2}$ keV, where r_s is the Schwarzschild radius and M the mass of the central black hole, and can be of the order of 1-2 keV. In fig.1 line profiles in the case of fluorescence from an X-ray illuminated accretion disk around a Schwarzschild black hole are shown (Matt, Perola, Piro 1991; Matt et al. 1992). Note the red tail due to gravitational redshift for a face-on configuration and the appearance of two Doppler horns at higher inclination angle, with the blue horn brighter than the red one due to Doppler boosting.

3.3 Equivalent width

The iron K line equivalent width depends, first of all, on the physical state of the medium, that determines the main process of emission. We discuss in more detail the case of fluorescent emission by a photoionized medium.

The fluorescence efficiency (that is the probability that a ionizing photon with $E > E_K$ produces a fluorescent photon escaping from the medium) can be expressed by:

$$P_f(E) = [(1 - \exp(-\tau_{tot}(E)(\tau_{Fe,K}(E)/\tau_{tot}(E))] \, \omega_k P_{esc}(E) \tag{1}$$

where each term represents the probability of the following processes: photoelectric absorption of the incident photon in the iron K shell (τ_{tot} is the total optical depth; $\tau_{Fe,K}$ is the optical depth for photoelectric absorption in the iron K shell); de-excitation by emission of fluorescence rather than ejection of Auger electrons (whose probability is the fluorescent yield ω_k); escape of the line photons from the medium, that depends of the geometry and the optical depth of the medium and that will be specialized in some particular case described below.

The equivalent width is then given by:

$$EW = \frac{\Delta\Omega}{4\pi} \int_{E_k}^{\infty} dE \, P_f(E) \, F(E)/F(E_\alpha) \tag{2}$$

where F(E) is the incident (photon) spectrum, E_K and E_α the iron edge and line energy and $\Delta\Omega$ the solid angle subtended from the medium to the X-ray source.

In the optically thin case ($N_H < < 10^{24}$ cm^{-2}) and $F(E) \simeq E^{-\alpha}$ eq (2) reduces to

$$EW = \frac{\Delta\Omega}{4\pi} \, \omega_k \, N_{Fe} \, \sigma_{K0} \, E_\alpha/(\alpha+2) \, (E_\alpha/E_K)^{(\alpha-1)} \tag{3}$$

where the iron cross section $\sigma_{Fe,K} = \sigma_{K0} (E/E_K)^{-3}$. Thus only photons with $E > E_K$ give a significant contribution to the fluorescent line.

Note that the EW is directly proportional to the cross section, the fluorescent yield and the column density of iron and weakly increasing with the hardness of the spectrum. The fluorescent yield raises slowly from 0.34 for FeI to 0.49 in FeXXII, then in FeXXIII drops to $\omega_k = 0.11$ and in FeXXIV increases back up to 0.75 (Krolik and Kallman 1987 and references therein). For the last two stages, FeXXV and FeXXVI, line photons are produced by recombination but it is possible to define an effective yield, $\omega_k = 0.5$ and 0.7 respectively (Osterbrock 1974). Note however that when ξ increases above $10^{3.5}$ the EW reduces as iron atoms become fully ionized (Kallman and McCray 1982). The increasing trend of the fluorescent yield is somewhat compensated by the decreasing cross section σ_{K0}, that goes from $3.2 \ 10^{20} \ cm^{-2}$ for neutral iron to $2.2 \ 10^{20} \ cm^{-2}$ for FeXXIV (Reilman and Manson 1979).

For a medium in a neutral state equation (4) becomes

$$EW = \frac{\Delta\Omega}{4\pi} Y_{Fe} (N_H/10^{24}) \ keV. \tag{4}$$

where Y_{Fe} is the iron abundance normalized to the solar value.

Let us now consider an optically thick slab seen from the side illuminated by incident X-rays (reflection). Equation (1) becomes:

$$P_f(E) \simeq \frac{Y_{Fe}\sigma^*_{Fe,K}(E)}{Y_{Fe}\sigma^*_{Fe,K}(E) + Y\sigma^*_L(E) + 1.2\sigma_T} \ 0.5 \ \exp(\frac{Y\sigma^*_L(E_\alpha)}{Y_{Fe}\sigma^*_{Fe,K}(E) + Y\sigma^*_L(E) + 1.2\sigma_T}) \tag{5}$$

where $\sigma^*_{Fe,K}$ and σ^*_L are the photoelectric cross sections of iron K shell and lighter elements per hydrogen atom for a medium with solar-like abundances ($Y_{Fe} = Y = 1$) and P_{esc} has been approximated assuming that half of the line photons escape the medium through a column density corresponding to $\tau_{tot} = 1$. At variance with the optically thin case, the EW is roughly proportional to the iron abundance only for $Y_{Fe} \simeq 1$ then, at $Y_{Fe} >> 1$, saturates. Another difference is the dependency on the optical depth for photoelectric absorption of lighter elements, that determines the fraction of photons photoabsorbed by the iron K shell and the escape probability. In fig.2 I show the dependence of $P_f(E_K)$ on Y and Y_{Fe}, computed taking into proper account the radiative transfer (Basko 1978). An EW of about 100-150 eV is expected to be produced by cold slab of solar-like composition seen face-on and intercepting half of the photons produced by the X-ray source.

4. X-RAY REPROCESSING IN AGN

Before the launch of the GINGA satellite iron K lines with were observed only in a few AGN which showed strong low energy absorption ($N_H \simeq 10^{23}$ cm^{-2}) like Centaurus A (Mushotzky et al. 1978) and NGC 4151 (e.g. Perola et al. 1986) and could be attributed to fluorescence from the same matter causing the absorption (eq. (4)).

The large area and small background afforded by GINGA allowed the detection of an iron line at 6.4 keV with EW $\simeq$ 150 eV and a new flat component above 8 keV in several bright Seyfert galaxies (Matsuoka et al 1990, Piro et al. 1990, Pounds et al. 1990, Nandra et al. 1990). The small low energy absorption ($N_H < 10^{22}$ cm^{-2}) could not account for the presence of the line (cfr. eq.(4))

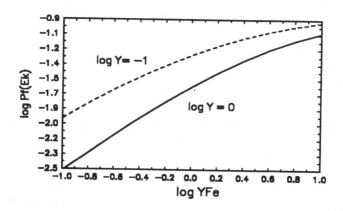

Fig.2 - Fluorescence efficiency $P_f(E_K)$ vs. Y and Y_{Fe}, after Basko (1978), when the incident beam of ionizing photons is normal to the surface of the medium.

Previous theoretical work by Lightman and White (1988) and Guilbert and Rees (1988) had pointed out that a cold thick medium in the environment of the central X-ray source would in fact produce two characteristic X-ray signatures by reprocessing the primary continuum. Along with the iron line we have discussed about in the previous section, a reflected continuum, produced by Compton scattering and photoelectric absorption of the primary radiation (whose albedo is shown in fig.3), was expected. More detailed models of a thick slab illuminated by an X-ray source have been more recently carried out by Matt, Perola and Piro (1991) and George and Fabian (1991). The X-ray spectra observed by GINGA (an example in fig.4) are in good agreement with that expected from reprocessing of the primary continuum by a cold thick slab subtending 2π sr to the central source and seen face-on.

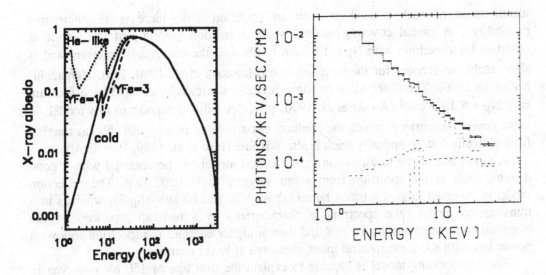

Fig.3 - X-ray albedo of a cold and highly ionized thick medium.

Fig.4 - X-ray spectrum of IC4329A fitted with a reflection model (from Piro et al. 1990). The different components (direct and reflected continuum, iron line) are shown with dashed lines.

5. OPEN ISSUES AND OBSERVATIONAL PROSPECTS

The most common scenario of the central power-house of an AGN and galactic black hole candidates encompass an accretion disk around a massive black hole. Detailed computation of the reprocessed spectrum have been carried out by Matt, Perola and Piro (1991) and Matt et al. (1992). The main difference from reprocessing by a slab are in the line profile, heavily affected by gravitational and Doppler shifts (fig.1).

X-ray mission based on solid state spectrometers like BBXRT or, in the near future, JET-X, SODART/SIXA, ASTRO-C have the capability of resolving the line profile sharply, thus opening the possibility of mapping the innermost regions of the medium in the environment of the central source. On the other hand the limitation of bandwidth to energy < 10-15 keV will hinder the collection of the other crucial piece of information, that is the reflected continuum. Note also that our present information on the reflected continuum, collected with GINGA, is restricted below 30 keV and is of limited statistical quality, since above 10 keV the efficiency of the detectors is rapidly decreasing. A good determination of both the iron line and the reflected continuum is decisive for a full understanding of several open issues regarding the reprocessing medium. I will describe a few of those in the following.

Although an explanation of the iron line and high energy bump in terms of reprocessing from a thick medium is widespread, particularly because of the tempting identification of such a medium with an accretion disk, there is an alternative possibility. A partial covering model, where a fraction ($\simeq 50\%$) of the source is absorbed by a medium with $N_H \simeq 10^{24}$ cm^{-2}, fits well the continuum spectrum and is also capable to account for the emission line (Matsuoka et al. 1990, Piro, Yamauchi, Matsuoka 1990). The observation of some Seyfert galaxies fully covered by a medium with $N_H \simeq 5 \cdot 10^{23}$ cm^{-2} (Awaki et al. 1990) provides indirect support to this model. A more general scenario in which the medium has a column density distribution ranging from optically thin to optically thick is also possible (Piro et al. 1989, Piro 1990).

A direct test of the transmission vs. reflection model can be obtained with a good determination of the spectrum from a few keV up to 100-200 keV. The reflection model is expected to give a broad hump between 20 and 60 keV (fig.3), whereas in a transmission model the spectrum is characterized by a turn up produced by the absorbed component around 10 keV and then at higher energies the spectrum follows a power law with the same spectral index measured at lower energies

The reprocessing model is capable to explain the iron line of 150 eV observed in the majority of unobscured Seyfert galaxies. On the other hand there are a few Seyfert galaxies like NGC 6814 (Kunieda et al. 1990), Mkn 841 (Day et al. 1990) and MCG-5-23-16 (Piro, Matsuoka, Yamauchi 1991 a,b) which show an iron line with an EW=300-400 eV. A greater solid angle of the reprocessing medium or a primary X-ray continuum beamed towards the reprocessor provide possible explanations to such strong lines. In all these situations the increment of the line EW is associated with a comparable increase of the level of the reflected component. However in MCG-5-23-16 and NGC 6814 Piro, Matsuoka and Yamauchi (1991a,b) obtained only an upper limit to the value of the reflected component, which is well below that predicted from the EW. Increasing the iron abundance gives a larger EW (and a lower albedo in the GINGA range -see fig.3), although the non linear dependence of EW on Y_{Fe} (fig.2) requires a very high value, $Y_{Fe} \simeq 10$, in order to have an upper limit on the reflected component consistent with the prediction. The other possibility explored to increase the EW is an highly ionized medium. In this situation the EW is larger (up to a factor of 4) than the cold case, due to the higher fluorescent yield and the lower photoelectric absorption of the line by elements lighter than iron, that are nearly or completely ionized (eq.5). The reflected continuum is now characterized by a very deep iron edge (fig.3) which is however not observed in the data. The conclusion is that in these objects X-ray reprocessing from a thick medium, although cannot be completely excluded, does not provide an explanation as direct and elegant as that offered in objects with EW$\simeq$150 eV. Clearly, a better determination of the spectrum above 10 keV along with a good measurement of line parameters, are necessary.

The SAX satellite, developed in Italy in collaboration with Holland and ESA (e.g. Butler and Scarsi 1990, Perola 1991), with its three orders of magnitude bandwidth, from 0.1 to 200 keV, will provide the unique opportunity, among the various satellites of the coming decade, to address those problems. To show the capability of SAX to determine with good accuracy both the line and the spectrum above 10 keV I have simulated a 30000 s. exposure of a bright Seyfert galaxy with reprocessing features like

IC4329A (spectral parameters as determined with GINGA by Piro, Yamauchi and Matsuoka 1990). In fig.5 the data are shown along with a best fit single power law., which is evidently unacceptable. Note the high statistical significance of the iron line and the excess at high energies.

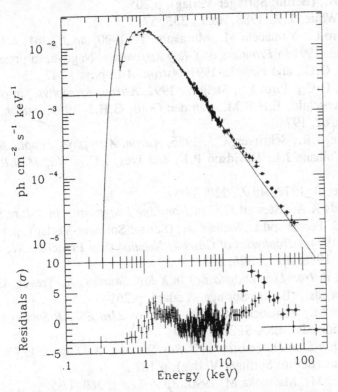

Fig.5 - Simulation of a 30000 s. observation by SAX of the Seyfert galaxy IC4329A. Note the excess at high energy and around 6 keV above the fit with a single power law represented by a solid line

6 REFERENCES

Awaki H et al. 1991, preprint
Basko M.M. 1978, *Ap.J.*, **223**, 268
Butler R.C., Scarsi L. 1990, SPIE proc., Vol.1344, p.465
Day C.R.S., Fabian A.C., George I.M., Kunieda H. 1990, *M.N.R.A.S.*, **247**, 15p
Fabian A.C., Rees M.J., Stella L., White N.E. 1989, *M.N.R.A.S.*, **238**, 729
George I.M., Fabian A.C. 1991, *M.N.R.A.S.* **249**, 352
Guilbert P.W., Rees M.J. 1988, *M.N.R.A.S.*, **233**, 475
Hayakawa S. 1991, *Nature*, **351**, 214
Kaastra, J.S., and Bleeker, J.A.M. 1990, in *Iron Line Diagnostics in X-Ray Sources*,
    A. Treves, G.C. Perola and L. Stella eds., (Berlin: Springer-Verlag), p.35

Kallman, T.R., and McCray, R. 1982, *Ap.J.Suppl. Ser.* **50**, 263

Kallman T.R., White, N.E. 1988, *Ap.J.*, **341**, 955

Kunieda H. et al. 1990, *Nature* **345**, 786

Laor A. 1990 in *Iron Line Diagnostics in X-Ray Sources*, A. Treves, G.C. Perola and L. Stella eds., (Berlin: Springer-Verlag), p.205

Lightman A.P., White, T.R. 1988, *Ap.J.*, **335**, 57

Matsuoka M., Piro L., Yamauchi M., Murakami T., 1990, *Ap.J.*, **361**, 440

Matsuoka M. et al. 1991 in *Frontiers of X-Ray Astronomy*, Nagoya, in press.

Matt, G., Perola, G.C., and Piro, L. 1991, *Astron. Astrophys.*, **247**, 25

Matt, G., Perola, G.C., Piro, L. , Stella L. 1992, *Astron. Astrophys.*, in press

Mewe, R., Gronenschild, E.H.B.M., van den Oord, G.H.J. 1985, *Astron. Astrophys. Suppl. Ser.*, **62**, 197

Mewe, R., Lemen, J.R., Schrijver, C.J. 1986, *Astron. Astrophys. Suppl.*, **65**, 511

Mitchell, R.J., Culhane J.L. Davidson P.J., and Ives, J.C. 1976, *M.N.R.A.S.*, **175**, 29p

Mushotzky R.F. et al. 1978, *Ap.J.*, **220**, 790

Nandra K., Pounds K.A., Stewart G.C.in *Iron Line Diagnostics in X-Ray Sources*, A. Treves, G.C. Perola and L. Stella eds., (Berlin: Springer-Verlag), p.177

Osterbrock D.E. 1974 *Astrophysics of Gaseous Nebulae* (San Francisco:W. Freeman)

Perola G.C. et al. 1986, *Ap.J.*, **306,** 508

Perola G.C. 1990 in *Iron Line Diagnostics in X-Ray Sources*, A. Treves, G.C. Perola and L. Stella eds., (Berlin: Springer-Verlag), p.261

Piro L., Matsuoka M., Yamauchi M. 1989, proc. of *23rd ESLAB Symposium*, J Hunt and B. Battrick eds., ESA-SP219, p.819

Piro L. 1990 in *Iron Line Diagnostics in X-Ray Sources*, A. Treves, G.C. Perola and L. Stella eds., (Berlin: Springer-Verlag), p.187

Piro L., Yamauchi, M., Matsuoka M. 1990, *Ap.J.(Lett.)*, **360**, L35

——————— 1991, in *Frontiers of X-Ray Astronomy*, Nagoya, in press

——————— 1991, in *Physics of AGN*, Heidelberg, in press

Pounds K., Nandra K., Stewart G.C., George I.M., Fabian A.C. 1990, *Nature*, **344**, 132

Pravdo, S.H., and Smith, B.W. 1979, *Ap.J.(Lett.)*, **234**, L195

Preite-Martinez, A., and Fusco-Femiano, R. 1983, in *Non Thermal and Very High Temperature Phenomena in X-Ray Astronomy*, G.C. Perola and M. Salvati eds. (Roma: Istituto Astronomico), p.93

Raymond, J.C., Smith, B.W. 1977, *Ap. J. Suppl.*, **35**, 419

Reilman R.F., Manson S.T. 1979, *Ap.J. Suppl.*, **40**, 815

Serlemitsos, P.J., Smith, B.W., Boldt, E.A., Holt, S.S., and Swank, J.H. 1977, *Ap.J.(Lett.)*, **211**, L63

Stella L., *Nature*, **344G**, 747

Tanaka Y. 1990 in *Iron Line Diagnostics in X-Ray Sources*, A. Treves, G.C. Perola and L. Stella eds., (Berlin: Springer-Verlag), p. 98

Tarter, C.B., Tucker, W., and Salpeter, E.E. 1969, *Ap.J.*, **156**, 943

van der Woerd H., White N.E., Kahn S.M. 1989, *Ap.J.*, **344**, 320

X-Ray Spectroscopy with AXAF

T.H. Markert

Center for Space Research, Massachusetts Institute of Technology

ABSTRACT

NASA's Advanced X-Ray Astrophysics Facility (AXAF), currently scheduled for launch in 1999, will include 5 instruments capable of moderate to high-resolution X-ray spectroscopy in the energy range 0.1 - 10.0 keV. These are two transmission grating instruments, a CCD imaging spectrometer, an X-ray calorimeter, and a Bragg crystal spectrometer. Here we describe the 5 spectrometers and their spectroscopic capabilities, including simulations of their orbital performances. Finally, we illustrate (with an example taken from data obtained by 3 spectrometers on the Einstein X-ray Observatory) how moderate and high-resolution spectra, taken together, can be used to study the properties of a cosmic X-ray plasma.

1 INTRODUCTION

NASA's Advanced X-Ray Astrophysics Facility (AXAF) is undergoing development and is currently scheduled for launch in 1999. Figure 1 is a cut-away view, showing the principal parts of the spacecraft. The heart of AXAF is the assembly of high spatial resolution X-ray mirrors. There are 6 nested grazing-incidence mirrors, each of which is a hyperboloid/paraboloid pair. The peak effective area of the High Resolution Mirror Assembly (HRMA) is 1500 cm^2, and the image quality is specified to be better than 0.5 arc seconds diameter (FWHM) over the central 5 arm minute radius field.

Much of the recent effort in AXAF development has gone into fabricating the very sophisticated X-ray mirrors, and in verifying their required performances. During the past year, the outer mirror pair (the so-called P1H1 for paraboloid/hyperboloid) was cut and polished at Hughes Danbury Optical Systems and assembled at Eastman Kodak. In August of 1991 P1H1 was shipped to the Marshall Space Flight Center for X-ray testing in the MSFC X-ray Calibration Facility. There the properties of this first element of the HRMA were measured. The spatial resolving power of P1H1 was found to be much better than the requirements: the FWHM image of a point source at the Al K line (1.49 keV), for example, was less than 0.25 arc seconds. This result is quite encouraging, since it suggests that there are no overwhelming technical problems to be anticipated in the fabrication of the flight mirror assembly.

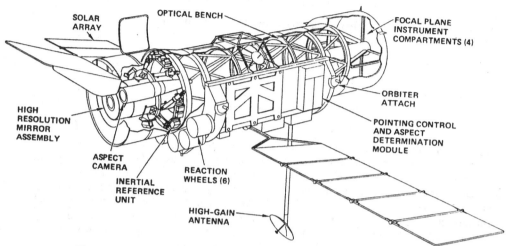

Fig.1—Cut-away view of the Advanced X-ray Astrophysics Facility (AXAF)

The high quality of the AXAF mirrors virtually guarantees that the images will be sharper than any previously obtained from X-ray telescopes, or from any currently being planned. Other important aspects of AXAF are its very long lifetime (the current design calls for 15 years, with occasional servicing and instrument replacement via the Space Shuttle), its usefulness at higher energies (it has an effective area greater than 10 cm^2 at energies as high as 9 keV), and its capability of extremely high spectral resolution measurements. This latter property results in part from the quality of the telescopes (for the transmission grating spectrometers) and in part from the inherent characteristics of the devices which are under development. In this paper we describe the spectroscopic instruments planned for AXAF, 4 of which are capable of extremely high spectral resolution (E/ΔE ~ 1000 at some energies).

2 SPECTROSCOPIC INSTRUMENTS

All of the AXAF instruments are listed in Table 1 We have divided the complement of devices into two categories, imagers and spectrometers, although the imagers also have spectroscopic capabilities. The High Resolution Camera (HRC), however, is a microchannel plate device and has a very small spectral resolution (E/ΔE ~ 1). It thus has a more limited spectroscopic utility than the other instruments. Consequently, we will not discuss it further here. More information about the HRC can be obtained in Murray *et al.*, 1987. We note that although the HRC has a rather small inherent resolution, it is one of the two readout devices for the Transmission Grating Spectrometers (discussed below) and thus plays an important role in AXAF spectroscopy.

TABLE 1 AXAF Instruments

	Principal Investigator
Imagers	
High Resolution Camera (HRC)	S. Murray (CfA)
AXAF CCD Imaging Spectrometer (ACIS)	G.Garmire (PSU/MIT/JPL)
Spectrometers:	
X-ray Spectrometer (XRS)	S. Holt (GSFC)
Bragg Crystal Spectrometer (BCS)	C. Canizares (MIT)
Low Energy Transmission Grating (LETG)	B. Brinkman (Utrecht)
High Energy Transmission Grating (HETG)	C. Canizares (MIT)

2.1 Transmission Grating Spectrometers (TGS)

There are two sets of transmission gratings on AXAF (Table 1). The gratings are assembled on support plates which can be rotated into place immediately behind the HRMA assembly (Figure 2). There the individual gratings intercept the X-rays from the mirrors and diffract them according to the grating equation ($n\lambda = p \times \sin\theta$, where λ is the X-ray wavelength, p is the period of the gratings, θ is the diffraction angle, and n, an integer, is the diffraction order). The diffracted X-rays then strike one of the two imaging detectors, from which the spectrum can be read. The gratings and the imaging detector approximate a Rowland circle of diameter 8.4 m. The Rowland circle geometry minimizes geometrical aberrations which can decrease the resolution of the instrument.

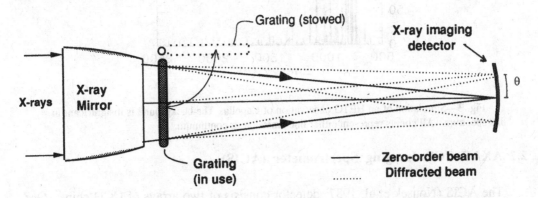

Fig. 2—Layout of the transmission grating spectrometers on AXAF. The grating assembly is a plate which can be swung into place behind the mirrors when in use. There are two such plates (only one is shown) for the LETG and the HETG assemblies. The detector array is an imager and, on AXAF, can be either the HRC or ACIS.

The Low Energy Transmission Grating (LETG, Brinkman et al. 1987) employs arrays of gold lines with period 10,000 Å (1 μm). The High Energy Transmission Grating (HETG, Canizares et al. 1987) has submicron structures of gold (p = 2000 Å = 0.2 μm) and silver (p = 6000 Å = 0.6 μm). These two sets of gratings are mounted on the same plate and diffract X-rays to different regions of the imaging detectors so there is no confusion between the two spectra. The LETG performs best with the HRC as its readout device, and operates over the range 0.08 - 1 keV, reaching a resolving power of ~2000 at the lowest energies. The HETG is well matched with a separate ACIS grating readout array (see section 2.2) in the range 0.4 - 8 keV, with resolving powers that reach ~1000 near 0.4 and 1 keV (see Figure 10). Although both transmission gratings are most effective for studying point sources, they can also be used as slitless spectrometers to make monochromatic images of the spatial structure in an extended source in the light of a strong emission line (similar to a spectroheliogram).

Figure 3 is a simulation of an observation performed with the HETG instrument. This figure illustrates the ability of a grating spectrometer to capture the spectrum of a stellar corona (in this case Capella).

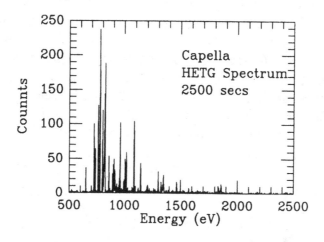

Fig. 3—Simulation of an HETG observation of Capella. The background is insignificant in this spectrum. All counts represent either line emission or continuum.

2.2 AXAF CCD Imaging Spectrometer (ACIS)

The ACIS (Nousek et al. 1987) detector consists of two arrays of CCD chips. One array is matched to the HRMA focal surface and is primarily intended for imaging. The second array, intended as a readout for the transmission gratings is a strip of 6 CCDs and lies approximately on the Rowland circle of the TGS. The imaging array has a field of view of 8 × 24 arc minutes, with individual pixels of dimension ~ 0.5 arc seconds.

The excellent imaging capability of the HRMA/ACIS combination is combined with remarkably good energy resolution (~50 eV at 1 keV and ~130 eV at 7 keV). This high spectral resolution allows ACIS to perform spatially resolved spectroscopy of high enough quality to isolate and measure individual line strengths. Figure 5 shows a simulation of an ACIS observation of a 4×10^7 K thin plasma. A number of lines and line are clearly discernable. It is important to remember that these kinds of spectra will be accompanied by sub-arc second quality imaging. One application of such spectral imaging will be to map out the temperature structure of cooling flows in clusters of galaxies.

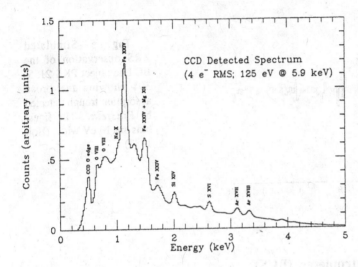

Fig. 4—Simulation of an ACIS observation of a thin plasma. The more prominent lines and line blends are noted

2.3 X-Ray Spectrometer (XRS)

The XRS is a microcalorimeter (see, for example, Holt 1987). The concept of microcalorimetry, and some recent advances in the field, are described elsewhere is this volume, and in particular in the article by Moseley. Briefly, an X-ray calorimeter measures the temperature increase in a mass as a result of the absorption of a single X-ray. For this scheme to work, the absorbing volume must be very small, and the temperature must be quite low, ~0.1 K. The low temperature, and the desire to operate the detector in space for several years, leads to a requirement for a large container of liquid helium and an additional refrigerator (to lower the temperature below the liquid helium level).

The XRS will employ an array of 36 rectangular calorimeter elements, each element having dimensions 0.25×1.00 mm (equivalent to a field of 5×20 arc seconds), arranged so that the XRS as a whole will have a 1×1 arc minute field of view. The energy resolution of a microcalorimeter of this design is, in theory, as small as 1 eV, and a practical XRS should have a ΔE (FWHM) better than about 10 eV (a sample element with

$\Delta E = 7.3 \pm 0.4$ K has been demonstrated). This high spectral resolution, coupled with the excellent sensitivity of the XRS to point sources, will make it an extremely valuable tool for studying the line emission from distant clusters of galaxies and from active galactic nuclei. It will also be sensitive to absorption features (Figure 5 shows a simulation of such an observation lasting only 100 seconds).

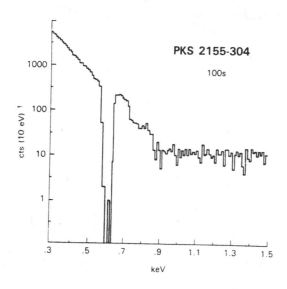

Fig. 5—Simulated XRS observation of the BL Lac object PKS 2155-304 showing the broad absorption trough detected by *Einstein*. The finest bins are 10 eV wide. (Holt 1987).

2.4 Bragg Crystal Spectrometer (BCS)

A schematic diagram of the Bragg Crystal Spectrometer (Canizares et al. 1987) is shown in Figure 6. X-rays from the HRMA pass through a small aperture (3 × 20 arc minutes), and strike one of three curved diffractors (particular diffractors can be selected by rotating the triangular crystal drum). Those X-rays which satisfy Bragg's condition ($n\lambda = 2d \times \sin\theta$, where d is the spacing of the crystal planes, n is the order of reflection [generally n = 1, or sometimes 2], λ is the wavelength and θ is the angle of incidence [the Bragg angle]) are reflected efficiently from the crystal surface and enter one of two redundant imaging gas flow proportional counters. A small wavelength (energy) is scanned by rocking the crystal over a range of ~1°.

The BCS is essentially an improved version of the Focal Plane Crystal Spectrometer (FPCS) that flew on the Einstein satellite between 1978 and 1981. The BCS achieves the highest resolving powers of any of the AXAF instruments (E/ΔE as high as 2000) and can be used effectively for observations of extended as well as point sources. It also has a fairly large field of view (3 × 20 arc minutes) within which it retains some imaging capabilities.

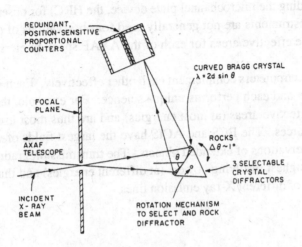

Fig. 6—Schematic diagram of the Bragg Crystal Spectrometer.

Figure 7 is a simulation of observations of the lines (resonance, intercombination and forbidden) of helium-like oxygen from a region in the interior of the supernova remnant Puppis A. Both the BCS and the XRS (assuming a 10 eV resolution) are simulated. The higher resolving power of the BCS clearly pays off here, since the three lines are cleanly resolved and the various plasma diagnostic tests can be made in a straightforward manner. Note also that the larger field of view of the BCS allows this instrument to accumulate a significant number of events in an observation time somewhat less than that of the XRS

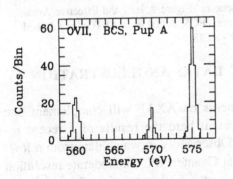 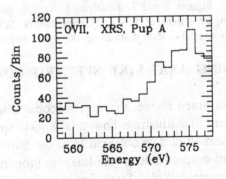

Fig. 7—Simulations of BCS and XRS (10 eV resolution) observations of lines from helium-like oxygen from the interior of the Puppis A supernova remnant. The XRS observation is 5000 s: the BCS observation is 3000 s.

2.5 Summary of Properties

Figure 8 shows the spectral resolving powers ($E/\Delta E = \lambda/\Delta\lambda$) of the various AXAF instruments (excluding the microchannel plate device, the HRC) for observations of a point source (the TGS instruments are not generally used for observations of extended objects). Figure 9 shows the effective areas for each of the AXAF spectrometers (again excluding the HRC).

The AXAF spectrometers complement each other effectively. Each instrument fills an observational niche and each performs unique science. For example, the XRS and ACIS have the highest effective areas (at most energies) and are thus most useful for studies of distant and dim sources. The BCS and ACIS have the largest fields of view and thus are appropriate for observations of extended objects. The transmission gratings, the BCS, and the XRS have the highest resolving powers (at different energies) and thus can be used for plasma diagnostics of different X-ray emission lines.

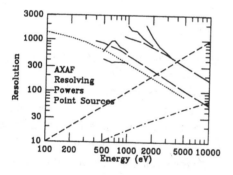 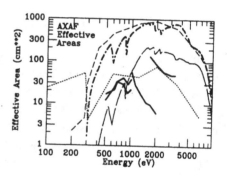

Figures. 8 and 9—Resolving powers for point sources (Figure 8, left) and Effective Areas (Figure 9, right) of the AXAF spectrometers. Solid lines - BCS; dotted line - LETG; long dashed lines - HETG; short dashed line - XRS; dot-dashed line - ACIS.

3 USING AXAF-LIKE SPECTROSCOPIC DATA: AN ILLUSTRATION

As noted above, the spectroscopic instruments on AXAF will complement one another. To illustrate how this works, we present here the results of a recent re-examination of data obtained from the *Einstein* Observatory in which data from a low-resolution spectrometer (the Imaging Proportional Counter, IPC), a moderate resolution spectrometer (Solid State Spectrometer, SSS) and a high-resolution Bragg crystal spectrometer (FPCS) are combined to reveal a clearer picture of the supernova remnant N132D, which is in the Large Magellanic Cloud (Hwang et al. 1992).

Using the observations made in the spectrometers broad-band instruments (the SSS and IPC), we obtained low- and moderate-resolution spectra, which we then fit to a model non-equilibrium plasma. The best-fit values of the plasma temperature, T, and ionization timescale, τ, ($\equiv n_e \times t$ = electron density $\times$ time since the plasma was initially shock-heated) were determined.

Figure 10 shows the FPCS scans of several lines or line blends in N132D. The much higher resolving power of the FPCS reveals a number of bright emission lines and line complexes which would not be discernable in the lower-resolution data.

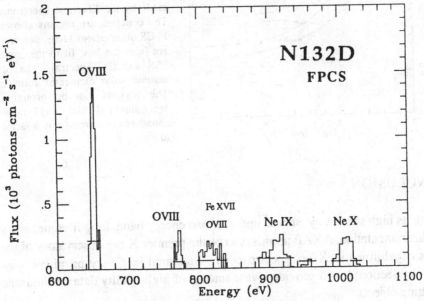

Fig. 10—Composite of several FPCS scans of N132D.

We used the non-equilibrium model and applied it to ratios of discrete line fluxes as measured by the FPCS. Flux ratios are useful since many of the unknown parameters of the plasma will cancel upon division. Ratios of ions of the same element are used to constrain T and τ jointly, independent of the elemental abundance. Having estimated T and τ, flux ratios of lines of similar energy in different elements can be used to constrain elemental abundances (relative to the solar abundances).

Figure 11 shows the regions of (T,τ) space allowed by two intensity ratio measurements made with the FPCS (Ne IX/Ne X and Fe XVII/O VIII) Also shown is the uncertainty region for the SSS/IPC measurements. We have also constructed similar plots for other line ratios. The net result is that all of the plots are in agreement, except for the iron to oxygen plots and, to a lesser extent, the neon to oxygen plots. These can be brought into agreement with the other observations by adopting an abundance ratio for oxygen (with respect to iron and neon) which is 5-7 times the solar value (for iron) or 1-4 times the solar value (for neon). Such relative abundances are consistent with some computations of the composition of the ejecta from a 20 M_O Type II supernova. Note that *all* of the data (SSS,IPC, and FPCS) was required for us to obtain this result.

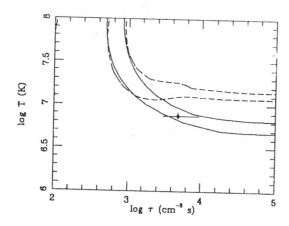

Fig. 11—Regions of the temperature-ionization time parameter space allowed by various observations made by the *Einstein* spectrometers. The contours are regions allowed by FPCS observations (2σ); the error bars are from the best fit to the combined SSS and IPC spectra. The fits all assume solar elemental abundances. The regions can be brought into agreement if the oxygen-to-iron abundance is increased by a factor of 5 to 7.

4 CONCLUSION

With its high sensitivity, superb optics, broad energy band, long lifetime, and state-of-the-art instrumentation, AXAF promises to be the premier X-ray observatory of the future. AXAF's capabilities in X-ray spectroscopy, illustrated in a very primitive way by the example in Section 3, will provide a vast amount of high-quality data on thousands of X-ray emitting objects.

I am grateful to all of the AXAF science teams for their contributions to this paper. I particularly acknowledge Prof. Claude Canizares. I also thank Una Hwang and Jack Hughes for the N132D analysis, and Mark Bautz, Dan Dewey, Ken Lum, and Brenda Parsons for assistance with the figures and the manuscript.

REFERENCES

Brinkman, A.C. *et al.* 1987, Astro. Lett. and Communications, **26**, 73..
Canizares, C.R., *et al.* 1987, Astro. Lett. and Communications, **26**, 87.
Holt, S.S. 1987, Astro. Lett. and Communications, **26**, 61.
Hwang, U., Canizares, C.R., Markert, T.H. and Hughes, J.P. 1992, in preparation.
Murray, S.S., *et al.* 1987, Astro. Lett. and Communications, **26**, 113.
Nousek, J.A., Garmire, G.P., Ricker, G.R., Collins, S.A., and Reigler, G.R. 1987, Astro. Lett. and Communications. **26**. 35.

Spectroscopy with XMM

A.C. Brinkman

SRON - Laboratory for Space Research, Sorbonnelaan 2, 3584 CA Utrecht,
the Netherlands

ABSTRACT

The paper describes the X-ray spectroscopic capabilities of the XMM-payload. XMM (X-ray Multi Mirror Mission) is the second 'cornerstone' project of the European Space Agency.

1   INTRODUCTION

The High-Throughput X-ray Multi Mirror Mission (XMM) is one of the four 'Cornerstone' projects in the ESA Long-Term Programme for Space Science. The satellite observatory consists of a high-throughput facility for X-ray spectroscopy coupled to dispersive and non-dispersive imaging spectrometers. The Observatory will be placed in a deep eccentric orbit allowing for sensitive, long, uninterrupted observations of a wide variety of targets ranging from nearby stars to distant quasars.

The 0.1-10 keV energy band (1-100 Å) to be covered by XMM contains the K-shell transitions of carbon, oxygen, neon, magnesium, aluminium, silicon, sulphur and calcium as well as both the L- and K-shell transitions of iron. Of particular interest, due to their high abundances, are oxygen and iron at temperatures of 10^6 K and $\sim 10^7$-10^8 K, respectively. Detailed analysis of these spectral features will permit the determination of the physical characteristics (density, temperature, ionisation state, element abundance, mass motions, deviations from Maxwellian distributions and red shift) of the emitting region and its surrounding environment. The spectral properties of a gaseous medium will depend on whether the ionisation process is caused by electron collisions (a thermal model) or by photo-ionisation (a nebular model), and on whether the plasma is optically thin or optically thick. It is envisaged that the Observatory will be operated as a long-duration facility available to the worldwide astronomical community.

The principal technical characteristics of the Observatory can be summarized as follows:
- Spatial resolution < 30" (half energy width, HEW)
- Effective area ~6000 cm² at ~2 keV and ~3000 cm² at ~7 keV
- Broadband (0.1-10 keV) spectroscopy with a resolving power $E/\Delta E$ (FWHM) of between

5 and 60

- High-resolution spectroscopy between 0.35-2.5 keV (5-35 Å) with a resolving power of about 400 at 0.5 keV
- Simultaneous optical monitoring between 2000 and 6000 Å to a limiting magnitude of 24.5 over part of the wavelength band.

Some of the distinguishing features of the XMM Observatory are:
- The unprecedented sensitivity to perform medium-resolution spectroscopy. [For a source with a flux of 2×10^{-14} erg cm^{-2} s^{-1} between 0.3 and 3.5 keV (a source at the limiting sensitivity of the Einstein Deep Survey), at O VII and Fe XX, lines of equivalent width 100 eV and 150 eV, respectively, can be detected in a single XMM orbit (6×10^4s).]
- The combination of high sensitivity over a wide bandwidth coupled to medium spatial resolution (<30" HEW) and modest spectral resolving power (50 at 7 keV). [For point sources a limiting sensitivity of 2×10^{-15} ergs cm^{-2} s^{-1} (0.5 nJy at 1 keV), i.e. one order of magnitude below the Einstein Deep Survey, can be reached in only a six hour exposure. This high sensitivity will provide a major contribution to our understanding of the origin of the diffuse X-ray background as well as establishing the characteristics of its principal components.]
- Simultaneous medium- and low-resolution spectroscopy. This capability will, for example, allow the XMM Observatory to perform time-resolved spectroscopy of sources such as stars with stellar flares.
- Imaging spectroscopy of extended sources. [The spectrum of a cluster of galaxies at a red shift of 0.1 having a luminosity of 4×10^{44} erg s^{-1} (10^{-11} erg cm^{-2} s^{-1}) can be measured at medium resolution and an O VII line detected with an equivalent width of < 10 eV, while 'spatially resolved' low-resolution spectroscopy can be performed simultaneously on an arc-minute scale.]
- Uninterrupted simultaneous optical coverage, the long uninterrupted observations possible with XMM allow the study of coordinated optical/UV and X-ray variations in a wide variety of objects, e.g. BL Lac objects, Seyfert galaxies, and X-ray binaries on all time scales from seconds to one day.]

2   PAYLOAD

XMM contains three X-ray telescopes supplied by ESA. Each telescope module consists of 58 thin concentric Wolter type I paraboloid-hyperboloid mirror shells with a 7.5 meter focal length. The sensitive area as a function of energy is plotted in fig. 1. Comparing the XMM mirror with AXAF (Markert, this conference) one notices the nice complementarity between XMM and AXAF. XMM has a much larger sensitive area (with 30" spatial resolution) whereas AXAF has sub-arcsecond spatial resolution.

In the focus of each of the three telescopes resides a CCD imaging camera called EPIC

(European X-ray Photon Imaging Camera, Bignami et al. 1989;1990). Behind two of the three telescopes a reflection grating module is placed in such a way, that half the light passes through the module, unaffected by the grating module and about half of the X-ray light will be dispersed and detected with two dedicated long narrow CCD-detectors (one for each grating module). Grating modules and CCD-detectors together make up the RGS (Reflection Grating Spectrometer, Brinkman et al. 1989a; 1989b). In addition there is an independent Optical Monitor (Mason et al. 1989).

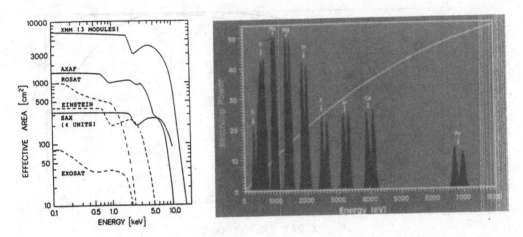

Fig. 1—The effective area of the XMM telescopes as a function of X-ray photon energy. A comparison with other missions is given.

Fig. 2—Resolving power of EPIC as a function of X-ray photon energy (solid line). The response to the H and He-like transitions of the principal elements is also shown.

2.1 The EPIC Camera

The three EPIC cameras in the foci of the three telescopes all contain CCD-arrays as detectors. CCD's (Charge Coupled Devices) provide the combined properties of high spatial resolution, wide spectral bandwidth, good spectral resolution and the ability to discriminate between X-ray photon and charged particle background events with very high efficiency. Different CCD-chips are under study and development at several institutes and industries; a final choice between the different alternatives will be made later. It is envisaged that two different types of CCD's will fly on XMM, one type based on the well known MOS technology, but optimized for X-ray response and the other one a novel type fully deep depleted pn CCD based on the semiconductor drift chamber principle (Strüder et al. 1990).

The CCD's are passively cooled in order to suppress the dark current with the aid of dedicated radiators looking towards cold space.

Of major importance while evaluating the X-ray spectroscopic potential, is the resolving power. In fig. 2 the resolving power is plotted as a function of energy. Also shown in fig. 2 are the H- and He-like transition complexes of the principle elements. All transitions were set at equal strength and folded through mirror and CCD-response. The CCD can resolve the H- and He-like complexes for elements with $Z > 8$ which are important diagnostic tools. The detector quantum efficiency (including visual light filter) as a function of energy is shown in fig. 3. The low-energy response is determined by the filter (2000 Å lexan + 1000 Å Al).

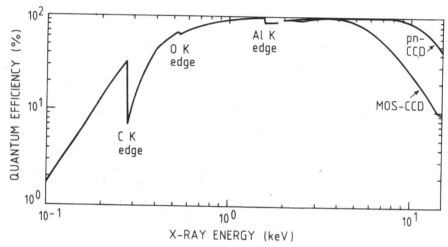

Fig. 3—The quantum efficiency versus energy for the MOS and pn CCD's (including filter).

The difference in depletion depth (over which photons can be absorbed) of the pn and MOS type CCD's is reflected in the different high energy responses of the two types.

2.2   The Reflection Grating Spectrometer (RGS)

With present technology, high-resolution X-ray spectroscopy can only be achieved with dispersive optical systems. Even the best available nondispersive spectrometers cannot yield adequate resolution at energies below about 1 keV. Since spectral resolution degrades linearly with the angular resolution of the focusing optics, however, the moderate resolution of the XMM telescopes imposes severe constraints on the design of potential dispersive spectrometers. For transmission gratings, which have been incorporated on Einstein and Exosat and will be used on AXAF, this requires very high line densities, which cannot be fabricated with existing technology. Reflection gratings, in contrast, offer the possibility of very high dispersion with modest line densities ($\leq$ 1000 lines/mm). With proper optimization of the grating parameters, they can also yield high diffraction efficiency in the desired spectral band.

The RGS design incorporates an array of reflection gratings placed in the converging beam at the exit from the X-ray telescope. The grating stack picks off roughly half of the X-ray light and deflects it to a strip of CCD detectors offset from the telescope focal plane (see fig. 4). The remaining light passes undeflected through the grating stack where it will be utilised by EPIC in the telescope focal plane.

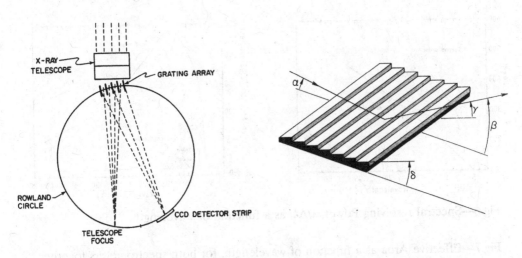

Fig.4—A schematic of the optical design of the RGS. An array of reflection gratings is oriented at grazing incidence to the beam at the exit of the X-ray telescope. The gratings intercept only a fraction of the beam; roughly half the light passes through the telescope focal plane. Rays which are intercepted are diffracted to a strip of CCDs which serve as the spectroscopic detector. The gratings, the telescope focus, and the CCD detector strip all lie on a large Rowland circle. This configuration eliminates aberrations associated with the arraying geometry.

Fig.5—A blow-up of a portion of one of the reflection gratings showing the orientation of the incoming and outgoing rays. The groove facets are triangular in shape and are tilted at an angle δ with respect to the grating plane.

The individual reflection gratings are mounted at grazing incidence to the beam in the in-plane or classical configuration, in which the incident and diffracted rays lie in a plane which is perpendicular to the grating grooves (see fig. 5). Because the projected line spacing is considerably smaller than the ruled line spacing in this geometry, the dispersion achievable at soft X-ray wavelengths can be relatively large (Kahn and Hettrick, 1985).

In order to maximize the diffraction efficiency in the relevant band, "blazed" reflection gratings have been incorporated which have the sawtooth groove profile illustrated in fig. 5

The reflection gratings diffract first and second order at high-to-moderate efficiency. The separation of the spectral orders is accomplished using the energy resolution of the CCD detectors. The energy resolution also provides a means for background suppression. A set of ten chips are arrayed lengthwise in a strip-oriented tangent to the Rowland circle. The cooling of the RGS-CCD strip detector is similar to the EPIC detector. So-called back-illuminated CCD's will be employed, to enhance the low-energy response. The calculated resolving power, $\lambda/\Delta\lambda$, of the RGS is shown as a function of wavelength in fig. 6.

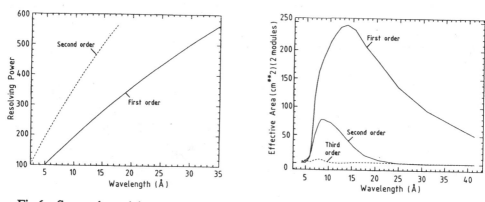

Fig.6—Spectral resolving Power, $\lambda/\Delta\lambda$, as a function of wavelength.

Fig.7—Effective Area as a function of wavelength, for both spectrometers together.

The main effect included in the calculation of $\lambda/\Delta\lambda$ is the telescope blur. The RGS design is specifically tailored so that the telescope blur provides the dominant contribution to the spectroscopic resolution. If the XMM optics achieve better than 30 arcsecond performance, the RGS system spectral resolution will improve accordingly. (ESA's telescope specification calls for 30 arcseconds, design goal is 20 arcseconds). Further factors taken into account are: alignment and flatness errors of the grating plates, line space errors of the grating plates, varied line-space aberrations and defocus errors.

Figure 7 shows the expected system spectroscopic effective area for the two reflection grating modules.

3   DIAGNOSTIC CAPABILITIES

Astrophysical plasmas are often discussed in terms of an optically thin thermal model, a nebular model, an optically thick model or combinations thereof.

3.1   The Optically Thin Thermal Model

An optically thin thermal model describes the X-ray emission from stellar coronae, clusters of galaxies, and the interaction of supernova remnants with the interstellar medium.

In general, a range of temperatures is present in these sources and a number of unique temperature diagnostics are necessary to construct accurate differential emission measures. For temperatures between 10^5 and 2×10^7 K, line emission from $Z > 8$ constituents dominates the total emissivity. These spectra are rich in emission lines whose relative strengths are dependent on the physical properties of the emitting plasma, such as the temperature, density and abundance. In particular for the iron spectrum, the L- and K-shell emission groups for stages from Fe XVII to Fe XXV are very prominent features. If present in the source, these give a powerful diagnostic of the distribution of ionisation stages. Above 2×10^7 K the emission is mainly Bremsstrahlung, although iron K emission around 7 keV remains a strong feature up to a few times 10^8 K. The main emission-line complexes in the X-ray regime will be from helium-like and hydrogenic ions; these become distinct from each other at a resolving power of >10. This can be achieved with the EPIC cameras for elements with Z greater than that of oxygen (see fig. 2). The ratio of the strength of the helium-like to the hydrogenic lines is a good measure of the plasma temperature. This does, however, assume equilibrium between ionisation and recombination, as well as certain limits on the departure from uniform temperature, which may not be true in some astrophysical plasmas (e.g. supernova remnants). For a detailed discussion on optically thin thermal models, see e.g. Keenan, this conference. Another temperature measure that is independent of this assumption, for the hydrogen and helium-like species, is the ratio of the K_α to K_β line intensities. In supernova remnants, the detailed study of these line ratios can lead to a better understanding of the non-equilibrium effects in a model-independent way.

At higher resolutions (of the order of 100 or greater), as given by the RGS, many new features are resolved. These include the so-called 'dielectronic satellite lines' due to inner shell transitions in partially stripped ions, as well as the detailed structure of the helium-like spectra. This higher resolution not only brings the advantage of access to new diagnostic possibilities, but is also necessary in many cases to avoid errors due to merging of lines, and failure to take account of the presence of such features. The relative strengths of the dielectronic satellite lines increase with Z. They are therefore of the greatest importance for iron, where they can contribute more than 50% of the line emission. Having two types of origin (dielectronic recombination and inner-shell excitation), they provide possibilities for measuring temperature, without many of the constraints of the methods already described, as well as ionisation stages and departures from ionisation equilibrium.

The intensities of the resolved components of the helium-like transitions also give information on the temperature, as well as the existence of departures from Maxwellian electron energy distributions. Such departures can be important for shockheated low-density plasmas, as for example in supernova remnants. At lower Z (e.g. in oxygen), the helium-like spectra can also be used for density measurements. Such detailed diagnostics have been applied with great success to the solar spectrum, and to laboratory tokamak devices. They have not in general been available to X-ray astronomy because of the lack of sufficient sensitivity combined with spectral resolution.

3.2   The Nebular Model

The nebular model is applicable to any case where gas is illuminated by an X-ray source. This material might be the flow surrounding an accreting compact object, the stellar wind/atmosphere of a companion star, or the interstellar medium in the line of sight. The ionisation and temperature structure of the illuminated gas are regulated by photoelectric absorption and Compton heating. The emerging spectrum will be dominated by absorption edges and emission lines that can be used to diagnose the properties and the geometry of the gas cloud. The strongest emission lines will be emitted at energies close to the low-energy X-ray cut-off. Because photo-ionisation, rather than electron collisions, dominates the ionisation structure, at any given temperature the atoms in a nebular model are more highly ionised than in a thermal model.

The X-ray spectrum of an X-ray photoionized plasma differs considerably from that of collisional plasmas with similar ion concentrations. Because the photoionized plasma is overionized relative to the electron temperature, the excitations of important lines are dominated by recombination, photoexcitation, and cascades as opposed to collisional excitation and dielectronic recombination. This effect on the emergent spectrum is illustrated in figures 8a and 8b (Liedahl, Kahn, Osterheld and Goldstein, 1990).

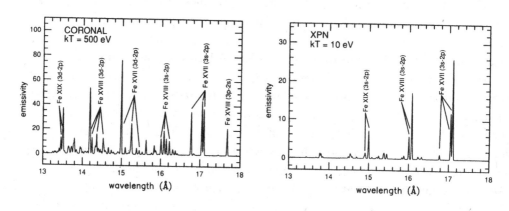

Fig.8a—Model emission rate spectrum for Fe XVI-XIX under conditions appropriate to coronal equilibrium. $kT_e = 500$ eV, $n_e = 10^{11}$ cm^{-3}. The line profiles are Gaussian with a FWHM of 0.025 Å. The emissivity scale is arbitrary.

Fig.8b—Model emission rate spectrum for Fe XVII-XIX under conditions appropriate to an X-ray-photoionized nebula. $kT_e = 10$ eV, $n_e = 10^{11}$ cm^{-3}. The line profiles and emissivity scale are the same as for Fig. 8a.

As fig. 8a shows, the coronal (optically thin thermal model) spectrum contains substantial contributions from 3d-2p transitions of the Fe XVII-XIX ion stages, which are formed by collisional ionization from the ground state. The spectrum of the X-ray photoionized nebula model however is almost entirely made up of transitions of the type 3s-2p, formed by recombination. Measurements of this spectral range with RGS therefore, will allow one to distinguish between coronal and photoionized plasmas.

The absorption-edge and emission-line energies are dependent on the ionisation state of the gas, with the energies increasing with higher ionisation state. A spectral resolving power of 10 is sufficient to distinguish between neutral and hydrogenic species. To resolve the ionisation state more precisely and to distinguish the K_α and K_β edges and lines requires a resolving power >30. The line emission will be from either fluorescence or recombination, depending on the ionisation state of the plasma. The former mechanism is particularly important for iron because it has by far the largest fluorescent yield (~30%) of all the elements with K edges in the X-ray band (e.g. oxygen has a yield ~3%). Photo-absorption only occurs in the line of sight to the X-ray source, whereas the fluorescent line comes from all the gas that surrounds the X-ray source. Comparison of the strength of the line with the corresponding depth of the edge provides a probe into the geometry of the gas surrounding the source.

The differences in the ionisation state of the various elements can radically effect the overall emerging spectrum. It is possible to find situations where the lower Z elements (e.g. oxygen) are fully ionised, but where iron still retains one or more electrons. Such effects can lead to strong soft excesses that can, in a low-resolution instrument, be indistinguishable from an additional spectral component or the effects of a partial covering of the X-ray source. In the interstellar medium, a sizeable fraction of elements such as oxygen and silicon may have formed grains. If this is the case then self-blanketing, depending on the grain size, can cause a large reduction in the depth of the K edges. To study these effects requires moderate-resolution (~10) measurements of the relevant edges in combination with a broadband measurement of the continuum shape.

3.3 Density Diagnostics

The determination of electron densities in hot cosmic plasmas provides one of the greater challenges for X-ray astronomy. Typically, one estimates densities by determining the emission measure of the plasma from spectral fits to the line and/or continuum emission. The emission measure is the density squared multiplied by the emitting volume.

Given the resolution of the RGS, it will be possible to measure electron densities using lines originating from metastable levels. The 5-30 Å wavelength region contains a number of electron-density-sensitive lines from the He-like ions and from Fe XVIII through Fe XXIII. As is well known from the early work of Gabriel and Jordan (1969; 1972), we can use the intensity ratios of the forbidden and the intercombination lines from the helium-like ions of nitrogen through sulphur to diagnose the plasma in the electron density range $n_e \sim 10^{10}\text{-}10^{15}$

cm⁻³ and corresponding temperature range T ~ 1-20 MK. Calculations by Mason and co-workers (Mason et al. 1984 and Mewe et al. 1985) on iron and nickel ions from a number of successive ionization stages can be applied to derive tools for diagnosing plasmas in the density range 10^{10}-10^{15} cm⁻³ and temperature range ~5-15 MK. These lines are concentrated in the wavelength region 7-13 Å and correspond to 2s-3p and 2p-3*l*.

An example is illustrated in figures 9 a, b, and c, where we have plotted the expected O VII emission line strengths, at the RGS resolution, for a coronal plasma with a temperature of 2 MK and densities of 10^{10}, 10^{11}, and 10^{12} cm⁻³, respectively. As can be seen, accurate constraints on electron density can be derived in this way. Since the range of densities over which this ratio shows discernible variation differs according to species, measurement of several such ratios for different He-like species provides a probe of the global density structure of the line emitting region of the plasma. Deviations from coronal equilibrium can alter these diagnostics through recombination and ionization effects, however the measured ratios remain density-sensitive even for these cases (Pradhan, 1985). Thus, this diagnostic can be quite generally applied to nearly all astrophysical sources.

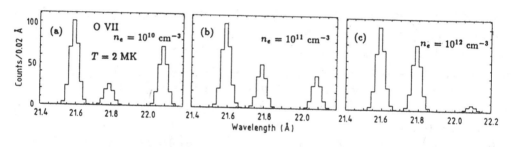

Fig.9—O VII triplet at 22 Å observed with the XMM-RGS in 10^3 s for electron densities $n_e = 10^{10}$, 10^{11}, and 10^{12} cm⁻³ and a reduced emission measure $\epsilon = 10^{50}$ cm⁻³ pc⁻².

4   SIMULATIONS

The detailed capability of XMM is shown for two simulations, one for an elliptical galaxy, the other a nearby star.

4.1   Study of Elliptical Galaxies

Fig. 10 shows the simulated spectrum (countrate in the EPIC-camera versus energy) for a 2 arcmin size element of hot gas in an elliptical galaxy in the Virgo cluster. Two gas temperatures of 0.5 and 2 keV were assumed for these elements which cover a diameter of 5 kpc. The luminosity of the galaxy as a whole was 10^{41} erg s⁻¹ and the exposure time 10^5 s. The gas distribution was taken as a worst-case uniform distribution over the whole galaxy.

The inset shows the simultaneously obtained RGS-spectrum of the low-temperature plasma of the cooling flow in the central 2 arcmin region.

XMM with its < 30" angular resolution will be able to determine the density and temperature distribution of the gas in systems out to a distance of ~ 100 Mpc, i.e. the distance of the Coma cluster, and thus derive the binding mass distribution of these objects.

If we assume a space density of ~0.002 per Mpc^3 for elliptical galaxies, we find that XMM will be able to study the mass distribution in over 1000 elliptical galaxies. At a distance of 50-100 Mpc, the 'average' elliptical, with L_x ~10^{41} erg s^{-1}, will have its integral spectrum determined in less than 10 000 s and a spectrum of high quality derived in 10^5 s (Fig. 10). If we take the XMM confusion threshold of ~2×10^{-15} erg cm^{-2} s^{-1}, it will be able to detect elliptical galaxies out to a distance of 500 Mpc, where their mean optical magnitude is only 18.

The most luminous ellipticals with L_x ~10^{42} erg cm^{-2} s^{-1} can be detected with XMM out to red shifts of 0.25, where there appears to be substantial evolution in their optical properties.

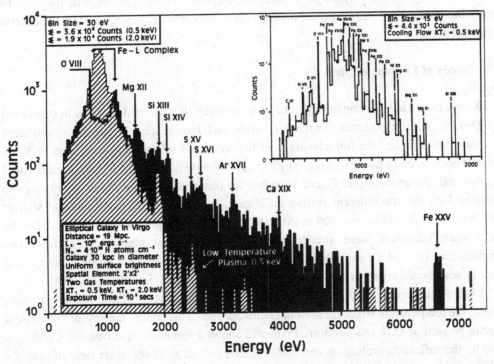

Fig.10—CCD spectra of a 2' element of hot gas in an elliptical galaxy in the Virgo cluster (19 Mpc) with a diameter of 30 kpc. For model assumptions see text. The inset shows the equivalent reflection-grating spectrum of the low-temperature plasma of the cooling flow in the central 2' region.

Theoretical arguments indicate that in the luminous systems the whole of the X-ray emission appears to be a cooling flow. The models developed so far indicate a substantial temperature gradient inside ~1 kpc. Thus for galaxies more distant than the Virgo cluster (~20 Mpc), the cooling part of the flow will be essentially a point source for XMM. However, this still means that XMM can determine the spatially resolved temperature structure of the cooling flows in the ~100 bright ellipticals within 20 Mpc. Because of the complex temperature structure in the central regions, where the gas is cooling rapidly, high-resolution X-ray spectroscopy is necessary to deconvolve the temperature structure. Here the grating on XMM, because it is able to do a good job on sources of angular size up to ~1'-2', will be of tremendous use in measuring the integral spectra of the central regions of the cooling flows, (cf. Fig. 10 (inset)). With its broadband spectral coverage, XMM should be able to distinguish, via their spectral signatures, galaxies whose X-ray emission is dominated by their X-ray binary population and those whose emission is dominated by hot gas.

In summary, for a very large number of elliptical galaxies XMM will be able to determine their mass profiles and the distribution of dark matter, as well as study the central cooling flow.

4.2   Study of Coronae of Stars

X-rays from stellar coronae originate from optically thin thermal plasmas in collisional equilibrium. The instruments flown on Einstein and Exosat had, in general, insufficient spectral resolution to take full advantage of the wealth of information contained in X-ray spectral lines. Even in the best cases, only a crude estimate of the relevant parameters could be obtained. As an example, Figure 11 shows an observation of the bright giant Capella obtained with the transmission grating on Exosat. Individual lines, particularly at shorter wavelengths, could not be resolved in this case; however, the complexes of lines revealed by the Exosat instrument were already sufficient to show that plasmas of at least two temperatures (one of a few million degrees and the other about one order of magnitude higher) were necessary to fit the data in a satisfactory way (e.g. Lemen et al. 1989)

With the improved resolution of the instruments on XMM, it will be possible to fully resolve the temperature structure of stellar coronae, and to determine the amount of coronal plasma present at each temperature. Figure 12 shows a simulated spectrum of Capella as seen by the reflection grating on two XMM telescopes, in an observation time of only 5×10^4 s. Individual lines of ions formed at different temperatures are clearly resolved, including the He-like triplets of abundant ions such as O VII and Ne IX. The inset shows an expanded view of the region of the He-like triplet of O VII. The resonance, intercombination and forbidden lines are clearly resolved. This is important, as was discussed before at this conference, since the ratio of the latter two lines is a diagnostic of coronal density.

Using the EPIC CCD array and the reflection gratings, XMM will have the sensitivity to observe and obtain good-quality spectra in 10^5 s of sources that are over three orders of magnitude weaker than a coronal source like Capella (cf. inset in Fig. 12, which also shows the simulated Capella spectrum obtained in only 1500 s by the CCD cameras, indicating that broadband time-resolved spectra can be obtained on such sources). There are thousands of such sources (either weaker or at greater distances than Capella) that could be observed with XMM.

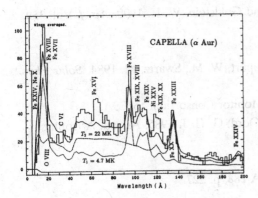 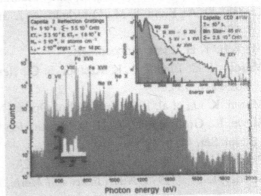

Fig.—11 The Exosat transmission-grating spectrum of Capella. The two temperature fits (T_1=4.7 Mk, T_2=22 Mk) are also shown.

Fig.—12 XMM reflection-grating spectrum of Capella obtained in an exposure time of 5×10^4 s. The oxygen He-like triplet is also shown (inset), along with the CCD spectrum. In the case of the CCD broadband spectrum, the exposure time was 1500 s.

5. ACKNOWLEDGEMENT

While preparing this paper, I have made extensive use of the XMM MISSION-SCIENCE REPORT (ESA SP-1097, March 1988) and the proposals of the EPIC and RGS consortia to ESA, in response to ESA's request for proposals, 1988.

6. REFERENCES

Bignami, G. F., et al. 1989, Proposal of EPIC-consortium to ESA.

Bignami, G. F., et al. 1990, *EUV, X-Ray and Gamma-Ray Instrumentation for Astronomy*, ed. Oswald, H. W. Siegmund and Hugh S. Hudson, in *SPIE*. Vol. **1344**, p. 144.

Brinkman, A. C., et al. 1989a, Proposal of RGS-consortium to ESA.

Brinkman, A. C., et al. 1989b, *EUV, X-Ray and Gamma-Ray Instrumentation for Astronomy and Atomic Physics*, ed. Charles J. Hailey and Oswald H.W. Siegmund, in

SPIE. Vol. **1159**, p. 495.

Gabriel, A. H. and Jordan, C. 1969, *MNRAS.* Vol. **145**, p. 241.

Gabriel, A. H. and Jordan, C. 1972, in *Case Studies in Atomic and Collisional Physics.* ed. E.W. McDaniel, M.R.C. McDowell (Amsterdam: N.H. Publ. Co.), Vol. **2**, p. 209.

Kahn, S. M. and Hettrick, M. C. 1985, *Proc. ESA Workshop on Cosmic X-Ray Spectroscopy Mission* (Denmark: Lyngby), *ESA SP-2.* Vol. **39**, p. 237.

Keenan, F., these proceedings.

Lemen, J. R, Mewe, R., Schrijver, C. J., Fludra, A. 1989, *Ap. J.* Vol. **341**, p. 474

Liedahl, D. A., Kahn, S. M., Osterheld, A. L. and Goldstein, W. H. 1990, *Ap. J.* Vol. **350**, p. L37-L40.

Markert, T. H., these proceedings.

Mason, H. E., Bhatia, A. K, Kastner, S. O., Neupert, W. M., Swartz, M. 1984, *Solar Phys.* Vol. **92**, p. 199.

Mason, K. O., et al. 1989, Proposal of Optical Monitor Consortium to ESA.

Mewe, R., Gronenschild, E. H. B. M., van den Oord, G. H. J. 1985, Astron. Astrophys. Suppl. Ser. **62**, p. 197.

Pradhan, A. K. 1985, *Ap. J.* Vol. **288**, p. 824.

Strüder, L. et al. 1990, *Nucl. Instr. and Meth.,* A **288**, p. 227.

X-ray Spectroscopy with the XSPECT/SODART Telescopes on SRG

H. W. Schnopper[1], C. Budtz-Jørgensen[1], F.E. Christensen[1], R. Mewe[2], H.U. Nørgaard-Nielsen[1] and N. J. Westergaard[1]

[1]Danish Space Research Institute
[2]SRON-Laboratory for Space Research

ABSTRACT

The XSPECT/SODART X-ray telescopes and their associated focal plane instrumentation will be launched on board the Russian X-ray mission SPECTRUM RÖNTGEN-GAMMA (SRG). Technical details of the mission and of the telescopes have been discussed elsewhere. This discussion will, therefore, first present the main properties of the instruments and then present simulations of some of the scientific questions which can be addressed by the observing program.

XSPECT/SODART consists of two, high-throughput telescopes. In the focal plane of each one, there are high- and low energy imaging proportional counters. One telescope also has an array of silicon detectors and the other a polarimeter. An imaging, objective spectrometer consisting of three different kinds of Bragg crystals and two different multilayers is mounted in front of one of the telescopes. The broad bands of the imaging X-ray detectors collectively cover the energy rangy from 0.2 to 25 keV. The objective spectrometer samples this range around the emission and absorption features from, among others, the cosmically important ions of Fe, S, Si and O. An optical/UV monitor, co-aligned with the X-ray telescopes and consisting of three separate telescopes, will support the X-ray observations and provide aspect information.

1 INTRODUCTION

SRG is to be the first of a series of new astronomical observatories to be launched under the sponsorship of the Russian Academy of Sciences. The expected launch date is mid-1995 and the observatory is intended to be operated for a period of 5 years. The satellite will be launched into a deep, highly eccentric orbit with a period of about four days from which long duration (up to about 80 h) observations can be made. Ground station operations which include up- and down-links for data and command transfer will take place approximately once every 24 hours. As experience grows, it should be possible to plan for between 10 and 15 reorientations of the spacecraft per 24 hour period. With all the operational and pointing constraints taken into account, there is access to approximately 80 per cent of the celestial sphere at any time. The major facilities on board are the XSPECT/SODART X-ray telescopes and their associated focal plane instruments. Instrumentation in addition to XSPECT/SODART will cover the energy range from EUV to gamma rays. A complete description of the SRG system (Figure 1) will be found elsewhere (Sunyaev 1990). This paper updates a previous version (Schnopper 1990).

2 X-RAY OPTICS

The XSPECT/SODART system consists of X-ray optics, thermal and structural elements, focal plane instruments, a focal plane transport assembly and an objective crystal spectrometer (Figure 2). Those elements provided by DSRI are called XSPECT. Brief descriptions of these elements follow.

SPECTRUM RÖNTGEN-GAMMA

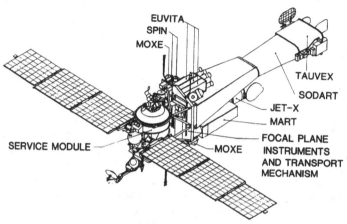

Fig. 1 SPECTRUM RÖNTGEN-GAMMA (R. S. Kremnev, Babakin Institute).

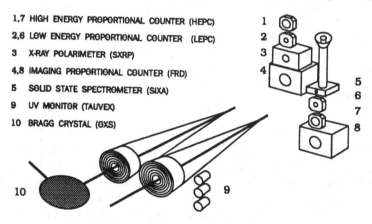

1,7 HIGH ENERGY PROPORTIONAL COUNTER (HEPC)

2,6 LOW ENERGY PROPORTIONAL COUNTER (LEPC)

3   X-RAY POLARIMETER (SXRP)

4,8 IMAGING PROPORTIONAL COUNTER (FRD)

5   SOLID STATE SPECTROMETER (SIXA)

9   UV MONITOR (TAUVEX)

10   BRAGG CRYSTAL (OXS)

Fig. 2 The XSPECT/SODART system: Telescopes and instruments.

Thin foil, conical approximations to Wolter 1 optics can fill a large fraction of the available aperture by densely nesting many shells of thin foils and can, therefore, yield a large collecting area and an extended energy range (Serlemitsos *et al.* 1984). Using this concept, the design goal for both of the XSPECT/SODART telescopes is a Half Power Width (HPW) of <2 arcmin of which <20 arcsec comes from the conical geometry. The parameters chosen for the telescopes are listed in Table 1 (Westergaard *et al.* 1990).

Figure 3a shows the effective collecting area *vs.* energy which can be expected for a single telescope. Measurements of reflection efficiencies at various X-ray energies for foil samples prepared at DSRI have been made with the DSRI X-ray Optics and Crystal Testing Facility and they agree very well with theoretical predictions. Both have been used to construct the XSPECT/SODART telescope curve. The parameters for the XSPECT/SODART instruments are given below in Tables 2 and 3 and, together with the data from Figure 1, they have been used to derive the instrument effective areas shown in Figure 3b.

TABLE 1. XSPECT/SODART: TELESCOPE SPECIFICATIONS

Field of view	arcmin	60
Half power width	arcmin	< 2
Focal length	m	8
Diameter: Inner Shell	cm	16
Outer Shell	cm	60
Reflecting surface	Å gold	350
Effective area	cm² @ 2 keV	1540
	cm² @ 8 keV	1200
	cm² @ 20 keV	60
Shell material	mm aluminum	0.3 (0.4)
Shell separation (min)	mm	0.5
Shell length	cm	20
Number of shells		154
Mass per telescope	kg	95
Total mirrored surface	m²	75

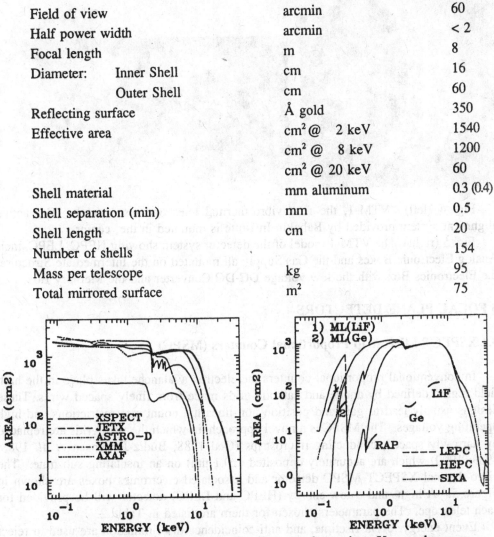

Fig. 3a (left) On-axis effective collecting area for various X-ray telescopes.
Fig. 3b (right) Effective area for various XSPECT/SODART instruments.

Each XSPECT telescope consists of two sections, 1α and 3α, each of which is divided into quadrants (Figure 4). Recent white light and He-Ne laser measurements on the DSRI Optical Test Stand show that an assembled, but not yet aligned, telescope is within assembly specifications. Once the alignment procedure is completed, the X-ray reflecting properties, including the HPW, can be measured using a specially constructed facility that can be used either with the DSRI X-ray Optics and Crystal Testing Facility with line energies available from X-ray tubes or at the Daresbury Synchrotron with monochromator selected energies (Christensen, *et al.* 1992). The Spare Model will be fully calibrated at Daresbury. Flight unit final calibration will take place in orbit.

Fig. 4 (left)   VTM-1, the first vibro-thermal telescope test model.   The active alignment system provided by Babakin Institute is mounted in the center.

Fig. 5 (right) The VTM-1 model of the detector system showing HEPC, LEPC, their Analog Electronic Boxes and the Gas Supply all mounted on the Intermediate Structure; the Electronics Box with the low voltage DC-DC Converter and the Memory Box.

3 FOCAL PLANE DETECTORS

3.1 XSPECT Microstrip Proportional Counters (MSPC)

In conventional proportional counters the electron avalanche takes place in the high field region defined by anode and cathode grids made from finely spaced wires. These designs usually lead to gain and position non-linearity, count rate restrictions and high operating voltages. The MSPC is a novel approach, in which the wire grids are replaced by narrowly spaced, conducting microstrips (Oed 1988; Budtz-Jørgensen *et al.* 1988; 1989; 1990) which are accurately deposited (±0.1 µm) on an insulating substrate. The vibro-thermal XSPECT MSPC detector and associated electronics boxes are shown in Figure 5. A high- and a low-energy (HEPC and LEPC) detector will be provided for each telescope. The parameters chosen for them are listed in Table 2.

Event energy, event risetime, and anti-coincidence discrimination are used to reject background. Measured results of energy- and position- resolution, and background rejection are well within the specifications given in Table 2.

3.2 Other focal plane detectors

Two additional imaging proportional counters (FRD), a Silicon X-ray Array (SIXA) and a Stellar X-Ray Polarimeter (SXRP) complete the complement of focal plane instruments. They are shown in place on the focal plane transport assembly (Figure 2) and are described by Sunyaev (1990), Vilhu (1990) and Kaaret (1990), respectively.

TABLE 2. XSPECT: HEPC, LEPC SPECIFICATIONS

Energy Resolution		$0.32\ [E/1(keV)]^{\frac{1}{2}}$
Background Rejection		> 99%
Time Resolution		< 5 µs
Maximum Count Rate		$< 5\times10^3$ ph s^{-1} mm^{-2}
Filling Gas		Xe: 90%; CH$_4$ 10%
Background Rejection		> 99%
Total Hard Disc Memory		> 128 Mbyte

		HEPC	LEPC
Field of View	arcmin	60	30
Active Area Dia.	mm	150	75
Window Diameter	mm	140	70
Position Resolution	mm	< 1	< 2
Energy Range	keV	2 < E < 25	0.2 < E < 8
Efficiency	%, (keV)	>70, (2<E<15)	>70, (1<E<8)
Gas Thickness	cm	4	4
Gas Pressure	atm	1	0.5
Window: Polyimide	µm	3	0.8
Al	nm	40	40

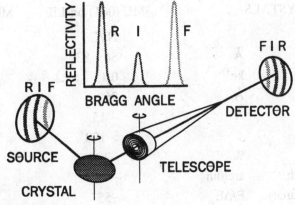

Fig. 6 The OXS concept.

4 OBJECTIVE CRYSTAL SPECTROMETER (OXS).

The OXS concept (shown in Figure 6) separates the processes of energy dispersion and imaging (Schnopper and Byrnak 1987; Christensen *et al.* 1990). The large mosaic panel of flat crystals in front of the telescope acts as a narrow pass filter and as a mirror. Each pixel in the reflected field of view of the telescope satisfies a specific Bragg angle on the crystal and, therefore, each pixel in the detector can be identified with a particular energy. Scans which involve repositioning of the telescope axis and the angle between the crystal panel and the telescope axis (45 ±15 deg) yield either the spectrum of a point source or energy resolved images of an extended source. Parameters chosen for OXS are given in Table 3 (Christensen *et al.* 1990). Multilayer (ML) structures will be deposited on the polished surfaces of the LiF and Ge crystals to allow simultaneous measurements of the Fe L and other prominent lines in the energy range below the C K-edge.

TABLE 3. XSPECT: OXS CRYSTAL SPECIFICATIONS

NATURAL CRYSTALS		LiF (220)	Ge (111)	RAP (001)
Lines: H- and He-like		Fe	S	O
Bragg Angle (He-like)	deg	~ 41	~ 50	~ 57
Wavelength Range	Å	1.7 - 2.5	3.8 - 5.7	15.4 - 22.6
Energy Range	keV	5.0 - 7.4	2.2 - 3.2	0.55 - 0.81
Max. Redshift (He-like)		0.32	0.12	0.05
Rocking Curve Width	arcmin	< 2	~ 1.6	~ 7
Resolution (point source)	E/ΔE	~ 3200	~ 2600	~ 770
Integrated Reflectivity	rad	~ 1.7×10^{-4}	~ 1.7×10^{-4}	~ 5×10^{-5}
Peak Reflectvity	%	> 25	~ 35	~ 2
Darwin Width	arcmin	~ 0.75	~ 1.6	~ 7
Mosaic Width	arcmin	~ 1.2	~ 0	~ 0

MULTILAYER CRYSTALS		ML (001) on LiF	ML (001) on Ge
Composition		Ni/C	Ni/C
Wavelength Range	Å	56 - 83	45 - 60
Energy Range	keV	0.15 - 0.22	0.21 - 0.28
2d Spacing	Å	96	77
d Spacing (Ni)	Å	~ 15	~ 10
Number of Layers		> 100	> 100
Peak Reflectivity	%	~20	~21
Rocking Curve Width	arcmin	~45	~40
Resolution (point source)	E/ΔE	~55	~80

5 SENSITIVITIES

Although the HEPC and LEPC detectors have unparalled energy resolution when compared with other proportional counter designs, it is their combined ability to establish the parameters of spectral shape over the full energy range of the SODART telescope for which they are best suited. In addition, HEPC and LEPC can fully sample the SODART HPW which, together with the large SODART telescope collecting area, makes them ideal for the study of the spectral characteristics of moderately distant objects which contribute to the diffuse X-ray background (DXRB).

OXS is suited for high resolution observations of both point and extended sources. Simulations of X-ray emission from optically thin plasmas with temperatures between 10^5 and 10^8 K which are typical of stellar coronae and the diffuse hot gas in clusters of galaxies are presented to illustrate both capabilities.

The SIXA sensitivities for line detection have already been presented elsewhere (Vilhu 1990) and are not discussed here.

5.1 Broad Band Detection (HEPC, LEPC)

The best available 2 - 10 keV spectra of active galactic nuclei (AGN), one of the main contributes to the DXRB, have been obtained with the large area proportional counters on GINGA. Pounds *et al.* (1990) have presented the mean spectrum of 8 AGNs each with flux levels of 10^{-11} erg cm^{-2} s^{-1} (0.5 - 3.5 keV). The combined spectrum corresponds to a sensitivity limit 200 times greater than the EINSTEIN Deep Survey Limit. That the SODART telescope is confusion limited at low energies (~1 keV) is shown in Figure 7. Confusion is defined here as there being more than 1 source in an area corresponding to 40 half-power beams. What is also shown in Figure 7, however, is that the confusion problem disappears at higher energies (~8 keV). It is clear, that for a great many sources, it will be possible to detect the continuum in 1 keV energy intervals. For the fainter sources, the energy interval can be broadened (Figure 7) to allow for more sensitivity. By exploiting the better angular resolution of the JET-X telescope, it should be possible to remove some of the confusion problem from the SODART analysis. With XSPECT/SODART it should, therefore, be possible to link the AGNs discovered with low energy telescopes (EINSTEIN, ROSAT) with those discovered with large area proportional counter arrays (HEAO-1, GINGA) and, therefore, to establish accurately their broad band spectral shapes.

Sources too weak to be resolved can also be studied through their contribution to the DXRB. For a typical observing time of 10^4 s, the spectrum of the DXRB can be obtained out to 7 keV for each SODART field of view. Studies of pixel to pixel fluctuations can be used to determine the epoch at which the DXRB was formed.

5.2 Narrow Band Detection (OXS)

Three prototypical stars which cover a temperature range of nearly two orders of magnitude (see Table 4) are chosen to illustrate how the high resolution data can be used for temperature, electron density and velocity diagnostics.

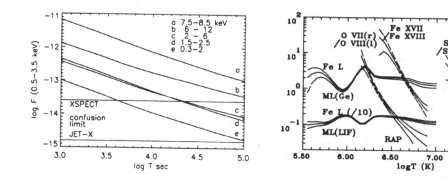

Fig. 7 (left) XSPECT/SODART point source detection sensitivities.
Fig. 8 (right) Line ratios are used to determine the ion temperature.

Table 4 Late-Type Stellar Parameters for OXS Observations

Object	Spectral Type	d	N_H	T_1	e'_1	T_2	e'_2	Ref
Capella	(G5III+)F9III	13	5	5	3	25	9	1
Procyon	F51V-V(+DF)	3.5	2	0.6	1	2	1.5	1
AR Lac[a]	K01V	40	10	20	3			2
AR Lac	G21V	40	10	7	3			2

[a]Eclipsing binary (P = 1.983[d])

[1]Mewe *et al.* (1990)   [2]Walter *et al.* (1983)

d = source distance (pc), N_H= interstellar column density (10^{18} cm^{-2}),T_i = temperature (10^6 K) and e'_i = reduced emission measure (10^{50} cm^{-3} pc^{-2})

The He-like : H-like resonance line ratio is a good diagnostic for ion temperatures over a broad range of temperatures and, as an example, Figure 8a and b show results calculated for S and O. For He-like OVII, the intercombination : forbidden line ratio is a sensitive function of electron density in the range $n_e = 10^{10}$- 10^{12} cm^{-3}. Simulations for Procyon are shown in Figure 9. Measurement of the Doppler shifted radiation from the orbiting component of a binary system can be used to separate the individual contributions to the total emission. In the RS CVn binary system Ar Lac , the K star orbits the G star with a period of 1.98 d (orbital velocity = 116 km s^{-1}). Figure 10 shows a simulation of the He-like S emission from the combined system as a function of orbital phase. The OXS wavelength resolution element is $\Delta\lambda = $ ~0.002 Å at $\lambda = 5$ Å.

Point source detection limits (5σ or 10 counts in the photon limited case) for each of the lines listed in Table 3 can be obtained from the parameters listed in Tables 1 - 3.

The expected line emission fluxes from the stars listed in Table 4 can be calculated from standard models using the parameters listed. The results for observing times of 10^3, 10^4 and 10^5 s per spectral resolution element are shown in Figure 11. Both particle and diffuse X-ray background have been taken into account.

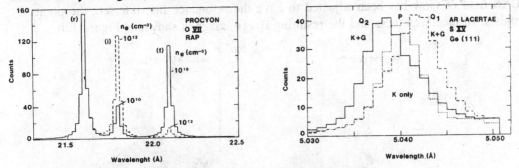

Fig. 9 (left) The effect of electron density on line strengths.
Fig. 10 (right) Doppler shifted lines from a binary system at different orbital phases.

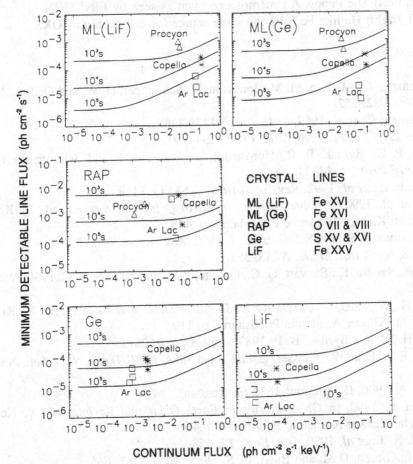

Fig. 11 OXS sensitivities for emission lines for various observing times.

The data from the EINSTEIN IPC observation of the Puppis supernova remnant shown in Figure 12a can be used to simulate an OXS observation. The raw data has been reprocessed using the parameters in Tables 1 and 2. The resulting image has been sampled by OXS using the parameters from Table 1 for the lines of He-like Fe. The OXS field of view has been adjusted to have the resonance line intersect the point of highest surface brightness and the resulting Bragg image is shown in Figure 12b.

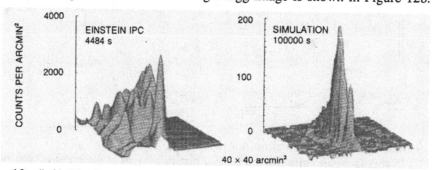

Fig. 12a (left) The Puppis A supernova remnant as seen by EINSTEIN.
Fig. 12b (right) He-like Fe from the same source for one setting of OXS.

6 REFERENCES

Budtz-Jørgensen, C., Madsen, M. M., Jonasson, P., Schnopper, H. W. and Oed, A. 1988, *SPIE Proc.*,**982**, 152.

Budtz-Jørgensen, C. *et al.* 1989, *SPIE Proc.*, **1159**,236.

Budtz-Jørgensen, C. *et al.* 1990, *SPIE Proc.*, **1344**, 91.

Christensen, F. E., Byrnak, B. P., Hornstrup, A., Shou-hua, Z and Schnopper, H. W. 1990, *SPIE Proc.*, **1344**, 14.

Christensen, F. E., *et al.* 1992, *Rev. Sci. Instrum.*, **63** (**1**), 1168.

Kaaret, P., *et al.* 1990, *in Observatories in Earth Orbit and Beyond*, ed. Y. Kondo (Dordrecht: Kluwer Academic Publishers), p. 443.

Mewe, R., Lemen, J. R. and Schrijver, C. J. 1990, *Adv. Space Res.*,**10**,129.

Oed, A. 1988, *Nucl. Inst. Meth.*, **A263**, 351.

Pounds, K. A.,Nandra, K., Stewart, G. C., George, I. M. and Fabian, A. C. 1990, *Nature*, **344**,132.

Schnopper, H. W. 1990, *in Observatories in Earth Orbit and Beyond*, ed. Y. Kondo (Dordrecht: Kluwer Academic Publishers), p. 119.

Schnopper, H. W. and Byrnak, B. P. 1987, *Appl. Opt.*, **262**871.

Serlemitsos, P. J., Petre, R., Glasser, C. and Birsa 1984, *IEEE Trans. Nucl. Sci.*, **NS-31**, 786.

Sunyaev, R. A. 1990, IKI preprint 1632 (in Russian).

Vilhu, O., *et al.* 1990, *in Observatories in Earth Orbit and Beyond*, ed. Y. Kondo (Dordrecht: Kluwer Academic Publishers), p.433.

Westergaard, N. J., *et al.* 1990, *Opt. Eng.*, **29**, 658.

Walter, F. M., Gibsen, D.M. and Basri, G. S. 1983, *Ap. J.*, **267**, 665.

Applications of CCD Detectors to Spectral Analyses of the X-Ray Emission from Tokamaks

A F Abbey[1], R Barnsley[1], J Dunn[1], S N Lea[1], N J Peacock[2].

1 Department of Physics, University of Leicester, University Road, Leicester LE1 7RH, England.

2 Culham Laboratory (UKAEA/EURATOM Fusion Association) Abingdon, Oxon. OX14 3DB, England.

ABSTRACT

Large area (~10^3 x 10^3 pixels), CCD detectors for direct imaging of X-ray sources are being developed in the X-ray Astronomy group at the University of Leicester. This paper reports on the incorporation of such a detector within a Johann crystal spectrometer for tokamak spectroscopy. With a pixel size of 22 μm, the CCD is an ideal detector for a relatively compact, 1 to 2 m Rowland circle, spectrometer devoted to X-ray line emission measurements. The use of a prototype CCD-Johann spectrometer for time-resolved temperature measurements of the $1s^2$ - $1s2p$ lines of Si XIII in the DITE (Diverter Injection Tokamak Experiment) is reported. In this case, the detector was a standard TV format CCD with video-tape data storage. Subsequently, a liquid N_2 cooled CCD detector system capable of accommodating chips with up to 1152 x 2186 pixels and with quantum efficiency ≥ 20% for photon energies between 700 eV and 12 keV, has been installed in the Johann spectrometer. Read-out modes can be programmed to suit the application. A 2-D single photon counting mode, with an energy resolution of 150 eV is appropriate to the study of low flux sources such as those associated with beam-foil or gas target experiments. By sacrificing the energy resolution, a 1-D mode with on-chip binning allows line spectra from high flux sources such as tokamaks to be recorded with a time resolution of ~ 1 ms.

1. Introduction

The use of CCDs for direct detection of X-rays from laboratory sources has certain advantages over more conventional detectors such as gas-flow proportional counters, scintillation converters, microchannel plates and other solid state devices.

The advantages include integrated electronics, a reasonably high quantum efficiency, ~50% in the X-ray region, Fig 1, a useful inherent energy resolution ~ 200 eV, a high spatial resolution with pixel sizes in 22 μm x 22 μm and most importantly, the flexibility using appropriate software options, to read out the X-ray signal electronically in a mode designed to optimise the information from the source. In the present context of measuring emission line profiles from fusion plasmas, then ideally, a finite spectral region, typically $0.01 \leq \Delta E/E \leq 0.1$, should be recorded and read out continuously by on-chip binning with a time resolution $\sim$1msec.

This paper describes, in the first instance, experiments on the feasibility of using a standard uncooled video read-out CCD for plasma X-ray spectroscopy on DITE. These experiments have led to the development of custom-designed, large area, cooled CCD's which have been used to record the X-ray spectra from JET.

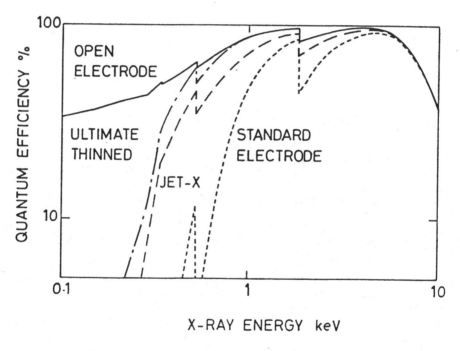

Fig.1   Quantum efficiency of CCD detectors.

2.   Use of Standard Visible TV Format CCD for X-ray Spectroscopy

In a trial experiment aimed at measuring the ion temperature of Si XIII in DITE, a standard uncooled, TV-format CCD was used to detect the diffracted light from a curved crystal spectrometer. The variable radius, Johann spectrometer has been described previously (Dunn et al 1988). In the present application, an ADP (101), 2d = 10.648 Å crystal was bent to a Rowland radius of 657 mm. The Rowland circle was placed orthogonal to the toroidal axis.

While the crystal rocking curve determined the ultimate resolution $\lambda/\delta\lambda \approx 4000$, Johann defocusing errors, arising from the finite 28 x 22 mm^2, width x height of the crystal, were responsible for an operating resolution $\lambda/\delta\lambda \approx 2,500$ which was still further reduced by avoidable focusing errors in this Si XIII study. The CCD chip active area of 8.5 x 6.5 mm^2 limited the spectral bandwidth to $\lambda/\Delta\lambda \approx 0.015$ at a Bragg angle, $\theta_B = 38.95^o$. This was sufficient to encompass the Si XIII resonance lines between 6.648 Å and 6.740 Å in each read-out frame, Fig 2 .

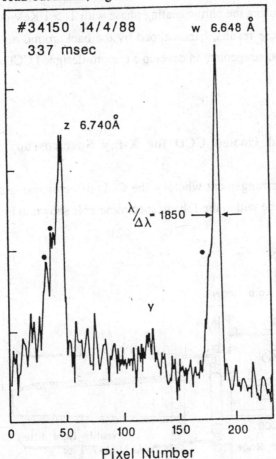

Fig.2 Wavelength region 6.65 = >6.74Å recorded with a commercial TV format CCD chip, in a 20 ms integration period, 337 ms into shot #34150 on the DITE tokamak. The crystal in the Johan spectrometer was PET (002). Features marked • on the resonance line 'w' and the forbidden line 'z' indicate defocused fringing due to disturbed crystal alignment during spectrometer installation.

The CCD chip was mounted on the Rowland Circle, but perpendicular to the diffracted X-ray beam instead of the usual tangential geometry. A transparent 0.5 μm polypropylene window was placed over the CCD aperture to protect it from dust. A light tight 2 μm Mylar filter with 1000Å Al coating was used to pass the diffracted rays.

The CCD was also useful for focusing the crystal in visible light. An optical focus of 60 μm FWHM was easily achieved. The visible and x-ray light images were recorded onto video tape using a commercial video cassette recorder. Individual frames were analysed with a frame store. The S/N ratio was improved by compressing the 2-dimensional height information into 1-dimension.

The experiments with the time resolved TV-format chip yielded reasonable results for T_i (Si XIII) ~ 700eV during the ohmic heating phase with $T_i > 1$ KeV during neutral beam auxiliary heating. These results, handicapped by the background noise in an uncooled chip, gave sufficient encouragement to develop a custom-designed CCD detector for the X-ray region.

3. Custom-Designed Cooled CCD for X-ray Spectroscopy

The mechanical arrangement whereby the CCD detector was accommodated on a rotation / translation stage within the Johann spectrometer is shown in Fig 3.

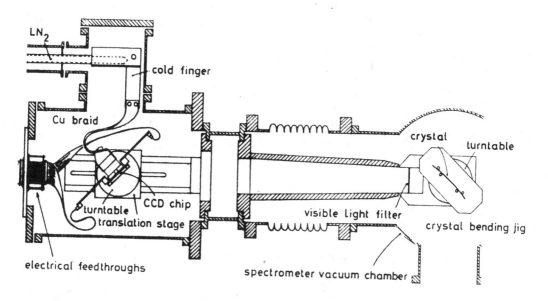

Fig.3   Plan view of diffraction arm of spectrometer.

The CCD detector and associated electronics were based on those already developed and in use for astrophysical studies, (Chowanietz et al 1985). The detector characteristics and read-out options are tabulated below.

3.1 Large Area (1152X2186 Pixels) Liquid N$_2$ CCD Detector

CCD Specifications:

Manufacturer.....	English Electric Valve Co (EEV)
Type.....	CCD 05-30-5 high resistivity (20Ωcm) silicon
Size (pixels)....	1152 X 1242
Pixel Size......	22.5 µm square
Active Area.....	27 X 25 mm
Quantum Efficiency.....	50% typically
Thermal Control........	Liquid Nitrogen to -100°C
Charge Transfer Efficiency......	0.999992/pixel (serial)
Energy Resolution (non-binned mode).....	≈ 200eV

3.2 Read-out Mode Specifications:

Visible Mode for focusing and alignment of spectrometer:

CCD mode.....	Frame transfer using 8-bit system:
Time/pixel.....	2 µs
Time/row.....	2487 µs
Frame rate....	576 X 2487 µs ≈ 1.4 µs, giving ≈ 0.7 fr s^{-1}

X-ray, single photon counting mode:

CCD mode.....	Full frame, 12 bit ADC
Energy resolution.....	150 eV
Time/pixel......	10 µs
Frame readout time....	10 X 1242 X 1152 = 14 s
Max integration time (at -100°C)......	300 s typically,
	(limited by dark current build-up)

X-ray, high time resolution mode:

(i) CCD mode(12 bit ADC)..... Full frame; on-chip binning ;
all vertical information lost.

Energy resolution......	none
Time/super-pixel.....	10µs
Transfer time, data to	
o/p register (dead time).....	1152 x 0.5 µs = 576µs.
Time resolution with max. line resolution.......	≈12ms
Time resolution with binned X10 horiz. pixels......	≈1.2ms
Time resolution with readout of 1/10 of horiz. pixels.......	≈1.2 ms

These figures could be improved x4 by utilising read-out at all 4 corners of CCD. (With additional hardware).

(ii) Alternative (using 8 bit ADC)higher speed mode..... Full frame, on-chip binning, all vertical information lost.

Energy resolution.....	none, unless count rate is low.
Time/super-pixel.....	2 µs
Transfer time of data to o/p register (dead time)....	1152 X 0.5 µs = 576 µs
Time resolution with max line resolution......	1242 X 2 + 576 µs = 3.1ms
Time resolution with binned X10 horiz. pixels.....	≈ 0.3 ms
Time resolution with readout of 1/10 of horiz. pixels.....	≈ 0.3 ms

4.   Results Using Johann-CCD Spectrometer on JET.

The Johann-CCD instrument was deployed on the same sight line as a 2-crystal Bragg spectrometer (Barnsley et al 1991) which viewed the JET plasma along a horizontal chord through the torus mid-plane. The results of monitoring the He-like Ni XXVII lines between 1.59Å and 1.60Å using a Ge 220 crystal are shown in Figs 4, 5 and 6. Fig 4 shows an example of the charge deposited on the pixel field due to line emission. Fig 5 shows five sequential time slices of the Ni XXVII spectra in second order, interspersed with the Cl XVII Lyman series in first order, taken at 5 sec intervals during a JET discharge. In this preliminary experiment, the signal intensity was near the upper limit of single photon counting. The first order peak at 3.884 keV, Fig 6, is produced by single photons and the remaining spectrum is composed of single photons plus pulse pile up. An analysis of the peaks, however, allows discrimination of the first and second order line intensities. The spectra are comparable to that taken on TFTR by Hsuan et al (1987) and show the intensity variations of the He-like resonance lines and their Li and Be-like satellites in response to T_e changes during the discharge.

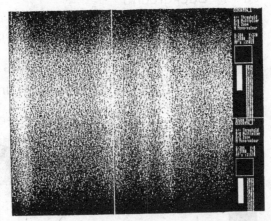

Fig.4   2-D Image (integrated throughout Tokamak pulse) of dispersed X-ray spectrum.

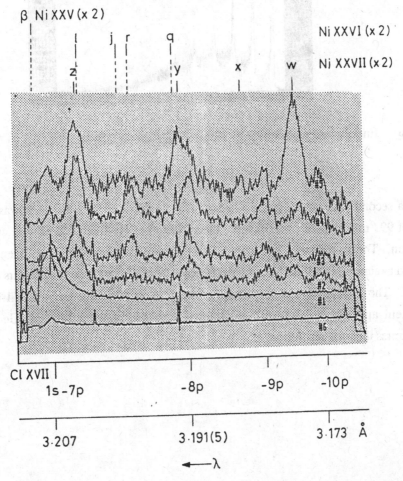

Fig.5   Sequence of spectra taken throughout JET discharge #26938 (CCD-Type 0530 EEV, 1152 X 1242 Pixels. Integration time ~2s. Ge (220), 2d = 4Å.).

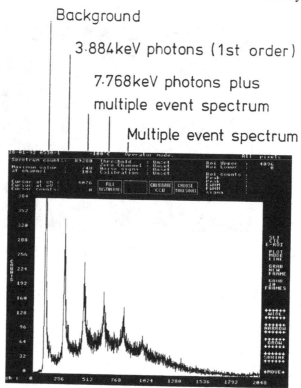

Fig.6 Energy spectrum on CCD chip for JET shot # 26938.

In a second experiment on JET, a Si (111) crystal, with 2d = 6.271Å, was bent to a radius of 997 mm and set up to observe the spectrum of He-like Cl XVI with 40 ms time resolution. The frame-store half of the CCD array was masked and the open area was protected against visible light by a graphite-coated 2μm Mylar filter. The results are shown in Fig.7. The increase in the intensities of the Cl XVI lines due to NBI at 14s and the subsequent appearance of additional lines due to the injected Mo and Ge at 15.5s, demonstrates the time resolution.

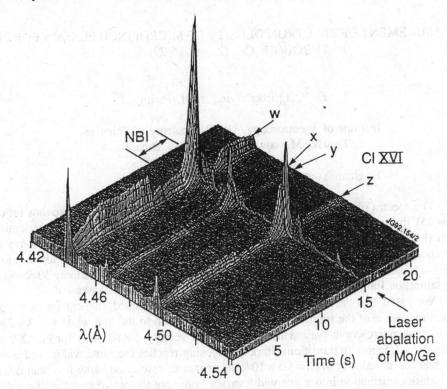

Fig.7 Isometric plot of Cl XVI n =2 resonance lines (w,x,y, z) during a JET discharge with a period of neutral beam injection (NBI) and with transient injection of Mo and Ge. Each time-slice results from compressing the 2-D image onto a single row of the frame store region of the CCD. The time resolution is 40 ms.

5.   References

1 Dunn J , Barnsley R , Evans K D and Peacock N J (1988) *Jnl. de Phys.* Colloque C1Supp.No3, 49 C1-91.

2 Chowanietz E G , Lumb D H and Wells A A (1985) *SPIE* Vol.597, 38.

3 Barnsley R, Schumacher U , Kallne E, Morsi H W and Rupprecht G (1991) *Rev. Sci. Instr.* **62** (4) 889.

4 Hsuan H et al *Phys Rev A* (1987) Vol.35 No.10, 4280.

MEASUREMENT OF ELECTRON DENSITY OF MICROPINCH PLASMA FOR ELEMENTS P THROUGH Cu (Z = 15-29)

E. V. Aglitsky and *A. M. Panin*[1]

*Institute of Spectroscopy, USSR Academy of Sciences,
Troitsk, Moscow region, 142092, Russia.

[1]Brigham Young University, Provo, UT 84602

The spectra of $1s^2$-$1snp$ transitions ($n = 3$-8) in highly charged He-like ions for elements P ($Z = 15$) throught Cu ($Z = 29$) were registered in a low-inductance vacuum spark micropinch plasma (MP) discharge. The spectra were rather weak and a highly sensitive x-ray imaging registration system based on a visible intensifier with fiberglass input-output windows was used [1]. The quantum efficiency of this system is about 50-80 %, noise sensitivity 30 photons/cm^2/c, and a saturation limit 10^6 photons/sm^2 (10^{-9} J/sm^2) in a 2-20 keV waveband.

We found that the registered spectral lines are broadened, depending on the principal quantum number n of the transition. The main contribution to the line width ($\sigma\lambda/\lambda = 2\times10^{-3}$) is a non-thermal macroscopic plasma motion (estimated velocity 3×10^7 sm/s for Fe XXV plasma). For the $1s^2$ - $1s5p,6p,7p$ transitions Stark broadening reaches the same value and a total line width reaches a value of $\sigma\lambda/\lambda = 3$-$5\times10^{-3}$, the apparatus resolution being less than $\Delta\lambda/\lambda = 10^{-3}$.

Stark contribution into a line width varies from shot to shot; thus reflecting a difference in the plasma ion density of each micropinch discharge due to different initial conditions. We found that the smaller Stark broadening, the "longer" $1s^2$ -$1snp$ series could be observed (in terms of maximum n).

Our estimation of plasma conditions showed that for higher n ($n = 5$-8) we can use a linear Stark approximation to simplify the line shape calculation. Electron density N_e corresponding to the observed broadening differs for different electrode matherials. Value of N_e lies within 10^{20} -10^{21} sm^{-3} for P ($Z = 15$), and rises substantially with Z: for Fe ($Z = 26$) $N_e = 10^{22}$ -10^{23} sm^{-3} , and for Cu ($Z = 29$) it reaches the value 3×10^{23} sm^{-3} (see Tabl.1). The dependence of measured density on Z is found to be as strong as Z^8. We couldn't estimate N_e for elements with $Z > 30$, although a resonant transition $1s^2$ -$1s2p$ was observed up to Zr ($Z = 40$). The spectra of these elements for the $1s^2$ -$1snp$ ransitions with $n > 3$ were too weak for broadening to be correctly measured. Assuming that the observed growth of N_e will take place for these elements too, one could expect the values N_e as much as 10^{25} sm^{-3} or more. MP plasma gives us the unique ability to investigate super-dense hot matter under laboratory conditions.

We also estimated the value of energy emitted by LIVS MP plasma in the resonance line of He-like ions as 10^{12} quantum/shot (for Fe, $Z=26$; $\lambda = 1.85$ Å). The dimensions of MP plasma as measured by specially designed pinhole were approximately 2-3 μm in diameter and 5-10 μm in length along an axis of MP plasma. The MP duration time was also estimated by measuring the plasma movement velocity and was found to be 20-50 psec for Fe and 40-60 ps for Ti plasmas.

Since one discharge of LIVS typically produces one micropinch (which is well seen from the pinhole picture), the power of linear K-radiation loses of MP was estimated as 10^8 Watt, or 10^{18} W/sm^3 , hence and a spectral brightness of the micropinch is 10^{18} W/sm pm (for $h\nu$=6.67 keV). This power of radiation is sufficient for photopumping of lasant ions in soft X-ray waveband, if the lasant ions are present into the MP plasma). Thus, a MP plasma of LIVS may serve as a small and bright "lamp" in X-ray region for photopumping. Unfortunately, a size of a "hot point" plasma is too small ($\approx \mu$m) to get reasonable gain.

Table 1. Measured density N_e of MP plasma for different elements.

Element	Z	Shot#	Measured density (cm^{-3})
P	15	1	2×10^{20}
		2	1×10^{21}
Ca	20	1	3×10^{21}
		2	5×10^{21}
		3	2×10^{22}
Fe	26	1	2×10^{22}
		2	1×10^{23}
Cu	29	1	8×10^{22}
		2	3×10^{23}

REFERENCES

1.A.M.Panin, "Sensitive x-ray imaging registration system" , Rev. Sci. Instr., 63, 1, IIA, 620, (1992).

2.E.V.Aglitsky, P.S.Antsiferov, A.M.Panin, "Electron density measurements of micropinch plasma", Plasma physics (Russian), 18, (1992), in press.

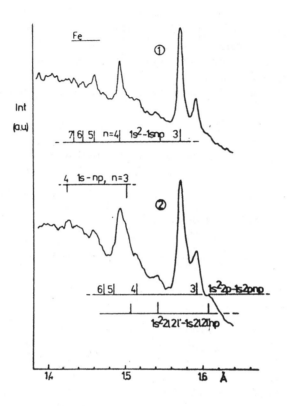

Fig. 1 . Spectra of $1s^2$-1snp transitions
in He-like Fe (Z =26).

Bayesian approach to soft x-ray line diagnostics

D. Alexander and A.J.M. Garrett

Department of Physics and Astronomy, University of Glasgow, Glasgow, U.K.

ABSTRACT

Bayesian analysis is applied to the problem of how to combine spectroscopic data, obtained from two different instruments observing the same localised region of plasma, in a logically consistent way. This approach promises a better determination of the line fluxes used as diagnostics for the physical parameters of the emitting plasma. A specific benefit would be in the assignment of the best probability distribution for the mean temperature in an observed region of the solar atmosphere (eg. an active region or solar flare). Two methods in common use for temperature determination are the analyses of the Differential Emission Measure distribution and of temperature-sensitive line ratios. A problem apparent in each of these methods is how to combine consistently two or more data sets obtained from distinct but complementary instruments. The Bayesian approach discussed here outlines a scheme which makes this possible.

1 BAYESIAN ANALYSIS

Rules for the manipulation of probabilities in a Bayesian analysis can be derived from consistency criteria and proceed via the sum and product rules (see Jaynes 1983).

To test a chosen hypothesis (H) in the light of a given set of data (D) we can use Bayes' theorem (cf. Loredo 1989)

$$p(H|DI) = K\, p(D|HI)\, p(H|I) \, , \tag{1}$$

where K is a normalisation factor, p(H|DI) denotes the probability that hypothesis H is true given the observed data, D and any additional information present, I, and p(H|I) is known as the prior distribution for the hypothesis being tested.

For two (or more) data sets we obtain

$$p(H|D_1 D_2 I) = K\, p(D_2|HI)\, p(D_1|HI)\, p(H|I) \, , \tag{2}$$

where we have used (1) with D now denoting all data present (i.e. $D = D_1 D_2$) and we assume that D_2 is independent of D_1 given H.

2 EXAMPLE

Consider the observation of an emission line by two different instruments (see Figure 1). Observations of emission line intensities from a region of the solar atmosphere

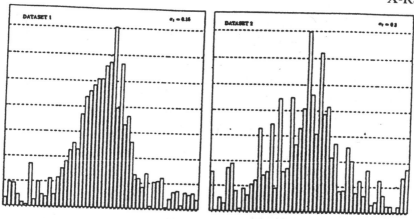

Fig. 1—Hypothetical observations by two distinct instruments

by the SOHO/CDS and SUMER instruments in a common wavelength range could provide a potential example (Harrison 1991). In producing thehypothetical data sets shown in Figure 1 we have assumed for illustration that each instrument is subject to Gaussian noise and that this noise is independent of intensity and wavelength. In principle, however, any instrumental properties may be included.

Our goal is to estimate from the combined data, using Bayesian analysis, the mean λ_0 and variance s^2 of the line profile, which is hypothesised to have a Gaussian form; this hypothesis is implicit in the information content, I. The introductory example presented here foreshadows our ultimate aim of using a non-parametric Bayesian approach to test different physical descriptions of the observed line profiles. In appropriate Bayesian language we wish to find the joint probability distribution for (λ_0, s^2) given H which is consistent with the data observed. Note that in this analysis there are two distinct Gaussians being discussed for each instrument; one relating to the instrumental noise with mean 0 and variance $\sigma_{1,2}^2$, the other to the hypothesised distribution for the observed line profile with mean λ_0 and variance s^2. Equation (2) thus becomes

$$p(\lambda_0, s | D_1 D_2 I) = K\, p(D_2 | \lambda_0, s, I)\, p(D_1 | \lambda_0, s, I)\, p(\lambda_0, s | I) . \qquad (3)$$

In order to solve for the joint distribution, $p(\lambda_0, s | D_1 D_2 I)$ we need to assign each of the probabilities appearing on the RHS of equation (3).

1. $p(\lambda_0, s | I)$: With *no additional information*, other than that the line has a well-defined peak and a definite spread, we assume initially that all values of (λ_0, s) within the observed range are equally likely and therefore our prior probability distribution is chosen to have the 'top-hat' form

$$
\begin{aligned}
p(\lambda_0, s | I) &= c & \lambda_1 \leq \lambda_0 \leq \lambda_2 \quad ; \quad s_1 \leq s \leq s_2 \\
&= 0 & \text{otherwise}
\end{aligned}
\qquad (4)
$$

where c is a constant, λ_1, λ_2 denote the lower and upper wavelengths observed by the instrumental channel being considered and s_1, s_2 are chosen from inspection of the

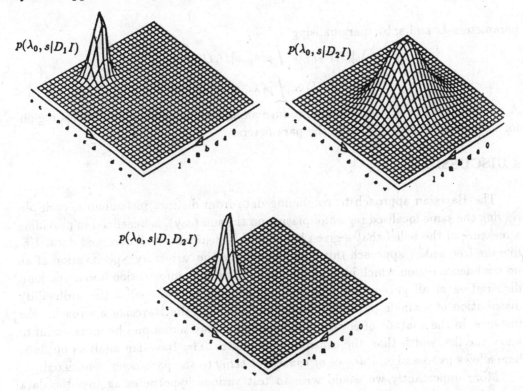

$p(\lambda_0, s|D_1 I)$ $p(\lambda_0, s|D_2 I)$

$p(\lambda_0, s|D_1 D_2 I)$

Fig. 2—Posterior probability distributions representing the 'truth' of the assumed hypothesis of a Gaussian line profile obtained from a Bayesian analysis (eq. 6) for the individual data sets and for the combined data

observed profile; $s_2 \leq (\lambda_2 - \lambda_0)$ when there is a definite line present. 2. $p(D_1|\lambda_0, s, I)$: Since the noise in each bin is assumed to be Gaussian we have

$$p(D_1|\lambda_0, s, I) \propto \prod_{i=1}^{n} \left[\int_{\text{bin}} exp\left\{ -\frac{(D_{1i} - N_i)^2}{2\sigma_1^2} \right\} d\lambda_i \right] , \qquad (5)$$

where i is the bin number, D_{1i} is the data observed in each bin of instrument 1 and N_i is the hypothesised Gaussian line distribution with mean λ_0 and variance s^2 evaluated at $\lambda = \lambda_i$, the wavelength corresponding to bin i. Likewise, $p(D_2|\lambda_0, s, I)$ has a similar form.

Therefore, we have

$$p(\lambda_0, s|D_1 D_2 I) = K \, exp\left[-\sum_{i=1}^{n} \frac{1}{2}\left(\frac{(D_{1i} - N_i)^2}{\sigma_1^2} + \frac{(D_{2i} - N_i)^2}{\sigma_2^2} \right) \right] p(\lambda_0, s|I) , \qquad (6)$$

where we have assumed that the function under the integral in equation (5) is constant across a bin. The probability distributions obtained are shown in Figure 2. The combined result is clearly weighted towards the 'better' data as would be expected. One advantage of the Bayesian approach over a frequentist approach, such as χ^2 testing, is that we are able to obtain probability distributions for the individual parameters λ_0 and s, by marginalising

$$p(\lambda_0|D_1D_2I) \propto \int p(\lambda_0, s|D_1D_2I)ds \; ,$$

$$p(s|D_1D_2I) \propto \int p(\lambda_0, s|D_1D_2I)d\lambda$$

(7)

A statistic, such as χ^2, would only produce an arbitrarily defined 'confidence region' for the *combined* distribution of the parameters.

4 DISCUSSION

The Bayesian approach to combining data from distinct instruments, each observing the same localised region of plasma on the sun (say), is beneficial in providing a measure of the belief that a given hypothesis is justified by all observed data. Unlike the frequentist approach this does not require the 'arbitrary' specification of an $n\sigma$ confidence region which in any case can only provide information about the joint distribution of all parameters required. If we wish to know what the probability distribution of a single parameter looks like, we need the Bayesian approach. For instance, in the analysis of emission line profiles it may sometimes be more useful to know the line width than the centroid wavelength. The Bayesian analysis outlined here allows us to assign a *degree of consistent belief* to the parameter considered.

More importantly we would wish to test various hypotheses against the data and compare them with each other in a consistent manner in order to determine which best fits the data. Bayesian analysis allows us to do this by comparing the probabilities assigned for each hypothesis (cf. Loredo 1989).

In the example given here a top-hat prior distribution was assumed for the parameters under consideration. Any additional prior information available, such as observations from other instruments, which cannot be directly introduced into the above analysis, should be included in the choice for the form of the prior distribution. For instance, observations of non-thermal processes may affect our decision about the expected shape of the profile. Any additional information we have should be included.

While we maintain that Bayesian analysis is the only logically consistent way to incorporate data into probability calculations, a more comprehensive study must be completed in order to assess any improvements over conventional sampling theory in the determination of physical parameters in the solar atmosphere.

REFERENCES

Harrison, R.A.: 1991, *The Coronal Diagnostic Spectrometer for SOHO*, RAL Scientific Report, SC-CDS-RAL-SN-91-0005.

Jaynes, E.T.: 1983, *Papers on Probability, Statistics and Statistical Physics*, ed. R.D. Rosenkrantz, (Dordrecht: Reidel).

Loredo, T.J.: 1989, in *Maximum Entropy and Bayesian Methods*, ed. P. Fougere, (Dordrecht: Kluwer).

ENERGY LEVELS 1s²2lnl' (n=2,3,4) OF NaVIII-SXIII IONS. COMPARISON OF TWO CALCULATION METHODS: MCDF AND MZ

K. Ando[1], U.I. Safronova[2], and I.YU. Tolstikhina[2]

[1]Riken, The Institute of Physical and Chemical Research, Japan
[2]Institute of Spectroscopy, Russian Academy of Science, Russia

1. Introduction

The present work is the direct continuation of paper (Safronova et al. 1990) where contributions of different effects to the first low excited states of Be-like ions were discussed. Use of the perturbation theory by 1/Z allows us to analyze the results of energy level calculations which were carried out on the basis of Dirac-Fock calculation method and to point the way to improve the accuracy of data obtained. In the present work we use the same method but we will consider the calculations of higher excited states that is 1s²2131' and 1s²2141'. For a more complete comparison with experiment we have chosen the ions in intermediate ionization stages, Na VIII-S XIII. We will give the results of two calculations and experiments by (Martin and Zalubas 1979, 1980, 1981, 1983, Martin and et al. 1985, 1990). The calculations on the basis of perturbation theory give the energy levels for the whole isoelectronic sequence in a wide interval of Z. To demonstrate the contributions of different effects we will use the results for another Z also.

2. 1/Z-Expansion (MZ-Method)

The MZ-program is based on perturbation theory by 1/Z (Safronova and Senashenko 1984). The energy matrix is represented by the sum of three terms:

$$E(a,a';J) = (E^N + E^R)\, \delta\,(LS,L'S') + E^S \tag{1}$$

where E^N is the nonrelativistic part, E^R is the relativistic term shift, defined by the Breit Hamiltonian and including the dependence of mass on velocity, contact term and orbit-orbit interaction. Finally, the third term E^S gives the fine structure splitting of terms and consists of three operators: spin-orbit (E_1^{SL}), spin-other-orbit (E_2^{SL}) and spin-spin (E^{SS}) interactions.

Each of these parts E^N, E^R, E^S can be represented as 1/Z-expansion and the calculation of every coefficient is carried out by means of the Feynman diagram technique. At present the MZ-code includes subroutines for the calculation of E^N with the precision up to the second order of the theory of perturbations and E^R and E^S up to the first order by 1/Z. There are no difficulties in including the new expansion terms as new coefficients will be calculated. Now the addition of the new terms is made by using the screening method. The constants of screening are the first order contributions of perturbation theory (it will be easily seen from formula given below). The core (C) of the Be-like system considered here is the three-electron system 1s² 21 with nl external electron. So for the diagonal matrix elements a=a' we have:

$$E^N = E^N(C) + E^N(n), \quad E^R = E^R(C) + E^R(n) \tag{2}$$

$$E^N(C) = Z^2 E_0(C) + Z E_1(C) + E_2(C) \tag{3}$$

$$E^N(n) = -[Z - \sigma(n)]^2 / 2n^2 + E_2'(n); \quad \sigma(n) = E_1(n)n^2 \tag{4}$$

After substitution of "n" for "C" the formula for E^R (C) and E^R (n) have the same form:

$$E^R(n) = \frac{\alpha}{4}\left(Z - \sigma^R(n)\right)^3 ZR_0(n); \quad \sigma^R(n) = R_1(n) / 3R_0(n) \tag{5}$$

The formulae for E^S (C) and E^S (n) are analogous:

$$E^S(n) = \frac{\alpha}{4}\left(Z - \sigma^S(n)\right)^3 \left(ZE_0(n)Q_1 + E_2(n)Q_1 + E_{SS}Q_2\right) \tag{6}$$

A more detailed description is in (Vainshtein and Safronova 1978) which also includes all formula for taking into account of the third order of nonrelativistic part of energy and high-order relativistic corrections E^D and radiative corrections E^L. It must be mentioned that the last two of the corrections above are one-electron corrections. To take into account the deviation of our system from the one-electron system we have inserted screening constants σ into these two corrections. The constants coincide with σ (n) and are defined in nonrelativistic approximation as:

$$E^D = \alpha^4[Z - \sigma(n)]^6 D, \qquad E^L = \frac{4}{3}\alpha^3[Z - \sigma(n)]^4 \Lambda \tag{7}$$

The method of the calculation of all the coefficients in (2) - (7) for the Be-like ions for the states $1s^22121'$ was given in [15]. The coefficients for more excited states was calculated in the present work.

To demonstrate the contribution of different corrections to the energy matrix the values of the corrections for the states $1s^22131'$ are represented in Table 1 for three different values of Z as an example. We have combined contributions of E^R and E^S and represented the sum of these corrections E^B. One can see from the Table that the two additional corrections E^D and E^L increase rapidly with Z so the calculations for the ions of high ionization degree will be not precise enough if these two are not taken into account.

TABLE 1.

Contribution of Different Effect for Energy Level

Level	Z	E^N	E^B	E^D	E^L	$E(\Sigma)$
$1s^22s3p$ $^3P_0$	10	1026929	1587	2	-57	1028461
	26	8972009	104501	1564	-2977	9075416
	42	24720125	787207	36799	-17629	25526501

In agreement with the general idea of the method of 1/Z-expansion all states are grouped in complexes which unite the states with definite values of J and parity (1+1'). Thus for the even complex of Be-like system we have the configurations 2s3s, 2p3p, 2s3d for n=3 and 2s4s, 2p4p, 2s4d, 2p4f for n=4. The energy matrix is diagonalized independently for each complex. After this procedure the value of aLS which corresponds to the maximum value of eigenvector component C^J (aLS) is assigned to the mixed level obtained.

3. Multiconfiguration Dirac-Fock Method (MCDF)

Multiconfiguration Dirac-Fock calculation of the energies of $1s^2 2lnl'$ (n=2,3,4) states for the ions with Z=11-16 was carried out on the basis of Grant code (Grant et al. 1980).

Let us now discuss the correlation corrections. Like in Z-expansion method in MCDF calculation we considered the complexes of states, i.e. the states with the same parity and the same values of zero-order energy. In this case the first order correlation correction theory that is proportional to Z was taken into account completely. The second-order correction was also taken into account in same way but the small part of it only. We also have performed the calculation with an extended basis: $1s^2 2l2l'+1s^2 2l3l'$, and $1s^2 2l2l'+1s^2 2l4l'$. In this case the energy is changed by 500-2000 cm^{-1}, that is 0.002-0.01 a.u.. For the states $1s^2 2l2l'$ E^{corr} is equal to 0.08-0.12 at.u. (Safronova and Weiss 1989). So, the inclusion of states $1s^2 2l3l'$ into the basis leads to the accounting of 10% of correlation correction of the second order.

4. Discussion of the Results of the Calculations

The results obtained by means of two codes -MCDF (c) and MZ(b)- and available experimental data (a) which were taken from reviews of the NBS are represented in Table 2. The table includes only 5 levels chosen by us for illustration from 10 levels of the $1s^2 2l2l'$ configuration, 36 levels for the $1s^2 2l3l'$ configuration and 52 levels for the $1s^2 2l4l'$ configuration. For all levels of $1s^2 2l2l'$ configurations we have a very good agreement of experimental data with the theoretical results obtained by the MZ- code. The MCDF-code does not give the same agreement because this method does not take into account the correlation corrections of the second order. The difference was considered in detail by (Palchikov and Safronova 1990). The agreement of experimental data with theoretical results obtained by the MZ-code for the configurations $1s^2 2l3l'$ is worse. In a number of cases it can be explained by low accuracy of experimental data (for the $1s^2 2l2l'$ levels we have used the smooth data from (Edlen 1983)). On the other hand in the MZ-code, the third-order correlation corrections are not sufficiently accurate. These corrections were evaluated by an empirical method. For the MCDF-code the difference is 1000-3000 cm^{-1}, which corresponds to the correlation correction of second order. This part was added to the energy matrix and have obtained better agreement with experiment for the levels of $1s^2 2l2l'$ (Safronova et al. 1990). At present we can not perform this in the general case. Unfortunately there are not enough experimental data for the configurations $1s^2 2l4l'$. In fact the whole configuration $1s^2 2l4f$ has not been studied and we hope that this paper will stimulate such experiments.

TABLE 2.

Energy Be-like ions (cm^{-1}), a-exp. :NBS data, b-MZ, c-MCDF

Level		NaVIII	MgIX	AlX	SiXI	PXII	SXIII
2s2p ^{1}P$_1$	a	243208	271687	300490	329679	359343	389583
	b	243132	271650	300482	329712	359415	389703
	c	252852	281346	310153	339355	369039	399298
2s3s ^{3}S$_1$	a	1239974	1532450	1855760	2210700	2595600	3011500
	b	1241027	1532524	1855857	2210101	2595334	3011647
	c	1237755	1530174	1853434	2207606	2592771	3009016
2p3s ^{3}P$_2$	a	1402200	1713900	2057140	2432037	2838800	
	b	1402430	1714236	2057379	2432037	2838402	3276683
	c	1401783	1713524	2056609	2431214	2837530	3275768
2p4p ^{3}P$_2$	a	1822880	2235350		3187370	3727900	
	b	1822376	2235250	2690239	3187535	3727347	4309913
	c	1820621	2234543	2689572	3186907	3726775	4309476
2p4d ^{1}D$_2$	a	1827570	2241210		3193530		
	b	1827817	2241063	2696253	3193452	3732705	4314087
	c	1826964	2240452	2695872	3193274	3732703	4314290

REFERENCES

Edlen, B. 1983, *Phys. Scripta,* **28,** 51.

Grant, I.P., McKenzic, B.T., Norrington, P.H., Mayers, D.F., and Pyper, N.C. 1980, *Comput. Phys. Commun.,* **21,** 207.

Martin, W.C., and Zalubas R. 1979, *J. Phys. Chem. Ref. Data,* **8,** 817.
1980, *J. Phys. Chem. Ref. Data,* **9,** 1.
1981, *J. Phys. Chem. Ref. Data,* **10,** 153.
1983, *J. Phys. Chem. Ref. Data,* **12,** 323.

Martin, W.C., Zalubas, R., and Musgrove, A. 1985 *J. Phys. Chem. Ref. Data,* **14,** 751.
1990, *J. Phys. Chem. Ref. Data,* **19,** 821.

Palchikov, V.G., and Safronva, U.I. 1 990, *Opt. and Spectr.,* **68,** 281.

Safronova, U.I., and Senashenko, V.S. 1984, *Theory of Spectra of Multicharged Ions* (Moscow: Energoatomizdat).

Safronova, U.I., and Weiss, A.W. 1989, *Opt. and Spectr.,* **66,** 2.

Safronova, U.I., Tolstikhina, I.Yu., and Chen, M.H. 1990, *Opt. and Spectr.,* **68,** 151.

Vainshtein. L.A.. and Saronova. U.I. 1978, *A.D.N.D.T.* **21,** 49.

Broadband (1 - 100 Å) Bragg Spectroscopy of Impurity Ions in Tokamak Plasmas

R Barnsley[1], S N Lea[2], A Patel[3], N J Peacock[3]

1 JET Joint Undertaking, Abingdon OX14 3EA, UK
2 Dept. of Physics and Astronomy, Leicester University, LE1 7RH, UK
3 Culham Lab. (UKAEA/Euratom Fusion Ass.), Abingdon, OX13 3DB, UK

ABSTRACT

The useful wavelength range for Bragg spectroscopy of Tokamak plasmas has been extended to 100 Å by using organic crystals, Langmuir-Blodgett films and multilayer mirrors. Theoretical work, confirmed by experiment, shows that the best resolving power between 25 Å and 100 Å is provided by true crystals such as OHM (octadecyl hydrogen maleate). A range of long wavelength diffractors, together with more standard crystals at shorter wavelengths, have been used to monitor all the main impurities in the JET Tokamak plasma during 1991 operations. Survey spectra from the COMPASS and JET Tokamaks of impurities ranging from Be to Zn are presented.

1. Introduction

There is an incentive to extend the practical upper limit for Bragg spectroscopy of Tokamak plasmas from about 25 Å to about 100 Å, since it would allow a single instrument to monitor the most highly ionised species of <u>all</u> impurities with $Z \geq 3$ (Li). This is important for the light impurities Be, B and C, which often dominates the radiated power, fuel dilution and effective charge state of Tokamak plasmas.

Above 25 Å, the low transmission of most window materials and the modest energy resolution of gas proportional counters both result in reduced sensitivity and signal-to-noise ratio. The main problem however is the low resolving power of available diffractors, which can result in severe line blending.

Langmuir-Blodgett films, (Henke 1964) such as Lead Stearate (Willingale 1979) are being superseded by sputtered or evaporated multilayer mirrors, which are physically more stable and offer high reflectivity in a wide range of 2d spacings (Huang et al 1989; Moos 1990). The main disadvantage of all these diffractors is their poor resolving power ($\lambda /\Delta\lambda$), which declines from about 100 at 20 Å to about 10 at 200 Å. Large organic molecules such as octadecyl hydrogen maleate (OHM) 2d = 62.5 Å and dioctadecyl adipate

(OAO) $2d = 91.2$ Å, have been crystallized (Luck and Urch 1990) and have been demonstrated in a fluorescent analysis spectrometer, with significantly improved resolution compared to commercially available multilayers.

Following a successful demonstration of a range of long-wavelength diffractors, together with conventional crystals, a composite set of diffractors was installed on a Bragg rotor spectrometer (Barnsley et al 1986), to cover the spectrum from about 1 Å to 100 Å. This instrument was then used to monitor the JET Tokamak plasma throughout 1991 operations.

2.   Selection of Bragg Diffractors for the Range 25Å to 200Å

At long wavelengths, where absorption dominates, the "perfect-latttice-with-absorption" model gives a good approximation to the resolving power (Henke and Tester 1975) of a Bragg diffractor:

$$\frac{\lambda}{\Delta\lambda} = \frac{\pi \sin\theta}{\mu_1 d} = \frac{\pi \lambda}{2\mu_1 d^2} \tag{1}$$

where d is the lattice spacing and μ_1 is the linear absorption coefficient. Calculations for the diffractors discussed here are shown in figure 1, together with the experimental results. The main result is that for all wavelengths theory predicts better resolution for true crystals than for currently available synthetic diffractors, and above the C-K edge predicts resolution comparable with that achievable at 20 Å. The upper envelope is the "perfect-lattice-with-absorption limit" for notional organic crystals having 2d values equal to the wavelength, by setting $2d = \lambda$ in eq (1) and this represents the effective upper limit for Bragg spectroscopy.

On the basis of the above calculations, the following diffractors were evaluated:
Lead Stearate: Langmuir-Blodgett film with $2d = 100.4$Å.
Multilayer Mirrors: several multilayers, manufactured by Ovonics, were tested:

OV-045A (W-Si) $2d = 44$Å   OV-151A (W-Si) $2d = 51$Å
OV-070A (W-Si) $2d = 70$Å   OV-120N (Ni-C) $2d = 117$Å

Octadecyl Hydrogen Maleate (OHM). This is a true crystal, $2d = 62.5$Å, and although not usually available, a sample of about 10×15 mm^2 was obtained from Quartz and Silice.

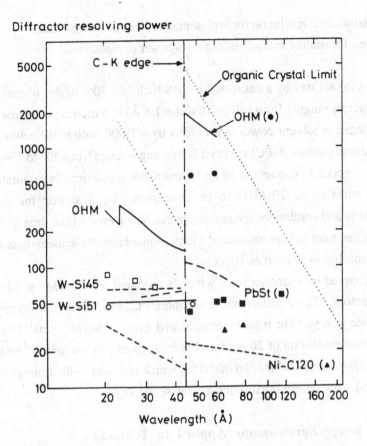

Fig.1 Diffractor resolving power $\lambda/\Delta\lambda$ estimated from the line-widths in the Tokamak spectra (points), compared with "perfect-lattice-with-absorption" curves for the diffractors used in these experiments. The PbSt curve is derived from double-crystal diffraction measurements.

3. Design Criteria for an Extended Wavelength-Range Bragg Rotor Spectrometer.

The design criteria for a tokamak impurity monitor based on Bragg diffraction are ;

    (a) Full coverage of the soft x-ray spectrum, to monitor a range of ionisation stages of any possible impurity.

    (b) High monochromatic sensitivity to monitor trace impurities and give good time resolution for the study of transient events such as impurity injection.

(c) Best possible time resolution for representative lines of the few main impurities, for routine analysis of radiated power components.

Aims (a) and (b) are met by a hexagonal rotor which fills 70% of the 40 cm^2 aperture, and carries diffractors ranging from LiF (420) (2d = 1.8 Å) to a multilayer mirror (Ni-C 2d = 117 Å). Moderate resolving power is provided by a 1:600 Soller collimator. A large area gas proportional counter (GPC) covers a Bragg angle range from 20° to 75°. Each of the ten detector anodes is connected to an independent amplifier-discriminator chain, allowing count-rates up to 20 MHz to be processed. By scanning the diffractors sequentially, a large and continuous spectral range can be covered with a time resolution of about 300 ms. If the rotor is kept stationary, a high monochromatic sensitivity is achieved, allowing a time resolution as short as 10 μs.

Aim (c) is covered by a smaller rotor which is mounted with a side-by-side array of four small diffractors. These reflect into a second GPC, which has four corresponding independent anode groups. The rotor is reciprocated over a relatively small Bragg angle range to give a time resolution of 20 ms for about 10 representative lines. The beamline provides a relatively coarse collimation of 1:150, which is suited to the line-widths of the multilayers and does not unnecessarily decrease the sensitivity.

4. Results of Bragg Spectroscopy Applied to Tokamaks.

In the COMPASS tokamak, the plasma-facing surfaces of which had been conditioned with boron compounds in order to reduce influxes of other less tolerable impurities such as oxygen, the use of extended range diffractors in monitoring the boron is illustrated in Figs 2 and 3.

In the JET tokamak the plasma-facing materials are predominantly carbon or beryllium tiles, the latter effectively gettering oxygen impurities and thus performing much the same function as the boron film in COMPASS. Apart from the the predominant light impurities Be and C in JET, the presense of other elements such as Cl ,O, Ni, Al, Zn can be detected, depending on the operating conditions, at concentrations as low as ~ 0.001% of the electron density. Typical JET data are shown in Figs 4 and 5.

5. Discussion

The resolving powers obtained from the tokamak results are shown as points in fig. 1,

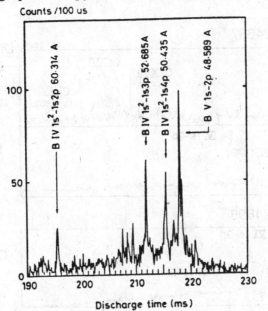

Fig.2 OHM crystal 2d = 62.5 Å used on the COMPASS tokamak . This crystal has a $\lambda / \Delta\lambda \sim 700$ which is very high in this spectral region and compares favourably with gratings. The unidentified lines are mainly from Fe XV and Fe XVI and show the value of the high resolution of OHM compared to PbSt.

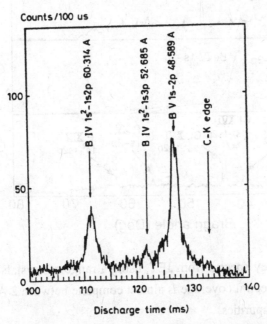

Fig.3   PbSt diffractor 2d = 100.4 Å used on the COMPASS tokamak, The peak reflectivity of PbSt is similar to that of OHM and though its integrated reflectivity is higher, the consequent loss of resolution is a disadvantage when observing a complicated spectrum.

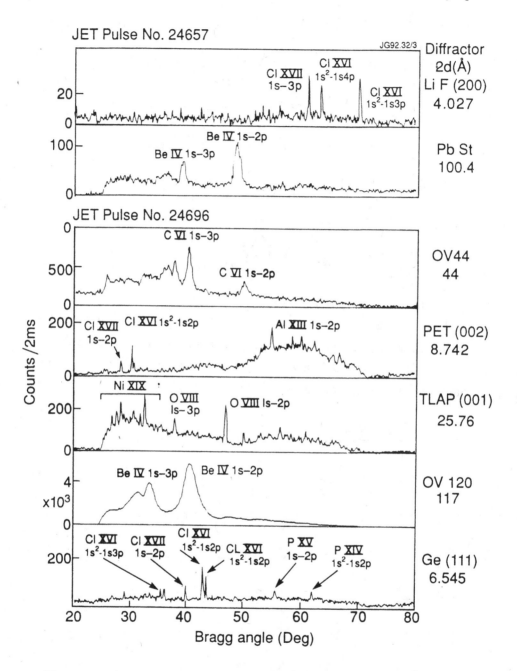

Fig.4 Typical survey spectra from JET, using a range of crystals and multi-layers on a hexagonal rotor. Spectral coverage is almost complete between 2 Å and 100 Å, thus monitoring all the main impurities.

together with the theoretical curves for crystals and multilayers, and two-crystal measurements for lead stearate. Good sensitivity was also achieved, giving a spectrum in about 100 ms from any of the diffractors.

The Lead Stearate spectra show $\lambda/\Delta\lambda$ between 40 and 50 which is slightly lower than the value of about 70 that would be predicted from the two crystal measurements, (Willingale 1979). When used on JET to monitor Beryllium, there was occasional contamination from strong higher order reflections.

The OHM Crystal exhibited a resolving power comparable with grating spectrometers in this spectral region. For B IV $1s^2$-$1s2p$ $^1P^0$ at at 60.31 Å, the overall instrumental $\lambda/\Delta\lambda$ was 700. This implies a value of about 1000 for the crystal alone, which is very close to the theoretical limit. This crystal was not used routinely on JET since its 2d value is less suitable for monitoring C and Be.

The W-Si 51 Multilayer, used on COMPASS, showed an almost constant $\lambda/\Delta\lambda \sim 50$ throughout its wavelength range, in close agreement with theory (fig. 1).

The W-Si 44 Multilayer, used on JET to monitor Carbon Ly α at 33.74 Å and C Lyβ at 28.47 Å, showed a better than predicted $\lambda/\Delta\lambda \sim 60$. The reversed Ly α:Lyβ ratio in the data, and absence of the C V $1s^2$-$1s2p$ $^1P^0$ line at 40.27 Å, are both due to the steeply-falling transmission of the 2 μm Mylar detector window.

The W-Si 70 Multilayer, was used on JET but not routinely, since it exhibited lower resolution than expected, and its wavelength range was not useful either for C or Be.

The Ni-C 117 Multilayer had a peak reflectivity about ten times higher than PbSt and about twice the line-width. Under most circumstances its $\lambda/\Delta\lambda \sim 35$ was adequate to monitor Be Ly α at 75.93 Å, and it was used throughout 1991 JET operations. Calculations suggest that a V-C multilayer with 2d $\sim$ 100 Å, $\lambda/\Delta\lambda \sim 80$, would be more suitable.

This work clearly demonstrates the feasibility of using a combination of crystals and multilayers in a single instrument for routine monitoring of all the impurities in a tokamak. plasma. The success of multilayers for monitoring C and Be was partly due to low concentrations in JET of metals such as Cr, Fe and Ni, and the resulting lack of line blending from their strong transitions between 25 Å and 100 Å. For this reason, the use of potential high-resolution crystals such as OAO will be investigated.

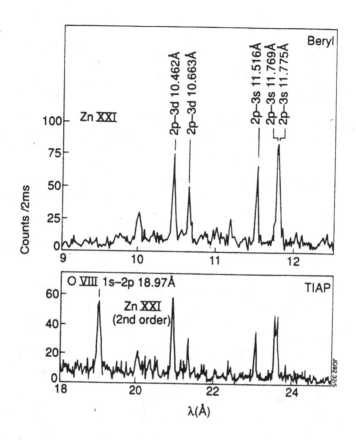

Fig.5 Spectra of the main 2-3 transitions of Zn XXI recorded from JET with a beryl (1010) 2d = 15.954 Å and TlAP (002) 2d = 12.88 Å.

6 REFERENCES

[1] Henke B L , (1964) *Adv. X-Ray Anal.* **7** 460

[2] Willingale R , (1979) *PhD Thesis,* Leicester University,

[3] Huang T C , Fung A and White R L , (1989) *X-Ray Spectrometry,* **18** 53

[4] Moos W et al (1990) *Rev. Sci. Instrum.* **61** (10) 2733

[5] Luck S and Urch D S (1990), *Physica Scripta,* **41** 749

[6] Barnsley R et al, (1986) *Rev. Sci. Instrum.* **57** (8) 2159

[7] Henke B L and Tester M A, (1975) *Advances in X-ray analysis,* **8**

EXPERIMENTAL STUDY OF X-RAY EMISSION FROM LASER IRRADIATED PLANAR TARGETS ON "MISHEN" FACILITY.

V.A.Bolotin, I.N.Burdonskii, V.V.Gavrilov, A.Yu.Gol'tsov,
S.V.Zavyalets, E.V.Zhuzhukalo, V.N.Kondrashov, M.O.Koshevoi[*],
M.I.Pergament, A.A.Rupasov[*], A.S.Shikanov[*]

Branch of Kurchatov Atomic Energy Institute,
142092 Troitsk, Moscow region, Russia
[*]P.N.Lebedev Physical Institute, 117942 Moscow, Russia

X-ray emission from planar targets irradiated by 1.054μm laser pulses was observed with temporal, spatial and spectral resolution. The main purposes of these measurements were the investigation of energy transfer in multilayer targets and X-ray conversion efficiency. A mass ablation rate was determined from temporal analysis of multicharge ions line emission and a key role of corona X-ray emission in accelerated material preheating was established.

1. Experimental arrangement

Experiments were conducted on "Mishen" facility described in details elsewere [1]. The laser system of "Mishen" consists of two channels with output beams parameters as follows: the main beam – output energy 100÷200J (λ=1.054μm) in 3-nsec pulse, divergence ~2×10^{-4}rad, contrast ratio ~10^{6}, power density at the target surface ~$10^{13}\div10^{14}$W/cm^2; the diagnostic beam – output energy 10÷20 J (λ=1.054μm) and 5÷10 J (λ=0.53 μm) in 0.3-nsec pulse, divergence ~10^{-4}rad, power density $10^{13}\div10^{14}$W/cm^2.

An X-ray diagnostic complex includes: calorimeters, pinholes, spectrographs with flat and curved crystalls, transmission grating spectrographs, vacuum X-ray diodes, time-analysing X-ray streak camera. To conduct a time-spatial or time-spectral measurements an X-ray streak camera was used in conjunction with set of slits, filters, flat crystalls or transmission grating, installed in front of a photocathode. Besides of listed X-ray diagnostic methods registration of plasma emission in visible range with temporal, spectral and spatial resolution was used in experiments.

2. Experimental results and discussion

Previously we have reported [2] the results of heat transfer investigations obtained in experiments with multilayer targets using time-integtated X-ray methods. The X-ray crystal streak camera (with two KAP crystals) enables us to study the time evolution of multicharged ion line spectra and to obtain a more detailed information on heat conduction dynamics, particularly, to measure a mass ablation rate. As an example, Fig.1 shows the time histories of the He-like aluminum ion resonance line emission (for aluminum and mylar+aluminum targets). A time history of continuous emission is also shown for mylar target. The time delay of intense aluminum ion emission allows to determine a mass ablation rate due to finite time of the plastic ablator burn-through. We have measured the mass ablation rate to be $\dot{m} \sim 2\times10^5 g/cm^2$ for plastic target and $\dot{m} \sim 1.8\times10^5 g/cm^2$ for aluminum one (power density was $\sim 5\times10^{13} W/cm^2$).

We have discussed a key role of X-ray preheating in our early work [3]. In some experimental conditions the preheating may be very large. Consider, for example, an experiment with multilayer target (0.1 μm Cu + 1 μm mylar + 1 μm Al, copper side being irradiated). Fig.2 shows that lines of H- and He-like aluminum ions are observed simultaneously with copper lines, i.e. with the beginning of the laser irradiation. So a time delay due to a plastic ablator burn-through is absent (compare with Fig.1). Optical measurements show that rear side emission in visible range also appears at the beginning of irradiation. When a copper layer is absent, this emission starts with time delay that equals to the shock wave transit time. According to calorimeter data the addition of thin ($\sim 0.1 \mu$m) copper layer in these experiments enhances an X-ray yield by the factor of 30÷50. Therefore the process of energy transfer in the target changes drastically with the increase of the corona X-ray emission.

The X-ray energy transfer is also of particular importance in indirect drive schemes. In this case the spectral region to be studied covers the interval 10÷100 Å. We try to model a different processes of indirect drive in flat geometry. In our experiments we used X-ray calorimeters, vacuum diodes and crystal spectrometers. Also we have employed spectrometers with transmission gratings of the two types: 1 μm period gold grating 1×0.1 mm^2 [4] and 1 μm period tungsten grating (circular aperture diameter ~25 μm). Both gratings were employed with film or MCP detectors. In the case of tungsten grating a spatial resolution transverse to the direction of dispersion was provided. The combination of the gold grating with an X-ray streak camera enabled to carry out time resolved measurements. Because of the overlapping of the different diffraction orders on the recorder in this case a special data processing is necessary to obtain a real spectral distribution. For this purpose an unfold procedure taking into account a detector spectral response was developed.

We studied a conversion efficiency of laser energy to X-ray emission varying a thickness and diameter of irradiated targets. In these experiments sandwich-type targets with copper layer of different thickness and diameter evaporated on 2 μm mylar support were used. Fig.3 and 4 show an X-ray transmission grating streak record of copper plasma emission and unfolded X-ray spectra at two different times. Note a pronounced increase of plasma radiation intencity in 40÷60 Å region approximately 1.5 nsec after irradiation start. Measured conversion efficiency dependence upon Cu layer thickness and diameter are presented in Fig.5,6.

3. Summary

The developed diagnostic methods are shown to permit a temporal-spectral measurements in a wide X-ray spectra range. A temporal analysis of the line emission has enabled to determine a mass ablation rate of the different target materials with better accuracy and reliability than in our earlier experiments. Experimental results show a key role of corona X-ray emission in accelerated material preheating. Experimental data on X-ray conversion efficiency and temporal evolution of X-ray spectra for different experimental conditions were obtained.

REFERENCES

1. Bolotin, V.A. et al. 1989, Moscow, Kurchatov Atomic Energy Institute, Preprint IAE-4967/7.
2. Bondarenko, Yu.A. et al. 1981, Moscow, X European conference on controlled fusion and plasma physics, 2, p.157.
3. Bol'shov, L.A. et al. 1987 Sov. Phys. JETP 65, 1160.
4. Alexandrov, Yu.A. et al. 1988 Laser and Particle Beams 6. 561.

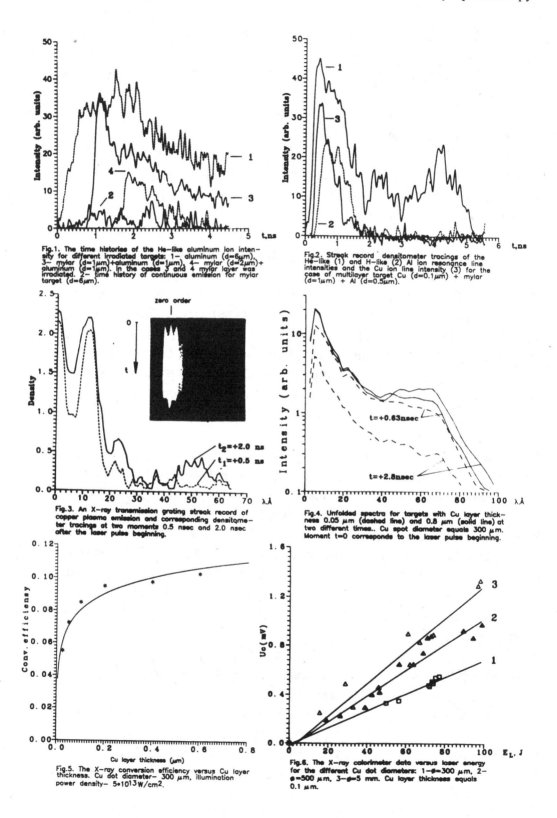

Fig.1. The time histories of the He—like aluminum ion intensity for different irradiated targets: 1— aluminum (d=6μm), 3— mylar (d=1μm)+aluminum (d=1μm), 4— mylar (d=2μm)+aluminum (d=1μm). in the cases 3 and 4 mylar layer was irradiated. 2— time history of continuous emission for mylar target (d=6μm).

Fig.2. Streak record densitometer tracings of the He—like (1) and H—like (2) Al ion resonance line intensities and the Cu ion line intensity (3) for the case of multilayer target Cu (d=0.1μm) + mylar (d=1μm) + Al (d=0.5μm).

Fig.3. An X—ray transmission grating streak record of copper plasma emission and corresponding densitometer tracings at two moments 0.5 nsec and 2.0 nsec after the laser pulse beginning.

Fig.4. Unfolded spectra for targets with Cu layer thickness 0.05 μm (dashed line) and 0.8 μm (solid line) at two different times. Cu spot diameter equals 300 μm. Moment t=0 corresponds to the laser pulse beginning.

Fig.5. The X—ray conversion efficiency versus Cu layer thickness. Cu dot diameter— 300 μm, illumination power density— 5•10¹³W/cm².

Fig.6. The X—ray calorimeter data versus laser energy for the different Cu dot diameters: 1—∅=300 μm, 2—∅=500 μm, 3—∅=5 mm. Cu layer thickness equals 0.1 μm.

X-Ray Spectroscopic Diagnostics of the Hydrodynamics of Flares on M Dwarf Stars

Chung-Chieh Cheng[1] and Roberto Pallavicini[2]

[1]Naval Research Laboratory, Washington, D.C.
[2]Arcetri Astrophysical Observatory, Florence, Italy

ABSTRACT

We calculated the X-ray spectra of the He- and Li-like Ca XVIII- XIX and Fe XXV-XXIV ions using the results from the numerical simulation of stellar flares on dMe stars. The line profiles can be used for the diagnostics of the flare hydrodynamics and the temperatures of the heated plasmas in the flaring loops.

1. Introduction

The realization that the atmospheres of late-type stars are solar like has stimulated efforts in theoretical interpretation of their atmospheric structures in terms of magnetic loop models (Linsky 1983; Antiochos and Noci 1986; Stern, Antiochos, and Harnden 1986). In particular, stellar flares observed in X-rays on dMe stars have been successfully interpreted, in analogous to solar flares, as sudden heating in magnetic loops (Haisch 1983). Many flares on dMe stars have been observed with good time and spectral coverage in X-ray by *EXOSAT* (for a review, see Pallavicini, Stella, and Tagliaferri 1990). The flaring plasmas have temperature ranging from 10^4 to 10^8 K, and emission measures ranging from 10^{50} to more than 10^{53} cm^{-3}, much larger than those from solar flares. To understand the diverse physical parameters of the observed stellar flares and the hydrodynamic evolution and energy transfer processes in the flaring loop, we have made a series of numerical simulations of loop models with different values of loop size ($2-4 \times 10^9$ cm), flare energy input, and initial loop conditions (Cheng and Pallavicini 1991). The hydrodynamic results are used here to calculate the expected X-ray spectra and X-ray count rates. The calculated Ca XIX and Fe XXIV line profiles show large blue shifted component, which could possibly be observed by a high resolution X-ray spectrometer such as the one planned for the *AXAF*. The ratios between lines from the He- and Li-like ions can be used to determine the temperature of the emitting plasmas.

2. Hydrodynamic Results

Using a one-dimensional hydrodynamic code (Cheng et al. 1983, Cheng, Karpen, and Doschek 1984), we have calculated 10 flare models (see Cheng and Pallavicini 1991). The flare energy is applied at the top of the loop at a constant input rate for 300 sec before decreasing. For the model dMe star, we have assumed a stellar gravity twice as the sun and a chromosphere at a temperature of around 9,000 K. The basic hydrodynamic results are similar and are illustrated in Fig. 1 for model 8 (large-loop). When the flare energy input is applied at the top of the loop, the temperature rapidly increases and a conduction front

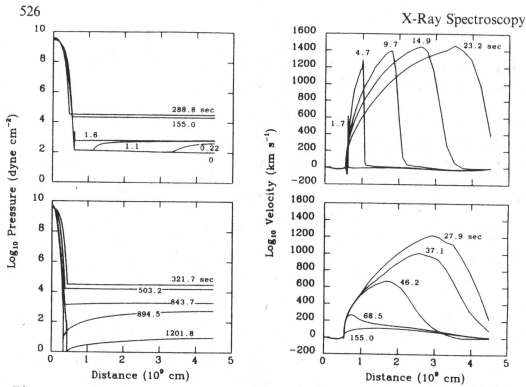

Figure 1. Gas pressure and velocity along the loop at various times (Model 8).

moves toward the chromosphere and reaches it within a few seconds. The chromosphere is heated then by the large heat flux, due to the inability of the chromosphere to radiate away the large deposited energy. And the gas pressure there becomes locally high, which acts like a pressure pulse that causes the gas to expand upward and also downward, and the density of the loop increases as a consequence. The downward moving front compressed the chromospheric plasmas ahead of it. Unlike the solar case, the resultant chromospheric condensation is short-lived in the stellar flare, owing to the much large gravity and a smaller chromospheric pressure scale height in the stellar atmospheres. As soon as the energy input stops, the loop temperature drops immediately, due to large radiative losses. The density in the loop reaches its maximum value after the temperature has reaches its maximum, similar to the solar case.

3. X-Ray Spectra and Plasma Diagnostics

We have calculated the X-Ray spectra in the 1–10 keV interval as a function of time during the flare evolution by folding the Raymond-Smith radiation code with the output from the hydro-code. Our simulated light curves (1-10 keV) reproduce the count rate and time characteristics of the observed stellar flares from *EXOSAT* (Cheng and Pallavicini 1991). The aspect ratio of the loop used for our models ranging from 0.25 and 0.3 and the distance to the star is assumed to be 10 pc.

The electron temperature in the flaring plasma can be determined from line ratios of He-like and Li-like resonance and satellite lines from highly ionized ions. We therefore have calculated detailed line profiles around 1.85 Å (Fe XXIV, XXV) and 3.2 Å (Ca XVIII, XIX) complexes at different phases of the flare evolution for our flare models. Fig. 2 shows an

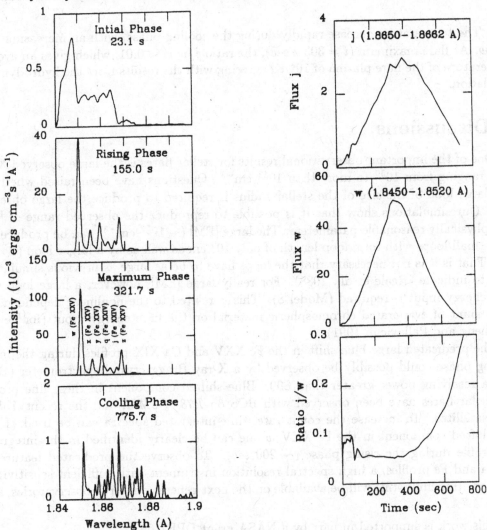

Figure 2.   *left*-Time Evolution of Fe XXIV-XXV line profiles.
Figure 3.   *right*-Fe XXV (w) and Fe XXIV (j) light curves and their ratio.

example of the Fe XXIV-XXV spectra for model 8 at selected times of the flare evolution. All the resonance and satellite lines from Fe XXIII to Fe XXIV have been included in the simulated spectra. We see from the figures that there is large blue shifts in the Fe XXV and Ca XIX line profiles during the initial heating phase. At $t = 23$ sec, the peak of the Fe XXV resonance line (w) is blue-shifted by about 1.7 mÅ, which corresponds to an upflow of about 300 km s^{-1}. Near the flare maximum and the decay phase, the blue-shifted component has disappeared. Note that, in the cooling phase, the intensity of the Fe XXIV j line increases relative to the Fe XXV w line and many Fe XXII-XXIII lines appears longer than 1.88 Å, indicating a decreasing temperature. Fig.3 shows the light curves of the resonance line w of Fe XXV at 1.85 Å and the j line of Fe XXIV at 1.87 Å and their intensity ratios throughout the entire evolution of the flare. The j/w ratio decreases with increasing temperature. The figure show that flare temperature increases to high value rapidly in the initial heating phase and stay high in the rising phase of the flare. The temperature decease rapidly during the cooling phase, with an increasing ratio of j/w. At flare maximum ($t = 305.2$ sec), the ratio j/w is ≈ 0.01, which gives an average temperature of the flare plasma of 10^8 K, agreeing with the results from the hydrodynamic calculation.

4. Discussions

One of the important observational results for stellar flares is the large observed X-ray [EM] ranging from 10^{50} to more than 10^{53} cm^{-3}. Questions have been raised whether a very large loop of the size of the stellar radius is required to produce the large observed [EM]. Our simulations show that it is possible to reproduce the observed range of [EM] with physically reasonable parameters. The large [EM] ($\sim 10^{52}$ cm^{-3}) can be produced by fairly small loops with total loop length of only 10^9 cm as well as by a larger loop (2×10^9 cm). That is it is not necessary that the loops have to be as large as previous simulations seem to indicate (Reale et al. 1988). For really large [EM], however, a large loop with large energy input is required (Model 8). This is related to the nonlinear dependence of the amount of evaporated chromospheric material on the flare energy input (for details, see Cheng and Pallavicini 1991).

The predicated large blue shift in the Fe XXV and Ca XIX profiles during the initial heating phase could possibly be observed by a X-ray Bragg crystal spectrometer (BCS) with a resolving power greater than 800. Blue shifted component in these line profiles from solar flares have been observed with BCS on *P78-1*, *SMM*, and the recent *Yohkoh* solar satellites. To increase the count rate, time-integrated spectra can be used. Large blue-shifted component in the Fe XXV w-line can be clearly identified in the integrated line profile during the rising phase (~ 200 sec). To observe the predicated features in the Ca and Fe profiles, a high spectral resolution instrument with sufficient sensitivity is required. Hopefully, this will be available on the next generation X-ray observatories, such as the *AXAF*.

This work is supported in part by a NASA grant DPR-1693.

References

Antiochos, S.K, and Noci, G. 1986, *ApJ*, **301**, 440.

Cheng, C.-C. et al. 1983, *ApJ*, **265**, 1090.

Cheng, C.-C., Karpen, J.T., and Doschek, G.A. 1984, *ApJ*, **286**, 787.

Cheng, C.-C., and Pallavicini, R. 1991, *ApJ*, **381**, 234.

Haisch, B.M. 1983, in *Activity in Red-Dwarf Stars*, ed. P.B. Byrne and M. Rodono, D. Reidel Publ. Co.

Linsky, J.L. 1983, in it Solar and Stellar Magnetic Fields: Origin and Coronal Effects, ed. J. Stenflo, D. Reidel Publ. Co.

Pallavicini, R., Stella, L., and Tagliaferri, G. 1990, *Astr. Ap.*, **228**, 403.

Reale, F. et al. 1988, *ApJ*, **328**, 256.

Stern, R.A., Antiochos, S.K., and Harnden, F.R. 1986, *ApJ*, **305**, 417.

Xe L and M X-Ray Emission Following Slow $Xe^{44+ \text{ to } 48+}$ Ion Impact on Cu-Surfaces

M. W. Clark[1], D. Schneider[1], J. McDonald[1], R. Bruch[2], S. Tanaka[2], F. Hao[2], R. Schuch[3], and U. I. Safronova[4]

[1]Lawrence Livermore National Laboratory, Livermore, CA 94550, USA
[2]Physics Department, University of Nevada, Reno, NV 89557, USA
[3]Manne Siegbahn Institute of Physics, Stockholm, Sweden
[4]Inst. of Spectroscopy, Russian Acad. of Sciences, Troitzk, Russia

ABSTRACT

The x-ray emission following the impact of highly charged Xe^{q+} (q=44-48) ions of 7 keV q energy on a Cu surface has been measured. Theoretically we have calculated $2\ell\text{-}n\ell'$ ($n \geq 3$), $3\ell\text{-}n\ell'$ ($n \geq 4$), $4\ell\text{-}n\ell'$ ($n \geq 5$), and $5\ell\text{-}n\ell'$ ($n \geq 6$) L, M and N x-ray transition energies and transition probabilities averaged over LS as a function of the electron occupation numbers (k_i) for the states $1s^{k_1}2s^{k_2}2p^{k_3}3s^{k_4}3p^{k_5}3d^{k_6}4s^{k_7}4p^{k_8}4d^{k_9}4f^{k_{10}}5s^{k_{11}}5p^{k_{12}}$. A comparison between measured peak positions and calculated transition energies reveals that by side-feeding of around 7 to 14 electrons into n=3,4, and 5 manifolds can explain the main features of the observed L and M x-ray emission spectra.

1 INTRODUCTION

Much effort is presently directed towards research on ion-surface interactions using very highly charged ions (Andrä 189). This research became feasible with the development of new and advanced ion sources such as Electron Cyclotron Resonance (ECR) sources, Electron Beam Ion Source (EBIS) and the Electron Beam Ion Trap (EBIT).

The recently developed technique (Schneider et al. 1991) to extract highly charged ions from EBIT, which has been originally designated as an ion trap, allows to use ions up to Th^{80+} e.g. for ion-surface interaction studies. These studies are aimed to understand the neutralization dynamics of very highly charged ions as they approach the surface and penetrate into the solid. Such highly charged projectile ions carry up to several hundred keV potential energy and x-ray emission studies are in particular suitable to illuminate the different interaction processes which lead to the transfer of this energy to the surface. The satellite intensities and energy positions of the emitted x-rays and Auger electrons provide important information on the history of the projectile ions interacting with the surface and the bulk material.

2 EXPERIMENT

In this study we report the first measurements on the ion-surface neutralization dynamics using highly charged Xe ions, characterized by an approach velocity of v ~ 0.22 a.u. corresponding to 7 Å/fs, where the number of L-shell vacancies in the incident Xe^{q+} (q:44 to 48) projectile is systematically changed. The Xe^{q+} ions have been extracted from the Lawrence Livermore EBIT source. After extraction the ejected ions are charge and momentum analyzed and focused onto a Cu surface (roughness: ~10μ), which is tilted by 45° with respect to the ion beam axis. The x-ray emission following impact of highly charged Xe^{q+} ions (7 keV x q energy) on a Cu surface has been measured by means of a Si(Li) detector. In order to identify the observed L-, M-, and N- x-ray spectra, we have performed extensive calculations of x-ray transition energies, including transition probabilities and autoionization rates.

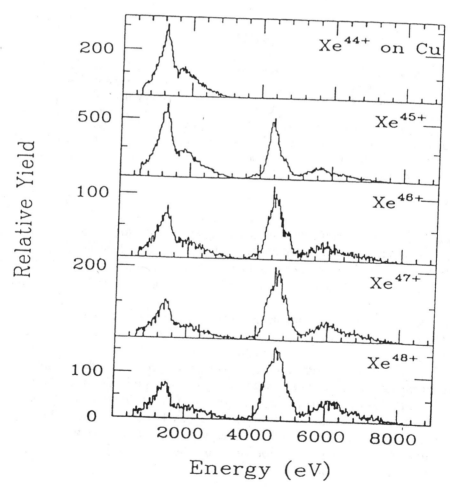

Fig. 1 - (a) - (e) M and L x-ray spectra following 7 kVxq Xe^{q+} (q = 44 to 48) ion impact on a Cu surface as a function of the projectile charge q.

3 THEORETICAL CALCULATIONS

The calculation of theoretical x-ray spectra for multiply excited Xe ions represents a challenging many-body problem (Safronova and Senashenko 1981, 1984). In order to reduce the vast number of possible states, we have averaged over the spin and angular moment quantum numbers. In this work we present for the first time comprehensive calculations based on nonrelativistic and relativistic perturbation theory for 2ℓ-$3\ell'$,2ℓ-$4\ell'$,2ℓ-$5\ell'$,3ℓ-$4\ell'$,3ℓ-$5\ell'$,4ℓ-$5\ell'$ x-ray transition energies and transition probabilities for highly excited Xe ions of the type $1s^{k_1} 2s^{k_2} 2p^{k_3} 3s^{k_4} 3p^{k_5} 3d^{k_6} 4s^{k_7} 4p^{k_8} 4d^{k_9} 4f^{k_{10}} 5s^{k_{11}} 5p^{k_{12}}$ with up to N=53 electrons occupying twelve different subshells where k_i (i = 1 to 12) are the electron occupation numbers. As an example we present here the analytic formula for LS averaged transition probabilities

$$W(Q,[n_1 l_1]^{-1} n_2 l_2 Q) = A_0 (n_1 l_1, n_2 l_2) Z^4 \times$$

$$\times \frac{k_1}{g_1}\left(1-\frac{k_2}{g_2}\right)\left[\frac{E(Q)-E([n_1 l_1]^{-1} n_2 l_2 Q)}{E_0(Q)-E_0([n_1 l_1]^{-1} n_2 l_2 Q)}\right]^3 \left(1-P(k_i)/Z\right)^2$$

where $g_i = 2(2\ell_i + 1)$ are the statistical weights and $P(k_i)$ the first order corrections for the dipole matrix elements. The hydrogenic transition probabilities A_0 (10^8 s^{-1}) are listed in Table 1.

Table 1.

nl-n'l'	A_0	nl-n'l'	A_0	nl-n'l'	A_0	nl-n'l'	A_0
2p-1s	37.575	3s-2p	0.1262	3d-2p	6.4625	4p-3d	0.02084
3p-1s	10.106	4s-2p	0.05154	4d-2p	2.0617	5p-3d	0.00897
4p-1s	4.0896	5s-2p	0.02576	5d-2p	0.9422	6p-3d	0.00469
5p-1s	2.0617	6s-2p	0.01402	6d-2p	0.4906	5p-4d	0.01130
6p-1s	1.1706	4s-3p	0.03669	4d-3p	0.7035	6p-4d	0.00565
3p-2s	1.3464	5s-3p	0.01809	5d-3p	0.3390	4f-3d	1.9296
4p-2s	0.5799	6s-3p	0.01014	6d-3p	0.1877	5f-3d	0.6356
5p-2s	0.2968	5s-4p	0.01290	5d-4p	0.1485	6f-3d	0.3003
6p-2s	0.1635	6s-4p	0.007162	6d-4p	0.08618	5f-4d	0.3617
4p-3s	0.1838			5d-4f	0.05046	6f-4d	0.1801
5p-3s	0.07770			6d-4f	0.02144	5g-4f	0.7654
6p-3s	0.05728					6g-4f	0.2470
5p-4s	0.04421						
6p-4s	0.02673						

It is evident from this table that the 3d→2p, 4d→2p, 4f→3d and 5g→4f transitions are characterized by large transition probabilities.

4 EXPERIMENTAL RESULTS AND DISCUSSION

Figure l(a) - (e) displays typical x-ray spectra originating from Xe^{44+} to Xe^{48+} ion impact on a Cu surface with the projectile charge as a parameter. For Xe (Z=54), q = 44+ represents the closed shell $1s^2 2s^2 2p^6$ ground state configuration. The other charge states (q = 45-48) are characterized by open L-shell configurations, namely $1s^2 2s^2 2p^5$ (q=45+), $1s^2 2s^2 2p^4$ (q=46+), $1s^2 2s^2 2p^3$ (q=47+) and $1s^2 2s^2 2p^2$ (q =48+) differing in the incident number of 2p-holes. As can be seen the Xe^{44+} spectrum (Fig la) consists only of lower energy M x-ray peaks due to the missing L vacancy states of the incoming projectile ion. From this figure we can also deduce that no L-shell vacancies are produced during the impact of the ion close to the surface. In contrast the Xe^{45+} to Xe^{48+} spectra show (Fig. 1b-1e), as expected, additional higher energy line structures, which arise from electric dipole transitions into empty 2p states (L-spectrum). In accordance with the increasing number of 2p vacancies, the measured L-lines energy positions shift towards higher energies with increased charge state q. The position of the main maximum in the L-spectrum for Xe^{45+} to Xe^{48+} falls well into the predicted range for 2p-3d transitions with innershell $1s^2 2s^2 2p^k$ (k=2-5) configurations.

A more detailed comparison between the measured and calculated x-ray transitions suggests the following picture: Following the classical overbarrier model the neutralization process starts at a critical distance $R_c \approx 30$ Å. We expect a "band" of Rydberg levels to be populated between about n=20 and n=40 during the approach to the surface. The interaction time ($\tau_c \sim 4 \times 10^{-15}$ s) is much too short for substantial cascading to lower states. Hence highly excited states consisting of ($1s^2 2s^2 2p^k$) inner shell configurations and a band of electrons in $n \approx 20$ to 40, "hollow atoms", (Briand et al. 1990) survive until the ion hits the surface. After the ion has penetrated into the solid, it experiences close binary collisions with Cu target atoms. The most significant radiative transitions observed arise from 3d, 4d, 4f, 5f and 5g states, which are most probably populated by direct capture into n=4, and 5 and cascade repopulation leading to a strong excitation of 3d levels ("side feeding") when the ions have already penetrated into the solid.

REFERENCES

Andrä, H. J. 1989, in Atomic Physics of Highly Charged Ions, ed. R. Marrus (New York: Plenum Press)

Briand, J. P., et al. 1990, Phys. Rev. Lett. **65**, 159.

Safronova, U. I., and Senashenko, V. S. 1989, J. Phys. B **14**, 603.

Safronova, U. I., and Senashenko, V. S., 1984, Theory of Spectra of Multicharged Ions (Moscow: Energoatomizdat).

Schneider, D., et al. 1991 Phys. Rev. A **44**, 3119.

DIFFUSION EFFECTS ON DIAGNOSTIC X-RAY EMISSION LINE RATIO MEASUREMENTS IN LABORATORY PLASMAS

I.H.Coffey,[2] R.Barnsley,[1] I.G.Hughes[1] F.P.Keenan,[2] K.D.Lawson[3] and N.J.Peacock[3]

[1]JET Joint Undertaking, Abingdon, Oxon, England, OX14 3EA, England
[2]Department of Pure and Applied Physics, Queens University, Belfast, BT7 1NN, Northern Ireland
[3]Culham Laboratory (UKAEA/EURATOM Fusion Association), Abingdon, Oxon, OX14 3DB, England

Abstract

The theoretical electron temperature sensitive X-ray emission line ratio $G = [I(1s^2\ ^1S_o - 1s2s\ ^3S_1) + I(1s^2\ ^1S_0 - 1s2p\ ^3P_{0,1,2})]/I(1s^2\ ^1S_o - 1s2p\ ^1P_1)$ for the He-like ionisation stage of chlorine is compared with measured ratios from the JET (Joint European Torus) plasma, where the electron temperature and density are measured by independent means. During the ohmic heating phase of a discharge Cl XVI is found to exist in plasma conditions approaching coronal equilibrium towards the plasma centre. However during additional heating the Cl XVI is approaching an ionising situation which results in significant changes to the measured ratios at a given temperature and density.

1 INTRODUCTION

The three main lines in the helium isoelectronic sequence, f, i and r, may be used to infer the electron temperature sensitive emission line ratio $G = (f + i)/r$, where f is the forbidden transition $1s^2\ ^1S_0 - 1s2s\ ^3S_1$, i is the intercombination transitions $1s^2\ ^1S_0 - 1s2p\ ^3P_{0,1,2}$ and r is the resonance $1s^2\ ^1S_0 - 1s2p\ ^1P_1$ transition. Although such ratios are not routinely used as electron temperature diagnostics on devices such as JET, non-spectroscopic techniques being preferred, they may be applied to the analysis of such sources for which no independent estimates of this parameter exists. However, comparison of theoretical ratios with experimental results from well diagnosed laboratory plasmas is important in confirming the accuracy of the atomic data used in the calculations. These calculations are generally carried out for coronal equilibrium conditions and hence it is important to consider effects that will cause departure from the coronal condition in laboratory plasmas.

2 DIFFUSION EFFECTS

Chlorine occurs as an intrinsic impurity in the JET plasma and the spectra were obtained with a double crystal X-ray spectrometer (Barnsley *et al.* 1991).

The atomic data used in the theoretical calculations was taken from Coffey *et al.* (1992) Experimental ratios were measured during both the ohmic phase of a JET discharge and an additionally heated phase (the plasma was heated by Ion Cyclotron Resonance Heating-ICRH). Typical central electron temperature values (measured by analysis of Electron Cyclotron Emission and in units of degrees Kelvin) were $\log T_e = 7.36$ for the ohmic phase (at time $t = 5s$) and $\log T_e = 7.93$ for the ICRH phase (at time $t = 9s$).

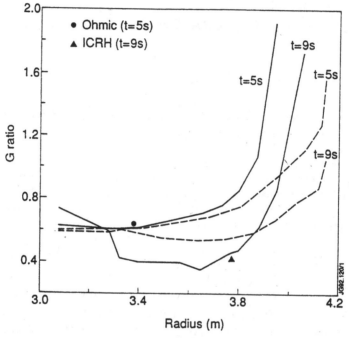

Fig.1 The theoretical emission ratio G plotted as a function of plasma radius Two cases are shown, the first taken at 5s during the ohmic phase of the discharge, and the second at 9s during the additionally heated phase. Each case is calculated using for a coronal ionisation balance (solid lines) or a diffusive ionisation balance derived from a transport code (dashed lines). Also shown are the measured values for the ohmic case (solid point) and for the heated case (triangle).

Fig.1 shows the radial dependence of the calculated ratios at both these times using a coronal ionisation balance and also a diffusive ionisation balance derived from a transport code (Lauro-Taroni) which calculated radial profiles of intensity density and radiated power for all the ionisation stages of an impurity element at given time during a discharge. In the ohmic case the Cl XVI emission peak occurs at a radius $R = 3.38m$. The difference between the coronal and diffusive cases is negligible. The experimental ratio at this time differs by approximately 4.5% which is within experimental errors. For the additionally heated case The

ClXVI emission peak occurs at R = 3.773m. Here the difference between the two cases is approximately 25%. The experimental point supports the diffusive calculation.

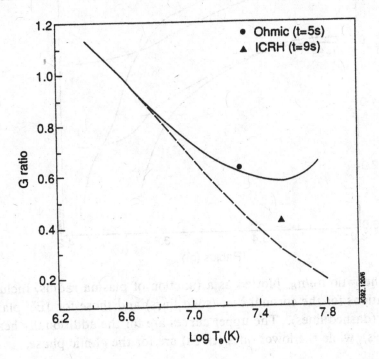

Fig.2 The theoretical emission Ratio G plotted as a function of temperature T_e, with dielectronic and radiative recombination included in (solid line) or excluded from (broken line) the calculations. Also shown are the JET experimental data at t = 5s (solid point) and at t = 9s (triangle).

Fig.2 shows the effect of diffusion in JET on T_e diagnostics. If coronal equilibrium were assumed then the spectrum taken during the ICRH phase would give a meaningless result. If non coronal conditions are assumed then the ratio indicates a lower H-like to He-like ratio (the broken curve in fig.2 is for $n_H/n_{He} = 0$). The variation of this ratio with plasma radius is plotted in fig.3. In the ohmic case the difference between the coronal and diffusive curves at the radius of peak Cl XVI emission (R = 3.38m) is approximately 6% and this relatively small difference is reflected in Figs.1 and 2. In the additionally heated case the difference (R = 3.77m) is now approximately 52%, resulting a lowering of the G ratio at this radius of 25%. In the extreme case the ratio $n_H/n_{He} = 0$, i.e. the Cl XVI is a completely ionising situation. This can occur for lower Z elements even during ohmic heating (see also Coffey *et al.* this colloquia).

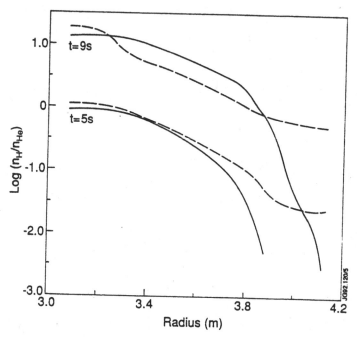

Fig.3 The ratio n_H/n_{He} plotted as a function of plasma radius. Included are calculations for the coronal case (solid lines) and those for JET plasma conditions (dashed lines). The upper curves are for the additionally heated phase (t = 9s), while the lower ones (t = 5s) are for the ohmic phase.

3 CONCLUSIONS

Significant changes to the values of the electron temperature G ratio from the coronal equilibrium case are found as a result of diffusion effects in the JET plasma. For CL XVI this effect is negligible during ohmic plasmas. In the additionally heated phase the n_H/n_{He} ratio is reduced which decreases the measured G ratios. G ratios calculated using a diffusive ionisation balance derived from a transport code are in good agreement with the measured values.

References

Barnsley,R., Schumacher,U., Källne,E., Morsi,H.W. and Rupprecht,G., Rev.Sci.Inst., **62** 889 (1991)

Coffey,I.H., Keenan,F.P., McAdam,C., Barnsley,R., Dickson,W.J., Lawson,K.D. and Peacock,N.J., JET Report JET-P(91)49 1991 submitted.

Lauro-Taroni,L. JET Laboratory, private communication.

HELIUM-LIKE Ne IX IN THE JET TOKAMAK

I.H.Coffey[2], R.Barnsley[1], F.P.Keenan[2], K.D.Lawson[3] and N.J.Peacock[3]

[1]JET Joint Undertaking, Abingdon, Oxfordshire, OX14 3EA, England
[2]Department of Pure and Applied Physics, Queens University, Belfast, BT7 1NN, N.Ireland
[3]Culham Laboratory (UKAEA/EURATOM Fusion Association), Abingdon, Oxon, OX14 3DB, England

Abstract

New calculations are presented of the electron density sensitive emission line ratio $R = I(1s^2\ {}^1S_0 - 1s2s\ {}^3S_1)/I(1s^2\ {}^1S_0 - 1s2p\ {}^3P_{1,2})$, and the electron temperature sensitive emission line ratios $G = [I(1s^2\ {}^1S_0 - 1s2s\ {}^3S_1) + I(1s^2\ {}^1S_0 - 1s2p\ {}^3P_{1,2})]/I(1s^2\ {}^1S_0 - 1s2p\ {}^1P_1)$ and $R_1 = I(1s^2\ {}^1S_0 - 1s3p\ {}^1P_1)/I(1s^2\ {}^1S_0 - 1s2p\ {}^1P_1)$. They are compared with measured ratios from the JET (Joint European Torus) plasma where the electron temperature and density have been determined by independent means. Consideration is given to diffusion effects which cause a departure from coronal equilibrium conditions, resulting in significant changes to the G ratio. These measurements are made in the absence of blending with Fe XVIII, Fe XIX and Na X lines which are sources of error in solar measurements. Good agreement with theory is found, with discrepancies of typically less than 10%. This implies that the theoretical results may be applied to the analysis of remote plasma sources for which no independent electron density and temperature estimates exist, such as solar flares and active regions.

1 INTRODUCTION

Emission lines due to transitions in He-like ions in the soft X-ray spectrum are readily observed in high temperature laboratory plasmas as well as the solar corona. The electron temperature and density may be found through the ratios, $G = (f + i)/r$ and $R = f/i$ respectively, where f is the forbidden $1s^2\ {}^1S_0 - 1s2s\ {}^3S_1$ transition, i is the intercombination $1s^2\ {}^1S_0 - 1s2p\ {}^3P_{1,2}$ transitions and r is the resonance $1s^2\ {}^1S_0 - 1s2p\ {}^1P_1$ transition. Also used to obtain the electron temperature is the ratio $R_1 = I(1s^2\ {}^1S_0 - 1s3p\ {}^1P_1)/I(1s^2\ {}^1S_0 - 1s2p\ {}^1P_1)$ Determination of these theoretical ratios is critically dependent on the atomic data adopted, especially for the electron impact excitation between the ground state and the $n = 2,3$ levels.

2 ATOMIC DATA

The model ion for Ne IX consists of the 23 lowest $1snl$ states with $n < 6$ and $l < 3$, making a total of 37 levels when the fine structure splitting in the ^{3}P and ^{3}D terms is included. The atomic data is taken from Keenan *et al.* (1987b), where the electron impact excitation rates from the ground state to the $1s2l$ and $1s3l$ levels have been interpolated (Keenan *et al.* 1987a) from the R-matrix calculations of Tayal and Kingston (1984; 1985) and the distorted wave calculations of Pradhan (1985). The effect of dielectronic and radiative recombination of H-like Ne X on the Ne IX level populations was included by using the recombination coefficients of Mewe and Schrijver (1978) and the ionisation balance calculations of Arnaud and Rothenflug (1985).

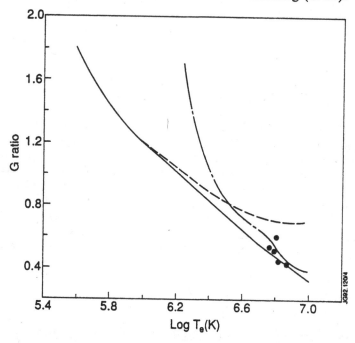

Fig.1-The theoretical emission ratio G plotted as a function of temperature T_e (in units of degrees Kelvin) at an electron density of log $n_e = 13$ (in units of cm^{-3}), with dielectronic and radiative recombination included in (solid line) or excluded from (broken line) the calculations. Also shown are the JET experimental data (solid points) and the theoretical G ratio (dashed line) calculated using an ionisation balance calculated for typical JET conditions.

3 EXPERIMENTAL

Neon was introduced into the JET plasma by gas puffing and the spectra were obtained with a double crystal X-ray spectrometer (Barnsley *et al.* 1991). It was used in general survey mode with TLAP crystals, providing coverage of the wavelength region from 11.3Å to 22.4Å, and was calibrated for absolute wavelength and intensity measurements with a resolving power ($\lambda/\delta\lambda$) of $\simeq$ 500. The radial position of maximum emissivity for Ne IX was obtained using a transport code (Lauro-Taroni) which calculated radial profiles of intensity, density and radiated power for all the ionisation stages of a particular impurity element at a given time during a discharge. T_e and n_e measurements at the predicted positions were found from radial profiles measured by analysis of Electron Cyclotron Emission and Thomson scattering of a ruby laser respectively.

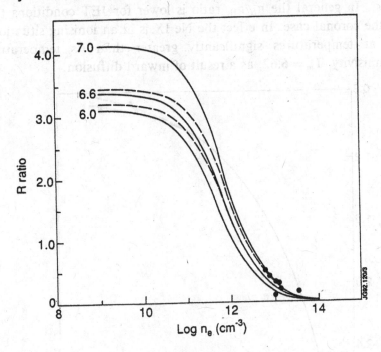

Fig.2-The Theoretical emission ratio R plotted as a function of density n_e, plotted at electron temperatures, log T_e = 7.0, 6.6 and 6.0, with dielectronic and radiative recombination included (solid lines) or excluded (dashed lines). Also shown are the measured values from JET which have an average electron temperature of log T_e = 6.85.

4 RESULTS AND DISCUSSION

Using the statistical balance population code of Summers *et al.* in conjunction with the atomic data discussed above, coronal Ne IX excited state populations were calculated for a range of electron temperatures and densities. Photoexcitation and de-excitation processes are negligible in comparison with the corresponding collisional processes and all transitions were considered optically thin. The temperature sensitive G ratio is plotted in fig.1.

Recombination into the triplet states results in the flattening of the theoretical curve as T_e increases. The model calculations assume coronal equilibrium which is generally valid in the centre of a tokamak, however the Ne IX emission shell is located close to the plasma edge. A diffusive ionisation model calculated using the transport code produces the curve also shown in fig.1 and is in good agreement with experimental values. The average difference between the experimental values and the theoretical G ratio calculated for JET conditions is typically 10%. In general the n_H/n_{He} ratio is lower for JET conditions than it would be in the coronal case. In effect the Ne IX is in an ionising situation and is occurring at temperatures significantly greater than its temperature of maximum emmisivity, $T_m = 6.62$, as a result of inward diffusion.

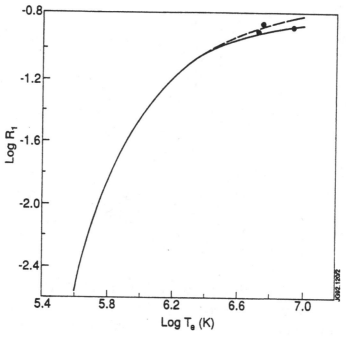

Fig.3-The theoretical emission ratio R_1 plotted as a function of temperature T_e, with dielectronic and radiative recombination included (solid line) or excluded (dashed line). The measured values from JET are also plotted.

Fig.2 shows the density sensitive R ratio. Inclusion of recombination effects increases the R values at low n_e. This increase is negligible for JET plasma densities, and hence the departure from coronal conditions has little effect. The R ratio varies strongly with n_e between log $n_e = 10.5$ to log $n_e = 12.5$, which is relevant to coronal diagnostics. The JET experimental points lie in the density range log $n_e = 12.8$ to log $n_e = 13.5$, which is outside the coronal interest region, but are in good agreement with theory over these densities. In fig.3 the temperature sensitive R_1 is plotted. Recombination is of lesser importance in determining R_1 than for the G ratio. The temperature range log $T_e = 6.0$ to log $T_e = 7.0$ comprises the temperature range in which the Ne IX ratios are useful as temperature diagnostics for the solar corona.

5 CONCLUSIONS

Comparison of the theoretical calculations with the JET results shows good agreement for the R, R_1 and G ratios providing support for the atomic data adopted in the calculations and the analysis techniques employed to derive the emission line ratios.

REFERENCES

Arnaud,M. and Rothenflug,R., Astron.Astrophys.suppl., **60** 245 (1985)

Barnsley,R., Schumacher,U., Källne,E., Morsi,H.W. and Rupprecht,G., Rev.Sci.Inst., **62** 889 (1991)

Keenan,F.P., McCann, S.M. and Kingston, A.E., Physica Scripta, **35** , 432 (1987a)

Keenan,F.P., McKenzie, D.L., McCann, S.M. and Kingston, A.E., Astrophys.J., **318** , 926 (1987b)

Lauro-Taroni, L. JET Laboratory, private communication.

Mewe,R. and Schrijver,J., Astron.Astrophys., **65** 99 (1978)

Pradhan,A.K., Astrophys.J.Suppl., **59** 183 (1985)

Tayal,S.S. and Kingston, A.E., J.Phys.B., **17** , 1383 (1984)

Tayal,S.S. and Kingston, A.E., J.Phys.B., **18** , 2983 (1985)

Summers,H.P., Briden,P.B., Dickson,W.J and Lang,J., "An Atomic Data & Analysis Structure for Spectral Emission in Laboratory and Astrophysical Plasmas" -to be published.

Determination of Element Abundances Using the Yohkoh Bragg Crystal Spectrometer

A. Fludra[1], J.L. Culhane[1], R.D. Bentley[1], G.A. Doschek[2], E. Hiei[3], K.J.H. Phillips[4], A. Sterling[2], T. Watanabe[3]

[1]Mullard Space Science Laboratory, University College London, UK
[2]US Naval Research Laboratory, Washington D.C., USA
[3]National Astronomical Observatory of Japan, Mitaka, Japan
[4]Rutherford Appleton Laboratory, Chilton, Didcot, UK

ABSTRACT

The Yohkoh mission was launched on 30 August 1991 by the Institute of Space and Astronautical Science from the Kagoshima Space Centre in Japan. There are four instruments on board which observe solar activity from soft X-ray to gamma-ray wavelengths with special emphasis on solar flares. The Bragg Crystal Spectrometer has four high resolution channels which register the resonance and satellite lines of S XV, Ca XIX, Fe XXV and Fe XXVI.

Preliminary results of Sulphur and Calcium abundance measurements made with the Yohkoh Bragg Crystal Spectrometer are presented. Theoretical spectra are fitted to the observed X-ray flare spectra to derive electron temperature and intensities of lines and continuum. Abundance values are obtained from the observed intensity ratio of the resonance lines of the Helium like ions to the adjacent continuum, by comparing it with the line-to-continuum ratio predicted from theory as a function of temperature.

Previous observations with the Bent Crystal Spectrometer on the Solar Maximum Mission (Sylwester, Lemen and Mewe, 1984, Nature, 310, 665) established that the Calcium abundance varied from flare to flare. First Yohkoh data also indicate variation of the Calcium abundance, ranging from the photospheric value to about a factor two above the photospheric value in a sample of seven flares. The Sulphur abundance in this sample of flares is smaller than the photospheric value by about a factor two and exhibits much less variation.

Femtosecond Laser-Induced Plasma X-Rays and Ionization Dynamics of High-Z Materials

J.C. Gauthier[1], J.P Geindre[1], A. Rousse[1], F. Falliès[1], P. Audebert[1], A. Mysyrowicz[2], J.P. Chambaret[2], A. Antonetti[2], A. Mens[3], R. Verrecchia[3], R. Sauneuf[3], and P. Schirmann[3]

[1]Laboratoire PMI, Ecole Polytechnique, 91128 Palaiseau (France)
[2]Laboratoire d'Optique Appliquée, ENSTA, 91120 Palaiseau (France)
[3]Centre d'Etudes de Limeil-Valenton, 94195 Villeneuve St Georges (France)

ABSTRACT

Hydrodynamic simulations taking into account radiation transport have been performed to investigate the factors controlling the time duration of x-ray emission from 100fs laser-produced plasmas. Maximum conversion efficiencies below 2% of the incident laser energy are calculated for high Z materials. The time duration of keV radiation is comparable to the laser pulse. Preliminary measurements of the x-ray pulse width for photon energies above 1.0 keV with a 2ps time resolution streak camera show qualitative agreement with hydrocode predictions.

1 HYDROCODE CALCULATIONS

In recent years, the technology of short laser pulse generation has progressed to the point that optical pulses of 100 femtoseconds in duration are routinely produced (Shank 1988). Such pulses can be amplified with relatively modest means to energies of a few tens of mJ corresponding to focused intensities exceeding 10^{18} W/cm^2. One of the interesting prospects with such short and intense light pulses is the generation of x-ray radiation in the subpicosecond time range (Murnane et.al. 1989,1991).

We have performed hydrodynamic simulations, taking into account radiation transport, to investigate the factors controlling the x-ray pulse duration in a 100 fs laser-produced plasma. We have used the XRAD hydrocode (Gauthier and Geindre 1988) including photon transport by a multigroup radiative transfer method. Emission and absorption data for the different elements studied here ($13 < Z < 78$) have been calculated using a non-LTE, average atom model. All simulations have been made at 10^{17} W/cm^2 and laser prepulse effects have not been taken into account. In Fig.1, we show the time-resolved emitted spectrum in the 0-3 keV x-ray range for Al, Cu, Sm, and Pt. The duration of sub-keV emissions is greater than 1 ps while K-shell (Al at 1.5 keV), L-shell (Cu at 1.2 keV), and M-shell emissions (Sm at 1.5 keV and Pt at 2.6 keV) last no longer than the laser pulse.

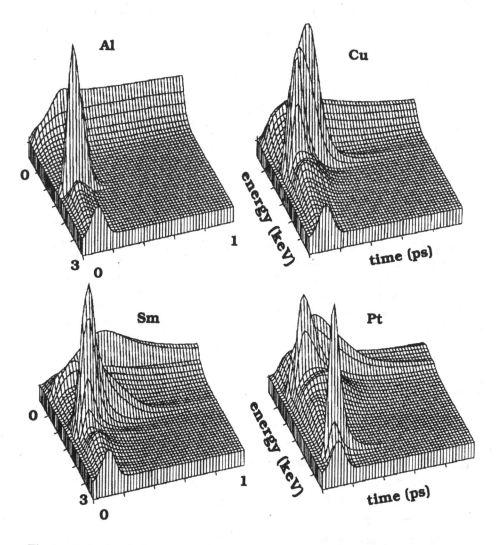

Fig.1 - Calculated time-resolved spectra of 0-3 keV x-ray radiation from Al (Z=13), Cu (Z=29), Sm (Z=62) and Pt (Z=78). The laser intensity is 10^{17} W/cm^2 and the laser pulse shape is indicated at the edge of each spectrum.

Figure 2 shows the variation of the conversion efficiency in the 0.01-10 keV range for Cu as a function of laser intensity and the evolution of the conversion efficiency with the nature of the material at a fixed laser intensity of 10^{17} W/cm^2. Conversion efficiencies never exceed 2 % for high-Z materials. This results from the low absorption rate of laser radiation, from the limited spatial range of the emission zone and from the high recombination rates at near solid densities which suppress the emitting ions. Even small, hydrodynamic expansion plays also a role in the ionization dynamics (Milchberg et.al. 1991).

2 MEASUREMENT OF THE X-RAY PULSE DURATION

Mesurements of a pulse duration below 2 ps for 500-1000 eV XUV radiation from a Si target have been previously reported (Murnane et.al. 1991). Here, we extend these results by measuring the pulse width of x-ray ($h\nu > 1$ keV) emission from a copper target at 10^{16} W/cm^2 laser intensity. We have used a new streak camera (Mens et.al. 1990) built around a P-850X bi-lamellar streak tube equipped with a CsI photocathode for this particular experiment. With an extracting field of 45 keV/cm, its dynamic spatial resolution is better than 20 lp/mm and its temporal resolution measured in the UV is about 2ps. A 2.3 fiber magnifier, an image intensifier, and a film holder have been used as a readout system for the camera. Film density was converted to intensity using a wedged neutral density filter. Figure 3 shows a line out of the signal recorded on a 2485 Kodak film. The FWHM of the pulse is about 2.1 ps, a value which is comparable to the signal already recorded by the camera in the UV. This shows that the x-ray pulse width of the Cu plasma emission above 1.0 keV (mostly L-shell radiation at 1.2 keV) is in the subpicosecond range. These preliminary results are in good agreement with our code predictions.

3 REFERENCES

(1)C.V. Shank in *Ultrashort Laser Pulses and Applications*, ed. W. Kaiser, Springer-Verlag, 1988. Numerous references can also be found in the Spinger-Verlag series of "Ultrafast Phenomena", vols. 46, 48 and 53.

(2) M.M. Murnane, H.C. Kaypten, R.W. Falcone, Phys. Rev. Lett. 62, 155 (1989) and Science, 251, 531 (1991)

(3) J.C. Gauthier and J.P. Geindre, PMI Reports n° 1971 and 1974, (1988).

(4) H.M Milchberg et.al., Phys. Rev. Lett. 67, 2654 (1991).

(5) A. Mens et.al. 19th International Congress on High Speed Photography and Photonics (Cambridge 1990).

* Work partly supported by the Stimulation Program of the European Economic Community and by the Direction des Recherches et Etudes Techniques under contract 90/057.

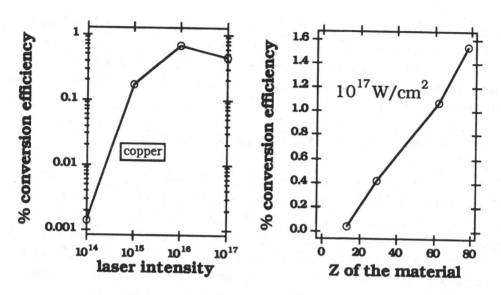

Fig.2 - Conversion efficiency in the range 0.01-10 keV as a function of laser intensity for Cu (left) and as a function of the material Z at $10^{17}W/cm^2$ (right).

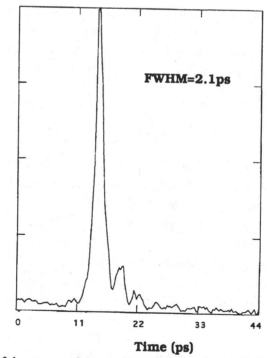

Fig. 3 - Lineout of the screen of the streak camera recording photons above 1.0 keV energy from a Cu target in a single shot. The measured laser intensity is about 1.-2. $10^{16}W/cm^2$.

ROSAT Observations of the Stellar Coronal Dividing Line

BERNHARD HAISCH[1] and J.H.M.M SCHMITT[2]

[1] Lockheed Solar and Astrophysics Laboratory, Palo Alto, California and Max-Planck-Institut für Extraterrestrische Physik, Garching, Germany
[2] Max-Planck-Institut für Extraterrestrische Physik, Garching, Germany

ABSTRACT

We present an update on the results of the ROSAT X-ray All-Sky Survey observations of stellar sources presented by Haisch, Schmitt and Rosso (1991). In that paper the presence of a coronal dividing line in the H-R diagram at approximately spectral type K3 II to K3 IV was established by the clear difference in distribution of the 65 ROSAT detections vs. the 868 non-detections of BSC stars in the 70 percent-complete survey. The remaining 30 percent of the survey has now been processed resulting in 31 additional detections of stellar coronae, all of which lie to the left of the dividing line.

1. THE ROSAT OBSERVATIONS

The concept of a dividing line in the H-R diagram was proposed by Linsky and Haisch (1979) to describe the apparently abrupt disappearance of coronae of stars evolving into the giant branch at early K spectral type. The ubiquitous nature of stellar coronae having been only just discovered with the *Einstein Observatory*, the non-existence of coronae of such stars became a tantalizing enigma, especially since this seemed to occur rather quickly, on an evolutionary time scale, and would almost certainly be an important clue concerning changes in stellar magnetic fields. Naturally the existence of a dividing line was critically scrutinized and, the number of stellar sources and non-sources being modest, this phenomenon was not universally accepted. Every last *Einstein* and *EXOSAT* X-ray photon was brought to bear on this problem (Haisch 1987, Maggio *et al.* 1990), along with many years worth of accumulated *IUE* spectra (Haisch *et al.* 1990), until ROSAT finally entered the scene on June 1, 1990.

The All-Sky Survey achieved approximately the same level of sensitivity, over the entire sky, as typical *Einstein* IPC fields, $f_x = (1-2) \times 10^{-13}$ ergs cm^{-2} s^{-1}. Almost 900 non-detections of single *Bright Star Catalog* GKM giants/supergiants were found to be distributed as shown in Figure 1 by Haisch, Schmitt and Rosso (1991). In sharp contrast, we found only one star out of 65 detected in the 70 percent-complete survey lying to the right of spectral type K3: HR 4289, a not-well-studied K5 III. The complete survey has now been examined for single GKM I to IV stars, yielding 31 new detections, all of which lie to the left of the dividing line. The entire sample of 96 detections is shown in Figure 2.

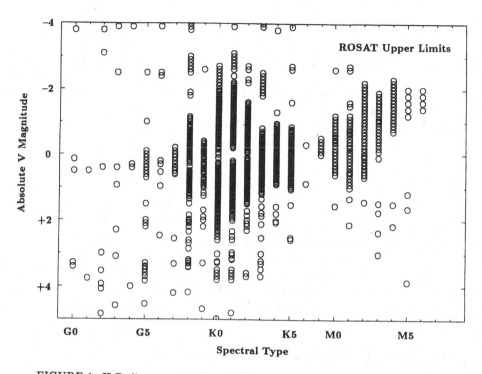

FIGURE 1. H-R diagram of $\approx 1 - 5\ M_\odot$ evolved stars. None of these single GKM (I to IV) stars were detected in the All-Sky Survey. Of these non-detections, 604 are K3 or earlier, 264 are K4 or later. The luminosity class spread is a plotting artifact.

2. CAN STELLAR WINDS ATTENUATE CORONAL X-RAYS?

The possibility of a connection between the disappearance of coronae and the onset of massive winds has been discussed for many years. The recent compilation of stellar wind properties by Judge and Stencel (1991) shows how closely the two situations may be juxtaposed, as shown in Figure 3. It is logical to ask whether attenuation of soft X-rays in such a wind might be significant. The density of a spherically symmetric wind is:

$$\rho(r) = \frac{\dot{M}}{4\pi r^2 v(r)}. \tag{1}$$

The mass column density along a line-of-sight for a constant outflow, v, is therefore:

$$\int_{r_0}^{\infty} \rho(r)dr = \frac{\dot{M}}{4\pi v}\int_{r_0}^{\infty}\frac{1}{r^2} = \frac{\dot{M}}{4\pi v}\left(\frac{1}{r_0} - \frac{1}{\infty}\right) = \frac{\dot{M}}{4\pi v}\frac{1}{r_0}. \tag{2}$$

Converting this to units of $\dot{M}(M_\odot/yr)$, $r_0(R_\odot)$ and $v(km/s)$, and assuming the outflowing matter is neutral hydrogen, the hydrogen column density is:

$$N_H = \frac{7.2\times 10^8}{1.66\times 10^{-24}}\frac{\dot{M}}{vr_0} = 4.3\times 10^{32}\frac{\dot{M}}{vr_0}. \tag{3}$$

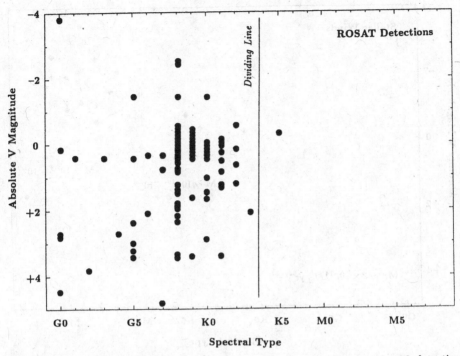

FIGURE 2. Same region of the H-R diagram as Fig. 1, now showing the 96 *detections of single giants and supergiants.*

Table 1 shows the calculated column densities, N_H, for typical stellar parameters and mass loss rates as compiled by Judge and Stencel (1991). For N_H in the neighborhood of 10^{20} soft X-ray attenuation will be significant. This would affect the interpretation of the coronal dividing line (Haisch, Schmitt and Rosso 1991).

Table 1. Stellar Wind Hydrogen Column Densities

SpTy	r_0	v	$\dot{M}$	N_H
K5 III	25	50	10^{-10}	3.4×10^{19}
K5 III	25	25	10^{-9}	6.9×10^{20}
K5 I	400	50	10^{-10}	2.2×10^{18}
K5 I	400	25	10^{-9}	4.3×10^{19}
M2 I	800	50	10^{-8}	1.1×10^{20}
M2 I	800	25	10^{-7}	2.2×10^{21}

While this implies that soft X-ray attenuation could become significant, we think it unlikely at this time that X-ray extinction is the entire story behind the dividing line. Ultraviolet transition region spectral lines also disappear (Haisch *et al.* 1990). We also point to the extremely low upper limit on soft X-rays from Arcturus determined by Ayres,

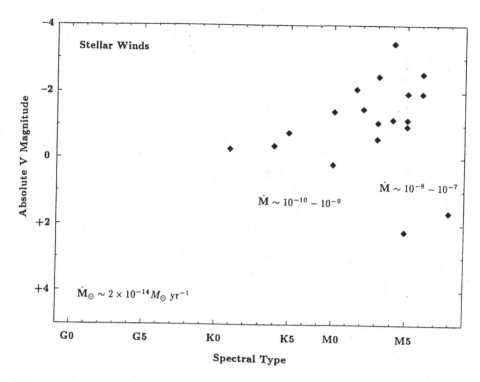

FIGURE 3. H-R diagram of the Judge and Stencel compilation of data on massive, cool stellar winds.

Fleming and Schmitt (1991) in a deep ROSAT exposure. The Alfvén wave reflection model in fact looks like a very promising way to bring about a true phase transition from coronae to winds (Rosner *et al.* 1991).

REFERENCES

Ayres, T.R., Fleming, T.A. & Schmitt, J.H.M.M. 1991, ApJ, 376, L45.

Haisch, B. 1987, in Lecture Notes in Physics, Vol 291, 269

Haisch, B. 1990, Bookbinder, J.A., Maggio, A., Vaiana, G.S. & Bennett, J.O., ApJ, 361, 570

Haisch, B., Schmitt, J.H.M.M. & Rosso, C. 1991, ApJ, 383, L15.

Judge, P.G. & Stencel, R.E. 1991, ApJ, 371, 357

Linsky, J.L. & Haisch, B. 1979, ApJ, 229, L27

Maggio, A., Vaiana, G.S., Haisch, B., Stern, R.A., Bookbinder, J., Harnden, F.R., Jr., & Rosner, R. 1990, ApJ, 348, 253

Rosner, R., An, C.-H., Musielak, Z.E., Moore, R.L., & Suess, S.T. 1991, ApJ, 372, L91

The Determination Of Solar Coronal Electron Temperatures From Mg XI Emission Lines in *SMM-FCS* Spectra Of Flares And Active Regions

L. K. Harra[1], K. J. H. Phillips[3], F. P. Keenan[1], E. S. Conlon[1] and A. E. Kingston[2]

[1]Department of Pure and Applied Physics, The Queen's University of Belfast, Belfast BT7 1NN, Northern Ireland, UK
[2]Department of Applied Mathematics and Theoretical Physics, The Queen's University of Belfast, Belfast BT7 1NN, Northern Ireland, UK
[3]Astrophysics Division, Rutherford Appleton Laboratory, Chilton, Didcot, Oxfordshire OX11 0QX, England, UK

ABSTRACT

Recent atomic physics calculations for Mg XI are used to derive the electron temperature sensitive emission line ratios G = $[I(1s^2\ {}^1S - 1s2s\ {}^3S) + I(1s^2\ {}^1S - 1s2p\ {}^3P_{1,2})]/I(1s^2\ {}^1S - 1s2p\ {}^1P)$, $R_1 = I(1s^2\ {}^1S - 1s3p\ {}^1P)/I(1s^2\ {}^1S - 1s2p\ {}^1P)$, and $R_2 = I(1s^2\ {}^1S - 1s4p\ {}^1P)/I(1s^2\ {}^1S - 1s2p\ {}^1P)$, which are found to be significantly different from earlier results. Values of T_e deduced from G, R_1 and R_2 ratios measured from solar flare and active region spectra obtained with the Flat Crystal Spectrometer (FCS) on board the *Solar Maximum Mission* (SMM) satellite are consistent. This provides support both for the validity of the theoretical G, R_1 and R_2 diagnostics, and for the FCS calibration curve in the wavelength region covering the Mg XI transitions, 7.472 – 9.314 Å.

1 Introduction

Emission lines arising from transitions between the $1s^2\ {}^1S$ and $1s2l$ levels in helium-like Mg XI are prominent features of high temperature solar and laboratory plasmas (Phillips *et al* 1982, Barnsley *et al* 1986). They may be used to infer the electron temperature of the emitting plasma through the well-know ratio G = $(f + i)/r$ (Phillips *et al* 1982), where f is the forbidden $1s^2\ {}^1S - 1s2s\ {}^3S$ transition at 9.314 Å , i the intercombination $1s^2\ {}^1S - 1s2p\ {}^3P_{1,2}$ lines at 9.232 Å, and r the $1s^2\ {}^1S - 1s2p\ {}^1P$ resonance transition at 9.169 Å (Blumenthal, Drake and Tucker 1972). Two additional temperature diagnostics for Mg XI in solar flares and active regions are used, $R_1 = I(1s^2\ {}^1S - 1s3p\ {}^1P)/I(1s^2\ {}^1S - 1s2p\ {}^1P) = I(7.850\ \text{Å})/I(9.169\ \text{Å})$, and $R_2 = I(1s^2\ {}^1S - 1s4p\ {}^1P)/I(1s^2\ {}^1S - 1s2p\ {}^1P) = I(7.472\ \text{Å})/I(9.169\ \text{Å})$, as they have been shown to be more sensitive to variations in T_e than the normal diagnostic for He-like ions, G. The theoretical determination of these ratios is critically dependent on the atomic data adopted in the calculations, especially for the electron collisional excitation rates between the ground state and $1s2l$ levels (Gabriel and Jordan 1972).

Using the most recent atomic calculations T_e-sensitive line ratios in Mg XI have been derived (see, eg, Figs 1 and 2) in this paper and compared then with observational data for solar flares and active regions (tables 1,2).

2. OBSERVATIONAL DATA

The observational data were obtained by channel no.3, with ADP crystal on the Flat Crystal Spectrometer (FCS) on board the *Solar Maximum Mission* (SMM). Although numerous flares and active regions were observed during the lifetime of SMM, few of these have Mg XI line spectra suitable for comparison with the present calculations, as the Mg XI $1s^2$ ^{1}S – $1s3p$ ^{1}P and $1s^2$ ^{1}S – $1s4p$ ^{1}P lines are often weak in active regions (leading to poor photon statistics). In addition, there may be significant variations in the flare emission during the time FCS takes to scan a spectrum. However we have identified 3 C1 and C2 flares whose X-ray fluxes determined from the companion Bent Crystal Spectrometer (BCS) show that the line intensities do not vary significantly during the time needed to scan the spectrum , and have also found 3 active region scans where the Mg XI $1s^2$ ^{1}S – $1s3p$ ^{1}P and $1s^2$ ^{1}S – $1s4p$ ^{1}P lines could be reliably measured.

3. RESULTS and DISCUSSION

The observed values of G, R_1, and R_2 are given in Table 1 together with the estimated errors in the line ratios. Table 2 lists the derived electron temperatures. They are similar to what is expected for hot active regions or weak flares. It is noted that the T_e sensitivities of R_1 and R_2 are much greater than for G, the former varying by factors of 1.7 and 2.1 respectively, between log T_e= 6.5 and 7.0, while the latter only changes by 34 % . Thus the use of R_1 and R_2 in the future should lead to better estimates than the more familiar G ratio.

The good agreement between theory and observation for R_1 and R_2 also provides support for the FCS channel 3 calibration curve in the wavelength region of the Mg XI transitions, 7.472 - 9.314 Å.

REFERENCES

Barnsley,R., Evans,K.D., Peacock,N.J., and Hawkes,N.C. 1986, Rev. Sci.Instrum.,57,159.

Blumenthal,G.T., Drake,G.W.F., and Tucker,W.H. 1972, ApJ, 172, 205.

Gabriel,A.H., and Jordan,C. 1972, Case Studies in Atomic Collision Physics, Vol. 2,ed E.W.McDaniel and M.R.C.McDowell (Amsterdam:North-Holland), p. 209.

Keenan,F.P.,Kingston,A.E., and McKenzie,D.L. 1986, ApJ, 303, 486.

Keenan,F.P.,Tayal,S.S., and Kingston,A.E. 1984, MNRAS, 207, 51P.
Phillips,K.J.H., et al. 1982, ApJ, 256, 774.

Table 1
Observed Mg XI Emission Line Ratios

Solar Feature	G	R_1	R_2
1986 July 13 C1 flare: 0242 UT	0.916 ± 0.035	0.117 ± 0.012	0.024 ± 0.005
1987 April 17 C2 flare: 0238 UT	0.723 ± 0.058	0.143 ± 0.016	0.024 ± 0.005
1987 April 17 C1 flare: 1955 UT	0.966 ± 0.070	0.051 ± 0.014	0.026 ± 0.007
1987 April 18 active region: 2237 UT	0.733 ± 0.061	0.103 ± 0.016	0.038 ± 0.009
1987 April 18 active region: 2247 UT	0.877 ± 0.079	0.123 ± 0.019	0.050 ± 0.006
1987 April 18 active region: 2257 UT	0.765 ± 0.082	0.112 ± 0.021	0.026 ± 0.009

Table 2
Derived Mg XI Logarithmic Electron Temperatures

Solar Feature	G	R_1	R_2
1986 July 13 C1 flare: 0242 UT	$6.33^{-0.08}_{+0.07}$	$6.75^{+0.11}_{-0.09}$	$6.49^{+0.06}_{-0.05}$
1987 April 17 C2 flare: 0238 UT	$6.80^{-0.18}_{+0.19}$	> 7.0	$6.49^{+0.06}_{-0.05}$
1987 April 17 C1 flare: 1955 UT	$6.24^{-\infty}_{+0.12}$	$6.32^{+0.08}_{-\infty}$	$6.51^{+0.10}_{-0.07}$
1987 April 18 active region: 2237 UT	$6.77^{-0.18}_{+0.20}$	$6.64^{+0.13}_{-0.10}$	$6.69^{+0.14}_{-0.14}$
1987 April 18 active region: 2247 UT	$6.40^{-0.15}_{+0.18}$	$6.80^{+0.20}_{-0.15}$	$6.90^{+\infty}_{-0.12}$
1987 April 18 active region: 2257 UT	$6.66^{-0.20}_{+0.29}$	$6.72^{+0.18}_{-0.21}$	$6.50^{+0.14}_{-\infty}$

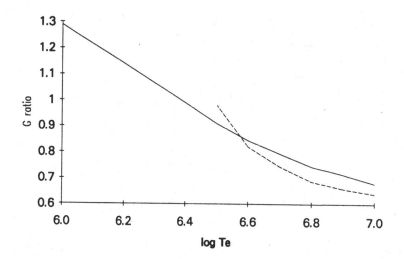

Figure 1. The theoretical Mg XI emission line ration G = [I(1s² ¹S - 1s2s ³S) + I(1s² ¹S - 1s2s ³P$_{1,2}$)]/I(1s² ¹S - 1s2p ¹P) where I is in energy units, plotted as a function of electron temperature at a density of $N_e = 10^{11}$ cm^{-3} , with : *solid line*- the present calculation; *dashed line* - the results of Keenan,Tayal & Kingston (1984).

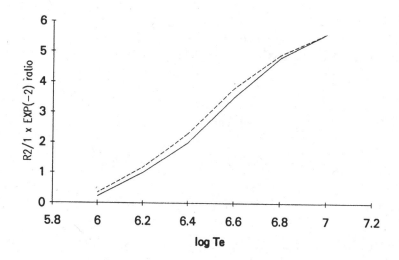

Figure 2. The theoretical Mg XI emission line ration R2 = I(1s² ¹S - 1s4p ¹P) / I(1s² ¹S - 1s2p ¹P) = I(7.472 Å)/I(9.169 Å), where I is in energy units, plotted as a function of electron temperature at a density of $N_e = 10^{11}$ cm^{-3} , with :*solid line* - the present calculation ; *dashed line* - the results of Keenan, Kingston & McKenzie (1986).

Accretion Disk Corona Modelling

Yuan-Kuen Ko[1,2], and Timothy R. Kallman[1]

[1]Laboratory for High Energy Astrophysics, NASA/Goddard Space Flight Center
[2]Department of Physics, University of Maryland

ABSTRACT

We investigate the vertical structure of an accretion disk corona (ADC) which will be formed by the illumination of a disk by X-rays from a compact object. The vertical temperature profile of the ADC is calculated including a large number of atomic processes for thirteen elements in the calculation of statistical and thermal equilibrium. Transfer of cooling radiation is treated using the escape probability formalism. The emission spectra formed in the corona are also calculated.

1 INTRODUCTION

Low mass X-ray binaries (LMXB's) consist of a late type star in orbit with a compact object, likely to be a neutron star. X-rays are generated as gas is transferred from the normal star and falls into the deep gravitational potential of the compact star. An accretion disk forms as this material loses its angular momentum and spirals inward to the surface of the compact star; disks have been identified in the spectra of other compact X-ray sources, notably cataclysmic variables, and also possible active galactic nuclei. The accretion disks in LMXB's are likely to be strongly heated by X-rays generated near the center of the disk. This can occur either due to the 'flaring' of the disk height with radius, allowing the disk surface to see the compact object directly (Vrtilek, et al. 1986), or by the scattering of X-rays in a corona or wind emanating from the disk (London 1982).

Illumination can affect the disk structure in a variety of ways, for example, heating of the disk interiors and associated effects on the vertical structure of the disk (Meyer and Meyer-Hofmeister 1982, Ko and Kallman 1991), production of a hot Compton-heated accretion disk corona (ADC) and wind (White and Holt 1982, Begelman, McKee and Shields 1983) and suppresion of the thermal instability which may be responsible for variability in dwarf novae, soft X-ray transients and AGN (Tuchman, Mineshige and Wheeler 1990).

Separating the hot Compton-heated corona and the possibly heated disk photosphere will be a transition region with temperatures ranging from 10^4 K to 10^7 K. This region is a likely site for the soft X-ray and UV emission lines which are observed from several LMXB (Vrtilek, et al. 1990). The goal of the work described here is to construct numerical models for this region.

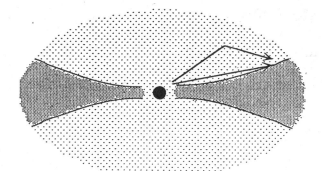

Fig. 1—Examples of the possible ways by which X-rays from the central compact object can back-illuminate the outer disk regions.

2 CALCULATION OF THE VERTICAL STRUCTURE OF ADC

Our model assumptions are as follows: We take the disk structure below the photosphere to be similar to that described by Shakura and Sunyaev (1973), modified to account for the effects of heating by hard X-rays (Ko and Kallman 1991). The X-ray luminosity generated near the compact object by disk accretion with accretion rate $\dot{M}$ can be expressed by:

$$L_x = \eta \dot{M} c^2 \tag{1}$$

where η is the efficiency of accretion and is ~ 0.1 for neutron star binary systems. The amount of X-ray flux that shines back onto the accretion disk depends on poorly known quantities, such as the degree to which the disk is flared, the shielding of the outer disk by a possible ADC in the inner disk regions, and the scattering in the ADC itself. We use a correction factor f, which we allow to be a free parameter, to reflect the efficiency with which the total X-ray flux available at a given radius is used to illuminate the disk surface. Therefore the incident X-ray flux onto the accretion disk is:

$$F_x = \frac{f L_x}{4\pi R^2} = \frac{f \eta \dot{M} c^2}{4\pi R^2} \tag{2}$$

where R is the distance to the central compact object. In our models, the incident X-ray spectrum is chosen to be a 10 keV Bremsstrahlung with total flux as shown by equation (2). The total external radiation field is thus the downward incident X-ray flux plus the upward black body radiation from the disk photosphere.

The state of the gas is defined by the ionization state, atomic level populations, density and temperature. The level populations and ionization balance are determined by assuming a balance between the various ionization, recombination, excitation and decay processes. All thermal processes are assumed to be on a steady-state balance, and the density is determined by the hydrostatic condition in the disk vertical direction. We use 168 ions, 1631 lines and 1649 levels from 13 abundent elements, H, He, C, N, O, Ne, Mg, Si, S, Ar, Ca, Fe, Ni, in the calculations of statistical and thermal equilibrium. The atomic data for lines are from Raymond and Smith (1977) and Mendoza (1986). The computational techniques are similar to those described in Kallman and McCray (1982), except that the 'nebular approximation' is not used; populations for all excited levels associated with line emission are explicitly calculated. Transfer of lines and recombination continua is treated using escape probabilities.

The simultaneous solution of the temperature, ionization and density of the gas in the disk atmosphere is done using an iterative procedure, since the attenuation of the incident X-rays requires integration from the top of the atmosphere, while the hydrostatic equation requires integration from the bottom of the atmosphere. In general 10-15 iterations are required to achieve convergence to within 10^{-2} in temperature and density.

3 RESULTS AND DISCUSSIONS

Two models are caluculated: (1)Model 1: $\dot{M} = 10^{17}$ g/sec, $f\eta = 0.01$, $R = 10^{10}$ cm, (2)Model 2:$\dot{M} = 10^{16}$ g/sec, $f\eta = 0.01$, $R = 3 \times 10^{10}$ cm. The boundary conditions at base of the corona are for (1)Model 1: $n_s = 2.53 \times 10^{15} \text{cm}^{-3}$, $T_s = 6.26 \times 10^4$ K, (1)Model 2: $n_s = 2.77 \times 10^{14} \text{cm}^{-3}$, $T_s = 1.9 \times 10^4$ K.

Fig. 2 and Fig. 3 show the vertical temperature profiles and the emission spectra for Model 1 and Model 2 respectively. A resolution of $\varepsilon/\Delta\varepsilon = 1000$ is assumed for line fluxes in both models.

The transition region between the disk photosphere and the top of the corona consists of three zones:(1) the low-T zone $\sim O(10^4$ K) which extends from the disk photosphere with decreasing T upwards because of the decreasing black body radiation flux. This is the region where most UV emission lines form. (2) the mid-T zone $\sim O(1 - 3 \times 10^6$ K) which is the main part of the transition zone. (3) the high-T zone where the gas is Compton heated to $T \sim 10^7$ K. The discontinuities between the three zones are caused by a thermal instability associated with Bremsstrahlung and atomic cooling.

For Model 1 most of the strongest lines are in the UV range, among them are HeII Balmer lines. For Model 2, because of lower temperatures in the line formation zone, most of the strongest lines are optical lines, among them are hydrogen Balmer lines and FeII lines. Those lines, however, are likely to be difficult to observe with low sensitivity instruments because of high continuum flux at that energy.

We find that level populations approach LTE for most abundant ions (e.g.HI and HeI) in $T \sim 10^4$ K gas. Model 2 is closer to LTE than Model 1 owing to lower temperatures and greater optical depths.

In this proceedings, we have shown the vertical structure and emission spectra of an accretion disk corona under the illumination of X-rays from the central compact object. A detailed description of the models and the results will be presented in a future paper.

REFERENCES

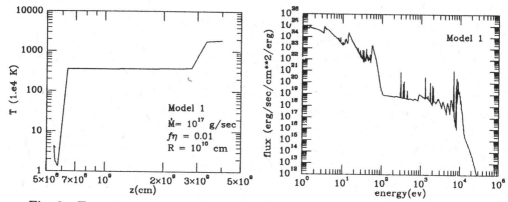

Fig. 2—Temperature profile and emission spectra for Model 1.

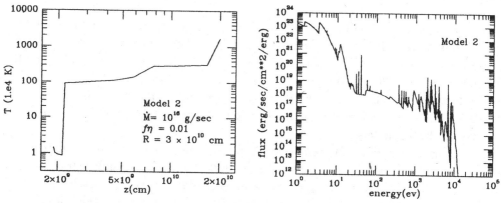

Fig. 3—Temperature profile and emission spectra for Model 2.

Begelman, M. C., McKee, C. F., and Shields, G. A. 1983, *Ap. J.*, **271**, 70.

Kallman, T. R., and McCray, R. 1982 *Ap. J. Suppl.*, **50**, 263.

Ko, Y., and Kallman, T. R. 1991, *Ap. J.*, **374**, 721.

London,R.A. 1982, in *Cataclysmic Variables and Low-Mass X-ray Binaries*, ed. D. Q. Lamb and J. Patterson, (Boston: D. Reidel Publishing Company), p121.

Mendoza, C. 1982, in *Planetary Nebulae*, ed. D. R. Flower, (Boston: D. Reidel Publishing Company), p143.

Meyer, F., and Meyer-Hofmeister, E. 1982, *Astr. Ap.*, **106**, 34.

Raymond, J. C., and Smith, B. W. 1977, *Ap. J. Suppl.*, **35**, 419.

Shakura N. I., and Sunyaev, R. A., 1973, *Astr. Ap.*, **24**, 337.

Tuchman, Y., Mineshige, S., and Wheeler, J. C. 1990, *Ap. J.*, **359**, 164.

White, N. E., and Holt, S. S. 1982, *Ap. J.*, **257**,, 318.

Vrtilek, S. D., Kahn, S. M., Grindlay, J. E., Helfand, D. J., and Seward, F. D. 1986, *Ap. J.*, **307**, 698.

Vrtilek, S. D., Raymond, J. C., Garcia, M. R., Verbunt, F., Hasinger, G., and Kurster, M. 1990, *Astr. Ap.* , **235**, 162.

Emission Line Polarization in SS433 ?

J. M. Laming

Naval Research Laboratory Code 4174L, Washington DC 20375-5000
(also SFA, Inc., Landover, MD 20785)

ABSTRACT

The possibility of detecting polarized line radiation from the X-ray jets of SS433 is discussed. If the observed radiation at 6.7 keV, currently interpreted as Fe XXV, is directly excited by particle impacts as the jets pass through a wind emitted from the companion, then due to the apparent monochromaticity and collimation of the jets, relatively high degrees of polarization (around ±50%) could be present in the resonance and magnetic quadrupole lines of Fe XXV.

1. INTRODUCTION

The jets of SS433 have stimulated much observational and theoretical work. In this admittedly more speculative paper, the possibility that line emission from these jets might be polarized is explored.

In the so-called kinematic model, the object is presumed to be a binary star with a 13 day orbital period, emitting two jets of material in a direction perpendicular to the plane of its orbit (Margon 1984). The temporal variation of Doppler shifts of lines in the optical spectrum (principally Hα and He I) shows that these jets precess with a period of 164 days. The precession axis is oriented at 78.2° to the direction of observation, and the precession cone angle is 19.8°. The jets appear to be highly monoenergetic, with a velocity of 0.26c, where c is the speed of light, and highly collimated.

X-ray observations of these jets with proportional counters on *EXOSAT* (Watson *et al.* 1986), *Tenma*, (Matsuoka, Takano, and Makishima 1986), and *Ginga* (Brinkmann *et al.* 1991) reveal a hard X-ray continuum with an emission line, interpreted as the 6.7 keV line complex in Fe XXV Doppler shifted in accordance with the precessional phase. Only a blue shifted component is generally seen, as distinct from the optical spectrum where blue and red shifted components are visible. Watson *et al.* (1986) attribute this to obscuration of the red shifted component by a thick accretion disk. Thus while the optical emission extends for a distance up to 10^{15} cm from the orbital plane, the X-ray emission must be confined to distances less than 10^{12} cm. It is a long standing problem to understand the nature of these jets, and their acceleration mechanism.

2. POLARIZATION OF IMPACT RADIATION

The directed nature of the jets makes it possible that line emission from them will be polarized. Atomic line emission excited in a plasma with an anisotropic electron distribution will in general be polarized. This has been modeled for laboratory beam-foil experiments (Laming 1989), solar flares, (Laming 1990a*, Haug 1979, 1981), and for non-radiative shocks in supernova remnants (Laming 1990b), and observed in the laboratory using the EBIT (Electron Beam Ion Trap) by Henderson *et al.* (1990). This can be understood for the case of impact excitation as follows. Let the initial bound electron have quantum numbers $L_i = 0$ and $M_{L,i} = 0$, where these are the orbital angular momentum and its z-component respectively. Taking the z-axis along the wavevector of the incident particle, its initial angular momentum z-component, denoted by $m_{l,i} = 0$ also. Conservation of L_z after the collision requires

$$M_{L,f} + m_{l,f} = M_{L,i} + m_{l,i} = 0.$$

Close to threshold, s-waves dominate in the partial wave expansion, requiring $m_{l,f} = 0$ for the scattered electron, and hence $M_{L,f} = 0$ for the bound electron. The preferential excitation of $M_L = 0$ states over others will lead to the emission of polarized light when observed at some angle to the z-axis, the degree of polarization being maximized at 90°. For neutral atoms on threshold, the cross section is zero. For ions, the cross section is non-zero on threshold, but some population of $M_L \neq 0$ states exists due to the depolarizing effect of the Coulomb field on the projectile. The polarization of radiation excited by electron impacts for Fe XXV has been calculated by Inal and Dubau (1987). For an impact velocity corresponding to the jet velocity in SS433, polarizations of about +50, -50, and +30 should be present in the w, x, and y ($1s^2\ {}^1S_0 - 1s2p\ {}^1P_1$, $1s^2\ {}^1S_0 - 1s2p\ {}^3P_2$, and $1s^2\ {}^1S_0 - 1s2p\ {}^3P_1$) lines respectively, observed perpendicularly to the z-axis.

3. THE JETS OF SS433

Could such a polarization be observed in the jets of SS433? First let us attempt to say what we can about whether such an effect is likely to be there in the first place, leaving the observational difficulties to the end. The optical emission has been modeled by Brown, Cassinelli, and Collins (1991) in terms of bullets of cold material ($\sim 10^4$ K) being collisionally heated by a stellar wind emitted from the companion. The abrupt turn off of this emission at 10^{15} cm is attributed to the bullets becoming collisionally thin at this distance, rather than being collisionally thick as postulated at smaller distances. These optical bullets must somehow start life as hot X-ray emitting material, and if collisionally thick would emit unpolarized radiation. However the

---

* The y-axis of figure 1 in this reference is a factor of 27 too big. The ensuing discussion applies to an electron beam energy of 27×10^{10} ergs/cm²/s, still well within typical flare energies for the cases considered.

temperature dependence of the collisional depth is approximately $\ln \Lambda / T^{3/2}$, where $\ln \Lambda$ is the Coulomb logarithm. Thus hot bullets with temperatures greater than 10^7 K will have 10^{-5} to 10^{-4} the collisional depth of their cold counterparts, which will possibly obviate this objection. Moreover, modeling of a thermal instability of the hot jets at the their base (Brinkmann et al. 1988) shows cold blobs forming there, making possible the existence of a medium of cold bullets surrounded by hot X-ray emitting gas. In such a scenario, the bullets could be collisionally thick, but the X-ray emitting material may not.

The next worry is that thermally excited radiation from the jet itself will dominate over the non thermal component of interest coming from the interaction between the jet and the wind. Taking the wind density to be 3×10^{10} electrons/cm^3 at a distance of 10^{12} cm (Brown et al. 1991), and the excitation cross section for the resonance transition to be 5×10^{-19} cm^2 (Inal and Dubau 1987), gives an excitation rate in this transition of order 10^2 per ion per second. Assuming that the thermal component comes from temperatures where the emitting efficiency is maximized requires a jet electron density of less than 5×10^{12} cm^{-3} for the non-thermal component to be dominant. Densities of up to an order of magnitude greater or less than this at the base of the jets have been suggested, (Brinkmann et al. 1991) for a thin and thick jet models respectively. Assuming this falls off as 1/distance^2, it is quite reasonable to suppose that the thermal component might not be significant compared to the non-thermal radiation.

It appears plausible then to expect polarized light to be emitted from the jets of SS433. It remains to consider processes which might depolarize it. The first of these is the effect of optical depth. According to the jet models of Brinkmann et al. (1991), and assuming cosmic elemental abundances, one should expect the w line to be optically thick in the radial direction from the jet axis, making it unlikely that polarized radiation in this transition will be emitted. The x, and quite possibly the y lines will however be optically thin in this direction. The next question is that of magnetic field within the jet. These are completely unknown, but fields of order $10^7 - 10^8$ G would be required to depolarize the w and y transitions in Fe XXV by the Hanlé effect.

4. DETECTION?

We should now bite the bullet and discuss whether it is feasible to detect such a polarization from SS433. It appears likely that the polarization will be large, but the big problem will be that SS433 is rather faint as an X-ray source. A flux of 2.4×10^{-3} photons/cm^2/s is reported by Matsuoka et al. (1986) in the 6.7 keV line complex. This flux is also expected to vary with the precessional phase (Brinkmann et al. 1991). Canizares et al. (1987) show a simulated spectrum of Fe XXV with a line flux of 7×10^{-4} photons/cm^2/s for a 10^4 sec exposure with the BCS on AXAF. Although AXAF will not carry polarimetric instruments, this demonstrates that future satellite missions will approach the sensitivity required to observe these transitions. Since the various lines have different (*i.e.* opposite sign) polarizations it is important to resolve them, making polarimeters designed for studies of continuum radiation generally unsuitable. As discussed by Laming (1990a), the intensity ratio between two transitions of different polarization will change with observation angle, due to the anisotropic nature of the emission. The intensity ratio between the resonance and magnetic quadrupole lines should change by around 20% with the precessional phase. However it is unlikely that this polarization signature will be detectable, since the change in observation geometry with the precession will make different parts of the jets visible at different times, which could well produce changes in line intensity ratios of a completely different nature.

REFERENCES

Brinkmann, W., Fink, H. H., Massaglia, S., Bodo, G., and Ferrari, A. 1988, *Astron. Astrophys.*, **196**, 313

Brinkmann, W., Kawai, N., Matsuoka, M., and Fink, H. H. 1991, *Astron. Ap.*, **241**, 112

Brown, J. C., Casinelli, J. P., and Collins, G. W. 1991, *Ap. J.*, **378**, 307

Canizares, C. R., *et al.* 1987, *Astro. Lett. and Communications*, **26**, 87

Henderson, J. R., *et al.* 1990, *Phys. Rev. Lett.*, **65**, 705

Inal, M. K., and Dubau, J. 1987, *J. Phys. B.*, **20**, 4221

Laming, J. M. 1989, *Nucl. Instr. Meth.*, **B43**, 359

Laming, J. M. 1990a, *Ap. J.*, **357**, 275

Laming, J. M. 1990b, *Ap. J.*, **362**, 219

Margon, B. 1984, *Ann. Rev. Astron. Astrophys.*, **22**, 507

Matsuoka, M., Takano, S., and Makishima, K. 1986, *M. N. R. A. S.*, **222**, 605

Watson, M. G., Stewart, G. C., Brinkmann, W., and King, A. R. 1986 *M. N. R. A. S.*, **222**. 261

Monte-Carlo calculations of X-ray spectra for a source with spherical or planar circumstellar matter

D.A. Leahy, J. Creighton

Dept. of Physics and Astronomy, University of Calgary

ABSTRACT

The propagation of X-rays through a cloud of material surrounding a point X-ray source is studied with Monte Carlo methods. Spherical and planar matter distributions, and the cases of a beamed or unbeamed X-ray source are considered. Absorption, scattering and iron and nickel K-shell fluorescence processes are included. Emergent X-ray spectra are calculated and equivalent widths of the fluorescent lines derived.

1 INTRODUCTION

X-ray spectra serve as a diagnostic for many properties of the source of the X-ray emission. Among the diagnostic tools available is the use of fluorescent line strengths to probe the properties of matter surrounding the X-ray source, where the line production occurs. Monte-Carlo methods have been used to study cases where spherical symmetry is not applicable, and some results are presented here.

2 THE MODELS

2.1 Source Models

The X-ray source was taken to have a continuum spectrum over the energy range 1-30 keV. The spectrum was assumed to be a power law of the form

$$N(E)\,dE = const.\,E^{-\alpha}dE$$

Results are presented here for an energy spectral index of -1. The source can be either isotropic, or beamed. For the beamed source the angular distribution, in spherical polar coordinates (θ_s, ϕ_s) with polar axis along the source axis, was taken as:

$$N(\theta_s, \phi_s)\,d\omega = f(\theta_s)\,d\omega$$

For the calculations here, we used:

$$f(\theta_s) = \cos^4(\theta_s)$$

2.2 The Surrounding Matter Distribution

The Morris and McCammon (Morris and McCammon 1983) cross-section was adopted for the absorption cross-section of the matter. The total cross-section included absorption and electron scattering. For electron scattering the Thomson cross-section was used. The number of electrons per H atom was 1.2131, determined using the element abundances given in Dalgarno & Lazer (1987).

Iron and nickel K-shell fluorescence was included. The iron Kα line energy was taken as the weighted mean (6.399 keV) of KαI and KαII line energies. Similarly, the iron Kβ energy is 7.085 keV, and the nickel Kα energy is 7.472 keV (nickel Kβ was not included as its intensity is less than nickel Kα by a factor of 0.135). Atomic data for the line energies and cross-sections was taken from CRC Tables. The fluorescent yields for iron and nickel were taken as 0.34 and 0.41 (Bambynek et al. 1972).

The matter distribution around the compact source was taken to have either spherical symmetry or planar geometry. For the case that the matter was spherically symmetric, three different distributions were used: a constant density (from r=0 to r=R); a $1/r^k$ distribution with zero density inside a specified inner radius (calculations were done for k=2); and an exponential ($n(r) = \exp(-\beta r)$). For the case of planar geometry, the matter distribution was taken to be exponentially decreasing on both sides of a central plane: $n(z) = \exp(-\beta |z|)$, with z= height above the midplane.

3 RESULTS

3.1 Isotropic Source and Spherically Symmetric Matter

The equivalent widths for the three fluorescent lines was calculated from the simulated spectra. There is a significant difference between source spectral index -1 and -2 since there are more photons above the Fe or Ni absorption edges for the -1 case. The $1/r^2$ and exponential matter distributions produce nearly the same results. They both give similar equivalent widths to the uniform distribution at low optical depth, but higher equivalent widths than the uniform distribution at large optical depth. In all cases the slope of the emergent spectrum is more positive than that of the source due to the preferential absorption of lower energy photons.

3.2 Beamed Source and Spherically Symmetric Matter

The outgoing photons were collected into spectra over ranges of polar angle. Equal cos(θ) intervals were taken between θ=0 to π/2. The limits on polar angle were chosen to give equal solid angle to the bins. The emergent spectra for optical depth of 0.2 are shown in Fig. 1.

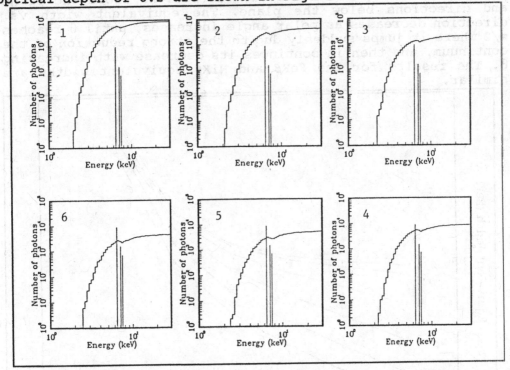

Fig.1 Monte Carlo calculated spectra for the case of a uniform density sphere, Thomson optical depth of 0.2, and a $cos^4(\theta')$ beamed source with photon spectral index of -1. The energy bin width increases logarithmically with energy. Spectra 1 through 6 are for viewing angle (θ) ranges of cos(θ)=1.0 to 0.833, 0.833 to 0.667, 0.667 to 0.5, 0.5 to 0.333, 0.333 to 0.167, and 0.167 to 0.

The main feature is a decrease of continuum intensity with polar angle, compared to nearly constant intensities for the fluorescent lines. This results in large equivalent widths near 90° for small to moderate optical depths (see Fig. 2), and large equivalent widths which are not as strongly angle dependent at large optical depths.

3.3 Isotropic Source and Planar Matter

Spectra were calculated for equal cos(θ) intervals between θ=0 to π. Emergent spectra were calculated for different viewing angles and for various source locations above the mid-plane. A sudden decrease in intensity is found between spectra for viewing angles just above the plane and just below the plane, resp. The equivalent width vs. optical depth for FeKα shows a clear separation between directions above the plane and directions below the plane. The equivalent width vs. direction decreases as polar angle increases, until θ reaches π/2 where it jumps suddenly due to the strong reduction of the continuum, and then it continues its decrease with increasing θ. The results for the FeKβ and NiKβ equivalent widths are similar.

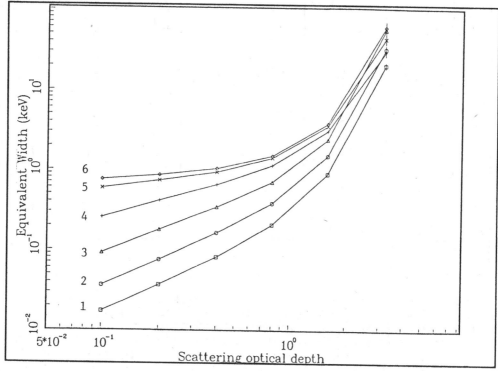

Fig.2 Equivalent width of Fe Kα flourescent line vs. Thomson scattering optical depth, for the case of a uniform density sphere and a cos⁴(θ') beamed source with photon spectral index of -1. (Cos(θ) ranges as in Fig.1).

REFERENCES

Bambynek,W., Crasemann,B., Fink,R., Frend, H., Mark,H., Swift,C., Price,R., and Rao,P. 1972 Rev.Mod.Phys., 44, 716.
Morrison, R., and McCammon, D., 1983, ApJ, 270, 119.
Spectroscopy of Astrophysical Plasmas, ed. Dalgarno & Lazer (Cambridge Univ. Press, Cambridge Mass., 1987).

STUDY OF MICROPINCH PLASMA PARAMETERS WITH TEMPORAL AND SPATIAL RESOLUTION BY MEASUREMENTS IN SOFT X-RAY SPECTRAL REGION

B.N.Mironov

Branch of Kurchatov Atomic Energy Institute,
142092 Troitsk, Moscow region, Russia

1. INTRODUCTION

The goal of this work is to determine some high temperature plasma parameters by soft X-ray measurements with spatial and temporal resolution. The experiments were carried out on Z micropinch discharge (low inductance vacuum spark-LIVS). The idea of these measurements is to study a temporal structure of X-ray pulses. Experimentally observed periodical changes of X-ray intensity $I(t)$ and transverse sizes of micropinch are associated with magnetic sound oscillations in the plasma.

An attempt is made to answer the following questions:
1. What are the transverse sizes, sound speed and ion density of plasma?
2. Is it important or not the self-absorption of line emission?

The data obtained in experiment are compared with the calculations [1]. It should be emphasized that the measurements have been made in soft X-ray region with the energy of quantum of $E_{h\nu} \simeq 1 keV$.

2. EXPERIMENTAL CONDITIONS

The experiments were performed with LIVS which is analogous to described in [2]. The typical discharge current was $I \simeq 200$ kA, current rise time $-\tau \simeq 1.2$ μs, energy stored in capacitor bank – $W_e \simeq 1.5$ kJ. Plasma was formed from the iron anode material. The X-ray emission of micropinch plasma was registered by means of high speed X-ray streak camera with resolution time of $\tau \simeq 20$ ps. The camera photocathode is sensitive in the spectral range 100eV ÷ 10^4 eV. The slit-shaped photocathode was placed perpendicular to Z-pinch axis. Three slits (their widths are 120 μm, 200μm and 250 μm) were mounted parallel to pinch axis between micropinch and camera. The mylar foil with thickness of 4μm and Al-coating of approximately 0.3μm was used as a filter and placed over all slits

(see Fig.1). A spatial resolution perpendicular to pinch axis was
obtained by this configuration.

3. EXPERIMENTAL DATA TREATMENT

In order to reduce the probability of errors all obtained
slide images were treated by two different methods. In the first
case the images were traced along time axis with relatively large
read-out window 200×400 μm^2, it was made for better signal to
noise ratio. Results are presented in Fig.3 (in Fig.3a different
slits and in Fig.3b different filters have been used to obtain
$I(t)$). In the second case the preliminary frequency filtering
of signals was made in 15 time points. The intensity distributions $I_i(x)$ behind all slits were obtained at the moments t_i,
where i=1,...,15. The spatial intensity distributions $I(x)$ at the
moment $t_1 = 1,17$ ns behind all slits are shown in Fig.2, where x-
the widths of intensity contours at half of maximum. The plots of
$I(t)$ ($I(t)$ - intensity values at half of maximum) for two slits
of 120 μm and 250 μm are presented in Fig.4, and the plots of
intensities contours widths $X(t)$ are displayed in Fig.5. In Fig.6
are shown the time histories of ratios I_3/I_1 and X_3/X_1 (where
I_3, X_3 and I_1, X_1 are values of intensity and contour widths at
half of maximum behind the slits of 250 μm and 120 μm respectively).

4. DATA INTERPRETATION

Let us consider two assumptions:

a) All the periods of intensity variations (T,T/2), along
with the period T_1 of the X-ray pulse as itself, are determined
by the sound wave double pass time (to the pinch axis and back,
Fig. 3).

b) We assume the Alfven velocity $v_A(r) \simeq const$. The validity
of these assumptions and the estimates of introduced errors are
beyond the scope of this paper. In our case $\beta \simeq 1$, so $V_A \simeq C_s$. Knowing
the oscillation period and plasma radius one can estimate V_A from
the formula $V_A = 2r/T$. The plasma transverse size R is obtained
from plots in Figs. 5,6. R is equal to $100 \div 150 \mu m$. Using assumption (a) we obtain $T_1 \simeq 2$ ns, $v_A \simeq (1 \div 2) 10^7$ cm/s, where T_1 is typical X-ray pulse duration. Then the observed period of T=0.5 ns,
in accordance with the assumption (b), must correspond to plasma

size $r_1 \simeq 30 \div 35$ μm, and the period T/2=0.25 ns – to $r_2 \simeq 15$ μm. Let us estimate plasma ion density. We shall substitute experimental data into equation

$$n_i = 4.84 \cdot 10^{20} \cdot I^2 \cdot A_i^{-1} \cdot r^{-4} \ [sm^{-3}]$$

where I – current (amps), T – period (s), A_i – mass number of element, r – pinch radius (cm). Taking into account that there are no substantial density changes during the pinch confinement stage, as evidenced by the constancy of periods T and T/2 (see Fig.3), let us estimate the ion density in the deep pinching stage. For r=1,5 μm the ion density equals to $n_i \simeq 4 \cdot 10^{23}$ cm^{-3}. This value is in a good agreement with calculations [1]. In Fig.6 the time dependence of ratio I_3/I_1 for slits of 250 μm and 120μm is depicted. The analysis of this ratio allows to distinguish three time phases I,II and III. For the time interval I when the source size is greater than the slit width (250μm), we believe that ratio $I_3/I_1 \simeq 4$ corresponds to a radiation from the plasma volume. When the ratio $I_3/I_1 \simeq 2$ (phase III) only plasma surface yields radiation. The deviations of measured ratios from theory are within the errors connected with the small signal amplitude at the beginning and at the end of process.

5. CONCLUSIONS

A relatively simple method of ion density measurements is demonstrated. The existence of several plasma regions with considerably different densities is shown. Also the magnetic sound wave speed is estimated. This method enables to obtain the nondiscrepant results, though the accuracy of ion density measurements is strong dependent on the accuracy of plasma transverse sizes measurements. The method may be used for inertial confinement systems with magnetic field at $\beta \simeq 1$ and $\tau > 2 \cdot r/V_a$, where τ – plasma confinement time.

REFERENCES

1. Koshelev K.N., Sidelnikov Yu.V., Vikhrev V.V. Micropinch as a spectral source of highly ionized atoms: Sov. Prepr. Institute for spectroscopy Acad. of Sci. №1, Troitsk, 1985.
2. Veretennikov V.A., Polukhin S.N., Semionov O.G., Sidelnikov Yu.V., Sov. J. Plasma Phys. 1981, v..7, №6, p.1199.

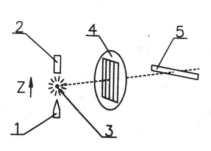

Fig.1. The experimental scheme: 1– anode Fe, 2– cathode, 3– X–rays from micropinch plasma, 4– the set of slits with d=100+250 μm, 5– the photocathode of X–ray streak camera.

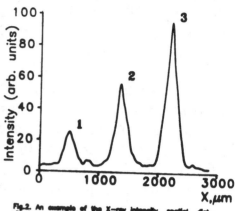

Fig.2. An example of the X–ray intensity spatial distribution at a streak camera photocathode behind slits with different widthes (1– 120 μm, 2– 200 μm, 3– 250 μm) at any moment.

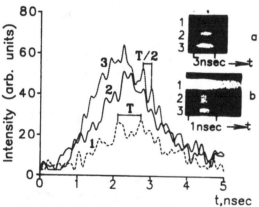

Fig.3. The time histories of the X–ray intensity from Fe micropinch plasma measured behind three slits with width 120 μm(1), 200 μm(2) and 250 μm(3); 3a– an X–ray streak camera records behind three slits; 3b– behind three filters: 1– mylar with thickness d=4 μm, 2– Be d=50 μm and 3– Al d=4.1 μm. Anode material – Al.

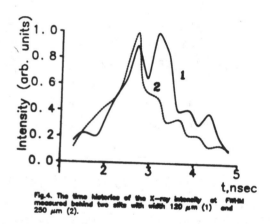

Fig.4. The time histories of the X–ray intensity at FWHM measured behind two slits with width 120 μm (1) and 250 μm (2).

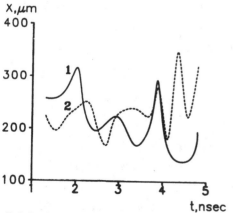

Fig.5. The time histories of a plasma size X(μm), measured at FWHM of X–ray spatial distribution behind two slits with width 120 μm (1) and 250 μm (2).

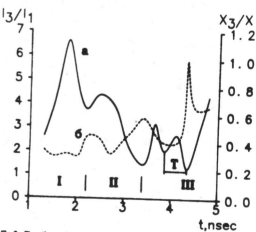

Fig.6. The time histories of values I_3/I_1 (a) and X_3/X_1 (b), where I_i is a X–ray intensity behind i^a slit and X_i =FWHM$_i$.

NEW TRENDS IN STARK BROADENING OF MULTICHARGED ION SPECTRAL LINES IN PLASMAS

Ya.Ispolatov, E.Oks

Physics Department, Auburn University, Auburn, Al 36849-5311, USA

Part 1. A consistent theory of spectral line broadening by dense plasmas.

There exist numerous papers [1-4] devoted to the theoretical description of spectral line broadening for multicharged ions in dense plasmas. However a significant ambiguity still remains being manifested as drastic discrepancies between the line shape calculations performed by different groups of scientists for the same plasma and ion parameters (as a result of a neglect or an incorrect treatment of important effects, some of which were indicated in [4,6]). To eliminate these shortcomings of previous theories we develope the following advanced model.

Interaction with plasma electrons in the vicinity of a radiating ion. A mean time τ_e of plasma electron relaxation to a local equilibrium state (of order of several plasma electron frequencies), a radiational decay lifetime τ_r of an excited state of a radiator and a characteristical time τ_i of varying of the ion microfield usually obey a relationship $\tau_e < \tau_r < \tau_i$ [7]. Besides, a number of plasma electrons simultaneously interacting with a radiating ion is much greater than unity. For example, a mean number of plasma electrons situating "inside" the 4s electron orbit of a hydrogen-like ion with a nuclear charge $Z_r \sim 10$ is of order of 10 for plasma parameters $N_e \sim 10^{22} cm^{-3}$, $T_e \sim 500\ eV$. In terms of the standard electron impact broadening theory it means that there exist multiply overlapped penetrating (monopole) collisions. These estimates justify our treatment of plasma electrons as a canonical ensemble (with fixed number of particles) in a state of local thermodynamical equilibrium under fixed states of bound electrons and stochastic configuration of plasma ions.

This "statistical" approach in explicit or implicit forms was used in many calculations [8,9] dealing with energy level shifts of a radiating ion due to a linear (Debye-like) or nonlinear screening of a nuclear charge by plasma electrons. These calculations are usually based on self-consistent determination of bound electron states and plasma electron densities in some region around a radiating ion ("ion sphere" model, "average atom" model).

A radiative lifetime of an ion is much smaller than a characteristic time of the plasma ion subsystem, and each radiating ion is perturbed by a stochastic (rather than equilibrium) ion configuration. Therefore while calculating a nonlinear screening of bound states and a ion-ion pair potential we use one- and two-centered models respectively. Plasma electrons belonging to one or two selected ions are separated from the bulk plasma electrons by a boundary (a generalized "ion sphere model") and are treated as a canonical ensemble (or grand canonical if not imposing the charge neutrality condition) in a bulk plasma thermostat.

An energy of a radiative transition (a photon energy) is calculated as a difference of free energies (rather than total energies) of this ensemble with bound electrons in excited and ground states, respectively.

To describe bound states and a distribution of free electrons we utilize the Density Functional Theory (DFT) for finite temperatures [9]. Bound state occupation numbers are fixed and correspond to an optical transition under consideration. Plasma temperature is high enough for nondegeneracy of free electrons, so that we treat them in the quasiclassical approximation. We use the Local Density Approximation (LDA) for finite temperatures to describe an exchange and correlations between electrons by some local potential [10]. The most important effect beyond the DFT calculation is a finite linewidth determined by an imaginary part of the RPA self energy of both excited and ground levels, and by cross-terms (vertex corrections) [11]. It can be shown that under some approximations this contribution can be identified with the classical expression for electron impact broadening operator.

In some papers a "quasistatic electron density fluctuations" approach is proposed for spectral line width calculations [12]. This approach might be justified for far wings of electron broadened spectral line where $\Delta\omega > \omega_{Pe}$. However even in these far wings the ratio of a contribution of quasistatic electron fluctuations to a quasistatic ion broadening is of order $Z^{-1} \ll 1$.

Interaction with plasma ions. An influence of plasma ions on a radiating ion is treated under the standard quasistatic microfield approach however including a spatial inhomogeneity of the ion microfield (a quadrupole interaction of a radiator with the ion microfield is taken into account in addition to a dipole interaction).

Ion-ion pair potentials and a single ion electric field strength are input data for calculations of a distribution function of the ion microfield and its spatial derivatives. We employ the DFT approach [13] to calculate pair ion-ion characteristics. It allows to avoid commonly used unjustified assumptions (Debye-like potentials, APEX scheme [14]). Another significant advantage of our approach compared to the spherically symmetrical electron pileup of the paper [13] is the two-centered ion model with a self-consistent spatial distribution of plasma electrons (see Fig.1).

Since the quadrupole interaction is smaller than the dipole interaction, we consider the former at the "constrained average" approximation: only average values of microfield spatial derivatives $\langle \partial F_i / \partial x_j \rangle_F$ are calculated for each microfield strength F

To take into account relativistic effects essential for multicharged ions we use the $1/Z_r$ expansion of the total relativistic Hamiltonian [15] in the basis formed by the Kohn-Sham radial and LS angular wave functions [11].

Details of calculations. Calculation of ion emission spectra including all aforementioned effects for a given microfield strength F are performed by a numerical diagonalization of the perturbed Hamiltonian in the finite basis of Kohn-Sham radial and LS angular wavefunctions corresponding to the same principal quantum number n. An influence of neighboring groups of levels with principal quantum numbers $\tilde{n} = n \pm 1$ is taken into account as a next order correction.

The last steps are convolutions with the plasma microfield distribution function W(F) and with the Maxwell distribution of ion velocities. The final line profiles obtained in this self-consistent approach differ significantly from the line profiles calculated under previous models (see Fig. 2).

Part 2. Intra-Stark spectroscopy of dense plasmas.

The term "intra-Stark spectroscopy" is used to describe a set of resonant phenomena in spectral line profiles which become effective when a radiating atom (or ion) interacts simultaneously with a static electric field $\vec{F}$ and a quasimonochromatic electric field $\vec{E}(t) = \sum \vec{E}_j \cos(\omega T + \varphi_j)$. The static (or more precisely, quasistatic) electric field $\vec{F}$ represent low-frequency plasma turbulence and/or ion microfields. The quasimonochromatic electric field $\vec{E}$ may represent Langmuir oscillations at a frequency $\omega \approx \omega_p = (4\pi e^2 N_e/m_e)^{1/2}$ where N_e is an electron density, or some other (e.g., lower-hybrid) waves developed in the plasma in such a way that their characteristic frequency ω is much greater than a characteristic width $\delta\omega$ of their frequency band. The field $\vec{E}$ may represent as well some laser or microwave radiation at a frequency $\omega > \omega_p$, which penetrates into a plasma from the outside or is generated within a plasma. In any case, it is essential that in the presence of such fields some local peculiarities arise within the quasistatic Stark profile: dips or depressions at definite separations $\Delta\lambda^{dip}$ from the line center λ_0 [16-22].

In various plasma experiments [16-20, 23-25] dips or depressions in hydrogen spectral line profiles were reliably identified with theoretically predicted ones and were successfully used for the measurement of parameters of quasimonochromatic electric fields and of plasmas.

Using the latest development of the theory of the dip effect [28] we re-analyze experimental line profiles of CVI and NeX from laser-produced plasmas of electron densities $10^{21} \div 10^{23}$ cm^{-3} published by various authors [27-29]. It turnes out that observed local peculiarities in these experimental profiles may be indeed identified with theoretically expected positions of dips (Fig. 12-14). In Fig.3 such an identification is shown for the NeX L_β - line registered in [28] at $N_e = 1.9 \times 10^{22}$ cm^{-3}; theoretical dip positions are indicated by vertical lines. The presence of the dips is a manifestation of a very strong oscillatory electric fields at the plasma electron frequency in the experiments [27-29].

Thus new <u>diagnostic methods based on intra-Stark spectroscopy</u> may work even <u>for super-high density plasmas</u>. In particular the experimental determination of a distance between dips provides an additional method for measuring high electron densities. The experimental determination of a dip halfwidth allows an estimate of an amplitude of the oscillatory (e.g. Langmuir) fields as well.

References.

1. H.R.Griem, M.Blaha, P.C.Kepple, Phys. Rev.A, v.41, 1990, p.5600.

2. J.W.Dufty, D.B.Boercker, C.A.Iglesias, J. Quant. Spectr. Rad. Transfer, v.44, 1990, p.115.

3. R.F.Jouce, L.A.Woltz, C.F.Hooper Jr., Phys. Rev.A, v.35, 1987, p.2228.

4. V.P.Gavrilenko, Ya.O.Ispolatov, Opt.Spectrosc.,v.68, 1990,p.583.

5. I.M.Gaisinsky, E.A.Oks, J. Quant. Spectr. Rad. Transfer, v.41, 1989, p.235.

6. I.M.Gaisinsky, E.A.Oks, S.E.Frid, J.Appl.Spectrosc.,v.52, 1990,p.128.

7. H.Nguen, M.Koenig, D.Benredjem, M.Caby, G.Couland, Phys. Rev.A, v.33, 1986, p.1279.

8. D Salzmann, H. Szichman, Phys. Rev.A, v.35, 1987, p.807.

9. W.Kohn, L.J.Sham, Phys. Rev., v.140, 1965, p.1133.

10. F.Perrot, M.W.C.Dharma-Wardana, Phys. Rev.A, v.30, 1985, p.2819.

11. F.Perrot, M.W.C.Dharma-Wardana, Phys. Rev.A, v.29, 1984, p.1378.

12. T.Błeński, B.Cichocki, Phys. Rev.A, v.41, 1990, p.6973.

13. M.W.C.Dharma-Wardana, F.Perrot, Phys. Rev.A, v.33, 1986, p.3303.

14. C.A.Iglesias, J.L.Lebowitz, D. MacGowan, Phys. Rev.A, v.28, 1983, p.1667.

15. L.A.Vainstein, U.I.Safronova, Physica Scripta, v.31, 1985, p.519.

16. E.Oks "Plasma Spectroscopy: The Influence of Microvave and Laser Fields", Springer, Heidelberg, 1992.

17. E.Oks "Spectroscopy of Plasmas Containing Quasimonochromatic Electric Fields", Energoatomizdat, Moscow, 1990.

18. E.Oks, 14-th Int.Symp. on the Physics of Ionized Gases. Invited Lectures. Sarajevo, 1988,p.435.

19. A.I.Zhuzhunashvili, E.A.Oks, Sov.Phys.JETP, v.46, 1977, p.1122.

20. E.A.Oks, V.A.Rantsev-Kartinov, Sov.Phys.JETP, v.52, 1980, p.50.
21. V.P.Gavrilenko, E.A.Oks, Sov.Phys.JETP, v.53, 1981, p.1122.
22. V.P.Gavrilenko, E.A.Oks, Sov.J.Plasma Phys., v.13, 1987, p.22.
23. K.H.Finken, R.Buchnwald, G.Bertschinger, H.-J.Kunze, Phys.Rev.A, v.21, 1980, p.200.
24. K.H.Finken, Forschr.Phys., v.31, 1983, p.1.
25. E.Bertshinger, Messungen von VUV Linien an einem dichten Z-Pinch-Plasma:
    Ph.D.Thesis, Bochum 1980, unpublished.
26. E.A.Oks, St.Böddeker, H.-J.Kunze, Phys.Rev.A, v.44, 1991, p.8338.
27. C.C. Smith, N.J. Peacock, J.Phys.B., v.11, 1978, 2749.
28. K.B. Mitchell, D.B.von Husteyn, G.H.McCall, P.Lee, H.R.Griem, Phys.Rev.Lett., v.42, 1979, p.232.
29. B.Yakobi, D.Steel, E.Thorsos, A.Hauer, B.Perry, Phys.Rev.Lett., v.39, 1977, p.1526.

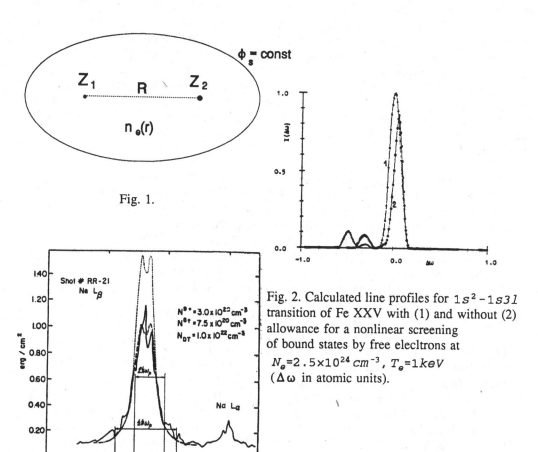

Fig. 1.

Fig. 2. Calculated line profiles for $1s^2-1s3l$ transition of Fe XXV with (1) and without (2) allowance for a nonlinear screening of bound states by free elecltrons at

$N_e = 2.5 \times 10^{24}\, cm^{-3}$, $T_e = 1 keV$

($\Delta\omega$ in atomic units).

Fig. 3.

SPATIALLY RESOLVED X-RAY SPECTRA OF MICROPINCH PLASMA

A. M. Panin

Brigham Young University, Provo, Utah 84604, USA

The spatially resolved x-ray spectra of a low-inductance vacuum spark micropinch plasma (LIVS MP) were registered with 5-7 μm resolution using a high apperture focusing Hamoss-type spectrograph [1] in the λ = 1.7-2.5 Å waveband (hν = 5-8 keV). We investigated the spectra of Fe and Ti ions. The spectral waveband covering the K_α and K_β regions is very suitable for the plasma diagnostics, as it contains both the lines of "hot" K ions (H-, He-like resonance transitions) and of "cold" L ions (K-shell transitions in Li...Ne-like ions), as well as the lines of inner shell transitions in "very cold" slightly ionized atoms being exited by electron beams.

The parameters of experimental LIVS set were the following: C = 16 μF, L = 30 nHn, U = 10-15 kV, I = 100-150 kA. The x-ray radiation was reflected by a cylindrical mica spectrograph (R = 60 mm, 2d = 19.9 Å, order of reflection n = 5) and was registered by a film placed on the spectrometer axis z. The detector plane coincides with both the mictopinch discharge axis x and the spectrograph axis z (see Fig. 1). Registered spectra resemble narrow (10-200 μm) strips along the dispersion axis z. Spectra were registered both for one discharge of LIVS only and with integration for 10-20 discharges. The images on the film did not overlap because of irreproducibility of "hot points" in space from discharge to discharge. At least 5-7 discharges could be integrated untill the spectrograms on the detector film began to interfere. Typical spectrogram of MP plasma in the vicinity of resoinance lines of H- and He-like Ti ions is shown in the Fig. 2.

At first sight, a spatial resolution in the described geometery is limited by aberrations of the spectrograph [1,2]. The aberrations could be too large (100-200 μm) as the "hot point" MP sometimes occurs too far from the discharge axis and from the detector plane (1-2 mm). Nevertheless, we found that the sharpness δ of the edge of the spectral image in the x direction was much smaller than the estimated aberration blur, and some spectra have as sharp an edge as $\delta \approx$ 4-6 μm which is close to the spatial resolution of the film used. That should take place only if the source is smaller than the measured sharpness δ. Thus, sharpness of the spectrogram edge in our geometry could serve as a measure to estimate plasma source size l_{pl} (along the x direction).

It was found that the bright lines of the K ions of Fe and Ti are regularly emitted from the small 5-10 μm "hot spots"(in the geometry of the experiment the axial dimension of the MP plasma was measured). The lines of F- and Ne-like ions are emitted from 70-200 μm "clouds" surrounding the appropriate "hot spots" (see Fig.3 for Fe ions). Sometimes these "clouds" of the relatively cold plasma are not followed by bright "hot spots". This fact reflects the lack of deep collapse of plasma. The spectrograms also show the presence of the emission of K_α and K_β lines of the slightly ionized plasma emitted by extensive (l_{pl} = 1-4 mm) clouds. This is surprising because the high energy (tens kV) electron beam required to excite K radiation in a cold plasma would be very wide and would penetrate too deep in the plasma. A more reasonable explanation is that this K radiation occurs at the very beginning phase of discharge, when the discharge gap is still nonconductive. The electrons from a trigger cathode plasma (we use traditional cathode trigger discharge to initiate main discharge) are accelerated by the full potential difference of the gap and strike anode, evaporating it and exiting K radiation in the expanding anode vapor.

The axial motion of MP plasma is free of magnetic braking. Thus we can estimate the lifetime of differently ionized plasma in the MP assuming the velocity v_i of a plasma motion to be thermal. The results of this estimation are shown in Table 1.

There is no direct measurement of these times in a MP discharge (especially for the "hot" plasma phase of a MP collapse). The problem of a direct temporal scanning of the micropinch x-ray radiation with a picosecond resolution is difficult to solve, because a MP plasma collapse time moment is not reproduced from discharge to discharge. A 50-100 nsec time scattering in collapse to occur is regularly reported by many authors. Some estimations of plasma dynamics show close times for different ions in LIVS plasma (see [3]). In [4] the lifetime of Ne-like Fe was measured to be less than the detector response time of 2 nsec.

Measured plasma dimensions reflect the evolution of MP plasma collapse; hence, the radiation of "hot" and "cold" plasma arises in different moments of time.

It would be interesting to measure the dimensions of a hot MP plasma in "light" of two types of lines: (i).those lines of [He]- or [Li]-like ions which are the transitions from the levels populated by dielectronic recombination, and (ii). those transitions from levels exited directly from the ground state by electron impact. This would give more information for understanding of MP plasma kinetics and plasma evolution. This problem still has more questions than answers.

REFERENCES

1. V. L. Hamoss, Ann. Phys. 17, 716 (1933).

2. B. Yaakoby, R. E. Turren, H. W. Schopper and P. O. Taylor, Rev. Sci. Instr., 50, 1609 (1979).

3. E. V. Aglitsky, P. S. Antsiferov, and A. M. Panin, "Experimental determination of the "hot point" plasma lifetime in the low inductance vacuum spark", Plasma physics (Sov.), 11, 10, 1266 (1985).

4. V. S. Veretennikov, Dissertation, Lebedev's General Physics Institute, Moscow (1988).

Ele-ment	Ion	l_{pl}	T_i (eV)	$v_i(10^6$ sm/s)	τ_i, ns
Ti (z=22)	[H],[He]	5-15μ	≅1200	25	.04-.06
	[F],[Ne]	200-400μ	≅ 70	5	3-10
	[Ar]...[Mn]	1-3 mm	≅ 10	2	40-100
Fe (z=26)	[H],[He]	4-10 μ	≅2000	32	.02-.05
	[F], [Ne]	70-100 μ	≅100	6	1-2
	[Ar]...[Sc]	1-3 mm	≅ 10	2	20-60

Table 1. Lifetime τ_i **for different ions in the micropinch plasma (for Fe and Ti plasmas) as estimated using measured plasma axial dimensions.**

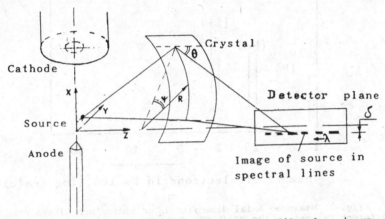

Fig. 1. Experimental arrangement. Detector film plane is *xz*. Spectrograph radius *R* = 60 mm, aperture 2ψ = 40° (mica, 2d/n = 19.9/5 Å). Spatial resolution δ ≈ 5 μ.

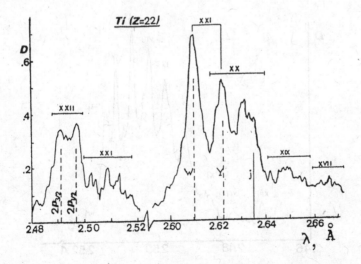

Fig. 2. Spectrum of Ti plasma being detected with the above Hamoss geometry. Resonant doublet of H-like Ti (3/2, 1/2) is not overlapped by K_β transitions in cold plasma (compare to Fig. 4 for Johann geometry spectra), because the latter are emitted by large (1-3 mm) plasma clouds.

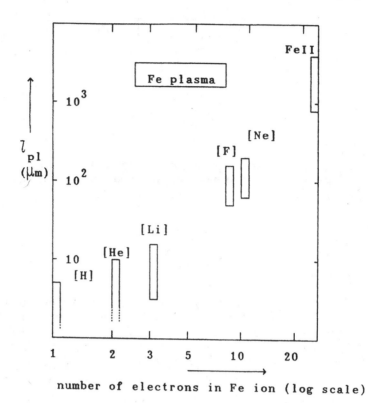

Fig. 3. Measured axial dimension l_{pl} of micropinch plasma for
different Fe ions in the MP (LIVS) discharge.

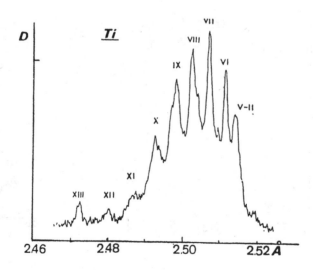

Fig. 4. Spectrum of Ti micropinch plasma being registered by
the Johann spectrometer in the same as in Fig. 1 waveband. Lack of
spatial resolution results in total blinding of Lyman α doublet of
H-like Ti.

Ar XVII X-ray Lines Emitted by Solar Flares

K. J. H. Phillips[1], F. P. Keenan[2], L. K. Harra[2], S. M. McCann[2]

[1]Astrophysics Division, Rutherford Appleton Laboratory, Chilton, Didcot, Oxfordshire OX11 0QX, England, UK
[2]Department of Pure and Applied Physics, The Queen's University of Belfast, Belfast BT7 1NN, Northern Ireland, UK

ABSTRACT

Recent calculations of electron impact excitation rates in helium-like argon (Ar XVII) have been used to derive emission line intensities for the resonance ($1s^2\,^1S_0 - 1s2p\,^1P_1$), intercombination ($1s^2\,^1S_0 - 1s2p\,^3P_{1,2}$) and forbidden ($1s^2\,^1S_0 - 1s2s\,^3S_1$) lines that appear in the X-ray region (~ 4 Å). These have been combined with calculations of nearby dielectronic satellites of Ar XVI to synthesize spectra that can be compared with observations. The synthetic spectra are sensitive to electron temperature T_e but not electron density unless extremely large ($> 10^{14}$ cm^{-3}). Comparisons have been made using observations taken during solar flares with the Flat Crystal Spectrometer (part of the X-ray Polychromator) on *Solar Maximum Mission* and with spectra from the Alcator tokamak. The observed spectra show good agreement with the theoretical spectra, and demonstrate the feasibility of using Ar XVII line ratios for determining T_e.

1 Introduction

Ar XVII emission lines in the spectra of solar flares and tokamaks arise from electron collisional excitation of $1s\,2l$ levels in helium-like argon ions from the ground ($1s^2\,^1S$) state. There are four prominent lines, due to transitions $1s^2\,^1S_0 - 1s2p\,^1P_1$ (resonance line), $1s^2\,^1S_0 - 1s2p\,^3P_{1,2}$ (intercombination lines) and $1s^2\,^1S_0 - 1s2s\,^3S_1$ (forbidden line); in the notation of Gabriel (1972: MNRAS, 169, 99), w, y, x, and z respectively. As with the corresponding lines of other He-like ions, electron temperatures T_e can be estimated from the intensity ratio, often called the G-ratio, defined by $[I(x) + I(y) + I(z)]/I(w)$. For low-$Z$ He-like ions, the so-called R-ratio, $I(y)/I(z)$, is sensitive to electron densities N_e but this is not the case for Ar XVII for the densities expected for solar flares or tokamaks.

For He-like ions with larger atomic number, dielectronic satellite lines due to Li-like ions make an increasingly important contribution to the spectrum of the He-like ion, and in general have sizable intensities for argon. The satellites can be formed both by dielectronic recombination and by inner-shell excitation; in either case, the intensity ratio to the w line of the He-like ion is a decreasing function of T_e. The theory is given by Gabriel (1972) and others.

In this work we outline calculations of the He-like Ar line intensities, derived from electron collision rates using the R-matrix code, and the Li-like Ar satellite line intensities, derived from the Cowan Hartree–Fock code. Synthetic spectra formed from these calculations are described as is comparison with observed spectra from solar flares and tokamaks.

2 The Calculations

The electron excitation rate coefficients were derived from collision strengths calculated using the R-matrix code (Keenan, McCann and Kingston 1987). The model Ar XVII ion used in this calculation had 23 $1s\,nl$ states with $n < 6$ and $l < 3$. Other atomic data needed to calculate level populations (e.g. energies of levels, A-values, recombination coefficients) were obtained from other published work, while a statistical equilibrium code was used to generate the line intensities and hence intensity ratios as a function of electron temperature T_e. There is no dependence on electron density N_e for $N_e \le 10^{14}$ cm^{-3}.

The wavelengths and intensity factors of dielectronically formed satellite lines were calculated using the Cowan Hartree–Fock code. The HFR version of this code, which includes relativistic corrections, was used with scaling of the Slater parameters, following standard usage of this code (see Cowan 1981; Fawcett and Wilson 1990). Relatively intense satellites with n up to 7 were included in this preliminary version of the calculation (in a final version of this work all allowed satellites up to $n = 11$ will be included). A few $n = 2$ satellites are also formed by inner-shell excitation, notably line q. Their intensities were calculated using scaled versions of the collision strengths given for Ca by Bely-Dubau et al. (1982).

Synthetic spectra were formed using these atomic calculations by adding together the intensities of all lines, broadened by thermal Doppler processes and an instrumental profile matching that of the Flat Crystal Spectrometer on *SMM*. Examples of spectra, with values of T_e that are widely different (8×10^6 and 20×10^6K) but for which the Ar XVII emissivity is a large fraction of its maximum value, are shown in Fig. 1. Over the range $8 \times 10^6 - 20 \times 10^6$K, the x, y and z lines all decrease in intensity somewhat relative to the w line; this may just be seen in Fig. 1 for the x and y lines, but for the z line the effect is masked by the fact that the intense dielectronic line j blends with z; the intensity ratio j/w (as with all dielectronically formed lines) decreases with increasing T_e.

3 Agreement with Observed Spectra

The Ar XVII X-ray lines have been observed during solar flares and from tokamak plasmas with high-resolution crystal spectrometers so that comparison with the calculated spectra presented here may be made.

The Ar XVII lines were observed in a wavelength scan performed by the Flat Crystal

Spectrometer on the SMM spacecraft during a flare on April 14, 1988. Although the best available spectrum from this or indeed any other solar spectrometer the observed spectrum is somewhat weak. However, general agreement of the principal Ar XVII lines and the more intense satellites was achieved with a synthetic spectrum calculated for a temperature $T_e = 8 \times 10^6$K.

Much better agreement was obtained with a spectrum observed with an instrument viewing the Alcator C tokamak. The figure shown here shows a spectrum from the paper by Källne *et al.* (1985). T_e and N_e were measured with diagnostic instrumentation, not from line ratios. Using a value of T_e very close to the measured one, a theoretical spectrum was obtained closely resembling the observed. In this case, several of the individual $n = 2$ satellites appear clearly, and their observed intensities are very close to those in the theoretical spectrum.

It is concluded that the present calculations of Ar XVII intensities closely match available solar-flare and tokamak observations.

REFERENCES

Bely-Dubau, F. et al. 1982, MNRAS, 201, 1155

Cowan, R. D. 1981, *The Theory of Atomic Structure and Spectra*, University of California Press

Fawcett, B. C. and Wilson, M. 1990, *A User's Guide for Atomic Codes on the Cray Supercomputer at RAL*, Rutherford Appleton Laboratory Report RAL-90-042

Gabriel, A.H. 1972, MNRAS, 169, 99

Källne, E., Källne, J., Marmar, E. S., and Rice, J. E. 1985, Phys Scr, 31, 551

Keenan, F. P., McCann, S. M. and Kingston, A. E. 1987, Phys Scr, 35, 432

Phillips,K.J.H., et al. 1982, ApJ, 256, 774.

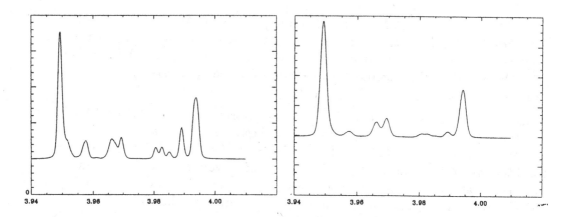

Figure 1. Calculated Ar XVII spectra, including Ar XVI dielectronic satellites, in the range 3.94 – 4.01 Å. The spectra were calculated for temperatures T_e of 8×10^6 (left) and 20×10^6K. There is a decrease in the G ratio over this range, but the more evident change is the increase (relative to the resonance line at 3.945 Å) in the satellite line intensities.

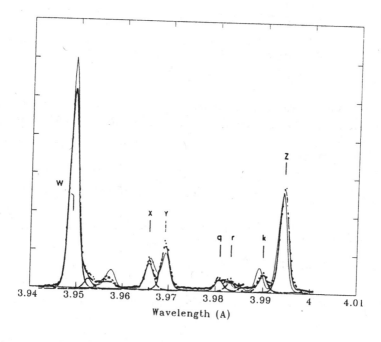

Figure 2. Comparison of theoretical Ar XVII spectrum (thin line) with an observed spectrum from the Alcator tokamak (Källne et al. 1985) (thick line with dots). Principal Ar XVII lines and dielectronic satellites are indicated, using notation of Gabriel (1972).

Gamma Ray and Neutron Spectroscopy with COMPTEL on the *Compton* Gamma Ray Observatory

J.M. Ryan[3], H. Aarts[2], K. Bennett[4], H. Bloemen[2], R. Diehl[1], A. Connors[3], H. Debrunner[5], C. deVries[2], W. Hermsen[2], G. Lichti[1], J.A. Lockwood[3], M. McConnell[3], D. Morris[3], V. Schönfelder[1], H. Steinle[1], A.Strong[1], B.N. Swanenburg[2], B.G. Taylor[4], W.R. Webber[3], C. Winkler[4],

[1]Max Planck Institut für Extraterrestrische Physik, Garching, Germany
[2]Research Organization of the Netherlands, Leiden
[3]Space Science Center, University of New Hampshire, Durham, NH 03824
[4]Space Science Division, ESTEC, Noordwijk, The Netherlands
[5]University of Bern, Bern, Switzerland

Abstract

COMPTEL on the *Compton* Gamma Ray Observatory in addition to producing images is also capable of measuring γ-ray spectra from cosmic sources as well as the Sun. By virtue of its imaging properties, the instrument has unique analyzing capabilities for a NaI based γ-ray detector. In constructing a γ-ray image of an astrophysical source it is possible to select only fully absorbed photons, thereby simplifying the instrument response and reducing the ambiguity of the spectral analysis. In a manner similar to that of γ-ray measurement, neutrons from the Sun can also be analyzed. The measured γ-ray (and neutron) spectra from astrophysical sources reflect upon the composition and spectra of energetic particles producing the γ-rays as well as the composition of the target medium. Recent measurements of solar flare γ-ray and neutron spectra illustrate the technique and the underlying physics probed through γ-ray and neutron spectroscopy.

Introduction

Spectroscopic studies of high energy radiation of astrophysical sources such as pulsars, quasars, black hole candidates, γ-ray bursters and solar flares provide information about the spectrum and composition of the energetic electrons and ions at these remote sites.

γ-ray spectra, depending on the source, are varying combinations of continuum and line spectra. Continuum spectra originate from energetic electron distributions undergoing either bremsstrahlung, synchrotron radiation or inverse Compton scattering. Line features in electron-dominated spectra can arise from annihilation radiation or through cyclotron lines, quantized synchrotron emission in intense magnetic fields ($\sim 10^{12}$ G).

Nuclear dominated spectra can either be composed of sharp or Doppler-broadened lines, or a combination of both. Narrow lines are present in the spectra of supernovae (e.g. ^{56}Co, ^{56}Fe) and solar flares. Doppler-broadened lines can be seen in solar flares. They arise from the interactions of accelerated heavy ions with the hydrogen in the solar atmosphere, whereas the narrow lines in solar flares originate from the collisions of accelerated protons with cold, heavy nuclei.

The Instrument

COMPTEL not only images cosmic sources, but measures their spectra in the difficult energy range of 1 to 30 MeV (Schönfelder *et al.* 1984). A schematic of the instrument is shown in Figure 1. The basic principle of both detections is the same; that is, the neutral particle elastically scatters in the low Z upper detector (NE213A, $CH_{1.286}$) and then scatters again in the lower detector (NaI). The energy deposit in the upper detector is measured with photomultiplier tubes, while the lower detector similarly records the energy of the scattered γ-ray. Recoil neutron energies are measured through the time-of-flight of the neutron. The energy deposits in the upper detector plus the energy measurements of the recoil γ-ray or neutron yields the scatter angle ϕ, when calculated with the appropriate kinematic formula (Compton or n-p scattering). The incident direction of the γ-ray or neutron is thereby restricted to a circle on the sky. Various statistical methods are employed to use this information to construct images (Diehl *et al.* 1992). However, it should be noted that as a consequence of the scattering law the spatial and energy information are interdependent, that is, the image data can be used to extract precise energy information. This can be seen in Figure 2.

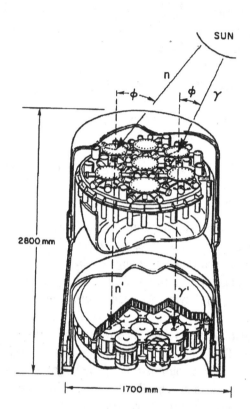

COMPTEL
IMAGING COMPTON TELESCOPE

Figure 1. Schematic of COMPTEL showing the interactions of γ-rays and neutrons. The scatter angle ϕ for γ-rays is given by $\phi = \cos^{-1}(1 - \varepsilon/E_2 + \varepsilon/(E_1 + E_2)$; where $\varepsilon = 511$ keV. The non-relativistic hard sphere, neutron-proton scattering formula is $\tan^2 \phi = E_1/E_s$.

In Figure 2 is a measured spectrum of a ^{24}Na calibration source accumulated from event whose the scatter angle is required to be consistent with the location of the source. The important feature to note is the strong suppression of the Compton tail in the spectrum, thereby yielding a spectral response more like X-ray scintillation detectors. Typically, γ-ray spectra are difficult to de-convolve at higher energies because of the more complicated scintillator response (escape peaks and Compton tails). This problem is ameliorated by utilization of the imaging information.

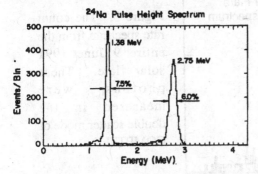

Figure 2. The resultant count rate spectrum for a ^{24}Na γ-ray calibration source when the measured scatter angle is required to be consistent with the true source direction. Most important is the significant suppression of the Compton tail.

Spectral information is also available for γ-ray bursts and solar flares through the use of the COMPTEL Burst system (Winkler *et al.* 1986). Two NaI D2 detectors are used in an single-detector mode to measure spectra of bursts and solar flares when triggered by the BATSE experiment on the Observatory. The energy range is 100 keV to 10 MeV with an effective field-of-view of ~ 2.5 sr and an energy resolution of ~ 10% at 662 keV.

The Solar Flare of 9 June 1991

The analyzing capability of COMPTEL is illustrated in the remaining figures with data from the X class solar flare which occured 9 June 1991. The flare was of ~ 10 minutes duration from 0136 to 0146 UT. Using only those events consistent with the solar direction, a count rate spectrum was created (Figure 3). Some important spectral features are the deuterium formation line at 2.223 MeV, a line at 1.6 MeV probably deriving from ^{20}Ne and a broad shelf above 4 MeV arising from both broad and narrow lines of C, N and O. Some undetermined fraction of the spectrum (mostly < 2 MeV) arises from bremsstrahlung of primary accelerated electrons.

In Figure 4 is the intensity-time profile of energetic neutron emission from the 9 June flare detected by COMPTEL. The neutron energy spectrum is a continuum which extends up to the greatest progenitor proton energy. The efficiency of COMPTEL in detecting neutrons falls off rapidly above 80 MeV, however.

Conclusions

The spectral analysis properties of COMPTEL have been demonstrated with the measurement of the complex spectra such as that of the solar flare on 9 June 1991. The spectra contain much diagnostic information about the spectra and composition of the parent proton, electron and ion population. In addition, the spectra can provide information on the composition of the solar atmosphere where the energetic particles interact. More observations of the Sun and other astrophysical sources continue into the second year of the mission.

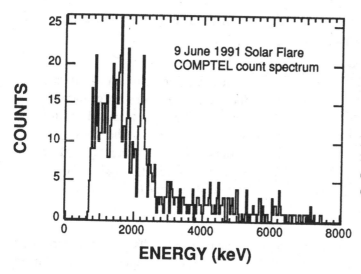

Figure 3. The count rate spectrum from the entire 9 June 1991 solar flare. These photons were measured in the double scatter mode of COMPTEL.

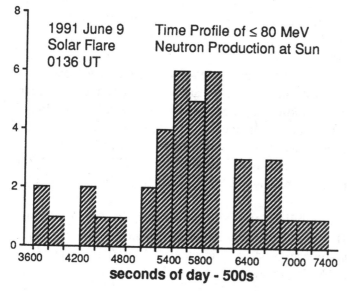

Figure 4. The intensity-time profile of neutron emission from the 9 June 1991 solar flare. The time axis represents the seconds of day of the neutron *emission* from the Sun. The corresponding time of the impulsive phase (0136UT to 0146 UT) as seen in γ-rays is from 5200 to 6000 s.

References

Diehl, R., *et al.* 1992, "Data Analysis of the COMPTEL Instrument on the NASA Gamma Ray Observatory", Proceedings of the Second GRO Science Workshop, Annapolis, Maryland:

Schönfelder, V., *et al.* 1984, *IEEE Trans. Nucl. Sci.*, **NS-31** (1) : 766-770.

Winkler, C., *et al.* 1986, *Adv. Space Res.*, **6** (4) : 113-117.

The Relative Coronal Abundance of Fe:Ne in Solar Active Regions Observed with the *Solar Maximum Mission* Flat Crystal Spectrometer

Julia L. R. Saba[1,2] and Keith T. Strong[1]

[1]Lockheed Solar & Astrophysics Laboratory, O/91-30, Palo Alto, CA 94304
[2]at Solar Data Analysis Center, Code 682.2, NASA/GSFC, Greenbelt, MD

ABSTRACT

We present early results from a study of the relative abundance of Fe:Ne in solar active regions, using high-resolution soft X-ray spectra acquired by the *SMM* Flat Crystal Spectrometer. To decouple the effects of temperature and abundance, we use the temperature-insensitive ratio of Ne IX at 13.45Å to Fe XVII at 15.01Å, and the sensitive temperature-diagnostic ratio of Fe XVIII blend at 14.22Å to the the Fe XVII line at 15.01Å. We find region-to-region changes in Fe:Ne of up to a factor of 7; for a given region, there are often factor-of-2 day-to-day variations and, occasionally, similar magnitude variations on time scales of less than an hour.

1. INTRODUCTION

There is a growing body of evidence for systematic differences in elemental composition between the photosphere and the corona, and from one type of coronal structure to another (see, *e.g.*, review by Meyer 1991 and references therein). Much of the observed variation appears associated with the first ionization potential (FIP) of the elements, suggesting that some differentiation process takes place in a temperature regime where some elements are ionized while others remain neutral.

There appears to be another layer of complexity to the coronal abundance picture. High-resolution soft X-ray spectra from the *Solar Maximum Mission (SMM)* Flat Crystal Spectrometer (FCS) show many examples of anomalous line ratios [see, *e.g.*, Fig. 1] whose only plausible explanation seems to be variability in the coronal abundances, from active region to active region and over time in a given region (Strong, Lemen, and Linford 1991; this work). Such variations could be related to variability reported between flares (see, *e.g.*, Lemen, Sylwester, and Bentley 1986) and during some flares (Sylwester, Lemen, and Mewe 1984).

2. ANALYSIS METHOD

To quantify the Fe:Ne abundance variations in FCS active region spectra, we use the ratio of Fe XVII at 15.01Å $[1s^2 2s^2 2p^6\ {}^1S_0 - 1s^2 2s^2 2p^5 3d\ {}^1P_1]$ to Ne IX at 13.45Å $[1s^2\ {}^1S_0 - 1s\ 2p\ {}^1P_1]$, an abundance diagnostic with a flat temperature response in the temperature range 2–6 MK relevant for active regions, and the ratio of the Fe XVIII doublet at 14.22Å $[1s^2 2s^2 2p^5\ {}^2D_{5/2},\ {}^2P_{3/2}]$ to Fe XVII at 15.01Å, a sensitive temperature diagnostic which changes by three orders of magnitude over

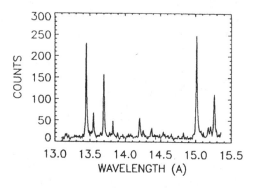 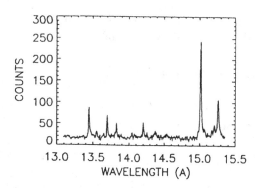

Fig. 1 – Comparison of FCS spectra from two active regions with about the same electron temperature ($T_e \sim 3MK$). Note the similarity in the relative intensities of the Fe XVIII lines at 14.22Å and the Fe XVII lines at 15.01Å. In the right panel, the Ne IX lines at 13.45Å to 13.7Å are considerably fainter with respect to all the other lines than in the left panel.

the same temperature range (See Fig. 2, *right*). An assumption that the plasma is isothermal is not required, provided the characteristic plasma temperatures fall in the regime where the abundance diagnostic ratio is temperature-insensitive and no substantial amount of material has a temperature outside this range.

3. WORK IN PROGRESS

We are examining the body of FCS spectral data acquired for quiescent and post-flare active regions during long scans of the wavelength drive, to determine the average value of the Fe:Ne relative abundance and to characterize the apparent variations in time and space. The study will be expanded to include O and Mg, which should allow detailed testing of FIP theories proposed for abundance variations. (The FIPs for O, Ne, Mg, and Fe are, respectively, 13.6 eV, 21.6 eV, 7.6 eV, and 7.9 eV.)

Our first results are shown in Fig. 3, which plots various samples of the abundance-diagnostic line ratio Fe XVII/Ne IX when the temperature-diagnostic line ratio Fe XVIII/Fe XVII falls in the 3-5 MK temperature range. Included are both one-hour averages of active region data and successive samples of regions if significant variation was found within the hour. For the latter, data are shown only if the recorded intensity was stable to about 10% during a given scan and if a contribution from transient, high-temperature plasma is ruled out by the absence of concurrent flux in high-temperature lines; the relevant lines are scanned within 2 minutes of each other, *i.e.*, fast compared to any observed changes. Most of both sets of data points are plotted as open diamonds. The extreme range in values of Fe:Ne appears to be about a factor of 7 in the data examined so far.

Repeated samples were taken of two active regions (NOAA AR 4901 and

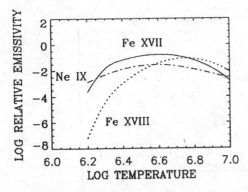
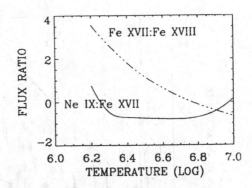

Fig. 2 – Line Emissivities and Flux Ratios versus log T. *Left:* Emission functions for the lines used in this project: Ne IX at 13.45Å, Fe XVII at 15.01Å, and the Fe XVIII doublet at 14.22Å. [from the calculations of Mewe *et al.* (1985).] *Right:* Ratios of the emissivity curves at left, for the sensitive temperature diagnostic, Fe XVII:Fe XVIII and the abundance-diagnostic ratio, Ne IX:Fe XVII.

AR 4787) denoted, respectively, by asterisks and open squares; both regions show changes in Fe:Ne of a factor of 2 or more. The solid squares show two samples of NOAA AR 4731, once near central meridian and once near the limb; the similarity in the deduced values of Fe:Ne suggests that the observed scatter is not simply a geometric effect from resonance scattering of Fe XVII. Most of the sequences of flare decay data show variation consistent with temperature evolution at a given Fe:Ne abundance (*i.e.*, they are roughly parallel to the dashed curve). However, some of them (*e.g.*, the two series of solid diamonds linked by straight line segments) show significant variations orthogonal to the dashed curve, indicating factor-of-2 variations in the Fe:Ne abundance on time scales less than an hour.

4. DISCUSSION

There is a large scatter of data points perpendicular to the dashed curve in Fig. 3 which gives the expected temperature dependence of the Fe XVII:Ne IX ratio. This scatter occurs in region-to-region samples, in day-to-day samples of a given region, and – in some cases – between 10-min samples of a region within a given orbit. Most of the variability is of order a factor of 2, but the extreme range in Fe:Ne between active region samples appears to be a factor of 7. The ten-minute or even one-day observed time scales for change appear to be too short for the diffusive time scales invoked by most elemental differentiation models. It is possible that different loops are being lit up within the 15 arcsec (FWHM) field of view of the FCS detector over time. We have not yet uncovered any pattern in the occurrence of abundance anomalies or variability in the data examined so far, but the work has just begun. We plan to search for a correlation between *SMM* BCS flare abundance anomalies and FCS abundances in the parent active region.

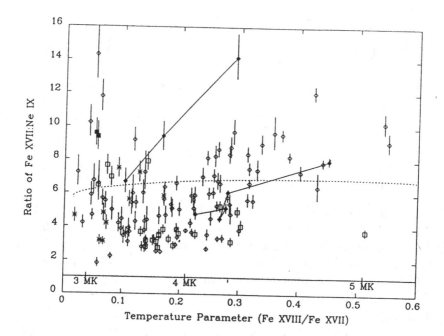

Fig. 3 – Fe XVII:Ne IX Flux Ratio vs. the Fe XVIII:Fe XVII temperature diagnostic for active regions sampled by the FCS. The (nonlinear) equivalent temperature scale is given at the bottom, inside of the plot. The shape of the dashed curve shows the expected temperature dependence of the Fe XVII:NeIX flux ratio, based on the tabulations of Mewe *et al.* (1985). The different symbols are discussed in the text. The error bars show 1σ statistical uncertainties.

ACKNOWLEDGEMENTS

This work has been supported by NASA contracts NAS5-28713 and NAS5-30431 and the Lockheed Independent Research Program. The data used here were made possible by the in-orbit repair of the *SMM* by the crew of the Space Shuttle *Challenger* on mission 41-C. This paper is dedicated to the memory of Francis R. Scobee, the pilot of mission 41-C and the commander of *Challenger*'s last mission.

REFERENCES

Lemen, J., Sylwester, J., and Bentley, R., 1986, *Adv. Space Res.*, **6**, No. 6, 245.

Mewe, R., Gronenschild, and van den Oord, 1985, *Astr. Ap. (Suppl.)*, **62**, 197.

Meyer, J.-P., 1991, *Adv. Space Res.*, **11**, No. 1, 269.

Strong, K., Lemen, J., and Linford, G., 1991, *Adv. Space Res.*, **11**, No. 1, 151.

Sylwester, J., Lemen, J., and Mewe, R., 1984, *Nature*, **310**, 665.

Timing of the Beginnings of the Solar Fast-Drift Bursts by H-alpha and X-ray solar flares

A. Tlamicha and Ladislav Krivsky

Astronomical Institute, Czechsolovak Academy of Sciences, 251 65 Ondrejov, Czechoslovakia

In the present paper we have statistically studied the problem of the onset of Fast-drift burst (FDP) in microwaves (2.0 - 4.5 GHz) in relation to the maxima of solar flares brightening in H-alpha and X-ray (GOES X-ray FLARES) emission. It is evident that the bursts are generated within most flares during the explosive phase of the flare, 1 to 6 minutes before the H-alpha flare maximum, since the relation to the X-ray maximum the occurrence of bursts appeared mostly in the interval 7 - 2 min, with number maximum 3 min before the X-ray maximum. 189 cases of FDB have been observed in 1989 - 1991 and used for our study.

In our case FDB covers the range 2.0 - 4.5 GHz. The recently published Catalogues by Tlamicha (1990) and Tlamicha et al (1991), enabled us to answer the question about the occurrence of Fast-drift bursts in the range of 2.0 - 4.5 GHz in relation to the development of the H-alpha and X-ray flares. In the past Svestka and Fritzova-Svestkova (1974) have studied the time connection between the onset of Type II bursts and the occurrence of microwave bursts, the microwave bursts precede the Type II bursts by 2 min. In the paper by Krivsky (1976) the time relation between the beginning of Type II bursts (start coronal shock) and the maximum brightening in H-alpha is treated. Type II bursts are centered around the maximum of the H-alpha flare but a high occurrence is found in the interval from 4 min before to the 1 min after the maxima of the H-alpha flares.

In order to treat the problem under study, it is necessary to start from physically significant and well-defined flare phenomena in the development of the flare which can be determined with sufficient accuracy and which can be evaluated from observations or records. As a result of extensive practical experiences we may say that this "time point" is not the beginning of the flare, but the maximum of the flare brightness of the H-alpha or of the X-ray. Both of these time values represent the typical processes mostly of a secondary nature, when the flare plasma has already been extensively thermalized and when the impulsive energetic phase of the flare (characterized by hard X-ray and FDB microwaves bursts) is already decreasing or passing.

The Catalogues published by Tlamicha (1990) and Tlamicha et al (1991) contain 189 cases of FDB. The time used for the statistical treatment is the time of start of the FDB. The data were processed as follows: at minute intervals before or after 0-moment (e.g. maxima of the H-alpha, Fig. 1 or of the X-ray flares, Fig. 2) the number of the occurrence of starting of the FDB were recorded and presented in histograms. FDB start time has been selected ± 10 minutes from the H-alpha or the X-ray flare maximum. Typical development of the H-alpha or the X-ray flares is demonstrated by full line taken as an average time from SGD (Boulder) for all our cases. The data of both maxima of the H-alpha or of the

X-ray emission (GOES X-ray FLARES) of flares were taken from Solar-Geophys. Data (Boulder). It is evident that the FDB are generated within most flares in the explosive phase of the flare between 1 and 6 minutes before the H-alpha flare maximum (max. occurrence 2 - 3 min), or between 2 and 7 minutes before the X-ray maximum (max. occurrence 3 min). Figures 3, 4 and 5 show some examples of H-alpha and X-ray Flares in the time development (H-alpha full line, X-ray Flare dashed line) together with Fast-drift bursts obtained by solar digital radio spectrometer.

The explosive phase of energetic flares and especially the impulsive phase as a part of the first one (defined by FDB and X-ray bursts), represents a short-lived complex of processes of flares connected with the main energy release.

Acknowledgements
The authors gratefully acknowledge the assistance of Z. Ambrozova, T. Sigmund, F. Zloch, L Zdarska in making the observations and with the analysis of the data. This work has been supported by the Czecholovak Academy of Sciences under Grant No. AI-GA-CAS 91/30317.

References
Krivsky L.: 1976 Bull. Astron. Inst. Czechoslov. 27, 374
Svestka Z.: Fritzova-Svestkova L.: 1974, Solar Phys. 36, 417
Tlamicha A.: 1990, Spectral Observations of Solar Radio Bursts in the Range 2.0 - 4.5 GHz.
        Publication of the Astronomical Institute of the Czechoslovak Academy of Sciences, No. 76
Tlamicha A.: Ambrozova Z, Sigmund T.: 1991, Spectral Observations of Solar Radio Bursts on the Range 2.0 - 4.5 GHz, Publication of the Astronomical Institute of the Czechosloval Academy of Sciences, No. 77 and No. 78
Tlamicha A.: 1991 Bull. Astrom. Inst. Czechoslov. 42, 257

Figures Captions

Figure 1
The occurrence of the beginning of the type Fast-drift bursts (FDB) before (-) and after (+) maxima of the H-alpha flare in 1989-1991).

Figures 2
The occurrence of the beginning of the type Fast-drift bursts (FDB) before (-) and after (+) maxima of the X-ray flare in 1989-1991).

Figures 3
Solar radion spectrum of September 2, 1989 with the time development of H-alpa (full line) and X-ray Flare (dashed line).

H–alpha Flare maximum

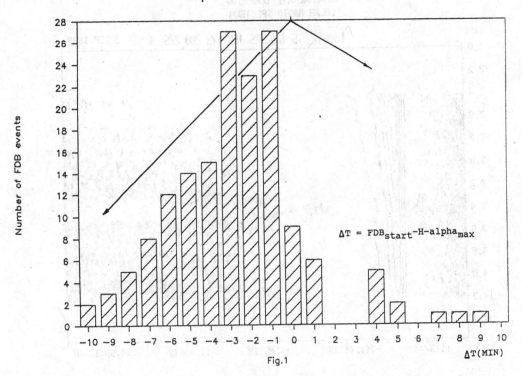

$\Delta T = FDB_{start} - H\text{-}alpha_{max}$

Fig.1

GOES X–ray Flare maximum
1–8 A

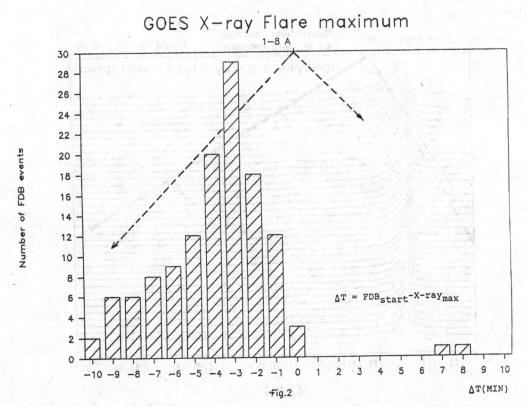

$\Delta T = FDB_{start} - X\text{-}ray_{max}$

Fig.2

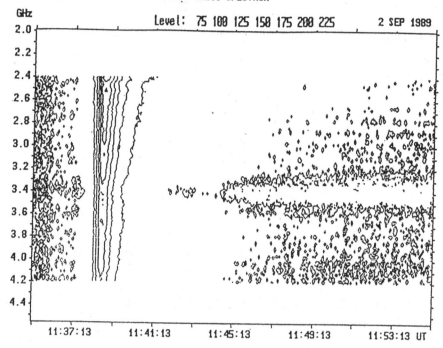

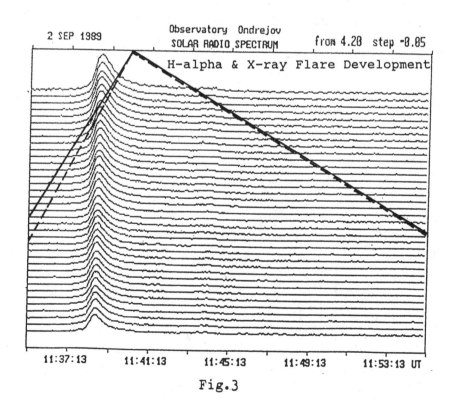

Fig.3

X-Ray Irradiation of Magnetic White Dwarfs

A. van Teeseling[1] and J. Heise[2]

[1]Sterrekundig Instituut, P.O.Box 80000, 3508 TA Utrecht, The Netherlands
[2]Space Research Laboratory, Utrecht, The Netherlands

ABSTRACT

We have calculated atmosphere models of hot white dwarfs whose stellar surface is irradiated by a bremsstrahlung spectrum (hard X-rays). Compton scattering is treated self-consistently as a noncoherent scattering process.

Because of an 'inverse' blanketing effect we find a very strong temperature inversion above the photosphere. A substantial fraction of the emerging soft X-rays is the result of absorbed and re-emitted hard X-rays.

Our models may be appropriate for the modelling of accreting magnetic white dwarfs. We compare some results with soft X-ray EXOSAT observations of AM Herculis.

1 IRRADIATION

A phenomenon that is important in several astrophysical objects is irradiation of a stellar atmosphere by X-rays. An example of such an object is a cataclysmic variable whose white dwarf has a very strong magnetic field (Liebert and Stockman 1985).

It is thought that in magnetic cataclysmic variables there exists column accretion onto the magnetic poles (Lamb 1985). If the accretion rate is sufficiently high the accretion flow passes a shock near the white dwarf surface. This shock heats the accreting plasma to a temperature of order 10 keV. Therefore, the matter radiates thermal bremsstrahlung observed as a hard X-ray spectrum (0.5-50 keV). If the shock is close to the surface half of these hard X-rays are scattered or absorbed by the white dwarf atmosphere. If the bremsstrahlung is absorbed, it will be re-emitted as soft X-rays.

There is still a discussion going on about the precise nature of the shock region and the hot spot on the white dwarf atmosphere and about the physical processes that contribute to the formation of the UV, soft X-ray and hard X-ray spectra. One of the problems is the effect of irradiation of the white dwarf atmosphere by hard X-rays (Williams et al. 1987).

Evidence for irradiation of the surface would be an emission feature at 6.5 keV observed in hard X-ray observations (Rothschild et al. 1981), if this feature is interpreted as an iron K-fluorescence line.

2 MODEL CALCULATIONS

We have calculated high-gravity, hot atmosphere models, including irradiation of the stellar surface by a bremsstrahlung spectrum (hard X-rays) and a self-consistent treatment of Compton scattering by calculating the redistribution of the scattered fraction of the intensity. Because the atmosphere is relatively cool compared to the hard X-rays we have only considered down scattering of photons and used the Klein-Nishina differential cross section to calculate the redistribution function (Pozdnyakov *et al.* 1983).

In the opacities we have included free-free absorption and the photoionization edges of all individual subshells of the 8 most abundant ions of the elements H, He, C, N, O, Ne, Na, Mg, Si, P, S, Ar, Ca, Fe and Ni (Kaastra 1991).

3 RESULTS

Because of the large bound-free absorption due to K and L-shell ionization of several abundant elements we find an 'inverse' blanketing effect: The incoming radiation is absorbed and leads to a backwarming of the outer layers of the atmosphere. The resulting temperature structures show a temperature inversion (fig. 1 and fig. 2). The consequence of this temperature inversion is that the limb darkening is smaller or even becomes limb brightening at certain frequencies in the case of strong illumination.

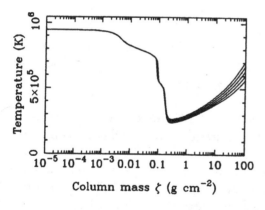

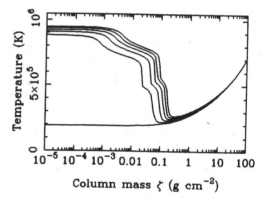

Fig. 1: Temperature inversion due to irradiation: The temperature at small optical depth is completely determined by the intensity (10^{-2}erg s^{-1} cm^{-2} ster^{-1}) and temperature (5×10^8K) of the irradiation flux. The flux coming from below the atmosphere is given by effective temperatures from 180000K to 260000K.

Fig. 2: Temperature structures of atmospheres with an effective temperature of 260000K and with different amounts of irradiation. The temperature below the photosphere is not affected by the amount of irradiation. The irradiation is increased in steps of 6.5×10^{16}erg s^{-1} cm^{-2}.

The spectrum of a model atmosphere with and without irradiation is shown in figure 3. For both spectra we have assumed an absorption column with $\log n_H = 19.8$ and a scaling factor 3.16×10^{-27} to simulate the incident photon spectra on the EXOSAT satellite. These parameters are appropriate for AM Herculis. Most of the soft X-ray energy from the irradiated atmosphere is due to irradiation.

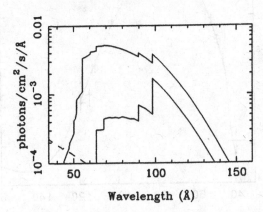

Fig. 3: Two model spectra of a white dwarf atmosphere ($\log g = 8$) with an effective temperature of 200000K: The spectrum from an atmosphere without irradiation (lower solid line) and the spectrum from an atmosphere with irradiation of $6 \times 10^{-3} \mathrm{erg\ s^{-1}\ cm^{-2}\ ster^{-1}\ Hz^{-1}}$. The dashed line gives the spectrum of the irradiation flux (the observed hard X-ray spectrum).

4 COMPARISON WITH OBSERVATIONS

The soft X-ray lightcurve of AM Herculis, observed with EXOSAT, is divided in minimum and maximum states due to occultation of the accreting pole(s) by this rotating magnetic white dwarf. The average spectrum of the maximum states, obtained with the $1000\ l\ mm^{-1}$ transmission grating when AM Her was in a high state, is shown in figure 4. This spectrum can be fitted to a black body with a temperature of 24.6 eV (285000K), but it reveals some additional features. At ~ 90Å the spectrum shows some edges, probably from oxygen. These features differ from cycle to cycle.

Fits with atmospheres without irradiation appear to be worse than the black body fit, because the temperature gradients are too steep and the calculated absorption edges in the soft X-ray spectrum are too big. The flux of model atmospheres, which include irradiation and Compton scattering, fits much better and requires a lower effective temperature than the black body fit. However, again the fit is not acceptable. This may be due to the finite size of the hot spot, that we observe at an angle. Then we must use the angle-dependent intensity instead of the angle-averaged flux to fit the observed spectrum. We are now investigating this (Van Teeseling and Heise 1992).

We conclude that it might be important to include irradiation of the white dwarf surface by X-rays in order to understand the X-ray spectra of accreting magnetic white dwarfs.

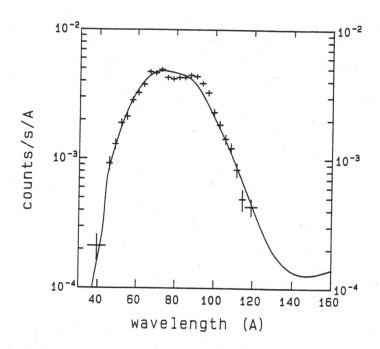

Fig. 4: The average spectrum of the maximum states of AM Her, obtained with the 1000 l mm^{-1} transmission grating on EXOSAT and the best fit to a black body (solid line) with a temperature of $\sim$24.6 eV, an absorption column with $\log n_H = 19.8$ and a scaling factor 3.16×10^{-27}.

REFERENCES

Kaastra, J. 1991, private communication

Lamb, D. Q. 1985 in *Cataclysmic Variables and Low-mass X-ray Binaries*, ed. J. Patterson and D. Q. Lamb (Dordrecht: Reidel), p. 179

Liebert, J. and Stockman, H. S. 1985 in *Cataclysmic Variables and Low-mass X-ray Binaries*, ed. J. Patterson and D. Q. Lamb (Dordrecht: Reidel), p. 151

Pozdnyakov, L. A., Sobol, I. M. and Syunyaev, R. A. 1983, Sov. Sci. Rev E2, 189

Rothschild, R. E., *et al.*, *Ap. J.* **250**, 723

van Teeseling, A. and Heise, J. 1992, *A. & A.*, in preparation

Williams, G. A., King, A. R. and Brooker, J. R. E. 1987, *M. N. R. A. S.* **266**, 725

Relative Elemental Abundances of a Solar Active Region

Katrina Waljeski [1] and Dan Moses [2]

[1] NRC Postdoctoral Associate, Naval Research Laboratory
[2] Naval Research Laboratory

ABSTRACT

The relative elemental abundances of iron, oxygen, magnesium, and neon in the solar corona are investigated with simultaneous broadband and spectral-line soft X-ray data. Coordinated observations of a solar active region were made by the American Science and Engineering (AS&E) High Resolution Soft X-Ray Imaging Sounding Rocket Payload, and by the X-Ray Polychromator Flat Crystal Spectrometer (FCS) onboard the Solar Maximum Mission spacecraft. By requiring mutual consistency between the broadband and emission line plasma diagnostics, the relative elemental abundances of iron, oxygen, magnesium, and neon are constrained. The relative abundances determined are found to be inconsistent with the photospheric abundances compiled by Allen (*Astrophysical Quantities*, London: Athlone Press, 1973) which have often been used in investigations of coronal plasma properties. The relative abundances from these observations are consistent with the "adopted coronal" abundances tabulated by Meyer (*Astrophysical Journal Supplement 57*, 1985), where the ratios of high first ionization potential (FIP) ions to low FIP ions are lower than the ratios observed in the photosphere.

1 OBSERVATIONS

On 11 December 1987, solar active region AR 4901 was observed simultaneously in soft X-ray wavebands (8-39 angstroms and 8-39,44-60 angstroms) by the American Science and Engineering (AS&E) High Resolution Soft X-Ray Imaging Sounding Rocket Payload, and in soft X-ray emission lines (Fe XVII at 15.01 angstroms, Fe XVIII at 14.22 angstroms, O VIII at 18.97 angstroms, Ne IX at 13.44 angstroms, and Mg XI at 9.17 angstroms) by the X-Ray Polychromator Flat Crystal Spectrometer onboard the Solar Maximum Mission spacecraft.

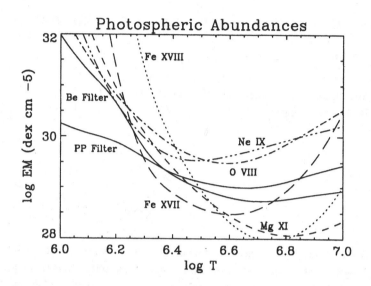

Fig. 1a - The emission measure is plotted as a function of the log of temperature, assuming the photospheric abundances of Allen.

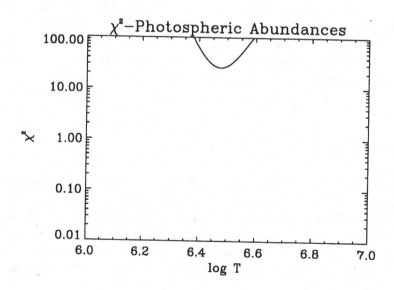

Fig. 1b - The χ^2 of all the lines (except Fe XVII) and both broadband measurements is plotted as a function of the log of temperature, assuming the photospheric abundances of Allen.

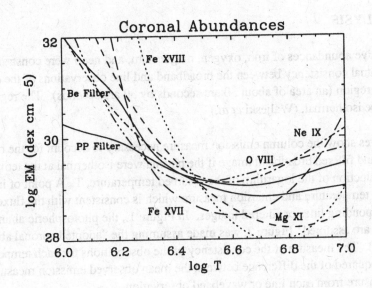

Fig. 2a - The emission measure is plotted as a function of the log of temperature, assuming the adopted coronal abundances of Meyer.

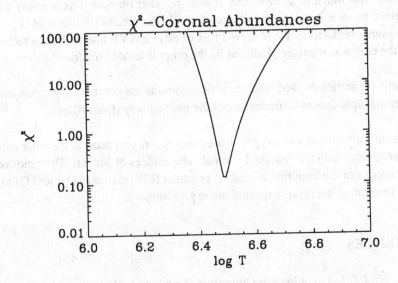

Fig. 2b - The χ^2 of all the lines (except Fe XVII) and both broadband measurements is plotted as a function of the log of temperature, assuming the adopted coronal abundances of Meyer.

2 ANALYSIS

The relative abundances of iron, oxygen, magnesium, and neon were constrained by requiring mutual consistency between the broadband and line observations of the central area of the active region (an area of about 30 arc seconds by 30 arc seconds). The region was assumed to be isothermal. (Waljeski *et al.*)

The figures show the column emission measure necessary to account for the observed flux of each line and filtered broadband image if the plasma were isothermal at the temperature, plotted as a function of the logarithm of the assumed temperature, T. A point of intersection is a solution - a temperature and emission measure which is consistent with the fluxes observed in all the corresponding lines or filtered images. In Figure 1., the photospheric abundances of Allen (1973) are assumed. Figure 2. was made assuming the "adopted coronal abundances" of Meyer (1985). The measure of the consistency of the observations for each temperature is the reduced chi-squared of the difference between the mean observed emission measure and the emission measure from each line or waveband observation.

3 RESULTS

The Fe XVII line at 15.01 Angstroms was not included in the calculations, as the observed flux for this line was found to be inconsistent with the other observations in a way which can not be explained by an anomalous abundance of iron in the corona (Waljeski *et al.*). The emission measure predicted by the observed flux of this Fe XVII line is about a factor of 3 lower than the emission measure predicted by the other line and broadband fluxes.

The plasma properties derived from these observations are not mutually consistent for the relative elemental abundances characteristic of the photosphere (from Allen).

The elemental abundances of oxygen, iron, magnesium, and neon in the solar corona were found to be consistent with the "adopted coronal" abundances of Meyer; The ratios of the coronal abundances of the high first ionization potential (FIP) elements (Ne and O) to the low FIP ions are lower than the ratios expected in the photosphere.

REFERENCES

Allen, C. W., 1973, *Astrophysical Quantities* (London: Athlone Press)

Meyer, J.-P., 1985, *Astrophysical Journal Supplement 57*

Waljeski *et al.*, 1992, in preparation

Flare Dynamics Observed In S XV

Dominic M. Zarro[1]

[1]Applied Research Corporation at the Solar Data Center (NASA/GSFC)

ABSTRACT

We have investigated the soft X-ray helium-like S XV (5.039 Å) emission line in a solar flare observed with the Flat Crystal Spectrometer (FCS) on the Solar Maximum Mission (SMM). Because the S XV emissivity function peaks at coronal temperatures of $\simeq 15 \times 10^6$ K, the S XV line provides a useful diagnostic of flare dynamics at relatively cooler temperatures than observed by Ca XIX and Fe XXV. Measurement of the S XV profile approximately 1 min after the peak of impulsive hard X-rays reveals a blueshifted component with a velocity of $\simeq 200$ km s^{-1}. We interpret this component as evidence for sustained chromospheric evaporation that is driven by thermal conduction.

1 INTRODUCTION

Most studies of the chromospheric evaporation process in solar flares have been based upon observations of the He-like Ca XIX and Fe XXV resonance lines (Doschek et al. 1980; Antonucci et al. 1982; Antonucci, Gabriel, and Dennis 1984). Since the emissivity functions of these ions peak at temperatures that are greater than about 20×10^6 K, the corresponding line emissions primarily diagnose physical conditions in the hottest regions of the flaring corona. Thus, high-temperature sensitive lines provide useful diagnostics during the flare impulsive phase when plasma temperature and mass motions are greatest. The present work extends the study of chromospheric evaporation to cooler temperatures below 20×10^6 K. As our diagnostic, we use the He-like S XV (5.039Å) line that we observed with the Flat Crystal Spectrometer (FCS) on the Solar Maximum Mission (SMM) X-ray Polychromator (XRP). This line has an emissivity function that is peaked at approximately 15×10^6 K. It therefore allows us to probe dynamical processes in cooler regions of the flaring corona.

After the impulsive phase, it is expected that supersonic mass motions will subside as hydrostatic equilibrium is restored in the corona. During this stage, soft X-ray blueshifts are difficult to detect with high-temperature lines such as Ca XIX since the line profile is usually dominated by emission from hot stationary plasma that has achieved thermal and hydrodynamic equilibrium in the corona (cf. McClements and Alexander 1989). The stationary component can potentially mask out weaker blueshifted emission that may persist after the main heating phase of the flare. By contrast, a low-temperature sensitive line such as S XV is less likely to be dominated by a stationary component since it is produced in cooler plasma that is still undergoing heating to flare temperatures.

2 OBSERVATIONS AND ANALYSIS

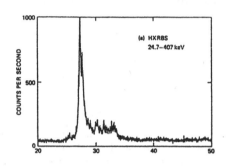 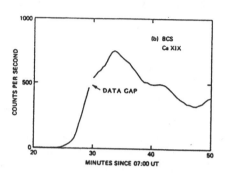

Fig. 1— (a) the HXRBS (24–308 keV) hard X-ray lightcurve at 128 msec resolution; (b) the BCS total soft X-ray Ca XVIII–XIX emission at $\sim$ 7s resolution.

A GOES class M1 flare occurred at 0725 UT on 1985 January 23. The flare was observed in hard X-rays with the SMM Hard X-ray Burst Spectrometer (HXRBS) and in soft X-ray Ca XIX with the SMM Bent Crystal Spectrometer (BCS). Figure 1 shows the temporal variations of the HXRBS and BCS data. The main impulsive phase was characterized by two intense bursts of hard X-rays lasting about 2 mins. The gradual phase of the flare was exceptional in that soft X-rays persisted at enhanced levels for up to 20 mins after the peak of hard X-rays.

Prior to the flare, the FCS was positioned in a part of the active region showing enhanced soft X-ray and UV emission. At 0732:28 (about 1 min after the peak of impulsive hard X-rays), the FCS commenced a series of rapid wavelength scans of the S XV line with 8 s time resolution and 0.6 mÅ spectral resolution. Figure 2a shows a 24 s average (sum of the first 3 scans) of the S XV profile. The profile shows evidence for a blue-asymmetry. A least-squares fit of a two-component gaussian function indicates a secondary component at a blueshift velocity $\simeq$ 187 km s^{-1}. Figure 2b shows a 24 s average of the S XV profile at 0743 UT (approximately 10 mins later). No significant blueshifted component is evident at this stage of the flare.

The blueshifted S XV component is indicative of line-of-sight upflows of coronal plasma. The question arises as to whether these upflows are physically related to the flare heating process. For any given heat input mechanism (e.g. thick-target electrons), it is expected that as the coronal plasma becomes hotter then an ever increasing fraction of the deposited energy will be conducted to cooler coronal layers, (i.e., the flare transition region) where it is channeled into radiation and conductively driven evaporation (Fisher 1986). The timescale for the domination of conductive cooling is dependent on the input heating rate and the peak temperature achieved by the flare plasma. It ranges typically between $\sim$ 2 and 20 s. Since the S XV blueshift

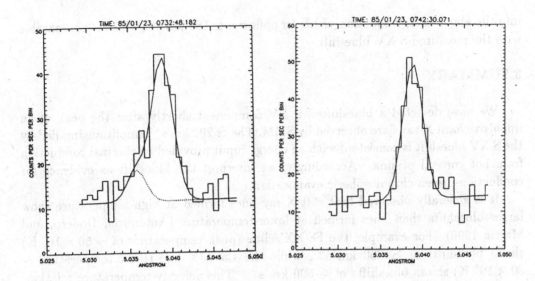

Fig. 2—(a) The S XV profile at 0732:28 showing evidence for a blue asymmetry. The solid line shows a model double gaussian that is fit by least-squares. The dashed line corresponds to the secondary component that is blueshifted by $\simeq 187$ km s^{-1} from the primary component; (b) The S XV profile at 0743 UT (approximately 10 mins later). The profile is fit with a single gaussian. No significant blueshifted is evident at this stage of the flare.

measurement was made approximately 1 min after the peak of impulsive hard X-rays (an indicator of nonthermal heating), we hypothesize that the upflow motion implied by the S XV blueshift is fueled by thermal conduction.

To test the above hypothesis, we perform a simple energy balance analysis to determine whether the amplitude of the S XV blueshift velocity is consistent with the energy deposited by thermal conduction. To a first approximation, we neglect radiative losses and solve for the upflow velocity by equating the mean downward-conductive flux from the corona with the upward-enthalpy flux transported by the S XV–emitting plasma (cf. Zarro and Lemen 1988). Thus,

$$10^{-6} T_0^{5/2} \frac{T_0}{L} \approx 5 k n_0 T_0 v,$$

where T_0 and n_0 denote the loop temperature and density, v is the upflow velocity, and L is the loop half-length. We infer the density from $n_0 \approx \sqrt{EM/V}$, where EM is the emission measure, $V = 2AL$ is the loop volume, and A is the loop cross-sectional area.

Zarro and Lemen (1988) used Ca XVIII–XIX line ratio observations of this flare to estimate $EM \simeq 10^{49}$ cm^{-3} and $T_0 \simeq 13 \times 10^6$ K at the time of the S XV blueshift measurement. For this same event, they used preflare Mg XI FCS imaging observations to derive $L \simeq 8 \times 10^8$ cm and $A \simeq 1.5 \times 10^{17}$ cm^2. Substituting these values into the above energy equation yields an upflow $v \simeq 180$ km s^{-1} which is comparable with the measured S XV blueshift.

3 SUMMARY

We have detected a blueshifted S XV component shortly after the peak of an impulsive hard X-ray flare observed by SMM. The $\simeq 200$ km s^{-1} amplitude implied by the S XV blueshift is consistent with an energy input provided by thermal conduction from hot coronal plasma. Accordingly, we interpret the blueshift as evidence for conduction-driven chromospheric evaporation.

It is generally observed that soft X-ray lines formed at high-temperature show larger blueshifts than lines formed at lower-temperature (Antonucci, Dodero, and Martin 1990). For example, the Fe XXV line (peak temperature of $\sim 50 \times 10^6$ K) shows blueshifts of $\lesssim 800$ km s^{-1}, while the Ca XIX line (peak temperature $\sim 30 \times 10^6$ K) shows blueshifts of $\lesssim 500$ km s^{-1}. This velocity-temperature relationship suggests a physical link between the plasma upflow motions implied by these blueshifts and the plasma heating that is occuring in flares. Our measurement of a blueshifted S XV component with an implied upflow velocity that is less than typical velocities implied by higher-temperature lines appears to extend this relationship to lower temperatures.

Because of its low spectral sensitivity and limited field of view (14 arcsec), the SMM-FCS could not observe S XV profiles during the critical impulsive phase of flares. This impasse is now remedied by new S XV observations that are being obtained currently at much greater sensitivity (factor of 5–10) with the full-Sun field of view Bragg crystal spectrometer on the Japanese Yohkoh satellite. The analysis of these improved observations will shed further light on the velocity-temperature relationship in flares.

REFERENCES

Antonucci, E. et al. 1982, *Solar Phys.*, **78**, 107.
Antonucci, E., Gabriel, A.H., Dennis, B.R. 1984, *Ap.J.*, **287**, 917.
Antonucci, E., Dodero, M.A., and Martin, R. 1990, *Ap.J.*, **73**, 147.
Doschek, G.A., Feldman, U., Kreplin, R.W, Cohen, L. 1980, *Ap.J.*, **239**, 725.
Fisher. G. 1986, *Ap.J.*, **317**, 502.
McClements, K.G. and Alexander, D. 1989, *Solar Phys.*, **123**, 161.
Tanaka, K., Watanabe, T., Nishi, K., Akita, K. 1982, *Ap.J. (Letters)*, **254**, L59.
Zarro, D.M and Lemen, J.R. 1988, *Ap. J.*, **329**, 456.

Carbon Transport Estimates in a Tokamak Plasma During Auxiliary Heating Experiments using Soft X-Ray CVI Emission

A. P. Zwicker[1], M. Finkenthal[2], S. Lippmann[3], H.W. Moos[1]

[1]The Johns Hopkins University, Baltimore, Md.
[2]Racah Institute of Physics, Hebrew University, Jerusalem, Israel
[3]General Atomics, San Diego, Ca.

ABSTRACT

Carbon, the dominant impurity in the DIII-D tokamak at General Atomics, in San Diego, originates from the graphite armour tiles. In the work presented here, a soft x-ray scanning monochromator that utilized a flat multilayer mirror (MLM) as the dispersive element was used to obtain the carbon emission in the 30-40Å region with a resolution of ≈1Å. A grazing incidence spectrometer measured the time history of the impurity lines above 100Å with a resolution of ≈1Å. During a series of ion cyclotron resonance frequency heating experiments (ICRF) the ratio of the Ly_α and H_α emissions of C VI (34Å and 182Å respectively) was monitored. Below a threshold ICRF input power level, this ratio was constant but, as the input power increased, the ratio changed. By modelling the change in this ratio with an impurity transport code coupled to a collisional-radiative model for C VI, it is possible to estimate the effect of ICRF injection on the transport of carbon.

1. INTRODUCTION

One possible scenario for a steady-state tokamak reactor is the use of high power radio frequency waves to drive a constant current through the plasma instead of, as is currently done, inducing a pulsed current through transformer action. The behavior of impurities in the plasma during radio frequency wave injection is important since accumulation of impurities in the center of the plasma can have a deleterious effect by, among other things, radiatively cooling the plasma. In this regard, an experiment was performed on the DIII-D tokamak in an attempt to study impurity transport across magnetic field lines in discharges with radio frequency wave injection. The impurity studied is carbon, the dominant low Z impurity in DIII-D. Carbon originates from the graphite tiles that are used to protect the vacuum vessel from the plasma. To infer information about the transport of carbon two soft x-ray emission lines of hydrogen-like C VI were monitored during a series of discharges with ion cyclotron radio frequency (ICRF) wave injection. Transport information can be inferred by comparing the ratio of these two lines to the ratio calculated from an impurity transport code coupled to a collisional-radiative model for CVI.

Two spectroscopic diagnostics were used for this experiment. One was a conventional grazing incidence spectrometer and the other was a multilayer mirror based soft x-ray monochromator. Multilayer mirrors (MLMs) are an attractive possibility for use as dispersive elements in soft x-ray spectroscopic diagnostics for the next generation of tokamaks (Moos, et al. 1990). A spectroscopic instrument based upon a MLM can be simple and have a large photon throughput, as compared to a conventional grazing incidence spectrometer. These devices may be a simple way to monitor impurity levels, measure soft x-ray radiative power losses or even produce soft x-ray images of the spatial distribution of an impurity in an extremely harsh environment.

2. EXPERIMENT

For this particular experiment, a soft x-ray scanning monochromator was built (Zwicker et al., 1992) that utilized a flat Monel/B_4C MLM (2d=84Å, number of layers=100, peak reflectivity= 7.5% [at 23.6Å], FWHM= 0.56Å [at 23.6Å]) as the dispersive element. The monochromator was mounted on the DIII-D tokamak at General Atomics and used to monitor hydrogen-like lyman alpha (Ly_α) emission (1s-2p) of carbon at 33.7Å. For the deuterium discharges discussed below, the plasma parameters were as follows: major radius, 1.67 m.; toroidal field, 2.1 T; plasma current , 1.0 MA; auxiliary heating, 0.3-0.5 MW of input ICRF power.

The spectrum above 100Å was monitored with a SPRED survey spectrometer (Fonck et al., 1982) SPRED provides time-resolved spectra between 100-1200Å with a resolution of ≈1Å. In this experiment the SPRED was used to monitor the emission from balmer alpha (H_α) of hydrogen-like carbon (n=2-3) at 182Å.

Fig. 1 is an example of the time history of the Ly_α signal of carbon at 33.7Å measured by the MLM-based monochromator, the time history of the H_α signal of carbon at 182Å measured by the SPRED spectrometer, and the ratio r, $r=Ly_\alpha/H_\alpha$. For this discharge, there were two ICRF pulses: 0.35 MW from 1600-2200 ms. and 0.45 MW from 2800-3400 ms and these are also shown schematically in the figure. Note that the ratio shown in the top trace (Fig. 1c) decreases soon after the ICRF pulse is turned on - the H_α signal is enhanced in relation to the Ly_α signal.

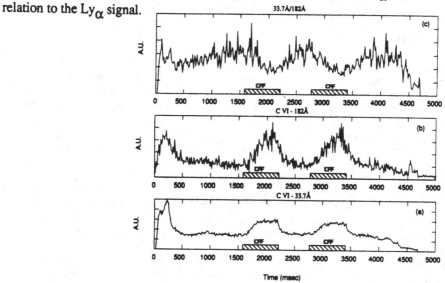

Fig.1: Time histories of (a) CVI Ly_α (33.7Å), (b) H_α (182Å) and (c) the ratio Ly_α/H_α during a discharge with ICRF (0.35 MW from 1600-2200 ms. and 0.45 MW from 2800-3400 ms)

3. RESULTS

To determine how this ratio, r, changed as a function of input ICRF power, a series of discharges with varying amounts of input ICRF power were studied. These results are presented in Fig. 2 where the value of the ratio during the ICRF pulse normalized to the value before the ICRF pulse is plotted as a function of input ICRF power. The deviation in the normalized ratio is approximately 10% and is primarily due to the low signal level of the H_α signal in the SPRED spectrum. Also shown in Fig. 2 are two calculated points (circles) assuming certain transport coefficients, as discussed below.

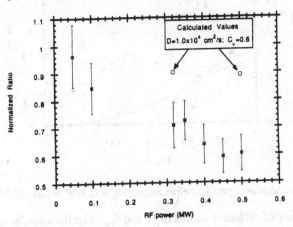

Fig. 2: The value of the normalized ratio as a function of input ICRF power. Also shown are two calculated points (circles) assuming certain transport coefficients, as discussed below.

4. DISCUSSION

The two carbon emission lines studied are temperature dependent and so, the question becomes, is this decrease in R during ICRF due to a change in the local electron temperature profile from the ICRF or is the transport of carbon, across the magnetic field lines and into the hotter region of the plasma, changing. In order to gain some insight into this question the MIST impurity transport code (Hulse, 1983) was used to calculate the hydrogen-like carbon distribution throughout the plasma given the independently measured temperature and density profiles (by Thompson scattering, electron cyclotron emission, and CO_2 laser interferometry) before and during the ICRF pulse (Fig. 3).

MIST assumes a constant anomalous diffusion coefficient, D and a radially dependent convective velocity, V determined by the expression,

$$V(r) = C_v D \left(\frac{\partial \ln n_e(r)}{\partial r} \right)$$ (1)

where C_v is a constant. Once the ion distribution was calculated, the results were input into a collisional-radiative model for C VI that assumes that the level populations of the ion are determined by electron collisions to an excited state while decaying to a lower level radiatively. The brightnesses of Ly_α and H_α are then calculated for a variety of different values of D and C_v.

In Fig. 2, two examples of the calculated normalized ratio are shown with the assumption that the carbon transport does not change during the ICRF pulse for typical transport coefficients: $D = 1 \times 10^4$ cm^2/s and $C_v = 0.6$. As is shown in the figure, the calculation does not match the experiment and in fact, if it is assumed that the transport coefficients are constant throughout the discharge, it is not possible to match the experimental values for any reasonable values of transport coefficients. Therefore, the ICRF pulse does effect the transport of carbon. However, for this particular experiment, it is not possible to determine exactly how the coefficients are effected since the experimental value can be

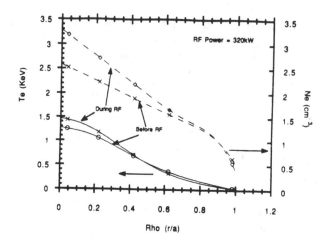

Fig.3: Temperature and density profiles before and during 0.32 MW of input ICRF.

matched with a variety of different values of D and C_v. For instance, at an input ICRF power level of 0.32 MW and an assumed initial value of the transport coefficients of D = 1×10^4 cm^2/s and C_v = 0.6 the experiment can be matched by a variety of combinations of transport coefficients (eg. D = 1.0×10^4 cm^2/s, C_v = 1.6 or D = 5.0×10^3 cm^2/s and C_v = 1.4). Therefore, additional information is necessary to limit the range of values of D and C_v. This can happen by determining, experimentally, the C VI ion distribution's width and peak location as a function of radius throughout the discharge or by making absolute rather than relative line brightness measurements (the MLM-based monochromator is absolutely calibrated, SPRED is not) and thus, determining the absolute value of the Ly_α/H_α ratio rather than a relative one.

5. CONCLUSIONS

A simple multilayer mirror monochromator was mounted on the DIII-D tokamak and used, in conjunction with a grazing incidence SPRED spectrometer, to study the ratio of Ly_α/H_α in C VI. During ICRF injection into the plasma, this ratio, which is temperature dependent, decreased. The results were modelled with an impurity transport code and a collisional-radiative model and the results indicate that the transport of C VI during the ICRF is changing. Plans are currently underway to mount an instrument on DIII-D that uses a curved multilayer mirror to image the plasma and determine the ion distribution of C VI. This should make the determination of specific changes in the transport coefficients, under a variety of different plasma conditions, feasible.

The authors thank the DIII-D team for their support during this experiment. This work was supported by DoE grant DE-FG02-86ER53214 and contract DE-AC03-89ER51114.

REFERENCES

Fonck, R.J., Ramsey, A.T., Yelle, R.V., (1982), Appl. Opt., 21, 2115.
Hulse, R.A., (1983), Nucl. Technol./Fusion, 3, 259.
Moos, W., et al., (1990), Rev. Sci. Instrum., 61, 2733.
Zwicker, A.P., et al., (1991), Bull. Am. Phys. Soc., 36, No. 9, 2493 and (1992), Rev. Sci. Instrum.. to be submitted.

Printed in the United States
By Bookmasters